Springer Series in Information Sciences 11

Editor: King-sun Fu

Springer Series in Information Sciences

Editors: King-sun Fu Thomas S. Huang Manfred R. Schroeder

Volume 1 **Content-Addressable Memories**
By T. Kohonen

Volume 2 **Fast Fourier Transform and Convolution Algorithms**
By H. J. Nussbaumer 2nd Edition

Volume 3 **Pitch Determination of Speech Signals** Algorithms and Devices
By W. Hess

Volume 4 **Pattern Analysis**
By H. Niemann

Volume 5 **Image Sequence Analysis**
Editor: T. S. Huang

Volume 6 **Picture Engineering**
Editors: King-sun Fu and T. L. Kunii

Volume 7 **Number Theory in Science and Communication**
With Applications in Cryptography, Physics, Biology
and Digital Information
By M. R. Schroeder

Volume 8 **Self-Organization and Associative Memory**
By T. Kohonen

Volume 9 **An Introduction to Digital Picture Processing**
By L. P. Yaroslavsky

Volume 10 **Probability, Statistical Optics, and Data Testing**
A Problem Solving Approach
By B. Roy Frieden

Volume 11 **Physical and Biological Processing of Images**
Editors: O. J. Braddick and A. C. Sleigh

Physical and Biological Processing of Images

Proceedings of an International Symposium
Organised by The Rank Prize Funds
London, England, 27–29 September, 1982

Editors: O. J. Braddick A. C. Sleigh

With 227 Figures

Springer-Verlag Berlin Heidelberg New York 1983

Dr. Oliver J. Braddick
Department of Experimental Psychology, University of Cambridge
Cambridge CB2 3EB, England

Andrew C. Sleigh
Royal Signals & Radar Establishment, St. Andrews Road,
Great Malvern, Worcs. WR14 3PS, England

Series Editors:

Professor King-sun Fu
School of Electrical Engineering, Purdue University,
West Lafayette, IN 47907, USA

Professor Thomas S. Huang
Department of Electrical Engineering and Coordinated Science Laboratory,
University of Illinois, Urbana, IL 61801, USA

Professor Dr. Manfred R. Schroeder
Drittes Physikalisches Institut, Universität Göttingen, Bürgerstraße 42–44,
D-3400 Göttingen, Fed. Rep. of Germany

ISBN-13: 978-3-642-68890-4 e-ISBN-13: 978-3-642-68888-1
DOI: 10.1007/978-3-642-68888-1

Softcover reprint of the hardcover 1st edition 1983

2153/3130-543210

The Lord Rank, J.P., LL.D., 1888-1972

Lord Rank, who died in 1972, was an outstanding industrial personality in the United Kingdom. He consolidated his family's flour milling and bread business, Ranks Hovis McDougall, into one of the major British food manufacturers. He also built up The Rank Organisation, starting in films and now incorporating a wide range of activities, as well as having a close relationship with the Xerox Corporation of the United States through its share holding in Rank Xerox.

Lord Rank was much concerned with the quality of life and provided for many charities during his lifetime. Just before his death, Lord Rank established The Rank Prize Funds to encourage scientific work in the fields of Nutrition and Opto-electronics.

Preface

This book consists of papers presented at an international symposium sponsored and organised by The Rank Prize Funds and held at The Royal Society, London, on 27-29 September, 1982.

Since the inception of the Funds, the Trustees and their Scientific Advisory Committee on Opto-electronics have considered that the scope of opto-electronics should extend to cover the question of how the eye transduces and processes optical information. The Funds have aimed to organise symposia on topics which, because of their interdisciplinary nature, were not well covered by other regular international scientific meetings. It was therefore very appropriate that the 1982 symposium should be on Physical and Biological Processing of Images.

The purpose of the symposium was to bring together scientists working on the physiology and psychology of visual perception with those developing machine systems for image processing and understanding. The papers were planned in such a way as to emphasise questions of how image-analysing systems can be organised, as well as the principles underlying them, rather than the detailed biophysics and structure of sensory systems or the specific design of hardware devices. As far as possible, related topics in biological and artificial systems were considered side by side.

The initial stages of image processing, whether physical or biological, consist of fairly simple, usually local, transformations of the array of intensity values in the image. However, to derive any useful information from the image, operations are needed which compute higher-level descriptions in terms of motion, stereo, and segmented regions, and ultimately in terms of the identity of objects and their layout in space. The topics of the papers were selected to span this whole range of image-processing problems.

If we can understand how to mechanise image interpretation, a rich harvest of applications will follow. Applications in medical imaging, remote sensing, and character recognition are explicitly considered here, and the possibilities for robotics frequently arose in discussion. The continuing fall in the cost of information processing and storage components makes it feasible to construct practical systems which would have seemed extravagant fictions only a few years ago. Research into the field which some term the 'Physics of Computational Systems' is likely to be one of the key topics of the 1980's. Insights into how the brain organises and relates data could catalyse the production of new and powerful computer architectures. Conversely, fundamental limitations which emerge from the study of computing and data organisation will help to confirm or reject models of human visual processes deduced by psychologists and physiologists.

The proceedings of the symposium reflected the very great diversity of scientific interest in these problems; artificial intelligence, computer science, vertebrate and invertebrate neurophysiology, optical and electronic engineering, visual psychophysics, pattern recognition, and cognitive psychology were represented. The symposium made participants from each of these fields aware of the questions posed by other disciplines and the kinds of answers they can currently offer. It was clear that they had much in common. It was also clear that their different vocabularies, methods and backgrounds were a hurdle yet to be fully surmounted. This volume, we hope, will broaden the horizons of a wider audience of research workers and advanced students in all disciplines concerned with the processing and interpretation of images, and stimulate a continuing interaction between these diverse groups.

We gratefully acknowledge the support of the Trustees and Staff of The Rank Prize Funds in the running of the symposium and the advice on its scientific content given by Professor Horace Barlow, Professor Cyril Hilsum and other members of the Trustees' Scientific Advisory Committee on Opto-electronics.

London, December, 1982 *O.J. Braddick · A.C. Sleigh*

Contents

Part VII **Postscript**

Part I

Overviews

Understanding Natural Vision

H.B. Barlow

Physiological Laboratory, Downing Street,
Cambridge CB2 3EG, United Kingdom

Abstract

Because many of you may not be familiar with the biological structures with which we ourselves process images, I shall start with a brief anatomical tour of the complex and incompletely understood visual system of primates. The biological components which perform the computations may also be unfamiliar, and it is interesting to compare them with physical devices: synapses and neurones are very slow and have a very limited dynamic range, but they are packed at a very high density per unit volume, are very freely interconnected over limited distances, and these interconnections possess self-regulatory properties that can compensate for errors in construction or changes of use.

The problems facing physiologists or psychologists studying natural vision are also very different from yours. First, it is technically very difficult to monitor the activity of just one or two elements at a time in a structure that contains 10^{10} neurones or more; second, with such incomplete information it is hard to interpret the results, and third, there is the constant danger of interfering with normal function. Also, we often fail to understand the goal or purpose of the neuronal interactions that occur in a biological structure; this should not be a problem for physicists because I cannot imagine one of your devices ever doing something clever that it's designer had not intended it to do!

The overall achievement of a good visual system is impressive. It does far more than simply provide us with a good picture to look at, for the looker is part of the system. It gives us most of our knowledge of the world, which requires organisation of the results to minimize redundant representation and statistical testing to ensure reliability. All this is done on the spot, in real time, by a system weighing about two kilograms (including power supplies). So far the flow of important concepts has been from physics to biology, but if we knew exactly how a real visual system worked the flow might be reversed.

What is Vision?

One tends to think of natural, biological, vision as the process by which an image of the world around us is formed optically by the eye, transduced into nervous activity in the retina and transmitted to the brain by its optic nerve fibres. But seeing is much more than a matter of getting an image upstairs and into the brain: it includes the process of interpreting the image and thereby gaining knowledge of the real world that gave rise to the image. It is especially important for me to make this distinction because I am sure most of you are more interested in the second process and that is what you would like me to tell you about; but it is only the first process that we are actually beginning to understand, so that is what I am forced to talk about. With guidance from physicists and engineers we are beginning to see how a real brain tackles a simple task of information engineering, moving an image from one place to another. For understanding the rest we also hope to benefit from those who actually perform the task of gaining knowledge from artificial images and that is one of the points of this symposium.

The following sketchy description of a very complex system will, I hope, give you some feeling for the results we have obtained and the problems that face us.

Macrostructure or Gross Anatomy

Figure 1 is a schematic view of the human visual system from above. The first point of functional interest is the rearrangement of nerve fibres from the two eyes at the optic chiasma; this has the effect of bringing the messages from the right hemifields seen by both eyes to the left cerebral hemisphere, those from the left hemifields of each eye to the right hemisphere. It is quite a shock to realise that the two halves of your visual field are represented in separate structures several centimetres apart. There is, however, a large bundle of nerve fibres called the corpus callosum joining the two halves, and this is in fact the largest tract of fibres in the body.

The structures shown here are composed of nerve cells and nerve fibres which influence each other through synapses. The approximate numbers of elements are indicated, and it also shows the order of the elements, which are synaptically connected in series. In the retina the light is absorbed in receptors (1st order) whose internal potential is thereby affected. These make a synapse with bipolar cells, and by graded release of chemical transmitter substances the internal potential of these bipolar cells is influenced, and they are therefore called 2nd order cells. These in turn influence the retinal ganglion cells whose long processes or axons constitute the optic nerve; these are 3rd order cells. This is much over-simplified, for there are two other types of cell in the retina, horizontal and amacrine cells, that provide other links between ganglion cells and receptors; in some cases amacrine cells may be interposed between bipolar cells and ganglion cells, which would make the ganglion cells and their axons 4th order. There is also some evidence for a direct connection from receptors to ganglion cells, which would make them 2nd order. In the cortex the pathways from cell to cell are so complex that it becomes even less possible to specify the order of a cell, but the numbers I have given do roughly indicate the position in the sequence.

Note that the number of components decreases from above 10^8 in the retina to about 10^6 in each optic nerve. The 10^8 does not represent the number of pixels, for the optical quality of the image in the periphery would not support that number. But it would support more than 10^6, the number of optic nerve fibres, and it is interesting to see how the retina deliberately degrades the quality of information transmitted from the periphery of the visual field: each fibre is excited by light falling in a large patch of roughly gaussian-shaped sensitivity profile, these patches getting progressively larger in size and fewer in number towards the periphery, so that resolution is deliberately degraded. The minimum resolvable angle is about 1 minute of arc at the fovea, and increases by about 1 minute for every 2 degrees of eccentricity, so is down to about 6 minutes at 10 degrees. Of course a shift of eye position changes the selection of information that is preserved or discarded, and it is important to realize that the human eye continually makes such a new selection by flicking from one place to another in the visual field. The frequency of these flicks varies greatly, from 2 or 3 times a second if you are reading or keenly interested in what you are inspecting, down to one flick every 2 or 3 seconds if you are in a more leisurely mood; but it is rare for 5 seconds to elapse without a flick. Of course the vast majority of shifts are made automatically without conscious effort; like breathing, in fact, but at a much greater repetition rate.

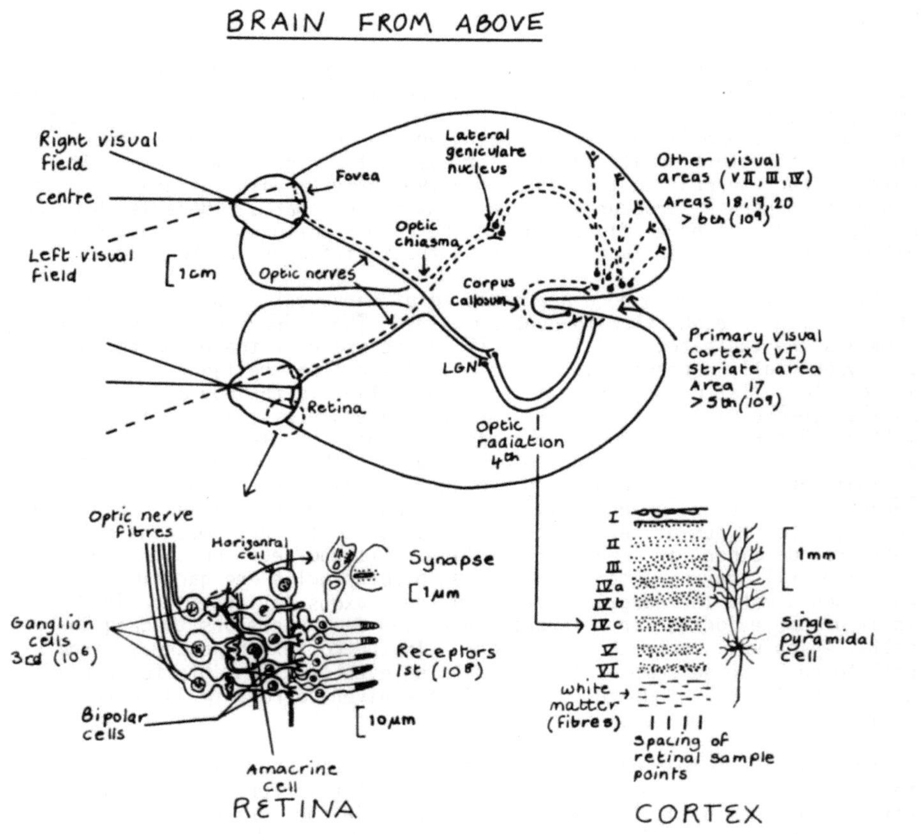

Fig. 1. Schematic view of the visual system viewed from above, with insets showing retinal neurones (left) and the laminae of visual cortex with a single pyramidal cell (right). The ordinal numbers indicate the serial order of the cell, those in parentheses the total number of cells of that type

In the cerebral cortex the numbers of elements increase to approximately 10^9 in the primary visual area (V 1) (also called the striate cortex or area 17) and another 10^9 (or possibly much more) in the surrounding secondary visual areas (also called parastriate, peristriate, V 2 to V 4 or areas 18, 19, 20). In the fourth layer of V 1 we know that there is a very precise representation of the visual field. It is grossly distorted because cortical area is apportioned according to the number of entering nerve fibres, rather than according to degrees of solid angle in the visual field; as we saw the resolution and density of ganglion cells decline rapidly away from the fovea. But it is precise in the sense that activity at one point corresponds unambiguously to something happening at one point in the visual field, and consequently it is possible to read backwards from the map to the real world: if there is activity at a particular spot in the map, this implies something happening at a particular point in the visual field of the eyes. Note, however, that there is more to be said about what that something is, and also that eye and head positions need to be known if you wish to point your finger to the position of the something in the field. Quite accurate topographical mapping also exists in the other layers of V 1, and there is evidence for quite good mapping in V 2, but in V 3 and V 4 it is very much less precise and appears to lack this unambiguous character that allows one to read back reliably from the map to the real world.

Retinal Components and what they do

Figure 1 also has a diagram of the retina, where it is possible to record from the nerve cells as they interact with each other [1]. Note first their size; though some are smaller, some larger, 10 µm is a reasonable figure for the size of a cell, and it is therefore possible to pack $10^6/mm^3$. This would leave little space between them, but packing densities close to this are certainly achieved, and may be exceeded in some places.

Now a word about how they work. The ionic concentrations of the fluid inside the cell are maintained by metabolically driven "pumps" at different values from those in the surrounding fluids. Electro-chemical gradients therefore exist across the cell membranes, and ions will pass down these gradients at a rate that depends upon the voltage difference between inside and outside, the concentration ratios for each particular ion, and the permeabilities of the cell membrane for each particular ion. Changes of these permeabilities in turn change the internal potentials of the cell. Take the receptor cell; without light the membrane is quite freely permeable to Na^+ ions, and since there is a strong inward electrochemical gradient for Na^+, there is a strong inward current in the resting condition. When light falls on the cell this inward current is reduced (by mechanisms that are not fully known) and this results in the inside of the cell becoming more negative. This change in potential spreads to the synapse, at the end of the cell remote from where light is absorbed, and there it causes a reduction in the release of a transmitter substance, whose nature is again unknown. This transmitter influences the permeability to ions of the next cell in the chain, the bipolar cell, with consequent effects on its own internal potential and the amounts of transmitter it releases at its own synapses.

And so it might continue, except for this difficulty: the potential at one end of a cell is not necessarily the same as that at the other, because if you do a simple calculation from the known ionic concentrations you will find that the resistance of the cytoplasm is considerable, and since that of the membrane is not infinite there will be substantial decrements of voltage along a cell. If the distances are only fractions of a millimetre this decay is tolerable, but passive conduction would be quite inadequate to convey potentials from retina to brain, let alone from your brain to your foot. Hence it is no surprise that in nerves longer than a millimetre or so information is transmitted by a different means: it is done by varying the number of all-or-nothing, self-sustaining, pulses transmitted along the axon. This is much more reliable, for the pulses almost always seem to reach their destination, but there is probably a considerable reduction in the dynamic range of signals. The maximum instantaneous rate of firing of impulses is about 1000/sec, but average rates rarely exceed 200/sec, so in a short period of, say, 1/20 sec only about 3 bits could be transmitted. This is obviously only a very small fraction of what you might expect from an electronic line.

Note that the potentials are slow also in their rise and fall times. The time to peak of a receptor in response to a brief flash is about 30 msec, during which time a fast-bowled cricket ball will have travelled over a yard. In spite of the slowness of the receptors the precision of judging the positions of moving objects corresponds to a time of about 1 msec or better, and it would be interesting to know how this is achieved.

So far I have outlined the biophysics of the retina, but what does it achieve in the way of information processing? Some of these steps are illustrated in Fig. 2, and they can be described very easily, though perhaps too simply, as 1) automatic gain control; 2) signalling deviations above and below the mean instead of absolute values; 3) high-pass temporal filtering; 4) high-pass spatial filtering; 5) conversion of graded signals into varying numbers of output pulses. The first occurs largely in the receptors: when the mean light level changes up or down, the input-output characteristic shifts bodily to higher or lower levels. The second occurs at the bipolar cells and is illustrated in the waveforms of Fig. 2. This also shows the transient character of the discharges from amacrine and ganglion cells. The fourth is brought about by a laterally connecting pathway through horizontal cells, which antagonizes the direct pathway through bipolar cells; the result is that the bipolars signal something approximating to the second spatial derivative of image luminance. The production of pulses is first seen in amacrine and ganglion cells.

Thus a network of elements such as exists in the retina can do tasks equivalent to spatial and temporal filtering, as well as automatic gain control and generation of a pulse-frequency modulated train of pulses. It is not too hard to see how this can be achieved by the interaction of graded signals, together with a threshold device for pulse generation, but we know from studies in other animals that such a network can achieve more. In the rabbit there are retinal ganglion cells that signal the direction of movement of the image over the retina, as well as the presence of oriented objects in the image [2, 3]. I shall not go into these here because there is no evidence for such pattern selectivity in the eyes of man or primates, but their presence in other animals raises an important point. So far I have spoken as if the cell was the important element, and mentioned that they could be packed at $10^6/mm^3$. But the evidence suggests that the crucial nonlinear interaction that gives rise to pattern selectivity can occur at each individual synapse [1,4]. In the layer where this occurs electron-micrographs show [1] that there are some $3 \times 10^8/mm^3$, and double this figure has been reported in the human cerebral cortex [5]. Synapses may be slow, but they are certainly compact.

There is one other point to discuss in connection with the retina: What is the need for the five operations I said it did? The need for pulse gener-

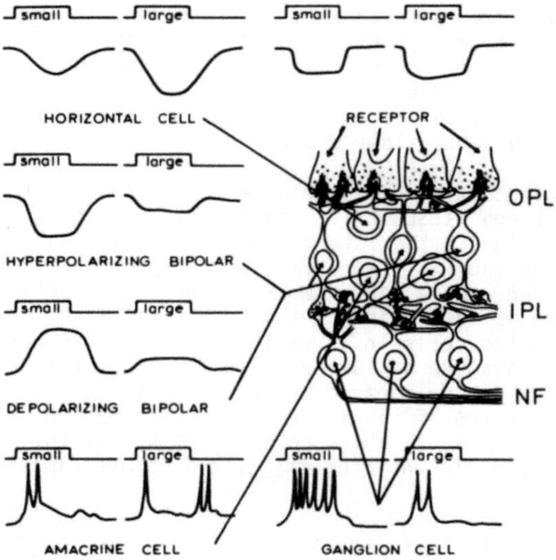

Fig. 2. Waveforms that are recorded from various cells in the vertebrate retina when a small stimulus spot is shone on to the retina centred on the cell, and when a larger spot that includes surrounding elements is used. The stimuli last about 1 second, and the responses are up to 30 mV in amplitude; in ganglion cells and amacrines the pulses would be larger if recorded with a full frequency response. Depolarization (i.e. reduction of intracellular negativity) is plotted upwards. (Adapted from ref. [2])

ation is clear enough, for graded potentials would decay to nothing before reaching the brain. The need for automatic gain control obviously arises from the enormous ($>10^{10}$) dynamic range of the input signals which have somehow to be passed down ouput channels of miserably narrow dynamic range, and signalling deviations from the mean also helps. The need for the other two can perhaps be explained in the same way, for it would be the low frequency signals that would tend to cause overload and saturation. But an explanation that is superficially quite different is sometimes given, namely the need to emphasize the borders of objects, or the temporal changes resulting from image movement, because these are "important" characteristics of images. Which of these explanations is correct? Personally I think that the two explanations are related: if one regards compression of the message while preserving its information, that is the reduction of redundancy,as the prime task, this will both tend to reduce the load on optic nerve fibres, and may also explain the psychological importance of transients. But we cannot go into predictive coding and such at this point.

Cortex

The retina performs an intelligible task, and we understand quite a lot about how it does it, but the situation is very different in the cortex, for it is an exceedingly complex structure and also very large. In the Rhesus monkey 60% of the whole surface area of the cerebral hemispheres is thought to be intimately concerned with vision [6]. One can divide these visual areas in the cortex into the primary area also called V 1, or striate area, and other areas called V 2, V 3, V 4, peristriate, parastriate, and so forth. We do not know for certain what tasks these areas perform, even in the case of V 1. Instead of burdening you with a mass of uninterpreted information I shall present a story which is fairly coherent, as far as it goes, but be warned that I have made a biassed selection of facts. What I shall say is certainly grossly incomplete and could be wildly wrong.

First look at the primary visual cortex (V 1), about which we know most, thanks mainly to the work of HUBEL and WIESEL [7]. The majority of the fibres from the retina, after relaying in the lateral geniculate nucleus, terminate on cells in layer IV. As you will see from the average separation of two fibres I have indicated, there are a vast number of cells available for each incoming fibre, and there is some evidence that the first task is to recreate among these closely packed cells in layer IV a very precise representation of the 2nd spatial derivative of luminance in the image [8]. This is a much simplified view, for I have ignored evidence about the different terminations of different classes of fibre, about the segregation of the endings from the two eyes in this layer, about colour, and so on, but this idea of the re-creation of the differentiated image, though oversimplified, will have to do for the present.

Now what is this map used for? The first step towards finding an answer is to look at the anatomy: Where does each cell pick up its information from? The answer for the cells representing the map is of course "from the incoming fibres of the optic radiation". But the cells in other layers must also be influenced predominantly from what is mapped locally, because the lateral connections to each cell are most dense in the immediate vicinity and are restricted to a total range of a few millimetres. We do not know the full story about the distribution of lateral spread of connections to a cell, but the total distance across V 1 is several cms, and few would deny that any given cell must be influenced predominantly by what is represented in the local parts of the map close to where it lies.

Next we need to see where the information is transmitted. It has emerged over the last ten years or so that the cells in different layers of the cortex send their messages to different parts of the brain [9]. The axons from the two lowest layers, VI and V, go back towards the region of the LGN relay and to a lower visual centre that I have not mentioned, the superior colliculus. The existence of a large population of fibres feeding back to the LGN is certainly interesting, but so far no-one has succeeded in finding out what they do or in making any theoretical sense of it. The fibres going to the superior colliculus make more sense, for this structure is concerned with the control of eye movements [10] and layer V is presumably important in helping to decide where the eye shall next fixate. However from the viewpoint of image processing the cells in layers II and III are most interesting, for they have the task of relaying information to the other visual areas. Thus the anatomy seems to be telling us that the first major step in image processing (after gain control, and spatial and temporal filtering) is to perform some kind of local analysis on each separate patch of image; this information is then passed on to other parts of the brain.

Now let us look at some physiological evidence (a film produced by HUBEL and WIESEL was shown at this point). Figure 3 shows diagrammatically what you find if you record from cells at various positions in the visual system and map what is known as their "receptive fields". This is the region of the visual field in front of the eye from which responses from a particular cell can be obtained. Initially one uses a small spot of light and records whether you get a response when it is turned on, when it is turned off, or

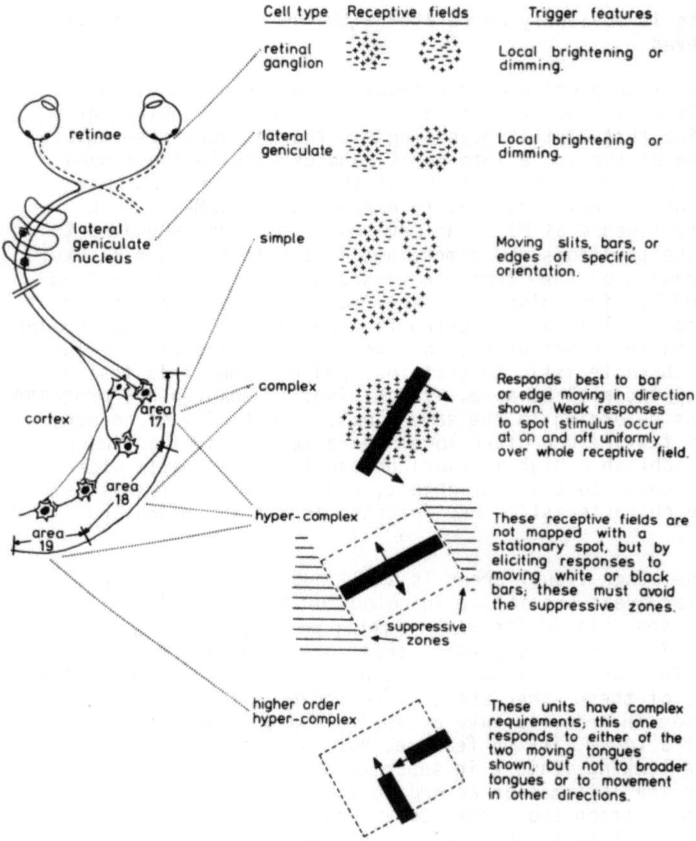

Fig. 3. Maps of the receptive fields that can be obtained from neurones at different stages in the visual pathway. (Adapted from ref. [2])

not at all. This figure shows the "centre-surround" arrangement found in the retina and LGN, which if you think it through you will see is what you would expect from a cell performing high-pass spatial filtering, or taking (on a rather coarse scale) the second spatial derivative of local light intensities. What is new in the cortex is the fact that the orientation of the stimulus becomes critically important, and this was one of the main findings made by HUBEL and WIESEL when they first recorded from cortical neurones [11] .

What cannot be shown in this figure are the results of more quantitative and extensive investigations of the properties of cells in V 1. They are selective for size as well as orientation, and some people [12, 13] believe that the activity of each cell in a set of cells represents the coefficient of one term in a local Fourier transform of the luminance distribution in a local region. We shall be hearing more about this view later on in this conference.

In addition to selectivity for orientation and size one finds that many cells require the stimulus to move in one direction rather than its opposite. Colour is also an important parameter in many cases [14] , and finally so is binocular disparity, that is the precise relative alignment of the images in the two eyes [15]; this is obviously correlated with the distance of a stimulus object from the eyes.

This list of the local properties of the image for which neurones in V 1 show selective sensitivity is interesting because it correlates quite well with the characteristics that Gestalt psychologists thought were responsible for causing segregation of the image into parts, and especially the separation of "figure" from "ground". The writings of this school are not easy to follow, but the principle was very clearly expressed in GUZMAN's work [16] in Artificial Intelligence at MIT. He devised a computer programme to derive an account of the positions and dimensions of a set of wooden blocks, given a single line drawing of them from one viewpoint. Since many of them were partially obscured by other blocks he required a means of identifying whether an edge seen to the left of an obstruction belonged to the same wooden block as an edge seen to the right of the obstruction. Co-linearity of the edges was the obvious characteristic to test for, and he named this test a search for a "linking principle" because, if positive, it enabled his programme to link two features as belonging to the same block. Similarity of colour, disparity, motion, and texture are other local characteristics of an image that can be used to establish a high a priori probability that particular parts belong, in some sense, to the same object, and it seems that this is likely to be why these characteristics are detected in V 1 and signalled elsewhere [17].

Our knowledge of the other visual areas is still rudimentary but there is evidence that particular areas specialize in colour or motion [18]; perhaps there is one such area specialized for each linking principle. This seems to make sense, and one is tempted to predict that the variables by which a particular linking feature is specified will turn out to be important in the anatomical arrangement of these other areas. The nerve cells of V 1 require an accurate topographic map in order to have access to the information that enables them to detect a <u>local</u> linking feature, but to detect the coincident occurrence of the same <u>linking</u> feature in separate parts of the visual field one needs to rearrange the information according to new principles: one needs to collect together information about particular characteristics regardless of whereabouts in the visual field they occur. At the moment there is little evidence that the brain makes such non-topographic maps [17], but time will tell.

Properties of Cortical Components

It is not easy to record the intracellular potentials of cortical neurones, but as far as is known they interconnect and communicate with each other in much the same way as the retinal neurones. However experiments have shown one important property, namely that these connections are not fixed solely by the mechanisms responsible for the development of the brain, but are subject to a limited amount of modification by the use, or lack of it, to which the connections are put during an early phase of maturation [19]. This is not too surprising, for something in the brain must be modifiable because we all learn and remember. Also, it has long been recognised that in young animals a function normally performed in one part of the brain can be taken over by another part following injury or experimental removal of the first part. This happens even with a very complex function such as speech, for injury to the dominant left hemisphere in a child does not prevent the acquisition of good capacity for language in the non-dominant hemisphere, whereas the equivalent damage to an adult would cause profound and permanent defects.

We have no idea how such things happen, but a simpler form of plasticity can be demonstrated in the neurones of the visual system by rearing animals in abnormal visual environments. I cannot go into these experiments here, but I think they require for their explanation a mechanism which was postulated by HEBB many years ago [20, 21]. If activation of a synapse is followed by activity of the postsynaptic neurone, then the effectiveness of that synapse increases. If the synapse fails to activate the neurone, then not only does it become less effective, but also the whole of the presynaptic neurone seems to be adversely affected: its cell body shrinks, and its terminal arborisations wither and become less extensive. Somewhat similar positive and negative effects are produced in peripheral nerves of the sympathetic nervous system by excess or deficiency of a well-characterised substance called nerve growth factor. It is therefore tempting to postulate that "synaptic rewarding factors" are released by activated neurones and picked up by the activating synapses; these would regulate the effectiveness of central synapses and promote growth of the presynaptic cell and sprouting of the successful terminal branches. Components with such adaptable properties would, I imagine, be exciting for electronic engineers, though also disconcerting.

The role of Psychophysics

What I have said so far may have given the impression that what we know about the visual system has been derived almost entirely from anatomical and physiological investigation. Nothing could be more misleading, for all the important properties of the visual system were first established by psychophysical and psychological observations made on the system working as a whole. For example, more than 100 years passed between the establishment of trichomacy as a fact of human colour vision, and the detection within the last fifteen years of three different cone photoreceptors containing three different photosensitive pigments. It is the same with sensitivity, for it was established forty years ago that a rod photoreceptor must signal the absorption of a single quantum, but these signals have only been directly recorded within the last couple of years. Anatomists and physiologists need to be told what the visual system does before they can set about the difficult task of finding out how it does it.

Unique Features of Natural Vision

Table 1 attempts to summarise what goes on in the early stages of natural vision, but what makes it different from artificial image processing? Table 2 attempts to summarize the unique features of natural image processing. I have already mentioned some properties of the components: they are very slow and have a narrow dynamic range, but they can be packed at very high density and have an interesting capacity for self-regulation. The parallel organization of elements is also unusual, and I am sure most of you would agree that nature must have some interesting programmes up her sleeve for handling images, but I do not think anatomists and physiologists are likely to discover what they are without the help of those who handle artificial images. DAVID MARR [22] has argued persuasively that the "computational theory of vision" must be understood before much progress can be made, and I think there is a possibility of fruitful interaction here, provided that we listen to each other. But I think there is a major difference between the aim or purpose of our own visual system and that of most artificial image processing, and I want to enlarge on this briefly.

Table 1. Suggested first steps in natural image processing

Retina	Lateral Geniculate Nucleus	Primary visual Cortex (V 1)	Other cortical areas
Automatic gain control Signals deviations from mean not absolute luminance High-pass temporal filtering High-pass spatial filtering	Aligns, but does not combine inputs from the two eyes More high-pass filtering Controls input to cortex	Detects local features of image that are likely to belong to same object – linking features:- Orientation Movement Colour Texture Disparity	? Specialize in individual linking features ? Mapping according to non-topographic variables

Table 2. Unique features of natural image processing

Components	Organization	Purpose
Very slow (<1000 Hz) Narrow range (<8 bits) Very many very small junctions ($10^9/mm^3$) Self-regulatory property	Junctions arranged in parallel Pattern of connections determines program Many aspects not understood	To represent visual scenes by activity of a sparse selection of reliable and non-redundant (i.e. independent) elements

What our own visual system does for us can be crudely described as follows. From a view lasting a second or two it provides reliable knowledge of the positions in space of some dozens of familiar or semi-familiar objects. Let us do a very approximate calculation on the amount of information this represents. Suppose there are 100 objects, each requiring two 16-bit words to identify in our internal catalogue of known objects. For each we might need 3 words for specifying position, and another 3 for orientation. Even with these quite generous figures we only seem to require about 800 16-bit words to specify a particular visual scene, though this would also require a very large store of information about familiar objects. A few more words would be needed if any objects were moving or had an unusual colour, and also for such items as illumination conditions, but in spite of the crudity of this estimate it really seems that such a representation of a whole visual scene would only require about 1000 words. This is quite remarkable, for it is two orders of magnitude below the requirements of a single raw image of any quality. Of course the representation is incomplete and much information will have been lost, but surely the main economy must be achieved be exploiting the redundancy of the population of images that our eyes interpret for us.

This is an idea with an ancient ancestry [23,24] and ATTNEAVE [25] applied it to the psychology of vision 3 decades ago, while I have argued [26] that the physiological facts fit the idea. I think it makes good sense to interpret the early transformations in the visual pathway as measures to reduce redundancy, or attempts to find symbols that are used with high relative entropy and are adequate to represent the image [27]. Gain control, subtracting off the mean value, frequency filtering in the temporal and spatial domains, all fit this interpretation, as does segmentation into sub-regions with more uniform properties. There must be many more stages we do not yet understand, such as the formation of a catalogue of familiar objects together with an adequate means of referring to it: however I think we can see the direction the route is heading even though we have not gone far along it.

One important point that arises from this description of the overall goal of perception is that it gives a means to test each step to see whether progress towards that goal is actually being made; indeed this yard-stick may be what enables the self-regulating properties I mentioned earlier to be effectively utilized. Perhaps if you were to set yourselves the goal of representing images as economically as our brains seem to, and if you were to reach it, then you might achieve a result that would be of very wide interest and importance.

Since this talk was supposed to be about understanding natural vision I must conclude by saying what should by now be obvious:– We do not understand much of it. But we are optimistic that we shall go on finding out more because we have a working system always available to study and we have the physical image processors to tell us what to look for.

References

1 J.E. Dowling, Investigative Ophthal. 9, 655 (1970).
2 H.B. Barlow and J.D. Mollon (Eds), *The Senses* (Cambridge University Press, 1982).
3 W.R. Levick, J. Physiol. 188, 285 (1967).
4 V. Torre and T. Poggio, Proc. Roy. Soc. B. 202, 409 (1978).
5 M. Collonier, "The electron-microscopic analysis of the neuronal organization of the cerebral cortex", in *The Organization of the Cerebral Cortex,* F.O. Schmitt, F.G. Worden, G. Adelman and S.G. Dennis, eds. (MIT Press, Cambridge, Massachusetts, 1981).
6 D.C. Van Essen and J.H.R. Maunsell, J. Comp. Neurol. 191, 255 (1980).
7 D.H. Hubel and T.N. Wiesel, Proc. Roy. Soc. B. 198, 1 (1977).
8 H.B. Barlow, Nature 279, 189 (1979).
9 C.D. Gilbert and T.N. Wiesel, "Laminar specialization and intracortical connections in cat primary visual cortex", in *The Organization of the Cerebral Cortex,* F.O. Schmitt, F.G. Worden, G. Adelman and S.G. Dennis eds (MIT Press, Cambridge, Massachusetts, 1981).
10 R.H. Wurtz and J.E. Albano, Ann Rev. Neurosci. 3, 189 (1980).
11 D.H. Hubel and T.N. Wiesel, J. Physiol. 160, 106 (1962).
12 J.G. Robson,"Neural images: The physiological basis of spatial vision", in *Visual Coding and Adaptability,* C.S. Harris, ed. (L. Erlbaum and Associates, Hillsborough, New Jersey, 1980).
13 B. Sakitt and H.B. Barlow, Biol. Cybernetics 43, 97 (1982).
14 C.R. Michael, J. Neurophysiol. 41, 572 (1978).
15 H.B. Barlow, C. Blakemore and J.D. Pettigrew, J. Physiol. 193, 327 (1967).
16 A. Guzman, Proc. Fall Joint Computer Conf. (San Francisco) Vol. 33, Pt 1, 291 (1968).
17 H.B. Barlow, Proc. Roy. Soc. B. 212, 1 (1981).

18 S.M. Zeki, Nature 274, 423 (1978).
19 D.H. Hubel and T.N. Wiesel, J. Physiol. 206, 419 (1970).
20 D.O. Hebb, *The organization of behaviour*, (Wiley, New York, 1949).
21 J.P. Rauschecker and W. Singer, Nature 280, 58 (1979).
22 D. Marr, *Vision*, (W.H. Freeman and Co., San Francisco, 1982).
23 E. Mach, *The analysis of sensations, and the relation of the physical to the psychical*. (Dover Publications Inc, New York, 1959; 1st edition 1886).
24 K. Pearson, *The Grammar of Science*, (Walter Scott, London, 1892).
25 F. Attneave, Psychol. Rev. 61, 183 (1954).
26 H.B. Barlow, "Possible principles underlying the transformations of sensory messages, in *Sensory Communication*, W. Rosenblith, ed. (MIT Press and John Wiley, New York, 1961).
27 H.B. Barlow, "Perception: What quantitative laws govern the acquisition of knowledge from the senses?", in *Functions of the Brain, Wolfson College Lectures 1982*, C. Coen, ed. (Oxford University Press, 1983).

Aspects of Visual Automation

J.R. Ullmann

Department of Computer Science, The University of Sheffield,
Sheffield S10 2TN, United Kingdom

Introduction

This has been written in response to an invitation to present a paper which would serve as a tutorial review of techniques in image analysis and understanding by machine. Accordingly this paper contains some very elementary tutorial material; but in limited space it is not possible to present a comprehensive distillation of dozens of books, e.g. [1,2,3,4,5] and thousands of papers, many of which are catalogued in ROSENFELD's annual surveys, e.g. [6]. The invitation also welcomed the illustration of such a talk with examples from my own work and current interests. Accordingly, this paper includes exploratory material concerning generalised correlation, which leads to a theory of surface representation of three-dimensional objects.

We begin with an elementary introduction to recognition. Recognition is not a fashionable topic in "computer vision" [7,8] but is essential groundwork for this paper.

2. Recognition based on Attributes

In the technical literature a pattern is essentially an array of values. If any of these values is changed, the pattern is thereby changed to a new pattern. Fig 1b, which is a digitised image of an apple, is an example of a two-dimensional pattern. The apple is shown in outline in Fig. 1a. Each value in Fig 1b is an example of a pixel or pel and is a measure of the average brightness over a small area in the image. Figure 1c is a slightly different image of the same apple. Because of this difference, although it is slight, Figs. 1b and c are said to be different patterns. Fig. 1d is yet another image of the apple, and this is identical to Fig. 1a except that the apple has moved parallel to the image plane. This shift makes Fig. 1d a different pattern to Fig. 1b. In Fig 1e the apple has moved towards us, and therefore Fig. 1e and Fig. 1b are different patterns. Associated with an apple there is generally not just one possible pattern, but many thousands, of which Fig.1 shows four examples.

These thousands of apple-patterns constitute a recognition class, that is, a class of patterns to which the same name should be assigned. By way of example we will consider a machine that is required to determine whether a given pattern belongs to the apple class or the banana class or the orange class, or perhaps to none of these. Recognition means making such a determination, assigning a pattern to a known class, thus associating a name with (at least part of) an image.

(a)

0	0	0	0	0	0	0	0	0	0	0	0
0	0	0	0	0	0	0	0	0	0	0	0
0	0	0	0	0	4	0	0	0	0	0	0
0	0	0	4	4	4	4	4	0	0	0	0
0	0	4	4	5	5	5	4	4	0	0	0
0	4	4	5	5	5	6	5	4	4	0	0
0	4	4	5	5	6	6	6	5	4	0	0
0	4	5	5	5	6	6	5	4	0	0	0
0	0	4	4	5	5	5	4	0	0	0	0
0	0	0	4	5	5	4	0	0	0	0	0
0	0	0	0	4	4	4	0	0	0	0	0
0	0	0	0	0	0	0	0	0	0	0	0
0	0	0	0	0	0	0	0	0	0	0	0
0	0	0	0	0	0	0	0	0	0	0	0
0	0	0	0	0	0	0	0	0	0	0	0
0	0	0	0	0	0	0	0	0	0	0	0
0	0	0	0	0	0	0	0	0	0	0	0
0	0	0	0	0	0	0	0	0	0	0	0

(b)

0	0	0	0	0	0	0	0	0	0	0	0
0	0	0	0	0	0	0	0	0	0	0	0
0	0	0	0	0	4	0	0	0	0	0	0
0	0	0	4	4	5	4	4	0	0	0	0
0	0	4	4	5	5	5	4	4	0	0	0
0	4	4	5	5	5	6	5	4	4	0	0
0	4	4	5	5	6	6	6	5	4	0	0
0	4	5	5	5	6	6	5	4	0	0	0
0	0	4	4	5	5	5	4	0	0	0	0
0	0	0	4	5	5	4	0	0	0	0	0
0	0	0	0	4	4	4	0	0	0	0	0
0	0	0	0	0	0	0	0	0	0	0	0
0	0	0	0	0	0	0	0	0	0	0	0
0	0	0	0	0	0	0	0	0	0	0	0
0	0	0	0	0	0	0	0	0	0	0	0
0	0	0	0	0	0	0	0	0	0	0	0
0	0	0	0	0	0	0	0	0	0	0	0
0	0	0	0	0	0	0	0	0	0	0	0

(c)

0	0	0	0	0	0	0	0	0	0	0	0
0	0	0	0	0	0	0	0	0	0	0	0
0	0	0	0	0	0	0	0	0	0	0	0
0	0	0	0	0	0	0	0	0	0	0	0
0	0	0	0	0	0	0	0	0	0	0	0
0	0	0	0	0	0	0	0	0	0	0	0
0	0	0	0	0	0	0	0	0	0	0	0
0	0	0	0	0	0	4	0	0	0	0	0
0	0	0	0	4	4	4	4	4	0	0	0
0	0	0	4	4	5	5	5	4	4	0	0
0	0	4	4	5	5	5	6	5	4	4	0
0	0	4	4	5	5	6	6	6	5	4	0
0	0	4	5	5	5	6	6	5	4	0	0
0	0	0	4	4	5	5	5	4	0	0	0
0	0	0	0	4	5	5	4	0	0	0	0
0	0	0	0	0	4	4	4	0	0	0	0
0	0	0	0	0	0	0	0	0	0	0	0
0	0	0	0	0	0	0	0	0	0	0	0

(d)

0	0	0	0	0	0	0	0	0	0	0	0
0	0	0	0	0	0	4	0	0	0	0	0
0	0	0	0	4	4	4	4	0	0	0	0
0	0	4	4	4	4	4	4	4	4	0	0
0	0	4	4	4	5	5	5	4	4	4	0
0	4	4	4	5	5	5	5	5	4	4	0
4	4	4	4	5	5	5	6	6	5	5	4
4	4	4	5	5	5	6	6	6	6	5	4
4	4	4	5	5	6	6	6	6	6	5	4
0	4	4	5	5	5	6	6	6	6	5	4
0	0	4	5	5	5	5	6	6	5	4	0
0	0	4	4	5	5	5	5	5	4	4	0
0	0	4	4	4	4	5	5	4	4	0	0
0	0	0	0	4	4	4	4	0	0	0	0
0	0	0	0	0	0	0	0	0	0	0	0
0	0	0	0	0	0	0	0	0	0	0	0
0	0	0	0	0	0	0	0	0	0	0	0
0	0	0	0	0	0	0	0	0	0	0	0

(e)

Fig.1. Examples of images of an apple

A recognition class must somehow be defined. In principle, a class can be defined by explicitly listing all its members: this is known as extensive definition. Alternatively a class can be defined by specifying an attribute, or a combination of attributes, possessed by all members and by no non-members of the class: this is known as intensive definition. For our fruit-recognition example, extensive definition would involve storing many thousands of digitised images of each fruit. Not only would this have formidable storage requirements, but also it would make the machine unable to recognise any new image that had not been stored. For practical pattern recognition we may anyway be unable to obtain complete extensive definitions.

Intensive definition is usually more practical. For our fruit-recognition example, two attributes might suffice. One of these attributes could, for instance, be a rough measure of colour. The machine could obtain two images of a fruit, one via a red filter and one via a green filter. For each such pair of images the machine could compute the ratio of the average red-image intensity to the average green-image intensity. We will call this the redness ratio. It should be useful for distinguishing between an apple and an orange that have the same shape.

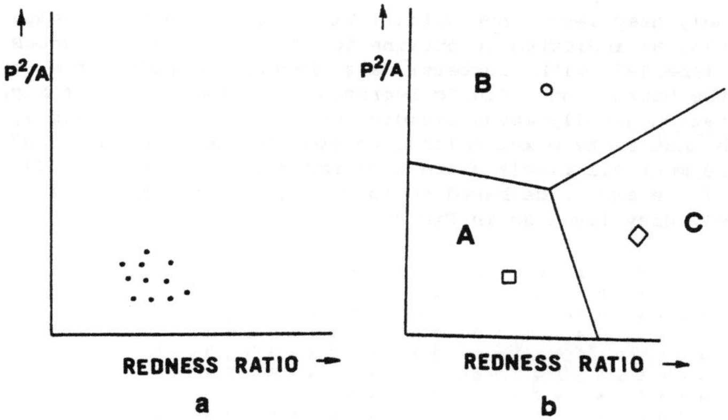

Fig.2. Representation space

To distinguish between an apple and a banana that have the same colour, we will measure perimeter-squared-over-area, which is a popularly-used attribute of shape. The dots in Fig.2a represent colour and shape attributes of a sample set of apple-patterns: there is one dot per pattern. The square in Fig.2b represents the average, or centre of gravity, of the Fig.2a dots. The diamond in Fig.2b denotes the average of a collection of dots, not shown in the figure, that represent a sample set of banana patterns. Similarly, the circle in Fig.2b represents the orange average. To recognise a given pattern, our machine could evaluate its redness ratio and perimeter-squared-over-area. These attributes would define a representative point in Fig.2b. The given pattern could be recognised as belonging to the class whose average was closest to the pattern's representative point. Boundaries between regions A, B and C in Fig.2(b) have been drawn so that region A contains all points that are closer to the apple average than to either of the other averages. Similarly, regions B and C respectively contain all points that are closest to the banana and orange averages. The use of our simple method of recognition implicitly subdivides Fig.2(b) into three regions A, B and C that respectively contain all apple, banana, and orange points.

Given a sample set of apple patterns, our machine could automatically average attributes, so as to fix the position of the square in Fig.2b. The machine could do the same thing for oranges and bananas, thus automatically determining inter-class boundaries like those shown in Fig.2b. This automatic determination is an example of learning; or, at least, of a computation that was called learning in the pattern recognition literature of the 1960's. Actually this example is unsophisticated: the literature of "statistical pattern recognition" is largely concerned with exploration of improved methods of automatic determination of inter-class boundaries [1,9].

In our example we chose two attributes quite arbitrarily. MUI and FU [10] and PAVLIDIS [35] have listed many attributes that may possibly be useful. A machine can automatically select, from a collection of candidate attributes, a "good" set of attributes for distinguishing between given recognition classes [9].

3. Segmentation

To measure an object's attributes such as perimeter-squared-over-area, a machine must first find and demarcate the object in a given picture. By way of example, using only grey level, not colour, Fig.3b is a digitised image of an apple and a banana as indicated in outline in Fig.3a. Fig.3c shows regions of Fig.3b labelled with letters: a's belong to apples, b's to bananas, and g's to the background. Fig.3c represents a segmentation of Fig. 3b. The term segmentation usually means dividing up a picture into regions, called segments, such that every pixel belongs to exactly one segment, and within any given segment all pixels share a characteristic property [20]. This property might, for example, be based on colour [11], motion [12,13], texture [14], or merely grey-level as in Fig.3b.

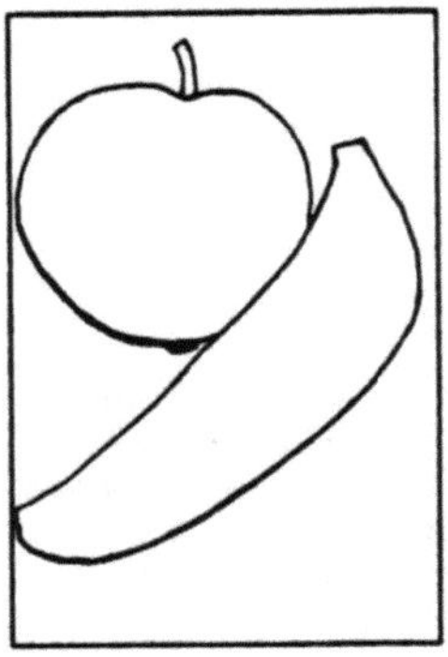

(a)

0	0	2	2	2	2	2	2	2	1	0	0
0	2	2	2	4	2	2	2	1	1	1	0
1	1	4	4	4	4	4	2	2	1	1	0
1	4	4	5	5	5	4	4	2	1	1	0
4	4	5	5	5	6	5	4	4	7	1	0
4	4	5	5	6	6	6	5	4	7	7	1
4	5	5	5	6	6	5	4	7	8	7	1
1	4	4	5	5	5	4	7	7	8	7	2
1	2	4	5	5	4	7	7	7	8	7	2
1	2	2	4	4	4	7	7	8	8	7	2
1	1	2	2	2	7	7	8	9	7	2	2
1	1	1	2	7	7	8	9	8	2	2	2
1	1	7	7	8	9	9	7	2	2	2	2
1	7	7	8	8	7	7	2	2	2	2	2
7	7	8	7	7	2	2	2	2	2	1	1
7	7	7	2	2	2	2	2	2	2	1	1
2	2	2	2	2	2	2	2	2	1	0	0
2	2	2	2	2	1	1	1	1	0	0	0

(b)

g	g	g	g	g	g	g	g	g	g	g	g
g	g	g	g	a	g	g	g	g	g	g	g
g	g	a	a	a	a	a	g	g	g	g	g
g	a	a	a	a	a	a	a	g	g	g	g
a	a	a	a	a	a	a	a	a	b	g	g
a	a	a	a	a	a	a	a	a	b	b	g
a	a	a	a	a	a	a	a	b	b	b	g
g	a	a	a	a	a	a	b	b	b	b	g
g	g	a	a	a	a	b	b	b	b	b	g
g	g	g	a	a	a	b	b	b	b	b	g
g	g	g	g	g	b	b	b	b	b	g	g
g	g	g	g	b	b	b	b	b	g	g	g
g	g	b	b	b	b	b	b	g	g	g	g
g	b	b	b	b	b	b	g	g	g	g	g
b	b	b	b	b	g	g	g	g	g	g	g
b	b	b	g	g	g	g	g	g	g	g	g
g	g	g	g	g	g	g	g	g	g	g	g
g	g	g	g	g	g	g	g	g	g	g	g

(c)

0	0	2	2	2	2	2	2	2	1	0	0
0	2	2	2	4	2	2	2	1	1	1	0
1	1	4	4	4	4	4	2	2	1	1	0
1	4	4	5	5	5	4	4	2	1	1	0
4	4	5	6	6	7	5	4	4	7	1	0
4	4	5	6	7	7	6	5	4	7	7	1
4	5	5	5	7	6	5	4	7	8	7	1
1	4	4	5	6	5	4	7	7	8	7	2
1	2	4	5	5	4	7	7	7	8	7	2
1	2	2	4	4	4	7	7	8	8	7	2
1	1	2	2	2	7	7	8	9	7	2	2
1	1	1	2	7	7	8	9	8	2	2	2
1	1	7	7	8	9	9	7	2	2	2	2
1	7	7	8	8	7	7	2	2	2	2	2
7	7	8	7	7	2	2	2	2	2	1	1
7	7	7	2	2	2	2	2	2	2	1	1
2	2	2	2	2	2	2	2	2	1	0	0
2	2	2	2	2	1	1	1	1	0	0	0

(d)

1	1	1	1	1	1	1	1	1	1	1	1
1	1	1	1	5	1	1	1	1	1	1	1
1	1	5	5	5	5	5	1	1	1	1	1
1	5	5	5	5	5	5	5	1	1	1	1
5	5	5	5	5	5	5	5	5	5	1	1
5	5	5	5	5	5	5	5	5	5	5	1
5	5	5	5	5	5	5	5	5	5	5	1
1	5	5	5	5	5	5	5	5	5	5	1
1	1	5	5	5	5	5	5	5	5	5	1
1	1	1	5	5	5	5	5	5	5	5	1
1	1	1	1	1	5	5	5	5	5	1	1
1	1	1	1	5	5	5	5	5	1	1	1
1	1	5	5	5	5	5	5	1	1	1	1
1	5	5	5	5	5	5	1	1	1	1	1
5	5	5	5	5	1	1	1	1	1	1	1
5	5	5	1	1	1	1	1	1	1	1	1
1	1	1	1	1	1	1	1	1	1	1	1
1	1	1	1	1	1	1	1	1	1	1	1

(e)

Fig.3. Images of an apple and banana touching

GURARI and WECHSLER [15] insist that each segment must be connected. Roughly speaking, a segment is connected if and only if it is possible to travel betweeen any two pixels within the segment without ever going outside the segment. ROSENFELD and KAK [2] define digital connectivity precisely. The requirement that each segment be connected does not appear useful in the segmentation of text into separate characters: a character that is broken may more conveniently be regarded as a single segment. An asbestos fibre, for purposes of measurement (for health protection) should preferably be deemed a single segment, even if its middle part is totally occluded.

The segmentation of Fig.3b that is shown in Fig.3c was obtained by applying the intensive definitions:

pixels in the range 0...2 are g's
pixels in the range 4...6 are a's
pixels in the range 7...9 are b's.
When appropriate intensive definitions exist but are not known a priori they can be determined automatically [16], for example by determining a histogram of grey levels and assigning values between two successive valleys in the histogram to the same segment.

Fig.3d is another apple-touching-banana image for which Fig.3c would be a plausible segmentation. For Fig.3d the a,b, and g regions do not correspond to distinct (non-overlapping) ranges of pixel values. For example the value 7 occurs both in the a's and b's regions. To deal with this we can arbitrarily select a pixel in Fig.3d and assign to it a label which might be one of the letters 'a','b','g'. We can find the difference in value between this pixel and each of its eight neighbours. We can assign the same label to each neighbour for which the difference is less than a threshold. Threshold=1 would be alright for Fig.3d. We can repeat the whole process for neighbours of newly labelled pixels, and stop only when no further pixels are given the same label. We can then repeat the whole process again, starting by assigning a new label to a remaining non-labelled pixel. We finally stop when no pixel remains unlabelled. This is a very simple example of a region-growing technique [16,17]. Regions may not end up with the specific labels that we would like them to have, unless we correctly assign the initial label in each region [18]. Region-growing automatically yields segments that are connected.

The methods of segmentation that we have so far mentioned are region-based. After accomplishing region-based segmentation it is easy to determine edges between regions, by picking out pixels where different segment-labels are adjacent. Instead of starting by determining regions and hence determining edges, we can, if we wish, adopt a dual approach, wherein we start by detecting edges and hence determine regions. A segment is now a region that is enclosed by edges and is not subdivided by any edge. Edge-detection is the subject of a substantial literature [2,8,16,19,20]. Region-based and edge-based methods can be combined to yield improved segmentation [21,22].

Fig. 3e. is a picture of an apple and a banana, as indicated in Fig.3c, but both of these objects have the same grey-level, and therefore they cannot be separated by region-based or edge-based segmentation. A method of BJOKLUND and PAVLIDIS [23] will separate a constant-grey-level region into maximal almost-convex segments; but we may not wish a very bent banana to be subdivided into more than one segment. When there is no contrast between segments, as in Fig.3e, we may only be able to achieve useful segmentation by correlation [2].

4. Correlation

4.1 Simple Correlation

A pattern that has not yet been recognised is said to be unknown. A classical method of recognition is to store one or more specimen patterns of each recognition class, and assign an unknown pattern to the class of the best-matching specimen pattern. We now explore the possibility of computing a matching score from the matching of corresponding pixels, instead of using attributes such as perimeter-squared-over-area.

We will use a simple one-dimensional example to illustrate this idea. In this example a 9-pixel unknown pattern $X=\{x_1,\ldots,x_i,\ldots,x_9\}$ is matched with a 9-pixel specimen pattern $Y=\{y_1,\ldots,y_i,\ldots,y_9\}$; and pixel-values are single digits. We compute a matching score for all plausible alignments of X with Y. Three such alignments are:

X = 0 0 0 3 4 6 3 2 0
Y = 0 4 4 5 3 2 0 0 0

X = 0 0 0 3 4 6 3 2 0
Y = 0 4 4 5 3 2 0 0 0

X = 0 0 0 3 4 6 3 2 0
Y = 0 4 4 5 3 2 0 0 0

We have drawn lines between corresponding pixels: a line joins pixel x_i to a corresponding pixel $y_{f(i)}$. The matching score might, for example, be

$$\sum_{\text{all } i} (y_{f(i)} - x_i)^2$$

The simpler score

$$\sum_{\text{all } i} (y_{f(i)} \text{ times } x_i)$$

is less satisfactory because, for example, its value is the same for

X = 0 0 0 3 4 6 3 2 0
Y = 0 0 0 4 4 0 0 0 0

as it is for

X = 0 0 0 3 4 0 0 0 0
Y = 0 0 0 4 4 0 0 0 0

The term correlation usually means evaluation of a matching score for each alignment. For two-dimensional patterns we evaluate a similar matching score for each of many two-dimensional relative displacements [24]. This involves considerable computation, which can be reduced by short-cut methods [25]. An unknown pattern can be assigned to the class of the specimen pattern for which, for some alignment, the matching score is higher than for any alignment with any other specimen pattern [24]. After we have generalised this idea we will explain some of its advantages over basing recognition on attributes such as perimeter-square-over-area.

4.2 Generalised Correlation

Suppose, for example, that Fig.1b is a specimen apple-pattern and that Fig.1d is an unknown pattern. Simple correlation between Figs.1a and c would give a good matching score, because simple correlation is designed to tolerate shift, that is, misalignment. Simple correlation cannot cope with dilation, and therefore would yield a poor score for Fig.1e matched with Fig.1b.

A practical recognition system may be required to tolerate shift and (at least limited) dilation, rotation, general perspective [26], and possibly also various non-analytic distortions of shape. A non-analytic distortion is one that cannot be formulated analytically, for example the distortion of a speaking mouth. In order to explore ideas we will make a simplifying assumption that in a given practical situation, there is a clearly-defined set of distortions that should be tolerated, and we will call these admissible distortions [27]. These are only distortions of shape, not variations in shading and colouring due, for example, to variations in conditions of illumination. Generalised correlation is a generalisation of correlation that is designed to provide tolerance for all admissible distortions but for no non-admissible distortions of shape. In principle, but usually not in practice, generalised correlation of an unknown pattern X with any specimen pattern Y could be accomplished as follows. For each admissible distortion, we apply this distortion to Y and evaluate a matching score for X with the distorted Y. The best of these matching scores can be used, in recognition, as a measure of the match between X and Y. If we know the distortion that yields the best matching score, then we know which point in X corresponds to which point in Y. In the special case when the only distortions that are admissible are shifts, generalised correlation is the same thing as simple correlation.

The term generalised correlation is new, but the idea, and its computational extravagance, are well known [28]. When, as is usually the case, the number of admissible distortions is large, then it is not practical to evaluate a matching score separately for each distortion in an extensively defined set of admissible distortions. If generalised correlation is to be of wide practical use it must be implemented using an intensive definition of the set of admissible distortions, but must still determine the optimal matching score and the optimal distortion (i.e. correspondence) that would have been found if extensive definition had been used. We do not yet know how to achieve this, nor do we know under what conditions this is achievable. Flexible matching methods of BURR [29] and TSUJI et al. [30] avoid extensive definitions of distortions, but perhaps provide insufficient control of discrimination between admissible and non-admissible distortions. Better control may be achievable at a cost that is currently unacceptably high [31].

In the present paper we will consider the question: if we had a black box capable of performing generalised correlation at acceptably low cost, what could this usefully achieve that could not readily be achieved otherwise? Exploration of this question has tutorial value and should be of interdisciplinary interest. Generalised correlation is a correspondence-determining process that has fairly obvious relevance to stereopsis [32] and interpretation of motion [33], which are areas of interdisciplinary research.

4.3 Segmentation by Correlation

We have mentioned previously that contrast-based (i.e. edge or region based) segmentation methods cannot straightforwardly be used to segment Fig.3e as shown in Fig.3c. If Fig. 1b is correlated with Fig.3e the best-matching alignment picks out the apple region in Fig.3e. This illustrates segmentation-by-correlation, which is used in practice for recognising touching characters in printed text [34]. As well as achieving segmentation, correlation yields a matching score that can be used in recognition, as we have explained previously.

If distortions other than shift are admissible, then we must replace correlation by generalised correlation. For example, if only Fig.1e, not Fig.1b, is available as a specimen of an apple-pattern, Fig.1e can be used to segment Fig.3e by generalised correlation, but not by simple correlation, because the apples differ in size in Figs.1e and 3e.

We have so far mentioned two methods of recognition. In one method, contrast-based segmentation is <u>followed by</u> determination of attributes such as perimeter-squared-over-area. In the other method, generalised correlation achieves segmentation <u>at the same time</u> as it yields a matching score; and this has the advantage that segmentation does not depend on the existence of contrast. An obvious disadvantage is that implementational problems are as yet unsolved. Even if they were solved, correlation would not be appropriate for matching textures that "look the same" but are not identical patterns. Automatic systems usually deal with texture by measuring attributes [14,36,37] rather than by correlation.

In Sections 5 and 6 we mention problems that may best be dealt with by some sort of correlation, rather than by measuring attributes.

5. Occlusion

<u>Occlusion</u> is the obstruction of our view of an object by another object. Occlusion has been investigated mainly in systems that attempt to describe the relative positions of points in a scene, without actually recognising objects [38,39].

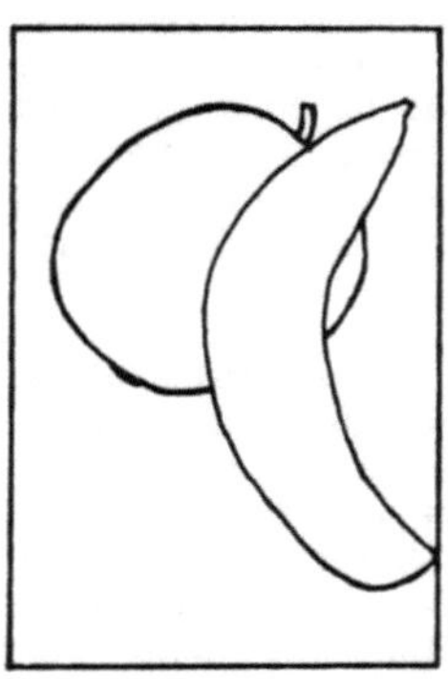

(a)

0	0	0	0	0	0	0	0	0	0	0	0
0	0	0	0	0	0	0	0	0	0	0	0
0	0	0	0	0	4	0	0	0	7	7	0
0	0	0	4	4	4	4	4	7	7	7	0
0	0	4	4	5	5	5	7	8	7	0	0
0	4	4	5	5	5	7	8	7	7	0	0
0	4	4	5	5	7	7	8	7	4	0	0
0	4	5	5	5	7	8	7	7	0	0	0
0	0	4	4	5	7	8	7	7	0	0	0
0	0	0	4	5	7	8	7	7	0	0	0
0	0	0	0	4	7	8	7	7	0	0	0
0	0	0	0	0	7	8	7	7	7	0	0
0	0	0	0	0	0	7	8	7	7	0	0
0	0	0	0	0	0	0	7	8	7	7	0
0	0	0	0	0	0	0	0	7	8	7	7
0	0	0	0	0	0	0	0	0	7	7	0
0	0	0	0	0	0	0	0	0	0	0	0
0	0	0	0	0	0	0	0	0	0	0	0

(b)

0	0	0	0	0	0	0	0	0	0	0	0
0	0	0	0	0	0	0	0	0	0	0	0
0	0	0	0	0	4	0	0	0	-	-	0
0	0	0	4	4	4	4	4	-	-	-	0
0	0	4	4	5	5	5	-	-	-	0	0
0	4	4	5	5	5	-	-	-	-	0	0
0	4	4	5	5	-	-	-	-	4	0	0
0	4	5	5	5	-	-	-	-	0	0	0
0	0	4	4	5	-	-	-	-	0	0	0
0	0	0	4	5	-	-	-	-	0	0	0
0	0	0	0	4	-	-	-	-	0	0	0
0	0	0	0	0	-	-	-	-	-	0	0
0	0	0	0	0	0	-	-	-	-	0	0
0	0	0	0	0	0	0	-	-	-	-	0
0	0	0	0	0	0	0	0	-	-	-	-
0	0	0	0	0	0	0	0	0	-	-	0
0	0	0	0	0	0	0	0	0	0	0	0
0	0	0	0	0	0	0	0	0	0	0	0

(c)

<u>Fig.4</u>. Images of an apple occluded by a banana

In Fig.4b, for example, a banana partially occludes an apple, as indicated in outline in Fig.4a. If the banana can be recognised as an occluding object, it can be deleted from Fig.4b leaving Fig.4c. In principle this might be done without recognising the banana as a banana but merely recognising it as an occluding object that is otherwise unidentified. Alternatively, the banana could be recognised specifically as a banana, perhaps by generalised correlation, and could then be deleted, again leaving Fig.4c. Either way, the value of perimeter-squared-over-area for the visible remaining part of the apple would be dangerously banana-like.

Because occlusion may spoil attributes of an occluded object's visible shape, it may be better not to base recognition on attributes, but to use correlation instead. To make correlation work, we treat deleted pixels (-'s in Fig. 4b) as <u>don't care</u> elements [40]. A <u>don't care</u> element is one to which we are free to assign whatever value will serve to maximise the matching score. Using this idea, there is an alignment of Fig.1b with Fig.4c that will give a good match, whereby the occluded apple can perhaps be recognised.

6. Finding Specific Points on an Object

A robot vision system may be required to identify a specific hole in an object, in order to put something in it [41]. If contrast makes the hole visible then it may be easy to find [42] and correlation may be unnecessary.

```
0 0 0 0 0 0 0 0 0 0 0 0     0 0 0 0 0 0 0 0 0 0 0 0
0 0 4 4 4 4 4 4 4 4 4 0     0 0 x x x x x x x x x 0
0 4 4 4 5 5 4 4 4 5 4 0     0 x x x x x x x x x x 0
0 4 4 4 5 4 0 0 4 5 4 0     0 x x x x x 0 0 x x x 0
0 4 4 4 5 4 0 0 5 4 4 0     0 x x x x x 0 0 x x x 0
0 4 4 4 4 5 4 4 4 4 0 0     0 x x x x x x x x x 0 0
0 4 4 4 4 4 4 4 4 0 0 0     0 x x x x x x x x 0 0 0
0 4 4 4 4 4 4 4 0 0 0 0     0 x x x p x x x 0 0 0 0
0 4 4 4 4 4 4 4 0 0 0 0     0 x x x x x x x 0 0 0 0
0 0 4 4 4 4 4 0 0 0 0 0     0 0 x x x x x 0 0 0 0 0
0 0 0 4 4 4 0 0 0 0 0 0     0 0 0 x x x 0 0 0 0 0 0
0 0 0 0 0 0 0 0 0 0 0 0     0 0 0 0 0 0 0 0 0 0 0 0

          (a)                         (b)
```

<u>Fig.5.</u> (a) Digitised image and (b) digitised map of an industrial component

To illustrate a further requirement, Fig.5a shows an image of an object. Fig.5b is a digitised map of the object, showing a point labelled p where a robot system is required to solder a wire onto the object. In Fig.5a the point p cannot be located by looking for contrast at or near p. If the object in Fig.5a is recognised by correlation, the correspondence that is thereby established shows which point in Fig.5a corresponds to point p in Fig.5b, and this shows where the wire should be soldered. The successful wire bonding technology of EJIRI et al [43] is based on correlation.

7. Representation of Three-Dimensional Objects

7.1 Introduction

We have considered the recognition of apples and bananas as if these were two-dimensional. Automatic recognition of blood cells may achieve clinically useful results even if the cells are regarded as two dimensional [10]. It is questionable whether objects such as aircraft can usefully be regarded as two dimensional [47].

A general idea is to store images of a three-dimensional object as it would be seen from many different viewpoints. An unknown object could in principle be recognised by matching it separately with all such stored images. This multiple-views idea would generally have excessive storage and computational requirements.

Efficient representation of three-dimensional data has been the subject of development for purposes other than computer vision. One very important purpose is graphical display: for example we may wish to display on a television screen a picture of a house, our viewpoint being controlled by twiddling knobs [44]. In the computer-controlled manufacture of complicated three-dimensional objects (such as car bodies) we may wish to display two-dimensional pictures, from various points of view, showing what an object will look like after manufacture. In computer graphics we may start from a three-dimensional representation [45] and compute from it, and display, a picture as seen from a desired viewpoint. Computer vision works in the opposite direction: it starts from an image and (if required) somehow finds a corresponding three-dimensional object.

In computer tomography; three-dimensional information is computed directly from a three-dimensional object and is assembled into a complete three-dimensional representation in computer memory [46]. Thereafter quantitative (e.g. volumetric) measurements can be made, and two-dimensional pictures can be displayed. Again, tomography differs from computer vision in that tomography does not start from views of surfaces.

MARR and NISHIHARA [49] and REQUICHA [45] have reviewed criteria that three-dimensional representations should satisfy. Generally the choice of an efficient representation of data depends on the manner in which the data is to be used. Computer vision differs from graphics and tomography in that it starts from images of surfaces, and we should therefore not be surprised if a three-dimensional representation appropriate for vision is inappropriate for other purposes. We now outline two types of representation for use in computer vision.

7.2 Generalised Cones

An ordinary simple cone can be generated by sweeping a dilating circle along a straight axis. A generalised cone can be generated by sweeping a two-dimensional shape, such as an ellipse, along an axis that need not be straight [52]. The dimensions of the two-dimensionsal shape may vary as a function of distance along the axis. If the axis, the two-dimensional cross-sectional shape, and the function of distance along the axis are defined analytically by means of fairly simple mathematical expressions, then these expressions constitute an intensive definition of a generalised cone, which is a three-dimensional shape. In computer memory this intensive definition can be expected to occupy far less space than would be required for explicit storage of images of a generalised cone as it would appear from a multiplicity of viewpoints. When required, any such image can be computed from an intensive definition of a generalised cone.

An object that is not easily represented by a single generalised cone may be represented by an assembly of generalised cones [50,52]. For example, a human hand may be represented by an assembly that includes a separate generalised cone to represent each finger. To recognise a two-dimensional image of such an assembly, one approach is to compute a synthetic image of the assembly as seen from the same viewpoint as the object in the image. The synthetic image is then matched against the actual observed image to check whether it is sufficiently similar to be deemed "the same" [52]. To produce an appropriate synthetic image, the system needs to know which generalised cones to generate by sweeping sections along spines, and which perspective transformation to apply. To obtain this knowledge without resorting to impractical trial-and-error, several short-cut strategies have been proposed.

For selecting appropriate cones, an hierarchical strategy may greatly reduce the search [49,52]. For example, to recognise a standing man we might first match his image with that of a single cone. If the degree of match was sufficiently good we would then attempt to match the man's arms, legs and trunk with smaller, more specific, cones; and it is important and helpful that we would know roughly where to look for these specific cones relative to the location of the initial very-rough-approximation cone. If we succeed in matching the specific cones with the picture then we may search for still more specific and detailed cones representing, for example, individual fingers. If at any stage of the hierarchical "detailisation" we find a sufficiently bad match then we may abandon the hypothesis that the initial cone represents a man, and instead try some other hypothesis, such as a banana.

For finding appropriate perspective transformations, published strategies have been centered on the use of invariants and almost-invariants such as the parallelism of parallel lines [53] and we will briefly discuss this in terms of polyhedra instead of cones. A change of perspective, if not accompanied by a change of occlusion, does not alter what-is-connected-to-what in a picture. For example, at a fully visible corner of a cube three edges meet, and this remains true when the viewpoint changes, so corners (ideally) possess some invariance. ROBERTS [54] matched corners in a picture only to corners in computer memory, thus avoiding futile trial-and-error matching of non-corners with corners. A change of perspective does not change the number of edges of a fully visible flat face of a polyhedron: ROBERTS [54] only matched faces that had the same number of edges. If we know which corners of a flat face in a scene match which corners of which face in computer memory we may be able to compute the appropriate perspective transformation to bring the faces into exact alignment.

It appears fair to suggest that the use of invariants has not been fully mastered except in very restricted circumstances; and the problems of finding perspective, and finding which cones to match, remain difficult. Generalised models are not good for representing surface-markings of objects, for example finger prints on fingers. Generalised cones "...have an ambiguity that an infinite number of them can represent a single object" [55], for example, a sphere. But "generalised cylinders appear to be quite useful for describing objects such as airplanes, animals, and dolls which may be decomposed naturally into a few tubelike components" [45].

7.3 Constrained-View Surface Representation

An image must be segmented into objects before these objects can be checked for the possession of invariant or quasi-invariant properties, such as more-or-less parallel edges. In this approach the initial segmentation of an image is contrast-based, not correlation-based, and may therefore fail if adequate contrast is not available, or if occlusion interrupts the view of an object. We now outline a method of three-dimensional representation that avoids this problem but presupposes that generalised correlation is practical, which is not yet the case.

In this constrained-view representation, a collection of normal orthographic images of flat fragments of the surface an object are stored explicitly in memory. Every point on the object's surface belongs to one of thee fragments. The surface of an orange would be subdivided into fragments small enough to be quasi-flat. A simpler example is the surface of a die: this could be stored as shown in Fig.6. Adjacent faces of the die need not be stored next to eachother in memory: indeed their placement in memory is unimportant.

The faces in Fig.6 could be matched against the faces in the perspective view in Fig.7, using generalised correlation, where the set of admissible distortions is exactly the set of all perspective projections. Fig.6 does not represent three-dimnensional shape: it does not tell us how the six faces are put together in space.

In constrained-view representation the vital idea is that the three-dimensional shape of an object is represented by a set of constraints between admissible distortions of faces that touch each other. For the Fig.6 example, let FA and FB be any two faces of the die that meet along an edge. There is a constraint that says that FA in Fig.6 can match a face in Fig. 7 only if FB in Fig.6 matches an adjacent face in Fig.7 as it would be seen from the same viewpoint as that from which FA is viewed in Fig. 7.A similar constraint applies to every pair of faces that meet, and all such constraints must be satisfied. If FA is viewed from a viewpoint such that FB is on the invisible side of the cube, and is therefore self-occluded, the constraints specify this naturally. Same-viewpoint constraints have been used centrally in various computer vision systems, e.g. [48,51]. In a constrained-view representation these constraints constitute the actual representation of three-dimensional shape [31].

To see why a constrained-view representation is not merely a camouflaged multiple-views representation, let us return to Fig.6 and temporarily remove constraints between distortions of faces. The crucial point is that the same set of admissible distortions will generate multiple views of any face. A multiple-views representation would require the storage of many views of each face, not just one view of each face as in Fig.6. If we reimpose constraints between distortions of adjacent faces, we note that, for example, the same constraints apply between any pair of mutually perpendicular rigid faces, regardless whether these faces belong to a cube, a wedge, or anything else.

Generalised cone representation achieves storage economy by intensively defining faces. Constrained-view representation uses storage-extravagant extensive definition of surfaces, but does not require a surface-generating phase in the recognition process. Most important, with constrained-view surface representation a system can in principle use segmentation-by-correlation to achieve segmentation when adequate contrast is not available for contrast-based segmentation, or when occlusion interrupts the view of an object. Furthermore, constrained-view surface representation can cope with surface markings, with the exception of texture, as we have remarked previously; and also with the exception of variable shading. Note that Fig.7 is purely black and white. Against constrained-view surface representation we must weigh heavily the disadvantage that the economic problems of generalised correlation are not

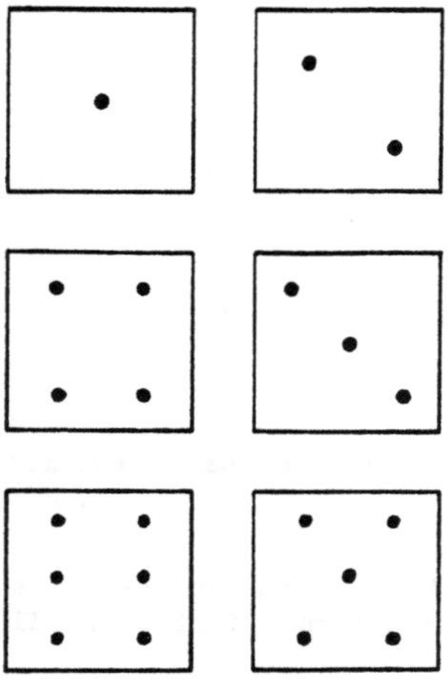

Fig.6. Faces of a die

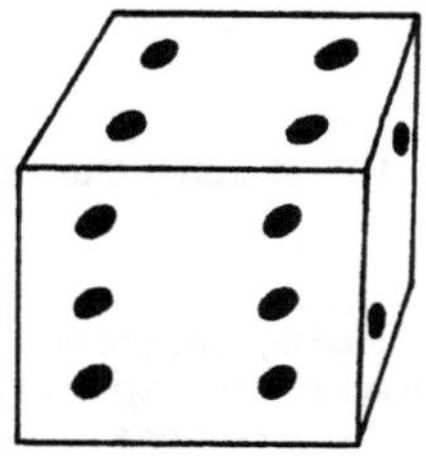

Fig.7. View of a die

only unsolved but are exacerbated by the cost of imposing constraints between distortions.

As a simple illustration of the practical meaning of imposing constraints between distortions, let us consider a two-dimensional example in which we view a V-shape from a point in its own plane at a distance that renders the projection approximately orthographic. Let the lengths of the arms of the V be A and B, and the acute angle between them be θ. Assuming that both are visible from our viewpoint, let the lengths of their images be A' and B' respectively. The relation

$$\mathrm{Cos}^{-1}(A'/A) + \mathrm{Cos}^{-1}(B'/B) = 180 - \theta$$

tells us that the contraction factor B'/B is constrained by A'/A and θ. There is a constraint between contraction distortions of A and B that depends on θ and can be regarded as a representation of θ.

Whatever type of representation we adopt, we may wish to ensure that a part that is common to many objects, e.g. a wheel common to different types of vehicle, has only a single stored representation, instead of a duplicated representation within each object, c.f. [56].

8. Computer Vision Without Using Models

If a vision system is intended to assign names to specific objects, possibly inferentially, then the system must include some sort of stored geometrical knowledge of these specific objects. There is, however, no need for a robot to name an object if the robot's only task is to "see" where to grasp the object, or to "see" where to move to avoid collision with the object. A man may be able to draw a picture of a new object that he is unable to name. Many contemporary computer vision systems do not aim to assign names to objects, and are not concerned with problems of representation in memory

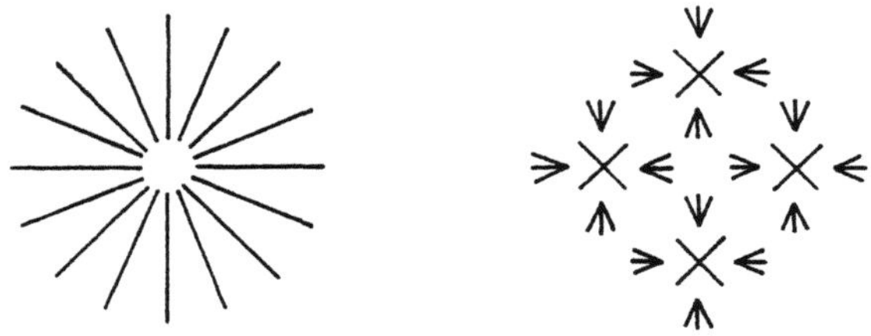

a b

Fig.8. Diagrams in which the total line-length is the same for all directions

[7,8]. Instead of naming objects, these investigations aim to describe objects, for example by specifying three-dimensional coordinates of all points on all visible surfaces.

A distinguished example is Horn's work on shape-from-shading [59], which, under restricted conditions of illumination and reflectance, may allow us to "reconstruct" a three-dimensional object from its two-dimensional image. More generally, this work has explored physical constraints between shape, illumination, reflectance, and image brightness.

Another possibility is to determine three-dimensional orientation from texture. For example, if Fig. 8a is held at an arbitrary angle to our line of sight some of the lines generally appear foreshortened, and from measurements of foreshortening we can determine the orientation of Fig. 8a as follows. In the image, we find all line segments. For each direction, we sum the total length of all line segments that are orientated in this direction. We thereby compose a histogram that shows the total line length at each orientation. This histogram has a single peak. The breadth of this peak, and the angle at which the peak occurs, tell us the three-dimensional orientation of the plane of Fig.8a relative to our line of sight [60]. This is still true if the lines in Fig.8a are chopped up into pieces and re-arranged to make a texture, for example as in Fig.8b, but without changing the total line-length in each direction.

As well as shading and texture, the shapes of junctions of lines in an image may yield valuable cues that constrain the possible shapes and dispositions of viewed objects. The use of junction-shape cues, which was originally confined to the polyhedral "blocks world" [38], has been extended to include a class of curved surfaces [62]. Another extension uses further cues, such as cues derived from symmetry and parallelism, to render more quantitative the constraints on possible reconstructed scenes [39].

9. Context

Visual interpretation may require the use of contextual knowledge, which is knowledge of interdependence of parts. As a very simple example, suppose that a picture is subdivided into (say) three subpictures (i.e. segments) that we will denote by X1, X2, and X3. We will denote possible names for subpictures by letters a,b,c,...,n. Any of our three subpictures may in principle have any of these names, but for a specific example picture we actually find, by methods such as those outlined previously in Sections 2 and 4, that the following names are the most likely for the three subpictures:

X1: a,c,e,h
X2: a,f,g
X3: b,d,h,k

Suppose we have a priori knowledge that X2 cannot have name f unless X1 has one of the names b,d,g. In our example X1 has none of the names, so we can delete f from the list of possible names for X2. This illustrates the use of contextual knowledge for narrowing down a list of possible interpretations; but sophisticated statistical techniques generally give better results [1,63,64].

Returning to our simple example, if we know that X3 cannot be a d unless X2 is an f or an e, then deletion of f from the X2 list will cause deletion of d from the X3 list. This knock-on effect is an example of discrete relaxation and is useful in the solution of a big family of contextual problems [65]. Discrete relaxation has been generalised so that instead of merely propagating deletion we propagate encouragement or discouragement for possible interpretations [61]. ROSENFELD [66] has surveyed practical uses of the resulting technology, for example in the contextual interaction of edge-detector outputs. Relaxation is not the only method that has been developed for the contextual editing of edge detector outputs [67,68].

In the opinion of MARTIN and AGGARWAL [69], "determining the proper interrelation of segmentation and recognition in the process of perception" is "the central problem in computer vision". In our simple example we assumed that the original picture had been correctly segmented into portions X1,X2 and X3. If segmentation is unreliable we can try various plausible segmentations and determine which of these are contextually credible [70,71,72].

Contextual techniques have been extended to allow for the possibility that at least one of the parts X1, X2 or X3 is accidentally missing; or a spurious extra part X4 has accidentally been inserted [73,81]. More generally, we can use contextual knowledge of syntax and semantics to restrict possible interpretations of a segment [74].

Direct constraints between possible names are not the only constraints that are actually used. Another important family of constraints are genre constraints: for example if we try to recognise ten human faces and we see that nine of these belong to politicians, then the tenth is likely to belong to a politician. Another example is that in a printed word we can reasonably expect all characters to be printed using the same font or type-face [75,76].

Yet another family of constraints are constraints between distortions [31]. For example we normally expect to find that all the characters in a given printed word have roughly the same size and orientation. If we see a man's hand, we expect his other hand to be roughly the same size.

Inter-distortion constraints are vital in constrained-view representation.

When one picture is a distorted version of another, we can correlate [77] or match [78] small regions of one picture with the other picture, seeking positions of best match, to tell us which point in one picture corresponds to which in the other. BARNARD and THOMPSON [79] apply contextual constraints whereby the correspondence between a pair of points depends on the correspondence between other pairs of points [31].

10. Learning

Hearsay II [74] is an archetypal example of a sophisticated system combining lexical, syntactic, semantic and other contextual constraints. Much of this contextual knowledge is not learnt, but is instead programmed into the system. When we compare a computer vision system with a brain, we should preferably compare not only functional capabilities (e.g. description and object-naming capabilities) but also learning and inferential capabilities (i.e. ability to work without specific programming). Otherwise our comparison may be misleadingly incomplete. We should remember that biological systems are subject to fundamental limitations, such as limitations of information capacity in the genetic code. This limitation is not an obvious problem for a programmed computer vision system.

Systems that can only read text [76] are obviously very limited functionally. Some such systems learn (in the sense of Section 2 of the present paper) to recognise characters [80] and others do not [56]. Even in this very limited domain it is not easy to compare the learning and inferential capabilities of man and machine. It is difficult to determine whether the machine, if it has less inferential power, needs to see and learn many more specimens of each character than the learning man requires.

We might wish a general automatic vision system to be capable of recognising new specific objects that belong to known generic classes of objects: for example new buildings, new tables, new human faces. A classifical approach is to attempt to find a representation space in which there are disjoint regions corresponding to generic objects. Fig.2 illustrates this approach, of which the fundamental limitations are not at all clear. Some of the difficulties are that occlusion and failures of segmentation may spoil the attributes used as coordinates of the representation space.

An alternative approach is to make a machine infer an abstract structural representation of the spatial arrangement of parts on an object, or parts of a scene. DIETRICH and MICHALSKI [57] have surveyed work on inference that is primarily symbolic, and presupposes that appropriate symbol-to-icon assignments have already been accomplished.

A classifical speculation is that low-level visual intelligence is a mechanism whereby the brain reassembles chunks of previous experience so as to solve current problems of recognition and description. Even if this speculation has some validity, the association principle [58], whereby chunks are reassembled correctly, remains a mystery. If we could understand the learning and inferential capabilities of a brain at a visual pictorial level, this understanding might perhaps subsequently percolate upwards to powerful abstract symbolic levels.

Acknowledgements

I am grateful to H.B. Barlow, L.G. Shapiro, J.P. Frisby and A.H. Watt for comments on a draft of this paper, but responsibility for shortcomings is entirely mine.

References

1. R.O. Duda, P.E. Hart :Pattern Classification and Scene Analysis (John Wiley and Sons, New York 1973)
2. A .Rosenfeld, A.C. Kak: Digital Picture Processing (Academic Press, New York, 1976)
3. E.L.Hall: Computer Image Processing and Recognition (Academic Press, New York, 1979)
4. W.K.Pratt: Digital Image Processing, (Wiley, New York, 1978)
5. T.Pavlidis: Structural Pattern Recognition, (Springer Verlag, Berlin, 1977)
6. A. Rosenfeld: Comp. Graph. and Image Proc, 13, 46 (1980)
7. H.G.Barrow, J.M. Tenenbaum: Proc IEEE, 69, 572 (1981)
8. M. Brady: Comp. Surveys, 14, 3 (1982)
9. P.A. Devijver, J. Kittler: Pattern Recognition a Statistical Approach (Prentice Hall, Englewood Cliffs, 1982)
10. J.K.Mui, K.S. Fu: IEEE Trans. PAMI-2, 429 (1980)
11. E.M. Riseman, M.A.Arbib: Comp. Graph. and Image Proc, 6, 222 (1977).
12. R. Jain, W.N. Martin, J.K.Aggarwal: Comp. Graph. and Image Proc, 11, 13 (1979)
13. B.E. Flinchbaugh, B. Chandrasekaran: Artificial Intelligence, 17, 387 (1981)
14. R.M. Haralick: Proc IEEE, 67, 786 (1979)
15. E.M. Gurari, H. Wechsler: IEEE Trans. PAMI-4, 304 (1982)
16. K.S.Fu, J.K. Mui: Pattern Recognition, 13, 3 (1981)
17. T. Kanade: Comp. Graph.and Image Proc, 13, 279 (1980)
18. C.A. Harlow, S.A. Eisenbeis: IEEE Trans. C-22, 268 (1973)
19. L.S. Davis: Comp.Graph. and Image Proc, 4, 248 (1975)
20. A. Rosenfeld: Proc IEEE, 69, 596 (1981)
21. D.L. Milgram: Comp. Graph.and Image Proc, 11, 1 (1979)
22. R.M. Haralick: Comp. Graph. and Image Proc, 12, 60 (1980)
23. C.M. Bjorklund, T. Pavlidis: IEEE Trans. PAMI-3, 144 (1981)
24. W.H. Highleyman: IRE Trans. EC-10, 501 (1961)
25. .J.K. Aggarwal, L.S. Davis, W.N. Martin: Proc IEEE, 69, 562 (1981)
26. R.M.Haralick: Comp. Graph. and Image Proc, 13, 191 (1980)
27. W. Doyle: J. Assoc. Comp. Mach, 9, 259 (1962)
28. M.A. Fischler, R.A.Elschlager: IEEE Trans. C-22, 67 (1973)
29. D.J. Burr: Comp. Graph and Image Proc, 15, 102 (1981)
30. S. Tsuji, M.Osada, M. Yachida: IEEE Trans. PAMI-2, 516 (1980)
31. J.R. Ullmann, to be published in Pictorial Data Analysis, Ed. by R.M. Haralick (Springer Verlag, Berlin, 1983)
32. W.E.L. Grimson: From Images to Surfaces (MIT Press, Cambridge, Mass, 1981)
33. S. Ullman: Computer, 14, 57 (August 1981)
34. Recognition Equipment Incorporated: U.K.Pat.1 115 909 (1968)
35. T. Pavlidis: IEEE Trans. PAMI-2, 301 (1980)
36. N. Ahuja, A. Rosenfeld: IEEE Trans. PAMI-3, 1 (1981)
37. L.S. Davis, S.A. Johns, J.K. Aggarwal: IEEE Trans. PAMI-1, 25 (1979)
38. K. Mackworth: Perception, 5, 349 (1976)
39. T. Kanade: Artificial Intelligence, 17, 409 (1981)
40. J.R. Ullmann: An Investigation of Occlusion in one Dimension, to be published in Comp. Graph and Image Proc.
41. C.A. Rosen: in Computer Vision and Sensor Based Robots, pp. 3–20, Ed by G.G.Dodd and L. Rossol (Plenum Press 1979)
42. M.Yachida, S.Tsuji: IEEE Trans. C-26,882 (1977)
43. S. Kashioka, M.Ejiri, K.Sakamoto: IEEE Trans. SMC-6, 562 (1976)
44. W.M.Newman, R.F.Sproull: Principles of Computer Graphics (McGraw Hill, New York, 1973)

45. A.A.G. Requicha: Comp. Surveys, 12, 437 (1980)
46. S.N. Srihari: Comp. Surveys, 13, 399 (1981)
47. T.P. Wallace, O.R. Mitchell, K. Funkunaga: IEEE Trans. PAMI-3, 310 (1981)
48. P.G. Mulgaonkar, L.G. Shapiro, R.M. Haralick: Proc. Workshop on Computer Vision, IEEE Computer Society Press, 116 (1982)
49. D. Marr, H.K. Nishihara: Proc. Roy. Soc. B200, 269 (1978)
50. H.K. Nishihara: Artificial Intelligence, 17, 265 (1981)
51. R.A. Brooks: "Model-based three dimensional interpretations of two-dimensional images", Proc. Image Understanding Workshop, April 1981
52. R.A.Brooks, R. Greiner, T.O. Binford: Proc ARPA Image Understanding Workshop, Cambridge, Mass, pp. 36-43, May 1978
53. R.A. Brooks, T.O.Binford: First National American Association for Artificial Intelligence Conference, Stanford, Ca, August 1980.
54. L.G.Roberts: in Optical and Electro-Optical Processing, pp159-197Ed.by J.T.Tippett (MIT Press, Cambridge, Mass, 1965)
55. Y. Shirai, in Computer Vision and Sensor Based Robots, pp 187-205, Ed.by G.G.Dodd and L.Rossol (Plenum Press, 1979)
56. J.L.Knab, J.J. McIntyre: U.S. Pat. 3 992 697 (1976)
57. T.G. Diettrich, R.S. Michalski: Artificial Intelligence, 16, 257 (1981)
58. E.G.Boring: Am. J. Psychol, 59, 173 (1946)
59. K. Ikeuchi, B.K.P. Horn: Artificial Intelligence, 17, 141 (1981)
60. P. Witkin: Artificial Intelligence, 17, 17 (1981)
61. S.W. Zucker, Y.G.Leclerc, J.L.Mahommed: IEEE Trans. PAMI-2, 117 (1981)
62. I.Chakravarty: IEEE Trans. PAMI-1, 202 (1979)
63. K.S. Fu, T.S. Yu: Statistical Pattern Classification Using Contextual Information (Research Studies Press, John Wiley, Chichester, England, 1980)
64. D.A. Landgrebe: Proc IEEE, 69, 628 (1981)
65. R.M. Haralick, L.G.Shapiro: IEEE Trans. PAMI-1, 173 (1979)
66. A. Rosenfeld: Pattern Recognition: 10, 181 (1978)
67. R. Nevatia, K.R. Babu: Comp. Graph. Image Proc. 13, 257 (1980)
68. P.V.Sankar, J.Sklansky: IEEE Trans. PAMI-4, 326 (1981)
69. W.N. Martin, J.K. Aggarwal: Pattern Recognition, 11, 169 (1978)
70. W.S. Rutkowski, S. Peleg, A. Rosenfeld: IEEE Trans. PAMI-3, 368 (1981)
71. E.W.Bollinger, A.M. Chaires, J.M. Ciconte, A.H. Elt, J.J. Hilliard, W.S. Rosenbaum: U.S. Pat. 3 969 700 (1976)
72. J.A. Feldman, Y. Yakimovsky: Artificial Intelligence, 5, 349 (1974)
73. P.A.V. Hall, G.R.Dowling: Comp. Surveys, 12, 381 (1980)
74. L.D.Erman, F.Hayes-Roth, V.R.Lesser, D.R.Reddy: Comp. Surveys 12, 213 (1980)
75. P.J.Mullan, W.S.Rosenbaum: U.S. Pat. 3 988 715 (1976)
76. J.R. Ullmann, in Applications of Pattern Recognition, pp. 197-236 Ed. by K.S. Fu(CRC Press Roca Baton, Florida, 1982)
77. E.L.Hall, D.L.Davies, M.E.Casey: IEEE Trans. PAMI-2, 313 (1980).
78. K.Price, R. Reddy: IEEE Trans. PAMI-1, 110 (1979)
79 S.T. Barnard, W.B.Thompson: IEEE Trans. PAMI-2, 333 (1980)
80. J.Sammon, J.Saunders; U.S. Pat. 3 755 780, (1973)
81. R.K. Moore: IEEE Trans. PAMI-1, 86 (1979)

Constraints, Descriptions and Domain Mappings in Computational Vision

A.K. Mackworth

Department of Computer Science, The University of British Columbia, Vancouver, B.C. V6T 1W5, Canada

1. Introduction

The central paradox of computational vision is that given only one or more images of a scene the set of possible scenes depicted is underconstrained; however, our subjective experience is the opposite: the scene appears to be heavily overconstrained. Every aspect of our own visual experience offers mutually confirming evidence for the existence of a single, specific, non-ambiguous scene. This paradox can only be resolved by postulating that any perceiving system must supply organized knowledge of the scene domain, the imaging projection process, the radiometric and geometric aspects of lighting, the reflectance, transmission and refractance properties of scene materials and many other relevant physical regularities. This knowledge spans a spectrum from general *a priori* knowledge of all scenes in the domain assumed to be imaged - what can be and what can not - to the specific, contingent knowledge of the actual scene and imaging situation involved - what is and what is not.

2. Perception

All perception necessarily involves three domains: W, a set of possible worlds; I, a set of possible images of the world; and P, a set of possible projections of W into I.

A particular world, w, leaves a trace of itself, i, in I that is determined by the particular method of projection p. The triple of configurations in W, P and I is related by a relation of representation [1] R(w,p,i) as shown in Fig. 1.

Parenthetically, this formalism should not be narrowly interpreted as applying to, say, single perspective views of a three-dimensional world of opaque objects. The image domain could include motion primitives, aural or visual stereo pairs, haptic images or depth maps - whatever is directly sensed by the perceiver's transducers. The projection domain could include holographic processes, x-ray, PET, NMR or CAT imaging or even free hand sketching. The world domain includes three-dimensional worlds, maps, electric circuits, text and the like.

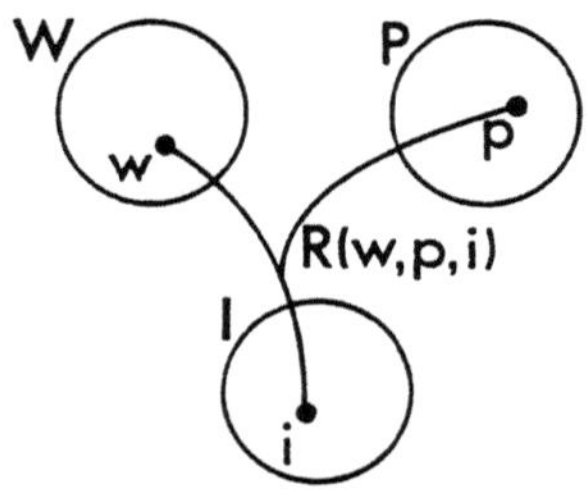

Fig. 1. The world, projection and image domains

The perceiver, minimally, has knowledge of i, I, W, P and R. Its task may be represented as finding a constructive proof for the imaging formula:

$$\exists w \exists p (w \in W) \wedge (p \in P) \wedge R(w,p,i) \quad . \qquad (1)$$

Any perceiver must be able to represent the potentially (and usually) infinite set of all solutions in a finite way. The representation must therefore be an intensional description of that set rather than an extensional list of its members.

In general the solutions will also be required to satisfy other constraints. Those constraints can be broadly categorised as a priori constraints on the nature of acceptable solutions in W or as arising from other images of the same w. Thus the representation must also allow specialization by additional constraints and intersection with other such representations. It might seem unnecessary to add the additional criterion that the representation include all and only those worlds that map to the given image i but many proposed representations have failed to satisfy this correctness criterion, in that they have excluded possible worlds for an image or included impossible worlds.

We assume that R(w,p,i) is restricted to the set of total many-to-one functional mappings from WxP onto I. This excludes for example, non-determinism from the imaging process (but not from the world). Thus, there is a mapping function,

$$f: WxP \to I \text{ such that } R(w,p,i) \text{ iff } f(w,p) = i \; ,$$

f does not have a function as an inverse. However it is possible to partition the set WxP into equivalence classes such that (w_1,p_1) and (w_2,p_2) belong to the same equivalence class WP_i iff $f(w_1,p_1) = f(w_2,p_2) = i$. We have:

$$WP_i \subseteq WxP$$

$$WxP = \bigcup_{i \in I} WP_i$$

and $\quad i_1 \neq i_2 \to WP_{i_1} \cap WP_{i_2} = \emptyset.$

We can project WP_i separately onto the set W as W_i and onto P as P_i where

$$W_i = \{w \mid (\exists p)(w,p) \varepsilon WP_i\}$$

$$P_i = \{p \mid (\exists w)(w,p) \varepsilon WP_i\} \quad ;$$

WP_i is a representation of the solution set of world-projection pairs for the imaging formula (1). Notice that if P is a singleton set $\{p\}$ (and the perceiver knows that) there is a one-to-one correspondence between members of WP_i and W_i. Since that assumption is often made, W_i is used to stand for the solution set of the simplified imaging formula:

$$(\exists w)(w \varepsilon W) \wedge R(w,p,i) \quad .$$

The history of computational vision research has been a search for good representations of the equivalence class W_i which must be correct, incremental, finite and efficiently computable from i. Using those criteria the most satisfactory representations of W_i are sets of constraints, induced by i and p, on the set of possible worlds.

3. Vision

The general view of perception given above can be applied to vision in various ways. Both the world domain and the projection domain can be refined or factored into sub domains. A possible factoring starts with the "universe" domain, U, in which each universe is a set of objects possibly moving in space over time. An instant selected from the time domain, T, selects out from a universe a particular world in W. A particular lighting chosen from the illumination domain, M, of that world produces a lit world in L. A view from a particular direction (in V) of that lit world produces a scene in S while a projection of that scene produces an image in I. These domains and their relational mappings are shown in Fig. 2.

The ordering of the domains shown is somewhat arbitrary and could be changed for some applications. Indeed, another version of this scheme could eliminate some of the arbitrariness of ordering by allowing mapping relations among more than three domains. However the scheme presented has the same structure as the archetypal perception model introduced in Section 2. Each of the relations constrains elements in three domains and has the same characteristics as the original relation of representation, R. The convention shown in Fig. 3 indicates that the domain of relation R_n is AxBxC but there is also a functional mapping f_n: AxB -> C such that $R_n(a,b,c)$ iff $f_n(a,b) = c$. Indeed all of computer graphics is concerned with good representations for specific configurations in the domains shown and the mapping functions, f_n. As before, the mapping function induces an equivalence class partitioning of

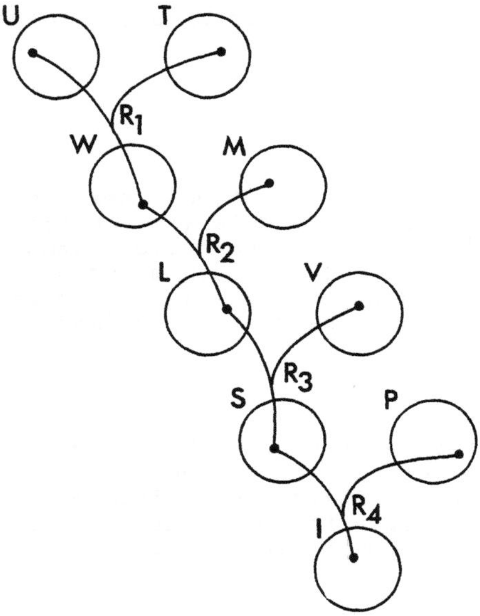

Fig. 2. The domains and relations for vision

AxB such that for any element $c \varepsilon C$ there is a class which we call $AB_c \subset AxB$ with $(a,b) \varepsilon AB_c$ iff $f_n(a,b) = c$. The fact that these equivalence classes are, in general, very large underlies tradeoffs such as surface photometry versus illumination or object shape versus projection technique that vary one factor while simultaneously varying another to produce identical images.

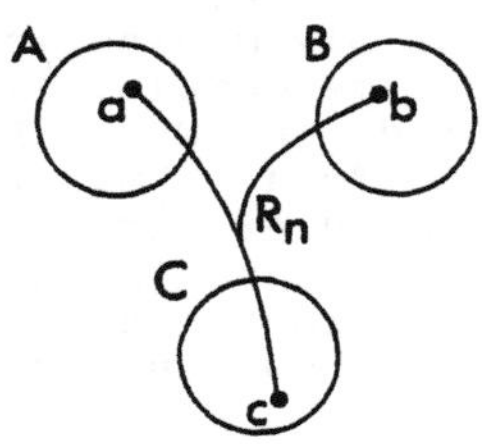

Fig. 3. The convention for showing R_n

Thus a given image, $i \varepsilon I$, induces an equivalence class partitioning $SP_i \subset SxP$. By projecting SP_i onto S and P respectively we have S_i and P_i. S_i in turn induces a partitioning of LxV, LV_i such that each pair (l,v) in LV_i maps into one member of S_i. LV_i can be similarly projected down into L_i and V_i and so forth.

Given the many-to-one nature of the mapping functions the size of the equivalence classes tends to increase with the level in the domain hierarchy unless other constraints on the classes are known a priori or available from other imagery known to relate to the same situation. Typical of the latter are motion, stereo and tomography, photometric stereo [2] and multispectral images which can each be illustrated as follows.

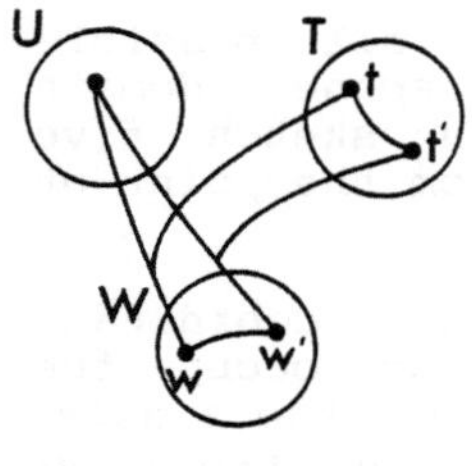

Fig. 4. Motion

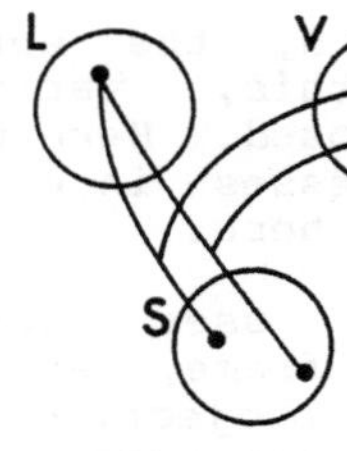

Fig. 5. Stereo

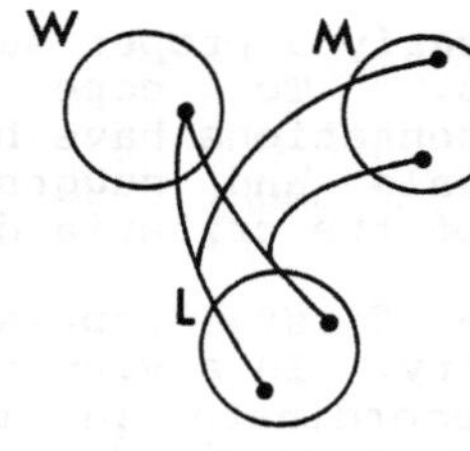

Fig. 6. Photometric stereo

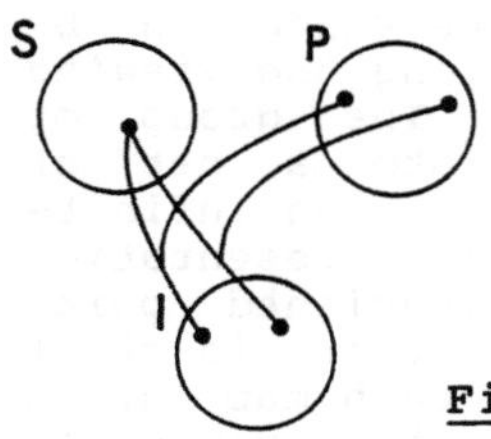

Fig. 7. Multi-spectral imaging

In multi-image situations that are known to arise from a single element of some higher domain one can exploit the extra constraints by intersecting the two (or more) equivalence classes that are induced in that higher domain: U for motion, L for stereo, W for photometric stereo, and S for multispectral imaging.

Most current theories of computational vision [3] distinguish between "viewer-centred" and "object-centred" representations. In this scheme the viewer-centred representations describe equivalence classes in S while object-centred representations describe equivalence classes in L and W.

The nature of each domain can be elaborated as much as necessary for any application. In a world of opaque objects, W, for example, could be composed of two domains, one for object geometry, the other for surface photometry, since they decouple into image geometry and radiometry.

In what follows, a brief sketch of some proposals for describing the equivalence classes in the scene domain, S_i, for orthographic line drawings of a world of opaque polyhedra will be given within this framework.

4. Some Scene Representations

When we say that a single image underconstrains the scene we mean simply that even given the knowledge that the projection is, say, orthographic the equivalence class S_i contains more than one member; in fact, it is obviously infinite and yet it is clearly a proper subset of S, the set of all possible scenes. To cope with this, various scene domain representations have been proposed. Here we can sketch five proposals and suggest weaknesses in each from the point of view of the criteria developed here.

The first representation uses Cartesian coordinate geometry. In a viewer-centred frame, let the image occupy the x-y coordinate plane and the projection be along the z axis. If the scene domain representation uses triples of reals as Cartesian coordinates, the scene representation fails the test of finiteness: the depth of the scene cannot be determined. To allow for that the second representation uses the equivalence class concept: represent all scenes which can be transformed into each other by a translation along the viewing direction by a single configuration. The group of translations along the z axis partitions S into a set of equivalence classes. Each equivalence class has an infinite number of members but it has a finite canonical representation as, say, the member of the class with a distinguished point having a z coordinate of zero. Unfortunately there is still an infinite number of equivalence classes that each map into a given image. We can now go on and look for another group of transformations which induce a further equivalence partitioning of our equivalence classes.

The third scene representation is the edge labelling approach [1,4]. The equivalence class descriptions are based on categorizing edges into four (or more) classes with respect to the dihedral angle the surfaces meet at and which view of the edge the viewer sees. Two scenes are then in the same equivalence class if they have the same visible edge and surface connectivitv and each edge has the same convex, concave or occluding label in each scene regardless of the actual dihedral angle between the two surfaces meeting at the edge. This equivalence class description is certainly finite; however, it is not correct. In general, not every member of the equivalence class produced can in fact map into the given image. This is always true but a convincing demonstration is given by "impossible objects" which are images that have no scene domain correspondent whereas the edge labelling algorithms return a non-empty equivalence class [4,5].

A fourth, more adequate, scene equivalence class representation is the gradient space representation of surface orientation [4,5]. This starts from the observation that equivalence classes induced by the translation group allow for one degree of freedom in the scene representation, but there are at least two more: the orientation of any surface in the scene can be arbitrarily specified. If there is a surface visibly intersecting that one another degree of freedom is introduced: the dihedral angle between them is unconstrained. However the locations of edge projections in the image supply a number of constraints in the scene that do constrain surface position and orientation in the scene thereby restricting the number of additional degrees of freedom in the scene description. A program POLY [5] based on this analysis, constructs equivalence class descriptions that are finite and eliminate many of the incorrect scene configurations allowed by edge labelling without, however, eliminating them all [6,7]. The descriptions are efficiently computable and are amenable to refinement or intersection with constraints available from a priori knowledge and multi-image constraints [8].

There is a general lesson in the usefulness of surface-based, viewer-centred intermediate scene representations. If a priori scene constraints such as "all surfaces are flat" or "all surfaces are smooth almost everywhere" are available. then it is necessary to find some configuration space in which such constraints are expressible. Minimally such a space must allow descriptions of the relevant objects (surfaces and edges, here) and constraints among the objects that are back-projected from the image description. Such a configuration space description is the requisite equivalence class description.

Another viewer-centred scene representation is the "intrinsic image" proposal [9]. Arrays of scalars are used to hold values at each image location for the various scene parameters that are confounded into image intensity such as surface illumination, albedo, orientation and depth. For our purposes here, the intrinsic image representation is not adequate. It does not satisfy the criteria of finiteness, correctness, refinability and efficient computability. It must be possible to construct finite equivalence class descriptions that describe the set of all scenes that could have produced the image and then refine that set as additional constraints become available. This process must use symbolic descriptions in the scene domain. Arrays of scalars can only represent one of the infinite number of members of the equivalence class. On the other hand, in an over-constrained imaging situation, where a particular member of several of the confounded domains is known and specified a priori, the intrinsic image representation is adequate.

5. Conclusion

The knowledge sources required to resolve the vision paradox can be constructed by understanding the confounding mapping processes discussed here. The needed equivalence class descriptions are best represented as sets of symbolic constraints on the allowable domain elements in a configuration space for the domain.

The general scheme is presented here to provide a framework for understanding theories of perception. It is not a theory of perception; however, any theory of perception must account for the fact that a perceiver does not perceive the world directly, pace Gibson [10]. The trace of the world available to the perceiver is a multiple confounding of configurations from many different domains. We see only the shadows dancing on the wall of Plato's cave.

6. References

1. CLOWES, M.B., On seeing things, Artificial Intelligence 2 (1971) 79-112.

2. WOODHAM, R.J., Analysing images of curved surfaces, Artificial Intelligence 17 (1981) 117-140.

3. MARR, D., Representing visual knowledge, AIM-415 (MIT, Cambridge, MA, 1977).

4. HUFFMAN, D.A. Impossible objects as nonsense sentences, in: B. Meltzer and D. Michie, Eds. Machine Intelligence 6 (Edinburgh University Press, Edinburgh, 1971) 295-323.

5. MACKWORTH, A.K. Interpreting pictures of polyhedral scenes, Artificial Intelligence 4 (1973) 121-137.

6. MACKWORTH, A.K. On the interpretation of drawings as three-dimensional scenes, D. Phil. Thesis (Sussex University, 1974).

7. DRAPER. S.W. The use of gradient and dual space in line-drawing interpretation Artificial Intelligence 17 (1981) 461-508.

8. MACKWORTH, A.K. Model-driven interpretation in intelligent vision systems, Perception 5 (1976) 349-370.

9. BARROW, H.G. and TENENBAUM, J.M., Recovering intrinsic scene characteristics from images, in: A. Hanson and E. Riseman, Eds., Computer Vision Systems (Academic Press, New York, 1978).

10. ULLMAN, S. Against direct perception, The Behavioural and Brain Sciences 3 (1980) 373-415.

Part II

Local Spatial Operations on the Image

Matching Coding to Scenes to Enhance Efficiency

S. Laughlin

Department of Neurobiology, Australian National University, Canberra

The father of the digital computer, J. VON NEUMANN, raised the problem of constructing a reliable brain from unreliable nerve cells [1]. One solution is to code the same information in many parallel channels, but a second possibility is to use the individual nerve cells as efficiently as possible by coding information in ways that minimise their faults. A first step towards efficiency is the elimination of redundancy from the incoming data [2,3,4,5,6] and a second is to code the essential messages in ways that fully exploit neuronal signalling capacity [5]; a procedure that often reduces the redundancy in a nerve cell's response.

It has been suggested that some of the neural interactions commonly observed during early visual processing (i.e. sensory adaptation and lateral inhibition) substantially improve coding efficiency [4,5,6]. We have recently examined early visual processing in some detail, using the fly's eye as a readily accessible 'model' system, and have analysed two procedures that enhance neuronal efficiency. Predictive coding removes redundancy [7], while matched amplification ensures that full use is made of a nerve cell's limited response range [8]. Both procedures have been developed by video engineers to increase the efficiency with which pictorial data can be transmitted and displayed, and both procedures necessitate the matching of coding to the statistical properties of the incoming signals. However, before presenting these procedures, they must be placed within the context of the limitations to neural coding, and the structure and function of the fly's eye.

Some Limitations to Neural Coding

Nerve cells usually transmit information as voltage signals generated across their membranes and as puffs of chemical transmitter molecule sent from one cell to another at specialised junctions called synapses. The voltage signals are either action potentials, brief unitary events of constant amplitude which frequency code information, or analogue signals, capable of adopting any voltage over a defined range, and thus called graded potentials [9]. The first limitation upon information transfer is the restricted range of response levels. Action potential rate has an upper bound (c. 1000 s^{-1}), set by event duration and a lower bound determined by spontaneous 'noise' events or by the requisite time resolution. Graded potentials are constrained to the range of -100mV to +50mV by the ionic gradients driving responses, although in practice they rarely exceed 60mV in total excursion. The second limitation is intrinsic noise, introduced at all stages of signal transmission. The ionic channels that generate electrical signals by conducting current across the membrane have a tendency to open and shut at random. Worse still, chemical transmission at synapses is a stochastic process. Small packets of transmitter molecules are emptied into the synaptic gap between cells at a mean rate which depends upon the input signal, but packet release is a binomial or Poisson process [10]. Such intrinsic noise, added during signal transmission, limits information capacity by determining the smallest resolvable difference in response. A neuron's response range is divided into a number of discriminable response states; a restriction that poses particular problems to a visual system operating over many orders of magnitude of light intensity.

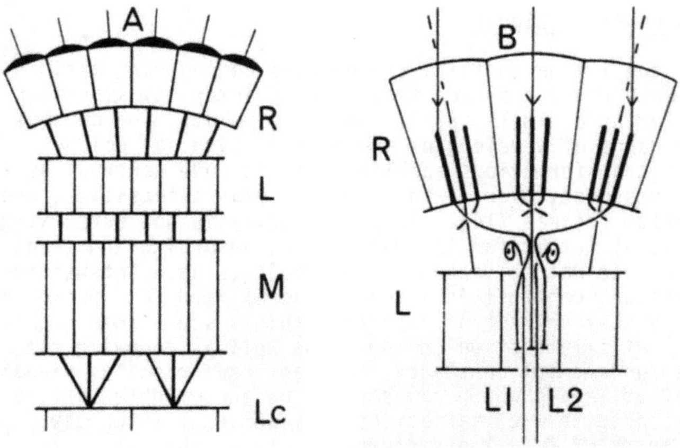

Fig.1. The orderly wiring of the fly's visual system. A. The array of directional light collectors and receptors, R, is mapped onto a corresponding matrix of neural subunits in the lamina L. The orderly spatial projection pattern is continued to the next 2 levels, M & Lc [12]. B. The receptors in adjacent light collectors, looking in the same direction, project to a single lamina subunit and connect with neurons L1 and L2 [13]

A Fly's Eye - an Ideal Subject for Studying Biological Image Processing

The fly's advantage is a structure and lifestyle conducive to several complementary lines of investigation [11,12,13]. The spatial intensity distribution is mapped onto an orderly radial array of light detectors (ommatidia) each having a restricted field of view (Fig.1a). A collector contains 8 photoreceptors looking out through its small lens in different directions. Because the visual angle between receptors within a collector equals the angle between collectors, there are several receptors located in neighbouring collectors but looking in the same direction (Fig.1b). The wiring of the visual system is precise and at the first level of information processing, the lamina, the outputs of the photoreceptors looking in the same direction are brought together from several collectors and connected to a small subassembly of neurons, known as a lamina cartridge (Fig.1b). There is one cartridge for every retinal collector so that the fundamental spatial sampling array of receptors is precisely mapped onto a corresponding array of neuronal sub-units. This orderly projection of one array of subunits onto another is continued at the next 2 levels of processing (Fig.1a). Because the system is a series of matrices, each composed of repeated neural sub-units, one can build up an unprecedently complete neural wiring diagram [12], and deduce the patterns of response set up in these arrays by different inputs to the receptors. One can record from photoreceptors and several classes of visual neuron, and relate these responses to both the wiring diagram and their function in generating visual behaviour [13]. Moreover, the behavioural studies are sophisticated enough to allow one to formulate and test models of visual processing [11]. A fly has form-perception, tracks objects, and visually stabilises its flight path, consequently it is eminently suited for investigating the fundamental principles of biological image processing. Its suitability has been developed and exploited by several groups [11,12,13,14] but discussion is limited here to the efficient coding of contrast at the first level of processing, the lamina.

Deriving and Processing Contrast Signals

Each neural sub-unit of the lamina contains two neurons coding contrast, designated L1 and L2. Both receive an identical set of intensity signals from the receptors monitoring a single spatial sampling point. The neurons are sensitive to a wide range of wavelengths and show no signs of colour coding [14]. The set of transformations enabling L1&2 to code contrast as the amplitude of graded response, over a wide range of mean intensities, are probably used in vertebrate retinas [14]. This similarity is not surprising because contrast is an ideal metric for the reflecting, absorbing and transmitting objects that dominate natural scenes. Such objects have intensities proportional to illumination, contrast is intensity normalised to a proportion of the mean, consequently the contrast of objects within a scene remain constant when the intensity of illumination changes. As well as endowing the visual system with this fundamental constancy, contrast coding scales signals down to a restricted set of levels which can readily be accommodated into a neuron's response range, despite wide variations in mean input intensity. Contrast is defined in terms of ΔI, the difference between the intensity at a point, I, and the mean intensity $\bar{I}$, because the formulation

$$\text{contrast} = (I - \bar{I})/\bar{I} = \Delta I/\bar{I} \qquad (1)$$

can be applied to aperiodic stimuli, and defines an object at the mean intensity as having zero contrast.

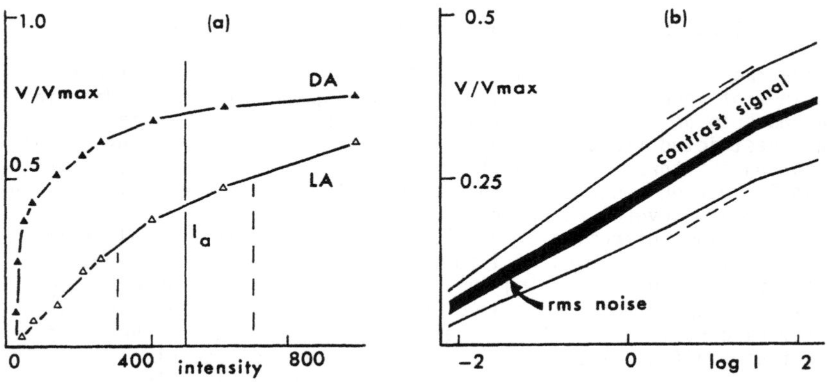

Fig.2. The coding of intensity and contrast by fly photoreceptors. (a) Response amplitude, V, vs. light intensity when cell has its highest gain (dark adapted state - DA) and when light adapted, LA, by background of intensity I_a. - - - = intensity excursion about I_a generated by average contrast. Data courtesy of T. MATIC. (b) Amplitude of response to average contrast signal and photon noise at different mean intensities. - - - - = modulation required for constant contrast to generate constant response. Responses normalised to maximum amplitude, Vmax

Contrast coding requires a sensitivity that varies inversely with the mean intensity and this is provided, albeit imperfectly, by the photoreceptors. The receptor's graded responses to light increase non-linearly with intensity (Fig.2) and, as in vertebrate cones [15], this non-linearity is combined with receptor gain controls to decrease the sensitivity as intensity is increased. Non-linearity ensures that the incremental sensitivity, dV/dI, drops with rising intensity. The gain controls prevent receptor saturation by decreasing the initial slope of the intensity-response function, thus reducing the response to a level where the incremental sensitivity is higher (Fig.2a). The effect of these sensitivity controls on contrast coding is shown in Figure 2b. The modulation of receptor potential induced by the intensity dependent photon noise plus the mean contrast signal occupies a narrow envelope, and is superimposed on a DC-bias. In the upper region of the curve, corresponding to normal light adapted operations, a constant contrast induces an approximately constant modulation, irrespective of mean intensity, but at lower intensities the sensitivity to contrast falls. Thus the receptors contain the contrast signal to workable proportions over a wide intensity range. About any one mean intensity the intensity-response function is approximately linear [15] over the average contrast range (Fig.2a), and one can regard the receptor's non-linearities as setting the gain on an essentially linear device. This type of sensitivity control is accompanied by two major deficiencies - contrast signals are small, and they often sit on a larger DC-bias. It is the task of the cells L1&2 in the lamina to remedy these shortcomings.

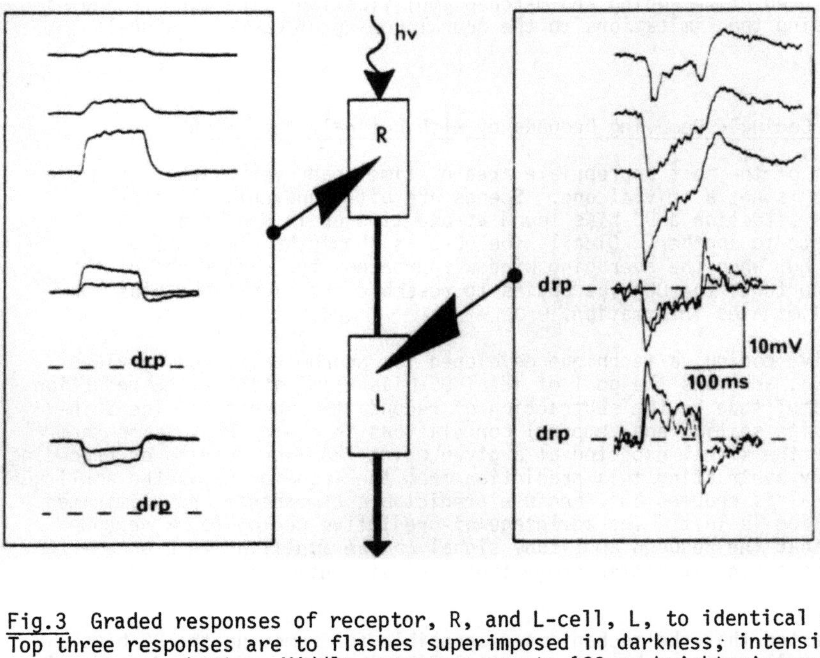

Fig.3 Graded responses of receptor, R, and L-cell, L, to identical stimuli. Top three responses are to flashes superimposed in darkness, intensity increases top to bottom. Middle responses are to 100 ms brightening pulses superimposed on steady background; contrasts = 0.2 & 1.0. Bottom responses are to dimming pulses; contrasts = 0.2 & 0.8. drp - resting potential level of membrane in darkness

A large number of chemical synapses, connecting the receptors to L1 and L2, invert and amplify the contrast signal (Fig.3). Nervous systems are adept at inverting graded potentials so that this change in signal polarity does not necessarily alter the information content. Amplification generates a robust contrast signal, suitable for transmission by L1&2 to the next processing level. A high gain is achieved by recruiting a large number of synaptic transmitter packets and this has the added advantage of improving the synaptic signal:noise ratio at a critical stage where the input signal is necessarily small [14].

The responses of L1&2 show no trace of the DC-bias. When intensity is elevated to a new constant level the receptors respond with a maintained increase in depolarisation (Figs.2 & 3). By comparison, L1&2 respond strongly to the change but then relax back to the original dark resting level of response (Fig.3). This relaxation is caused by a delayed inhibition which opposes the light response, and this inhibition effectively subtracts away the DC-bias. Although the mechanisms responsible for inhibition are unknown, they are driven by signals from two sources [14]. One is the time average of the receptor inputs to L1&2 (i.e. the time average at a single spatial sampling point) and the other a spatio-temporal mean averaged over a large circular field of surrounding spatial samples (Fig.4). By eliminating the DC-bias, inhibition removes redundancy while amplification makes full use of L1&2's limited response ranges. Subsequent analysis, comparing the properties of inhibition and amplification to the statistics of contrast signals, shows that this form of processing is carefully designed to increase neuronal information capacity. Thus DC-bias elimination and amplification are the foundations for predictive coding and matched amplification:- the two strategies for minimising the limitations to the neural coding of contrast signals.

Predictive Coding - Removing Redundancy with Retinal Inhibition

The problem of the most appropriate area or time span over which to compute the DC-bias is not a trivial one. Scenes are often unevenly illuminated and in this situation a DC-bias found at one time or in one area is quite inappropriate to another. Clearly the DC-bias should be a measure of the local mean but when the averaging window approaches the dimensions of the sampling aperture, the DC-bias begins to resemble the signal and bias subtraction eliminates information.

Predictive coding, a technique developed for economically coding video signals [16], achieves the goal of ideal DC-bias elimination - the reduction of signal amplitude by the subtraction of redundant components. The technique exploits spatial and temporal correlations in the data by using them to estimate the signal expected at a given point from surrounding or preceding signals. By subtracting this prediction from the incoming data, the amplitude of the signal is reduced but, because predictable components are eliminated, no information is lost. The advantage of predictive coding for a nervous system is that the reduced amplitude signal can be amplified to a greater extent, thus giving it better protection against contamination by intrinsic noise [7].

We suggested that the mechanisms responsible for removing the DC-bias from the L-cell response have the spatial distributions and time dependencies required for them to predict the signal arriving at L1&2 [7]. By assuming that the prediction is formulated from a weighted mean of samples, one can estimate the requisite spatial and temporal dependencies of inhibition from the first-order correlations between samples. Our analysis shows that the necessary functions are insensitive to the distances or times over which samples are correlated but are critically dependent on the signal:noise ratio. If the incoming data is decorrelated by accompanying photon noise, one must extend the weighting function to embrace a statistically significant sample. When input noise is reduced, the nearest data points are always the most closely related and one need look no further for an estimate, no matter how extensive the first order correlation. Because the input signal:noise depends on intensity (Fig.2b), one expects the weighting functions to be adjusted to the ambient light level.

This intensity dependence provides a good test for predictive coding in the fly's visual system because all the necessary parameters can be measured [7]. We obtained spatial autocorrelation functions from scans of natural scenes [8], and measured the temporal correlation introduced by the time course of the receptor's light response. The most important variable, signal: noise, was measured as follows. The signal is the mean contrast in natural scenes, measured from scans. At any given mean intensity, the noise was derived by measuring the fluctuations in L-cell voltage response, and then converting these into an equivalent masking contrast, using the empirical contrast-response relationship for the particular cell and intensity. From these measurements we estimated the weighting functions needed to implement predictive coding and compared these with the spatial and temporal characteristics of the L-cell response, determined under the same conditions (Fig.4). The impulse response shows that the delayed inhibition has the time and intensity dependence required for predictive coding. In the spatial domain, the inhibitory surround of the L-cell receptive field (Fig.4a), although intensity dependent, is always too wide and weak to formulate the best prediction. We suggested [7] that this widening is associated with the coding of moving stimuli. If unpredictable movements shift neighbouring samples into the centre of the L-cell receptive field before the inhibitory network has had time to use them for the formulation of a prediction, then inhibition must be extended to embrace a wider surround [7].

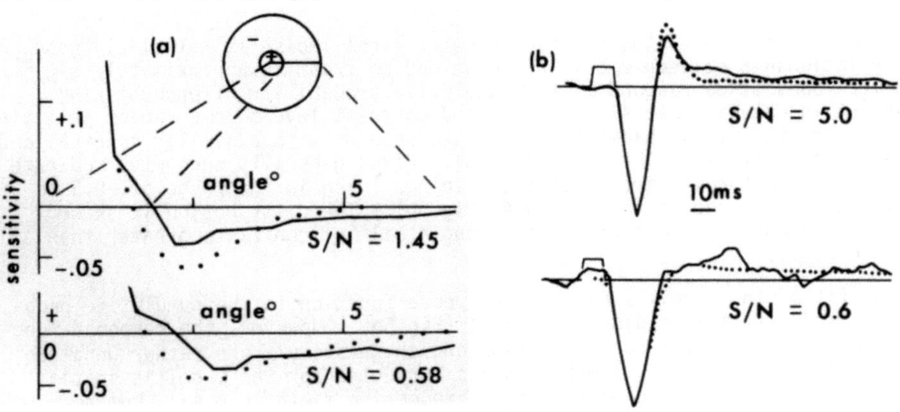

Fig.4. Intensity dependent inhibition and the implementation of predictive coding. (a) Sensitivity profile through circular receptive fields of an L-cell showing widening of outer inhibitory zone with decreasing signal:noise ratio, S/N. Spacing between sampling stations (light collectors) in fly = 1.5°. (b) Analogous weakening and lengthening of delayed inhibition illustrated by L-cell impulse response to 10ms brightening of background light. • • • • = inhibitory response required for predictive coding [7]

In conclusion, predictive coding is used to remove redundancy at the first level of processing in the fly's visual system. Predictions are formulated and then removed from the data stream by patterns of spatial and temporal inhibition, and these have to be adjusted to suit the quality of the incoming signals. Fortunately this adjustment depends almost entirely on intensity, and no account need be taken of the particular scene in which an animal operates. Thus predictive coding is a simple and robust procedure, well suited to removing redundancy in early visual processing [7]. Note that redundancy is not eliminated. Each individual scene contains non-random elements associated with features such as edges, contours and periodicities. This higher order redundancy is not specified by the autocorrelation function of the average scene but, because these features are the very characteristics that distinguish objects from noise [2,5], their elimination at a low level would be most undesirable.

Matched Amplification - Making Full Use of Response States

The efficiency with which amplification fill the L-cells' limited response range with the remaining signal can be gauged from elementary information theory. Neuronal response range is divided into a number of discriminable response states by intrinsic noise, and SHANNON showed that a communication channel's information capacity could only be maximised when all response states were used equally often [17]. This condition requires a coding procedure that is matched to the expected frequency of inputs and, for the general case of a neuron coding its input level as the amplitude of response, the necessary intensity-response function corresponds to the cumulative probability distribution of input intensity (Fig.5a). An identical coding procedure is often used in digital image enhancement to perform "histogram equalisation" [18]. By allotting sensitivity in proportion to the expectation of encountering a signal, the common measurements are enhanced at the expense of the rare, and in image processing this brings out the details contained in large areas of low contrast.

At high light levels, where the receptor signal:noise is relatively good, the relationship between L-cell response and contrast is approximately sigmoidal. Does it correspond to the cumulative probability of encountering contrast in natural scenes? I determined contrast levels in a random selection of appropriate natural habitats by using a scanner with a similar spectral and angular sensitivity to the fly's eye [8]. Flies generally encounter moderate contrast, with three-quarters of the readings lying in the range of -0.5 to +0.5, and an average contrast of 0.4 (Fig.5b). The distribution is skewed to negative contrasts and there is a long 'tail' of contrasts greater than +1.0 (Fig.5a).

The fit of the L-cell's contrast-response function to the cumulative probability suggests that coding is optimal (Fig.5b). However, the response function was measured using contrast steps and these stimuli are rather unnatural. Digital filtering of the intensity scans corrected for the L-cell's spatial filtering properties (the effect of its receptive field, Fig.4a), but no account was taken of the temporal filtering characteristics (the impulse response, Fig.4b). Because L-cell responses are transient, these neurons only respond to moving stimuli. Consequently the intensity scans should be convolved with an equivalent impulse response having an angular width proportional to speed. Unfortunately the critical filtering parameter, the statistical distribution of angular velocities encountered by the fly's eye in everyday life, has not been measured. Nonetheless, one expects the effects of temporal filtering to be small because, while the blurring effect of motion

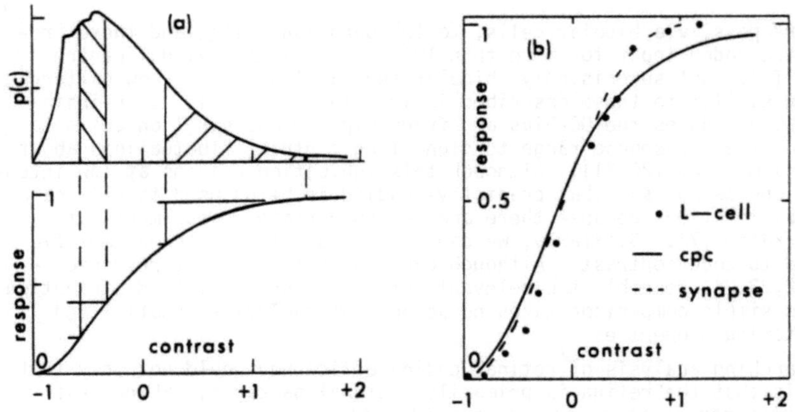

Fig.5. The matched amplification coding strategy. (a) Theory:- input levels occurring with unequal frequency (upper curve) are optimally coded by their cumulative probability function (lower curve) because equal response bands are occupied equally often. (b) Implementation:- cumulative probability curve for contrast, cpc, compared with response of light adapted L-cell to contrast steps, and to transfer function for chemical synapse [10]. Distributions of contrast determined from natural scenes using scanner with point spread function of 0.15° in (a) and 1.5°, equivalent to fly's light collector, in (b)

washes out contrast, the onesidedness of the impulse response increases it. Moreover, information capacity is rather insensitive to small inequalities in the utilisation of response states, so that simply giving a higher sensitivity to commoner contrasts achieves considerable coding efficiency.

Thus the study of contrast coding in the fly's eye demonstrates that neurons have at their disposal a simple means for enhancing their information capacity. The procedure of matched amplification adjusts sensitivity to the expectation of encountering an input. When inputs follow a bell—shaped probability distribution, matching is readily implemented by the sigmoidal transfer function of the typical chemical synapse (Fig.5b).

Finally, it should be noted that one does not need information or coding theory to appreciate the benefits of coding efficiency. Predictive coding and matched amplification convert small receptor modulations into bold excursions in membrane potential. The task of theory is to test hypotheses about the roles of L1&2 in early processing. Our analyses show that the strength of inhibition is more compatible with redundancy reduction than with edge detection, and small contrasts are not being enhanced because some later pathway is specifically interested in subtle fluctuations. L1&2 are generalised preprocessors, attempting to bundle as much data up the line as their limited resources permit.

Is Coding Efficiency Fundamental to Vision?

Do vertebrate visual systems improve their efficiency by employing similar matched coding procedures? The limited number of ganglion cells connecting the retina to the brain is an informational bottleneck that must encourage some degree of coding efficiency. In the monkey the outputs from about 10^8

photoreceptors pass, via bipolar cells, to 10^6 ganglion cells, and these provide the entire coded input for more than 10^9 neurons in the visual centres of the brain [6]. Not surprisingly, bipolar and ganglion cells show patterns of inhibition similar to those described in invertebrates [19]. Such inhibition inevitably reduces the DC-bias and frees bipolar and ganglion cells to devote more of their response range to signalling contrast fluctuations about a given mean luminance [20,21]. Although this inhibition weakens at low intensities [21], one cannot say that predictive coding is being used to eliminate first order correlations because there are no appropriate measurements of signal:noise ratio [7]. Similarly, we cannot conclude that matched amplification is used to code contrast. Although ganglion cell contrast-response functions [22,23,24] resemble the relevant cumulative probability distribution (Fig.6), this simple comparison takes no account of the cells' spatial and temporal filtering properties.

A more searching analysis of retinal coding efficiency would not only test the hypothesis that the retina is primarily a signal packager, making best use of the optic nerve, it would also provide additional examples of procedures by which a brain can minimise the deleterious effects of neuronal unreliability. Vertebrate retinas are more complicated than the fly's receptor-L1&2 system, and many of these complications, such as Y-cell non-linearity, may eventually show that coding is more subtle and sophisticated than was previously thought. Furthermore, retinal ganglion cells encode information as impulse frequency, not graded potential amplitude, and this may necessitate different coding strategies [24].

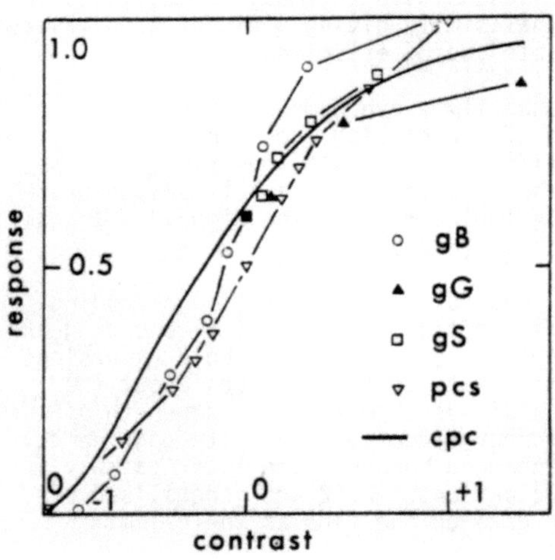

Fig.6. Does matched amplification operate in other visual systems? Cumulative probability distribution for contrast, cpc, compared with human perceptual contrast scale, pcs, [27], and contrast-response functions of light adapted retinal ganglion cells:- gB = cat, ON cell and OFF cell [22], gS = cat, ON cell [23], gG = rat, ON cell [24]. Response normalisations:- range of perceptual scale = 1.0, centred about zero contrast; ON cell response ranges from 0.6-1.0 and OFF cell from 0.6 to 0.0

With cells in the visual centres outnumbering retinal ganglion cells by 1000:1 there is little to be gained by continuously running these central neurons at full capacity. Efficiency is promoted by matching "filters", "channels", "primal sketchers" or "analysers" to the nature and quality of the objects they process. Nonetheless, when neurons respond, they should respond accurately and this entails matching their responsiveness to the levels of signals they encounter in their particular circuit. For example, some cortical neurons adjust their contrast-response function to the ambient level of contrast [25]. In the presence of low contrast stimuli they are extraordinarily sensitive, as if searching for detail, but exposure to higher contrasts depresses sensitivity appropriately. One can interpret this as adaptive matched amplification, whereby the probability of encountering a stimulus takes into account the immediate history. However, both the most strongly adapted cells, and the less responsive non-adapting cells, are far too sensitive to small contrasts for matched amplification. Are measurements of stronger contrasts unimportant to these circuits or did the sinusoidal gratings used as stimuli excite these circuits with a power that natural scenes cannot match?

Just as neuronal responses and sensitivities should be distributed and adjusted to the qualities of the signals they receive, so should our perceptual abilities. RICHARDS [26] independently formulated and investigated the perceptual equivalent of matched amplification. A limited range of percepts is used efficiently when perceptual scaling follows the cumulative probability distribution of the relevant stimulus set, so that all levels are perceived equally often. The relationship between perceived object lightness and object intensity follows the distribution of natural intensity levels [26], and further evidence for perceptual matching is presented in Figure 6, where our perceptual scale for contrast is compared to contrast's statistical distribution. The perceptual scale was derived by asking subjects to assign numbers to the contrast levels of sinusoidally modulated gratings [27] and some of the discrepancy between scale and probability distribution can be attributed to the fact that sinusoidal gratings treat positive and negative contrasts as equivalent, whereas their natural distributions are asymmetrical (Fig.5a). It seems unlikely that a matching of perceptual scales, or of sensitivity functions, to favour the inputs we most commonly use is confined to judging lightness or contrast. For example, the human eye seems most sensitive to spatial details when objects are moved across the retina at speeds corresponding to the drift of the eye during fixation [28].

In conclusion, there is every reason to think that the types of matching of neuronal responsiveness to natural stimuli seen in flies, are widespread among visual systems, and that the concept of matching can be extended from neurons to circuits, and onward to perception. I would like to stress that there is nothing particularly new about this concept of matching. It is a commonplace of comparative sensory physiology that animals put their sensitivity where they expect the signal to be. Crickets tune their ears to the carrier frequencies of their songs, fishes adjust the wavelength sensitivity of their photoreceptors to the colour of the waters they inhabit and fireflies match the sensitivity of their eyes to the colours of their flashes. Conversely, we do not put sensitivity where it is wasted. Photoreceptors are placed no closer together than the diffraction limit of a lens, and ganglion cell density is decreased in retinal areas where information is sparse [29]. The fly has demonstrated that these principles operate at a new level, information coding by single nerve cells, and this suggests new ways in which the brain can minimise the constraint of fallible neurons.

Acknowledgements

It was a privilege to be able to collaborate with M.V. SRINIVASAN and A. DUBS on the analysis of predictive coding. I am extremely grateful to P. COOTE for her careful preparation of camera ready copy, and to A.W. SNYDER and J. HOWARD for continuous and fruitful discussions.

References

1. J. von Neumann: *The Computer and the Brain* (Yale University Press, New Haven 1958)
2. F. Attneave: Psychol. Rev. 61, 183-193 (1954)
3. H.B. Barlow: In *Sensory Communication*, ed. W.A. Rosenblith (MIT Press, Cambridge, Mass. 1961) pp. 217-234
4. H.B. Barlow: In *Sensory Communication*, ed. W.A. Rosenblith (MIT Press, Cambridge, Mass. 1961) pp. 782-786
5. H.B. Barlow: In *Current Problems in Animal Behaviour*, eds. W.H. Thorpe, O.L. Zangwill (Cambridge University Press, Cambridge 1961) pp. 331-360
6. H.B. Barlow: Proc. R. Soc. Lond. B 212, 1-34 (1981)
7. M.V. Srinivasan, S.B. Laughlin, A. Dubs: Proc. R. Soc. Lond. B, in press (1982)
8. S.B. Laughlin: Z. Naturforsch. 36c, 910-912 (1981)
9. G.M. Shepherd: In *Neurones Without Impulses*, eds. A. Roberts, B.M.H. Bush (Cambridge University Press, Cambridge 1981) pp. 1-27
10. S.W. Kuffler, J.G. Nicholls: *From Neuron to Brain* (Sinauer, Sunderland Mass., 1976)
11. W. Reichardt: In *The Neurosciences: Second Study Programme*, ed. F.O. Schmitt (Rockefeller University Press, New York 1970) pp. 494-511
12. N.J. Strausfeld: *Atlas of an Insect Brain* (Springer, Berlin-Heidelberg-New York 1976)
13. K. Kirschfeld: In *The Neurosciences: Fourth Study Programme*, ed. F.O. Schmitt (MIT Press, Cambridge Mass. 1979) pp. 297-310
14. S.B. Laughlin: In *Handbook of Sensory Physiology*, Vol. VII/6B, ed. H. Autrum (Springer, Berlin-Heidelberg-New York 1981) pp. 133-280
15. R.M. Boynton, D.N. Whitten: Science 170, 1423-1426 (1970)
16. B.M. Oliver: Bell System Tech. J. 31, 724-750 (1952)
17. C.E. Shannon, W. Weaver. *The Mathematical Theory of Communication* (University of Illinois Press, Urbana 1949)
18. R.C. Gonzales, P. Wintz: *Digital Image Processing* (Addison-Wesley, Reading, Mass. 1977)
19. H.K. Hartline: Science 164, 270-278 (1969)
20. F.S. Werblin: Scientific American 228, 70-79 (1973)
21. H.B. Barlow, W.R. Levick: J. Physiol. 259, 737-757 (1976)
22. H.B. Barlow: Ann. N.Y. Acad. Sci. 156, 872-881 (1969)
23. B. Sakmann, O.D. Creutzfeld: Pflugers Arch. 313, 168-185 (1969)
24. D.G. Green, M.K. Powers: Vision Res. 22, 209-216 (1982)
25. I. Ohzawa, G. Sclar, R.D. Freeman: Nature 298, 266-268 (1982)
26. W.A. Richards: Applied Optics, in press (1982)
27. J. Gottesman, G.S. Rubin, G.E. Legge: Vision Res. 21, 791-800 (1981)
28. D.H. Kelly: J. Opt. Soc. Am. 69, 1340-1349 (1979)
29. A. Hughes: In *Handbook of Sensory Physiology*, Vol. VII/5, ed. F. Crescitelli (Springer, Berlin-Heidelberg-New York 1977) pp. 614-756

Neighbourhood Operators

M.J.B. Duff

Department of Physics and Astronomy,
University College, London, United Kingdom

1. Introduction

This paper considers a class of operations which can be performed on images. The original image will be assumed to be a continuously varying distribution of intensities ranging from black to white, covering a rectangular area. Before the processing operation, the image will be digitized both spatially and in intensity, usually by dividing it into an m by n array of square picture elements (*pixels*), each being of uniform intensity at a particular *grey-level*, the complete intensity range from black to white having been divided up into G equally spaced steps. Other digitization schemes have been proposed and are sometimes used for particular purposes (such as hexagonal tessellations and logarithmic grey digitization) [1].

If the digitized image is I, the pixel in column i and row j is $I_{i,j}$. As a result of applying a neighbourhood operator L_G to the image I, a new image I* is computed, where

$$I^* = L_G (I) .$$

Restricting the discussion to the class of parallel neighbourhood operators, then the global computation L_G can be decomposed into approximately (mxn) local computations to be performed at each of the (mxn) pixels. For these local computations, which can all be performed simultaneously, each pixel $I^*_{i,j}$ in the new image will have a grey level $g^*_{i,j}$ where

$$g^*_{i,j} = L\{N (g_{i,j})\}$$

given that $N(I_{i,j})$ is the set of pixels in the local neighbourhood N relating to the pixel $I_{i,j}$, and $N(g_{i,j})$ is the set of grey-levels of the pixels in N. In other words, each new pixel will have a grey-level which can be calculated from the grey levels in a *window* of pixels either surrounding or near to the corresponding pixel in the original image. The size and shape of the window N may be varied and it is not implied that all functions will make use of every pixel within the window; some may be ignored. In the set defined by N, those elements which are used will be labelled arbitrarily 1,2,3.....d,.... , reserving the label 0 for the element at (i,j). Later in this analysis, weighting factors w_d may be associated with the elements labelled d, so that $N (w\, g_{i,j})$ will imply the set of products of grey levels and the corresponding weighting factors in the local neighbourhood N. It should be noted that the function will not necessarily be computible for pixels near an edge of the image where the window overlaps the edge.

It can be shown [2] that any image processing operation can be achieved by an appropriate sequence of parallel (3x3) local neighbourhood operations, although some will be too cumbersome for serious consideration as a replacement for the more familiar serial techniques. Nevertheless, the method should be regarded as an important computational approach of complete generality.

2. Classification of Neighbourhood Operators

There are several ways in which neighbourhood operators might be classified. One possibility would be to draw up a functional classification in terms of the results obtained by applying each operator. Although this is worthwhile in that it provides a catalogue of useful functions, it is more informative to try to classify operators with respect to their construction, in the hope that rules may be perceived to assist in the design of new operators. The most significant factors to be taken into account are

1. The window dimensions (which pixels take part in the computation)

2. The computation (the details of the computation performed on each window of pixels)

2.1 Window Dimensions. A broad division would be as follows:

a) The central pixel and one adjacent to it,

b) All or some of the (3x3) local neighbourhood,

c) All or some of a (pxq) local neighbourhood,

in which at least one of p or q is greater than 3, but excluding (m x n),

d) The entire (m x n) array.

The last category may appear to be a global, rather than local neighbourhood, operator, but does in fact represent the limiting case of neighbourhood operation. In the same way, one could consider a point operator, in which

$$g^*_{i,j} = L\,(g_{i,j})$$

as representing the opposite and lowest limit of neighbourhood operation.

2.2 The Computation It is harder to find a systematic classification with respect to the form of the computation in the operator. One possibility would be:

a) Insertion and deletion operators. These are operators applicable to binary (black and white) images in which a pixel may be changed from black to white or from white to black on the basis of a decision depending on the exact composition of the binary image in the window. This is obviously a non-linear operation.

b) Linear grey operators. The new pixel grey-level is a weighted sum of the grey levels of pixels in the window.

c) Non-linear grey operators. In these, the new pixel value is computed by performing an appropriate set of arithmetic operations on neighbouring pixels, involving some non-linear operations such as thresholding, min-max functions and conditional operations.

It is convenient to adopt the computation classification as the prime classification, subsequently subdividing each class on the basis of window size. In the next sections, the characteristics of each class and subclass will be examined and discussed.

3. Insertion and Deletion Operators

As was stated earlier, these operators are only applicable to binary images. The convention adopted here is that white pixels will be assigned the value 1 and black pixels the value 0. Since many of these operations are implemented by Boolean functions, the values 1 and 0 should be treated as Boolean variables rather than as grey levels.

3.1 One adjacent pixel. The operations in this subclass are *Shifting, directional edge finding, directional shrinking and expanding*. Thus, for a window containing only the pixel $I_{i-1,j}$ immediately to the left of $I_{i,j}$, the pixel under consideration,

$$g^*_{i,j} = g_{i-1,j} \quad ,$$

causes a right *shift* of the image I by an array distance equal to one pixel. In the example

$$g^*_{i,j} = g_{i,j} \cdot \overline{g_{i-1,j}}$$

where the dot represents the Boolean AND and the stroke is the inverse function, the operator extracts as white pixels the left *edges* of white objects on a black background. *Expanding* white objects to the right is effected by

$$g^*_{i,j} = g_{i,j} + g_{i-1,j}$$

where + is the Boolean OR . Similarly, *shrinking* from the left results from

$$g^*_{i,j} = g_{i,j} \cdot \quad g_{i-1,j} \quad .$$

Each of the latter three functions can be further decomposed into a shift followed by a Boolean AND or OR. It is a property of all neighbourhood operators that the operation can always be compounded from as many shifts as there are pixels in the window, each of which is followed by a logical or arithmetic point function (i.e. a function in which $g^*_{i,j}$ is a function of only $g_{i,j}$). However, pixels which are displaced by p pixels from (i,j) will require a shift of amplitude p before applying the point function.

Regarding $g_{i,j}$ and $g_{i-1,j}$ as two Boolean variables, there are exactly 16 different Boolean functions which could be applied to compute $g^*_{i,j}$. All of them have effects similar to those described above, except for the four which, being independent of $g_{i-1,j}$, degenerate into point functions ($g^*_{i,j}$ equalling $g_{i,j}$, $g_{i,j}$, 1 or 0).

3.2 (3x3) local neighbourhoods. These operators generally prove to be more interesting than the one adjacent neighbour operators (which are themselves a subset of this wider class). The operators can be expressed in the form

$$g^*_{i,j} = B\ \{g_{i,j},\ A\ \{\ N\ (s\ g_{i,j}\ )\}\}$$

where B {x,y} is any Boolean function of the Boolean variable x and y

A {x,y,.....} is the Boolean AND of the variables x,y,......

s is a set of switching functions

$(s_0, s_1, s_2, \ldots\ldots s_d, \ldots\ldots)$ which

are applied to the grey-levels in the local neighbourhood N;

$s_d \, g_{i,j,d} = 1$ when

$$s_d = 1 \text{ and } g_{i,j,d} = 1$$
$$s_d = 0 \text{ and } g_{i,j,d} = 0$$
$$s_d = X \text{ and } g_{i,j,d} = 0 \text{ or } 1$$

otherwise

$sd \, g_{i,j,d} = 0$.

The function is then completely specified by listing the 9 values of s as d ranges from 0 to 8 in the (3x3) window.

Expressed more simply, the (3x3) window in the image is compared with a (3x3) mask of 0's, 1's and X's. If the 0's in the image match the 0's in the mask, and the 1's in the image match the 1's in the mask, then the mask matches the image and the central pixel is deleted, inserted or replaced as specified. The X elements in the mask represent 'dont care' states and corresponding pixels in the image may take either value without affecting the output image. For convenience, the expression can be written as

$$g^*_{i,j} = B \{g_{i,j}, M_{i,j}\}$$

where $$M_{i,j} = A \{N (s \, g_{i,j})\}$$

and for a (3x3) local neighbourhood, the set of values for s can be written in array format. For example

$$s = \begin{vmatrix} 0 & 1 & 0 \\ X & 1 & X \\ 0 & 1 & 0 \end{vmatrix}$$

would be a possible mask for detecting a particular feature in images.

Operations of this type will extract edges and other shaped features from binary images and can be used to delete small objects. Some examples are given in Fig. 1. Certain short sequences of these operators can be expressed in a more concise manner by using *counting*. Thus, the operator may specify that the central pixel be deleted if more than t of its 8 neighbours are 0's. A further complication may be that in the 8 neighbour ring of the (3x3) neighbourhood, the greater than t 0-neighbours must be in an unbroken sequence for deletion to occur. These operators are orientation independent and correspond to the so-called *Golay surrounds* [3] in hexagonally digitised images.

Operators in this class are of fundamental importance in the processing of binary images. As was implied in the introduction, any complex image function can be built up by using sequences of simple neighbourhood operators, but even the individual operators themselves prove to be very powerful. Also, the fact that a window is small implies minimal complexity in the hardware or software needed to implement the operator.

Of particular interest are the *shrink* and *expand* functions, about which much has been written elsewhere [4,5], sometimes using alternative names for the same functions (*erosion* and *dilation* [6], *reduction* and *augment-*

$$\begin{matrix} 0 & 0 & 0 \\ 0 & 1 & 0 \\ 0 & 0 & 0 \end{matrix}$$

Isolated Point Detector

$$\begin{matrix} X & 0 & 1 \\ 0 & 1 & 1 \\ 1 & 1 & 1 \end{matrix}$$

Diagonal Edge Detector

$$\begin{matrix} 0 & 1 & 1 \\ 0 & 1 & 1 \\ 0 & 1 & 1 \end{matrix}$$

Left Edge Detector

Fig. 1. Binary operations with 3 x 3 windows

ation [7]). A simple quasi-isotropic *shrinking* of a white object on a black background is given by

$$I^* = S(I)$$

in which $g^*_{i,j} = g_{i,j} \cdot M_{i,j}$

and where $S = \begin{vmatrix} 1 & 1 & 1 \\ 1 & X & 1 \\ 1 & 1 & 1 \end{vmatrix}$

Thus, any black neighbour of a white pixel deletes (i.e., changes to black) the white pixel. This operator therefore removes the outer edges of white objects. The complementary *expand* function is

$$I^* = E(I)$$

in which $g^*_{i,j} = g_{i,j} + \overline{M}_{i,j}$

and where $s = \begin{vmatrix} 0 & 0 & 0 \\ 0 & X & 0 \\ 0 & 0 & 0 \end{vmatrix}$

Here, note that a white pixel will be inserted unless all the neighbours of the pixel are black, but white pixels are always unchanged. The result is therefore the addition of a white layer around every white object.

The fuller significance of the shrink and expand operators becomes apparent when finite sequences of them are applied to binary images. Very small white objects are composed of pixels, so that the object is composed entirely of edge elements. A single shrink operation will therefore delete such objects irrecoverably. These objects are said to have an *image depth* of 1. A subsequent expansion will approximately restore the shapes and sizes of bigger objects (i.e. *deeper* objects) which were not totally deleted by the previous shrinking. The partially restored image

$$I^* = E(S(I))$$

although similar to the original image, will have lost all small details (objects of depth 1 and small projections on the larger objects). The parts which were lost can be formed into a separate image by using

$$I^* = I.\overline{E(S(I))}$$

The first operation is similar to a low pass, two-dimensional spatial filter and the second operation is similar to a high pass filter. Repetition of the functions results in filters which remove or extract objects of larger image depths, in which x shrinks are to delete an object of depth x

$$I^* = E^x\ (S^x\ (I))$$

Band pass filters can be formed by combining suitable high and low pass filters

$$I^* = E^x(S^x(I)).\ \overline{E^y(S^y(I))}$$

where $y > x$ (4)

An image which has been successively shrunk and expanded would be unchanged if subjected to the same treatment again:-

If $I_2 = E(S(I_1))$

then $I_2 = E(S(I_2))$

or, more generally

If $I_x = E^x(S^x(I_1))$

then $I_x = E^x(S^x(I_x))$

Similarly,

$$\text{If } I_y = S^y(E^y(I_1))$$

$$\text{then } I_y = S^y(E^y(I_y))$$

A second point to consider is that a better approximation to a low pass filter can be obtained by deleting both small white and small black objects (holes). Either of the following functions, which do not give exactly the same results, will serve:

$$I^* = E(S^2(E\ (I)))$$

$$I^* = S(E^2(S(I)))$$

A one-dimensional version of these operations is shown in Fig.2 .

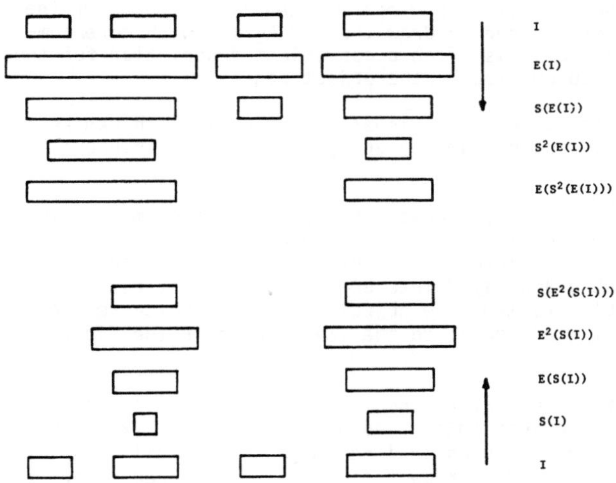

Fig. 2. One-dimensional shrink-expand operations

Unfortunately, a square (3x3) window does not produce truly isotropic operations, so it is necessary to combine sequences of expansions and shrinks into carefully constructed windows in order to obtain better directional uniformity in the filtering. This refinement has been studied by several authors [8].

A relationship which will be required later (in section 5.1) concerns the extraction of edges by using the shrink operation. Using the notation developed above, the operation

$$I_2 = S(I_1)$$

removes all the edges from white objects in I_1. These can be presented as a separate image I_3 by applying a pixel by pixel Boolean operation on I_2 and I_1, thus

$$I_3 = I_1 . \overline{I_2}$$

$$= I_1 . \overline{S(I_1)}$$

or, using the usual notation, for edges

$$I^* = I . \overline{S(I)}$$

An alternative numerical form of the operator (rather than Boolean) is

$$I^* = I - S(I)$$

A second type of significant operator in this class is the *skeletonizing* function. It is often found that the essential shape of an object can be represented by thin lines drawn centrally in the component parts of the objects. Alphanumeric characters provide good examples. The identifiable shapes of the characters are not noticeably dependent on the thickness of the pen or pencil used to draw them. Again, with more complicated shapes, the medial axis is often a convenient description for the object, especially when supplemented by the object edges.

As will be discussed later, the analysis of a single pixel thickness line is much more straightforward than the analysis of thick, 'blobby' objects.

Skeletons cannot be extracted by repeated shrinking of the image; this will almost inevitably break up the object into several parts before a single thickness residue is obtained. However, it is possible to devise a sequence of masks which gradually delete edge pixels whilst retaining pixels whose deletion would fragment the object. Fig. 3 shows one such sequence and the result of applying it to the image shown in the figure [9].

The third type of operator worthy of special consideration is the *line analysis* operator. Provided that an edge extractor or a skeletonizer has performed correctly, features in line images can readily be extracted by the use of the (3x3) window functions. The three principal features are line ends, junctions and crossings. As a first approximation, these features will be represented by white pixels having exactly 1, 3 or 4 white neighbours respectively. In cases of ambiguity, it is necessary to examine a wider neighbourhood, but the first indications obtained using the (3x3) window are usually sufficient. Some examples are given in Fig. 4. It should be noted that a (3x3) window has only 512 possible configurations, so that it is by no means out of the question to store binary coded interpretations in a RAM with 512 9 bit addresses, each address referencing, say, a byte of data, thereby allowing up to 256 distinct interpretations. This technique has been used in several specialised computers designed for image processing or pattern recognition [10,11 .]

3.3 Larger Neighbourhoods. Apart from the occasional need to resolve ambiguities in observed features, there is seldom cause to employ binary neighbourhood operators of larger size than (3x3). A few examples will be considered later as special cases of grey-level image processing.

0	0	X	X	0	0	X	1	X	X	1	X
0	1	1	1	1	0	1	1	0	0	1	1
X	1	X	X	1	X	X	0	0	0	0	X
	A1			A2			A3			A4	

0	0	0	1	X	0	X	1	1	0	X	X
X	1	X	1	1	0	X	1	X	0	1	1
1	1	X	X	X	0	0	0	0	0	X	1
	B1			B2			B3			B4	

Sequence = A1 B1 A2 B2 A3 B3 A4 B4 (repeated)

Fig. 3. Binary skeletonization

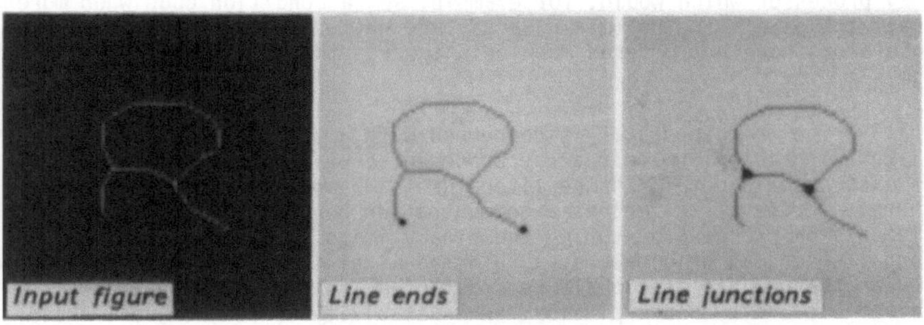

(1-points having one neighbour) (1-points having three or more neighbours)

Fig. 4. Features in line figures found by neighbour counting

3.4 Global Operators In this category, three operators will be discussed, each requiring special hardware if it is to be implemented efficiently, particularly in an array processor.

First, there is the *empty array test* operator which checks if at least one pixel is white in a binary image. This can be achieved very simply by means of an OR gate with (m x n) inputs, one for each pixel, and is used as the basic image content dependent test to terminate loop functions. For example, consider the short progamme:

```
        START

        I1= I0
LOOP:   I2= SKELETON (I1)
        I3= EXCLUSIVE OR (I1,I2)

        TEST I3

        IF I3IS EMPTY, GO TO END

        ELSE I1= I2  AND

        RETURN TO LOOP


        END
```

The function SKELETON produces no further change once the skeletonizing process is complete. This test is the most fundamental and therefore the most frequently used test in image processing when decisions are to be made depending on the state of an image.

The second operator merely *counts* the number of white pixels in the array. It is possible, but extremely inefficient, to implement this operator as a series of (3x3) windows. A better technique implies the use of special hardward counters, usually associated with the input or output circuits linking the computer to the image memories [12]. What seems to be lacking is a method which could be efficiently implemented on an array processor which would, for example, set a condition code when more than half the pixels in the (m x n) array are white. Local neighbourhood operators have not offered much help towards the solution of this problem.

Finally, the third operator is concerned with a process which is mainly associated with cellular logic array processors: *propagation*. This process is discussed in detail in the literature [13] and will not be emphasised here. Propagation implies a communication network between the memory elements or processors storing the binary image. Typically, a binary signal which may originate at a selected pixel in the image where a particular feature has previously been detected is passed on by white pixels and blocked by black pixels. When stability is reached, all those white pixels which have transmitted the propagation signal are known to be part of an object connected to the originating pixel. This operation is used in many ways, perhaps the most valuable being the segmentation from the background of connected objects in complex images. However, the advantage of having special hardware for this function is clearly implied.

4. Linear Grey Operators

In this class, pixels are taken to have grey levels in the range 0 to (G-1) where G>2.

4.1 One adjacent pixel. In principle, the operators relating to binary images and described in section 3.1 can be adapted to work with grey-level images, although in practice only the *shift* operators are particularly useful. Boolean functions are no longer applicable so some changes to the notation are necessary. The general expression for local neighbourhood operators.

$$g^*_{i,j} = L\{N(g_{i,j})\}$$

is still applicable and the function L for the class of linear grey operators performs a normalized weighted summation of the grey levels in N. A suitable form for the expression is

$$g^*_{i,j} = \frac{1}{F}\,\Sigma\{N(W_{i,j}\; g_{i,j})\}$$

where F is a constant factor, most usually chosen to be

$$F \quad = \Sigma\{N(w_{i,j})\}$$

so that the operators applied to uniform grey image will return a uniform grey image, of the same intensity,regardless of the weights used in the operator.

In the particular case being treated in this section, the operation

$$g^*_{i,j} = g_{i-1,j}$$

now right- shifts a grey image. Edge finding is better achieved using larger neighbourhoods and grey shrinking and expanding entail more complex functions (to take into account the missing Boolean operations).

4.2 (3x3) and larger local neighbourhoods. There is no important distinction between operators using (3x3) local neighbourhoods and those using larger neighbourhoods except that, as a matter of economy, the smallest effective neighbourhood would always be adopted. The algorithmic principles involved are similar regardless of the size of the neighbourhood. The main uses of these operators are *smoothing, edge finding,* or *enhancement, matched filtering,* and *convolution.*

Since these operations all require the calculation of a normalized weighted sum of the grey levels in a local neighbourhood (usually situated symmetrically about the pixel $I_{i,j}$), it is only necessary to specify the set of weights and the normalization factor, thus

$$I^* \quad = \frac{1}{F} \begin{vmatrix} w_{i-1,j-1} & w_{i,j-1} & w_{i+1,j-1} \\ w_{i-1,j} & w_{i,j} & w_{i+1,j} \\ w_{i-1,j+1} & w_{i,j+1} & w_{i+1,j+1} \end{vmatrix} I$$

for a (3x3) window.

The implementation of these operators depends strongly on the architecture of the computer. A valuable reformulation is possible when cellular logic arrays are used. The operation is carried out by performing a set of multiplications by the weights, each followed by shifts and summations. Thus, the entire image is multiplied by $w_{i-1,j-1}$, shifted diagonally south-east by one pixel, and added to $w_{i,j}$ times the original image. Similar contributions are added from the other 7 directions and the resulting image grey levels divided by F. This technique would most probably not be applicable in, say, a pipeline processor.

For particular sets of weights, it may occur that the process can be substantially shortened. For example, for the operator

$$\frac{1}{16}\begin{vmatrix} 1 & 2 & 1 \\ 2 & 4 & 2 \\ 1 & 2 & 1 \end{vmatrix}$$

the image is merely shifted, N,S,E and W, adding the shifted value to its previous value after each shift (see Fig.5). The normalization factor 16 is also effected very easily by means of left-shifting the binary numbers representing all the grey levels. It is worth noting that a careful choice of weights such as these might produce an adequate approximation to a more precisely specified operator with nearly similar weights, but which would be executed much more slowly in the particular computer being used. It is a property of many image operations that exact arithmetic is unlikely to be necessary to achieve a desired result, and a pedantic approach can be computationally expensive.

Smoothing or *blurring* results when the new pixel receives only positive contributions from neighbouring pixels and is therefore somewhat analagous to the expand function. The larger the window, the more extensive will be the blurring. If the weights for the more distant neighbours are of comparable size to or larger than the nearer pixels, then this also will accentuate the blurring. Small details will be lost in this process, but large areas with gradual intensity changes will be almost unaffected; hence the use of operators in this class for noise removal. Some examples are given in Fig. 6.

Edge enhancement or *edge finding* operators have almost the opposite effect to the smoothing operators. Their purpose is to simulate spatial differentiation of grey-levels. Thus, an approximation to the Laplacian

$$\nabla^2 f = \frac{d^2g}{dx^2} + \frac{d^2g}{dy^2}$$

is given by the operator

$$\begin{vmatrix} 0 & 1 & 0 \\ 1 & -4 & 1 \\ 0 & 1 & 0 \end{vmatrix}$$

and the result of applying this operator to the picture in Fig.7 (a) is shown in Fig. 7 (b).

I_1	a	b	c	
	d	e	f	Shift I_1 right and add to I_1
	g	h	i	
I_2	?	a+b	b+c	
	?	d+e	e+f	Shift I_2 left and add to I_2
	?	g+h	h+i	
I_3	?	a+2b+c	?	
	?	d+2e+f	?	Shift I_3 downwards and add to I_3
	?	g+2h+i	?	
I_4	?	?	?	
	?	a+2b+c +d+2e+f	?	Shift I_4 upwards and add to I_4
	?	d+2e+f +g+2h+i	?	
I_5	?	?	?	
	?	a+c+g+i +2(b+d+f+h) +4e	?	
	?	?	?	

Fig. 5. Shift and add sequence for weighted summation

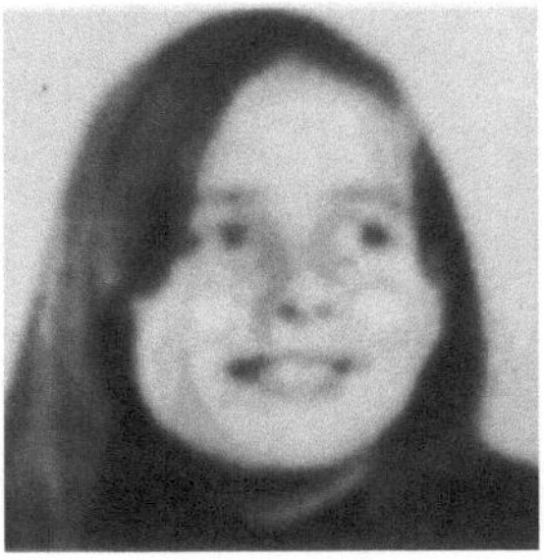

1	1	1
1	4	1
1	1	1

'Light' Smoothing

1	1	1
1	1	1
1	1	1

'Heavy' Smoothing

<u>Fig. 6.</u> Blurring by weighted summation of the photograph below (Fig. 7a)

(a)

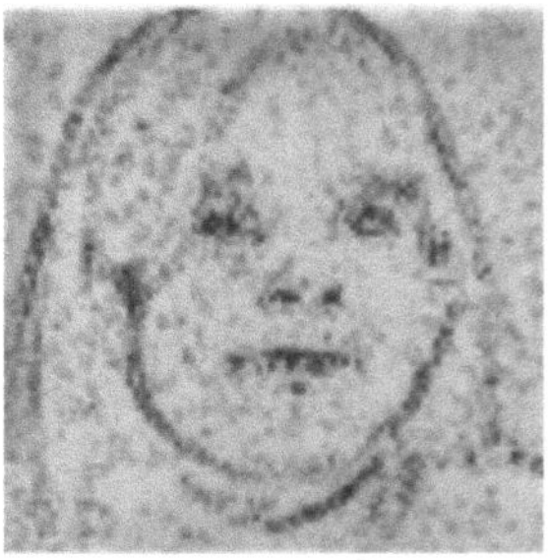

(b)

0	1	0
1	-4	1
0	1	0

<u>Fig. 7.</u> The Laplacian operator

Note that this operator is equivalent to taking the average with four neighbours and subtracting it from the original image, the result being multiplied by 5. This process (of adding a blurred negative to a sharp positive) is well known in photography and is called 'unsharp masking'. Whilst sharp edges are detected by (3x3) windows of this type, less pronounced edges require larger windows to reveal them, the optimum window width being approximately the width of the edge. In fact, edge operators are only special cases of the more general *matched filters* or *convolution operators*. The general form for a matched filter is

$$g^*_{i,j} = \frac{\Sigma \{ N (w_{i,j}\, g_{i\,j})\}}{\left[\Sigma \{ N (g^2_{i,j})\}\right]^{\frac{1}{2}}}$$

and this function maximises at locations (i,j) in the image where

$$\Sigma \{N(|w_{i,j} - g_{i,j}|)\}$$

is a minimum. In other words, when the pattern of weights in the window matches the grey-levels of pixels in the window, then the filter response is a maximum. Note that the denominator in the expression is the result of applying to the image a smoothing function in which all the weights are 1, which suggests a form for the computation.

The filter can be made to maximise at 1 by dividing by the normalization factor calculated from

$$F = \frac{\Sigma \{N(w^2_{i,j})\}}{\Sigma \{N(w_{i,j})\}}$$

It only remains to remark that 2-dimensional matched filters produce a translation independent response but are strongly influenced by re-orientation of the image.

4.3 Global Operators Brief mention should be made of the only important operator in this class, which is that which computes the mean grey level for the entire picture. As for the counting operator, there would seem to be no fast and efficient algorithm available for this operation.

5. Non-Linear Grey Operators

Unlike the operations in the previous class, the non-linear grey operations are usually completely irreversible and are often somewhat less predictable in their effect. They are characterised by conditional steps in the computation on the neighbouring pixels. As in section 4, virtually all the useful operators involve (3x3) or larger neighbourhoods, so this group will be subdivided in terms of their functions.

5.1 MIN and MAX operators. In the MIN operation, the central pixel is assigned the minimum grey level to be found in the window (the window usually being taken to include the central pixel itself)

$$g^*_{i,j} = \text{MIN}\ \{N\ (g_{i,j})\}$$

Implying parallel operation of this function over the whole image, the expression can be written

$$I^* = \text{MIN}\ (I)$$

Corresponding expressions for the MAX operation are

$$g^*_{i,j} = \text{MAX}\{N(g_{i,j})\}$$

and

$$I^* = \text{MAX}\ (I)$$

The use of these functions relates to fuzzy set theory in which the binary Boolean operations of OR and AND generalise to the MAX and MIN functions. To understand this relationship, consider the conventions following

BLACK pixels ↔ WHITE pixels		(BINARY images)
g= 0 ↔ g = 1		
DARK pixels ↔ LIGHT pixels		(GREY-LEVEL images)
g=low values ↔ g = high values		

In binary images, if any of the pixels in N is black, then

$$g^*_{i,j} = \text{AND}\ \{N\ (g_{i,j})\}$$

will cause $g^*_{i,j}$ to become black. In the same way, in grey-level images

$$g^*_{i,j} = \text{MIN}\ \{N(g_{i,j})\}$$

will cause $g^*_{i,j}$ to take on a value equal to the darkest pixel in N. The converse relations pertain for the OR and MAX functions. Another relationship is between the inverse of a binary image and its grey equivalent and is

$$\bar{I} \equiv (G-1) - I$$

Although these relationships are not exact, i.e. the binary operators and their grey equivalents do not form an isomorphism, many of the standard Boolean complex operations such as edge finding, skeletonizing, and band-pass filtering, can be translated into equivalent operations on grey-level images substituting the binary operators by their grey equivalents [14].

Comparing these relationships with those in section 3.2, it is clear that the functions which are in equivalence are (AND, MIN, SHRINK) and (OR, MAX, EXPAND). Just as various filters for binary were constructed by sequences of shrink and expand operations, so equally grey-level images can be *filtered* by using sequences of MIN and MAX operations. A *grey-edge*, in which the strength (or slope) of an edge is represented by the new pixel grey-level corresponding to the edge, can be extracted by translating the edge operation

$$I^* = I - S\ (I)$$

into

$$I^* = I - \text{MIN}\ (I)$$

It is interesting to note that the direct translation of the Boolean form of the binary edge operator

$$I^* = I.\overline{S(I)}$$

would translate to

$$I^* = \text{MIN}\ \{I,(G-1-\text{MIN}(I)\}$$

which does not reduce to the same expression and is considerably more complicated. On closer examination,it is found that the second expression does not produce an image representing edge gradients and is an example of the failure of the inversion equivalence. Nevertheless, this equivalence is found satisfactory elsewhere. The breakdown here is because of the degeneracy of the original binary expression which, it would seem, is in fact the limiting case of the difference between images rather than of an AND relationship between them.

The grey edge operator is illustrated in Fig. 8.

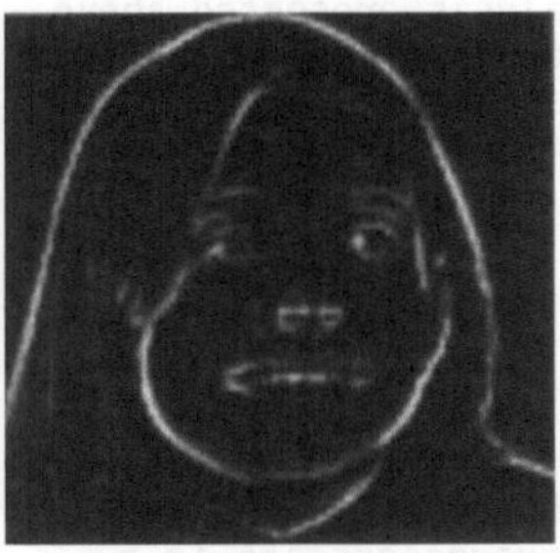

Fig. 8. Grey edge operator

5.2 Extremum filters Relating closely to the MIN and MAX operators is the *extremum* filter which is primarily used as an edge sharpener and noise filter, employing a form of smoothing which preserves edges in the image. The principle is that the new value of the pixel is either the MIN or the MAX of the local neighbourhood, whichever be the closer. Pixel values are thus 'pulled' into the dominant regions most closely resembling them. An example is given in Fig.9.

5.3 Median filters In this filter, also used to remove noise whilst preserving edges, the new pixel value is the median of pixels in the local neighbourhood. Fig.10 shows the effect of this filter, using the same original image as before.

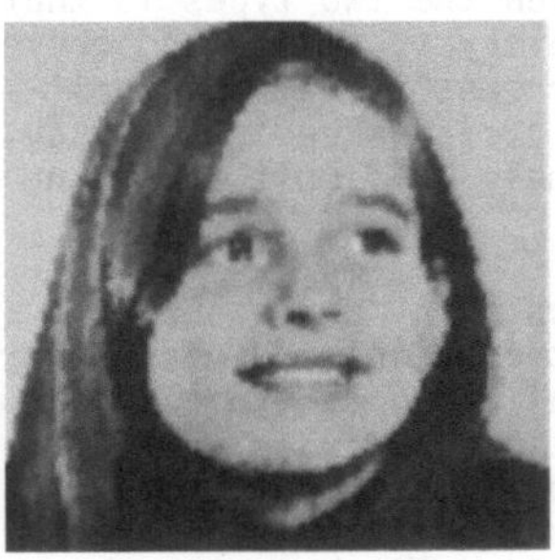

Fig. 9. Extremum Filter

Fig. 10. Median Filter

5.4 Local thresholding. It is often convenient to apply a threshold to an image so as to obtain from it a binary image in which the required information is represented by single bits in each pixel. Unfortunately, bad illumination conditions, uneven camera response and similar factors can produce sufficient image degradation to prevent the choice of a single threshold to be applied to the whole image. One solution is to find the average grey level over a window and then to use this as the basis for a threshold. The following expression is evaluated:

$$g_{i,j} - \frac{\Sigma \{N(g_{i,j})}{\Sigma \{N(1)\}}$$

and compared with a constant threshold. The denominator is just the number of pixels in N.

5.5 Other operators. This list of operators is far from complete; there is no end to the possibilities and many complex and ingenious operators have been devised for special purposes. The analysis presented above has introduced the more common operators and attempted to classify them in a helpful way. It should not be inferred that operators which have been omitted are of little value nor that the search for new and powerful operators need be unrewarding.

6. Discussion

It has been implied several times in this paper that a cellular logic architecture provides a particularly suitable structure for a computer intended to implement local neighbourhood operators. In such computers, an array of simple processors is assembled and, ideally, the processors are in a one-to-one relationship with the image pixels; each pixel has assigned to it one processor. Processors are interconnected so that each processor has inputs from and outputs to a picture store, and inputs from its immediate 4,8 (in square arrays) or 6 (in hexagonal arrays) neighbouring processors. For reasons of economy, the processors are simple in design and will be capable of performing single bit operations on binary data. Grey-levels will be therefore processed by sequences of identical operations on the consecutive digits of the binary numbers representing the grey-levels.

Two types of array [15] are currently being studied Single Instruction stream, Multiple Data stream [16,17,18] and Multiple Instruction stream, Multiple Data stream, although none in this latter class have yet been constructed. The essential difference between the two types is that in SIMD arrays, each instruction is broadcast to every processor and obeyed simultaneously, whereas in the second type, the processors may be executing different instructions in a given instruction cycle. Needless to say, the problem of programming MIMD machines is extremely difficult and by no means yet solved.

It is tempting to compare the circuitry in an array processor with the neural network in the brain or, at least, in the retina and visual cortex. One important difference must be that neural circuits exhibit *spatial layering* whereas the array processors utilise *temporal layering;* indeed, there would seem to be no concept of reprogrammable functions in neural information processing. Furthermore, it is uncertain whether the concept of a neural program is in itself meaningful. The 'programme' would appear to be replaced by structure in the interconnections between neurons.

The original work of HUBEL and WEISEL [19] stimulated much thought amongst those concerned with automatic pattern recognition, but it is not obvious that their findings have shown to be capable of translation into effective electronic circuits. Many of the most troublesome problems remain unsolved. For example, the use of local neighbourhood operators to find edges is now common practice but every new data set tends to bring with it special difficulties which invalidate previously well-tried methods for obtaining well-defined, continuous edges. It is frustrating, although challenging, that systematic approaches to pattern recognition do not usually produce satisfactory results. The best programs are often the product of intuitive design methods in which interaction can be all-important.

It might be argued that too much is being demanded of automatic pattern recognition systems. Perhaps it is unrealistic to expect fault-free recognition when the data itself may be ambiguous, and perhaps exaggerated claims are made for the apparent efficiency of neural recognition circuits. Attempts at automation too often set out to replace a human operator without allowing for the lapses in performance which are characteristic of the human and sometimes a direct and inevitable by-product of the task being carried out. Could it be that the unrivalled pattern recognition capability of the eye will have the unavoidable consequence that imprecision and error will also occur?

References

1. R.C.Gonzalez, P.Wintz: *Digital Image Processing* (Addison-Wesley Publishing Co., Inc., 1977) Chapter 2.
2. M.J.B.Duff: The Elements of Digital Picture Processing, in *Real-Time/ Parallel Computing*, ed. by Onoe, Preston and Rosenfield, (Plenum Press, 1981), p.1.
3. M.J.E.Golay: Trans. IEEE on Computers, C18 No.8, 733 (1969).
4. M.J.B.Duff: An Investigation of CLIP2 Programs Incorporating the SHRINK and EXPAND Functions, Int. Report No.73/2, Image Processing Group, Dept. of Physics and Astronomy, University College London 1973.
5. M.J.E.Golay: see ref.[3].
6. J.Serra: *Image Analysis and Mathematical Morphology*, (Academic Press, 1982).
7. K.Preston Jr. et al.: Proc. IEEE, 67 No.2, 826 (1979).
8. K.Preston Jr: Cellular Logic Algorithms for Graylevel Image Processing, in *Multicomputers and Image Processing*, ed. by Preston and Uhr, (Academic Press, 1982) p.135.
9. C.Arcelli, L.Cordella, S.Levialdi: Electr.Letters 11, 7 (1975).
10. F.A.Gerritsen, R.D.Monheimus: Evaluation of the Delft Image Processor DIP-1, in *Languages and Architectures for Image Processing*, ed. by Duff and Levialdi, (Academic Press, 1981) p.189.
11. K.-I.Mori et al: Design of a Local Parallel Pattern Processor for Image Processing, in *Special Computer Architectures for Pattern Processing*, ed. by Fu and Ichikawa, (CRC Press, 1982) p.191.
12. M.J.B.Duff: Proc. Nat. Comp. Conf. 1055 (1978).
13. M.J.B.Duff: Proc. Workshop on Picture Data Description and Management, Asilomar, California, 259 (1980).
14. V.Goetcherian: Pattern Recognition 12, 7 (1980).
15. M.J.Flynn: Trans. IEEE on Computers, C-21, 948 (1972).
16. M.J.B.Duff: see ref.[12].

17. S.F.Reddaway: in *Infotech State of the Art Report on Supercomputers*, 2, (1979).
18. K.E.Batcher: Trans. IEEE on Computers, C-29, 836 (1980).
19. D.H.Hubel, T.N.Wiesel: J.Physiol.160, 106 (1962).

Frequency Domain Visual Processing

J.G. Robson

Physiological Laboratory, University of Cambridge,
Cambridge CB2 3EG, United Kingdom

The sensitivity of human spatial vision has traditionally been characterised either by the minimum size of a high-contrast test figure which allows an observer to detect or recognize it, or by the minimum visible luminance difference between a large sharp-bordered target and the uniform background on which it appears. Information of this kind is not, however, adequate for determining what quality of reproduction of images is necessary for satisfactory human viewing, an important question to designers of, for example, photographic or television systems. Thus it is not surprising to find that a more powerful technique for characterising human visual sensitivity was initially introduced because of its potential application in these areas.

SELWYN [1] in 1948 and SCHADE [2] in 1956 were the first to report measurements of the minimum threshold contrast of a novel target, a *sine-wave grating* (Fig. 1). SELWYN, using photographic targets, and SCHADE using a television display, measured a subject's contrast threshold at a number of different spatial frequencies. Taking the reciprocal of these threshold values gives the subject's (sine-wave or spatial frequency) contrast sensitivity function (Fig.2).

Fig. 1. A sine-wave grating. The contrast *of the pattern is* $(L_{max} - L_{min})/2L_{mean}$ *where L is the luminance at a point.* Contrast *lies in the range 0 to 1. The* spatial frequency *of the grating is usually expressed as the number of periods subtending one degree at the eye*

While some use may be made of the measured contrast-sensitivity function without further interpretation, it has often in fact been supposed that it represents the modulation-transfer function of the visual mechanism (admittedly of both optical and neural mechanisms in cascade) on the assumption that each stimulus pattern at its own threshold contrast will generate the same small internal signal. On this basis it would be supposed that the detectability of patterns other than sine-wave gratings might be predicted by summing together the sensitivities for each of the Fourier components of the stimulus pattern. Of course this could only be expected to provide an exact prediction in so far as the visual mechanism operates in a linear manner, but it is not *a priori* unreasonable to expect it to do so for signals small enough for their sum to be at the threshold of detectability.

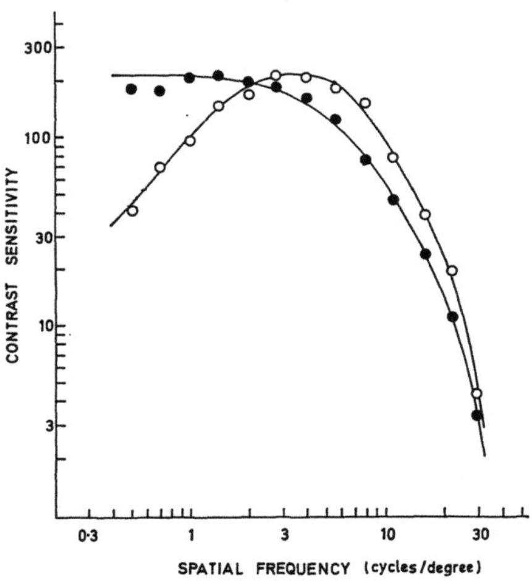

Fig. 2. The sine-wave contrast sensitivity of a human observer measured with a counter-phase modulated grating having a mean luminance of 20 cd/m². The attenuation of sensitivity at low spatial frequencies is only evident at low rates of modulation (open circles, 1Hz; Filled circles, 6Hz). (Adapted from ref. [3])

In fact the prediction does not work well even for such a simple pattern as a square-wave grating. Examining the relative visibilities of sine- and square-wave gratings, CAMPBELL and ROBSON [4] reported that the visual system seemed to behave as though the harmonic components which distinguish the square wave from the sine wave (3rd, 5th, etc. harmonics) were detected separately from the fundamental component. This finding led them to speculate that there might exist within the visual system a number of mechanisms each sensitive to only a restricted range of spatial frequencies. The existence of such "spatial-frequency selective channels" was further supported by the discovery by BLAKEMORE and CAMPBELL [5] of a spatial-frequency-specific adaptation effect (a reduction in contrast sensitivity over a limited range of spatial frequencies produced by exposure to a sine-wave grating of fixed spatial frequency).

Although these findings are obviously of some intrinsic interest, their main attraction has derived from their use as a basis for the suggestion that the visual system performs some kind of spatial frequency analysis of the visual input as a primary step in its image-processing function.

Because of the lack of unequivocal quantitative evidence, however, this idea was not initially made very explicit, though opponents of the notion have often imagined that it was proposed that the visual input is subjected to a global Fourier analysis with the implication that the visual scene is represented internally as an array of Fourier co-efficients. Such an idea borders on the ridiculous and it is hardly surprising that there is no real evidence to support it and much to rebut it. It has not proved possible, however, to counter the less implausible idea that (to give it in "physiological" terms) different spectral components (or perhaps more properly spectral regions) of the visual input are independently represented within the brain by the activity of different neurones. While this idea may now in fact be generally acceptable, there is still a good deal of controversy over its appropriateness or usefulness. Much of this argument has centred around the bandwidth of the spatial-frequency-selective visual mechanisms since it is maintained by some that a degree of spatial frequency selectivity would be shown by many mechanisms which could be much better understood in other terms. Higher degrees of selectivity would be more truly indicative of a frequency-analysing mechanism.

Despite a good deal of work it has proved peculiarly difficult to get acceptable estimates of the spatial frequency selectivity of underlying mechanisms from psychophysical experiments. The difficulty arises from a number of causes including 1) the dubious nature of the basic psychophysical linking hypotheses required for the interpretation of all psychophysical results, 2) the stochastic nature of the detection process which is involved in so many psychophysical studies and 3) the computational complexities introduced by the spatial non-uniformity of the visual process and the need to consider the contributions to vision of very many simultaneously active mechanisms. In any case it is inevitable that psychophysical experiments can never hope to provide an unambiguous picture of the physiological mechanisms of vision.

It may therefore be more rewarding to look at the results of neurophysiological studies. Though there are still many caveats to be borne in mind when interpreting neurophysiological results, at least the observations relate more directly to the physiological mechanisms and do not require the same detailed and model-dependent analysis as the psychophysical results.

I am therefore going to describe some of those neurophysiological findings which seem to me to be most relevant to an understanding of the image-processing role of the visual system particularly as revealed by studies mainly conceived, executed or interpreted in spatial frequency terms. It is worth emphasizing that while some of these studies have certainly been undertaken by people who have believed that the visual system is indeed performing some sort of spatial frequency analysis, it is certainly not necessary for this to be correct for their experimental approach to be useful or the results significant.

We may usefully start by a consideration of retinal function as a simple introduction to certain aspects of the operation of the visual nervous system. In fact for our purposes we can restrict our attention to the nature of the representation of the visual image at the output of the retina, that is to the behaviour of retinal ganglion cells.

The activity of individual nerve cells or axons can be recorded in anaesthetised animals using a microelectrode whose tip is either touching or at least close to the cell or axon. Such an electrode close to a retinal ganglion cell or an optic nerve fibre will generally record an irregular train of action potentials whose rate of occurrence is obviously modulated by visual stimulation.

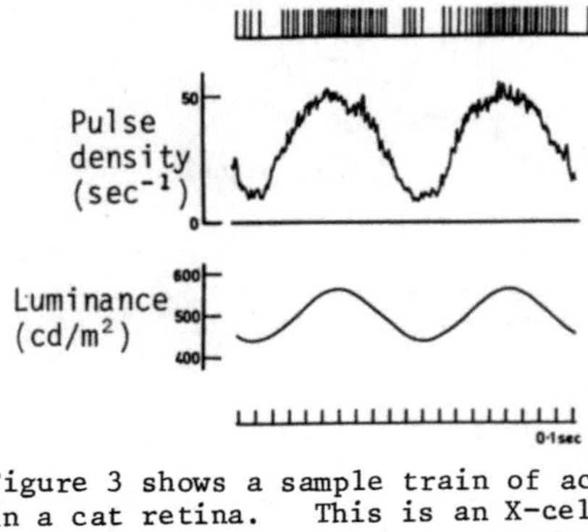

Fig. 3. Response of an on-centre X-cell in the cat retina to a drifting sine-wave grating (spatial frequency, 0.75 cycles/degree; drift frequency 1 Hz). Top record: a sample of the nerve impulse train; middle record: the average rate of discharge; lower record: the luminance at the centre of the receptive field. (From ref. [6])

Figure 3 shows a sample train of action potentials from a ganglion cell in a cat retina. This is an X-cell, the commonest type which comprises about one half of all the ganglion cells. The rate of discharge is varying above and below what it would be if the visual field were uniformly and steadily illuminated (about 30/sec in this cell) because the cat is in fact looking at a sine-wave grating whose bars are continuously drifting at a constant rate across the visual field. It is possible to quantify the magnitude of the cell's response by measuring the amplitude of the modulation of the discharge rate. This can be done by examining a "post-stimulus time histogram" which is generated by summing into a series of time bins related to the temporal period of the stimulus the action potentials occurring during many cycles of stimulation (such a record is shown in the middle of this figure). Alternatively the pulse train can be Fourier analysed (in time of course) to obtain the amplitude of the component at the stimulus frequency. When the contrast of the stimulus is fairly low and the response amplitude no more than 15/sec or so, the response is found to be very nearly sinusoidal and the response amplitude is proportional to the stimulus contrast (that is the cell behaves approximately linearly) [7]. In these circumstances the cell's sensitivity to the stimulus can be characterised by the ratio of the response amplitude to the stimulus contrast. This measure is the cell's *responsivity* and it can be used in determining the effect of changing the stimulus parameters. With a drifting grating as a stimulus the available stimulus parameters are the spatial frequency, orientation, temporal frequency (i.e. drift velocity x spatial frequency) and the luminance. The first two of these principally concern us here.

Figure 4 shows how the responsivity of an X-cell varies with spatial frequency. We may note an abrupt and profound fall in responsivity at high spatial frequencies and a limited and rather less rapid decline at spatial frequencies below the optimum. This sine-wave responsivity function is entirely typical of the behaviour of retinal X-cells when measurements are made at temporal frequencies of a few Hz or less. If the orientation of the grating stimulus is changed there may be some effect on the responsivity function which LEVICK and THIBOS [8] have recently shown can be described as a slight change in the spatial frequency scaling. However, this is a small effect and to a first approximation we may certainly assume that orientation is not a significant parameter for retinal X-cells.

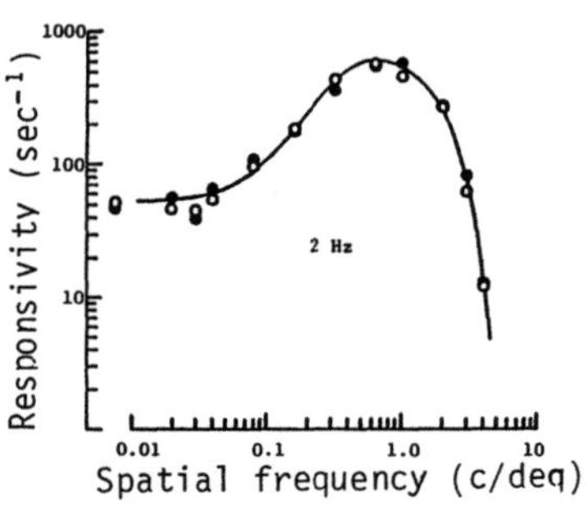

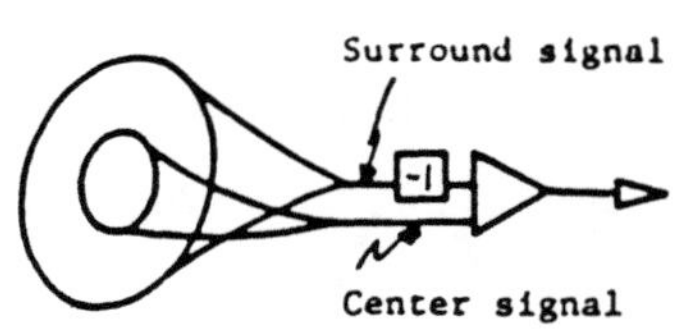

Fig. 4 (left). Sine-wave responsivity function of a retinal X-cell for both drifting gratings (open circles) and optimally positioned, counterphase-modulated gratings (closed circles). The temporal frequency was 2 Hz in both cases. Note the equivalence of both forms of inputs. (Adapted from ref. 7.)

Fig. 5 (right). The functional arrangement of the concentric centre *and* surround *regions of an X-cell receptive field. While the effects of light falling in these two regions is always antagonistic, signals from the centre may either increase the activity of the ganglion cell (*on-centre *cell) or decrease it (*off-centre *cell)*

How can we explain this behaviour? Let us briefly consider the ganglion cell's sensitivity not in the spatial frequency domain in which we have been making measurements but in the two-dimensional space of the stimulus screen at which the cat is looking. This is, of course, the domain in which KUFFLER [9] first observed that each ganglion cell in the cat retina only responded to visual stimuli in a restricted region of the visual field (the cell's receptive field). KUFFLER's experiments demonstrated that the receptive fields were roughly circular and organized into concentric antagonistic regions (excitatory centre and inhibitory surround or vice versa). This is sketched in Fig. 5. Following RODIECK and STONE [10] and ENROTH-CUGELL and ROBSON [11], we assume that signals from centre and surround regions are generated as a result of weighting the outputs of photoreceptors by two radially Gaussian functions of different radii (Fig. 6A). After inversion of the surround signal it is added to the centre signal and the resulting sum drives the ganglion cell. Given this model of the ganglion-cell receptive field we can predict how the cell will respond to any arbitrary stimulus pattern by convolving its luminance function with the "difference of Gaussians" weighting function of the cell's receptive field. To predict the responses to a set of sine-wave gratings of different spatial frequencies we could convolve each of the grating patterns with the weighting function or, more conveniently, recognise that this would be equivalent to computing a Fourier transform of the receptive-field weighting function. Even this can be simplified by realising that we can consider the overall sine-wave responsivity function of the cell to be determined by the difference between the separate responsivity functions of centre and surround. With this in mind we can understand how the bandpass form of a ganglion cell's overall responsivity function comes about (Fig. 6B). At low spatial frequencies the responsivities of the larger surround and smaller centre are nearly equal

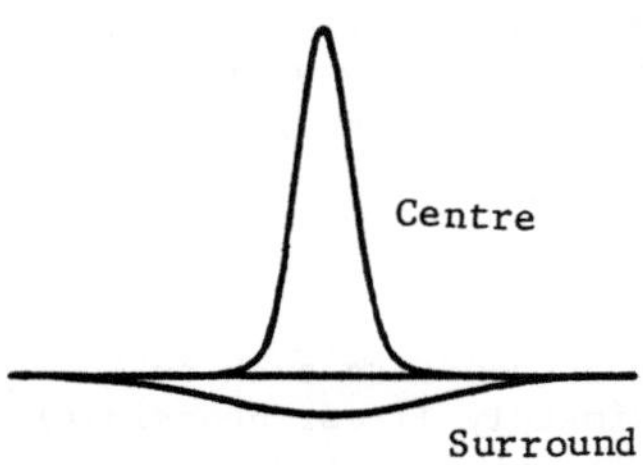

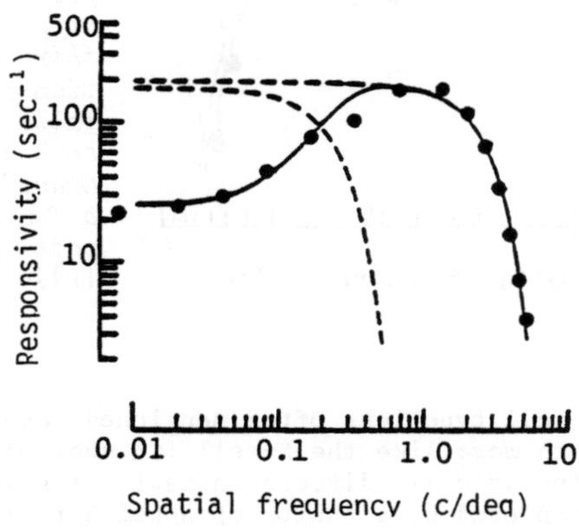

Fig. 6A (left). Gaussian weighting functions of centre and surround of a retinal ganglion-cell receptive field

Fig. 6B (right). The sine-wave responsivity function of a typical retinal X-cell fitted by the difference of two Gaussian functions. The responsivity of the surround mechanism is shown by the lower dashed line, that of the centre by the upper dashed line and its continuation to the right as a full line. In this example the diameter of the surround is about ten times that of the centre and its responsivity at zero spatial frequency about 0.88 that of the centre (from ref. [7])

and more or less cancel out to give an overall low responsivity. As the spatial frequency is raised, however, the responsivity of the larger surround falls off sooner than that of the smaller centre so that the overall response of the cell increases to become essentially that of the centre alone. At higher spatial frequencies even the response of the centre becomes attenuated.

I have somewhat laboured the relationship between spatial weighting function and spatial frequency response function because it is important to understand that these are simply alternative ways of describing the same thing. However I have glossed over the incompleteness of both descriptions as they are usually given. To describe a particular ganglion cell we must provide either the spatial weighting function relative to the midpoint of the receptive field together with the co-ordinates of this point in visual space or give the spatial frequency responsivity function including not only its magnitude but also its phase. These complete descriptions in space and spatial frequency would then be exactly equivalent.

We cannot leave the retina without mentioning two other cell types. The Y-cell (often confusingly and somewhat inappropriately described as a *transient cell*) does not conform to the simple linear model proposed for the X-cell [11] and though I am sure we shall hear more about this beast during the conference I shall not discuss it further now.

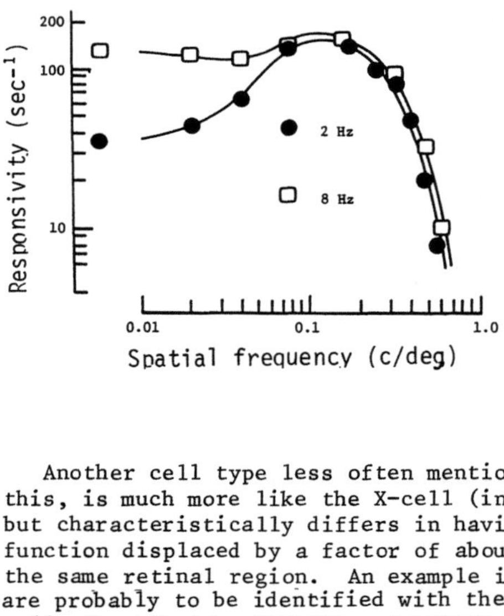

Fig. 7. The sine-wave responsivity function of a regularly firing "sluggish sustained" cell in the cat retina measured at two temporal frequencies. The descending high frequency limb of the function falls at spatial frequencies about one third of those at which the function for a neighbouring X-cell would fall. The experimental points are in both cases fitted by differences of Gaussians assuming an extra delay of a few milliseconds in the surround pathway (Fig. 5). (Adapted from ref. [7])

Another cell type less often mentioned, especially in contexts such as this, is much more like the X-cell (in showing approximately linear behaviour) but characteristically differs in having its spatial frequency responsivity function displaced by a factor of about 3 to the left of that of X-cells in the same retinal region. An example is shown in Fig. 7. Cells of this type are probably to be identified with the unfortunately named *sluggish sustained cells* described by CLELAND and LEVICK [12]. Their lower spatial frequency cutoff corresponds, of course, to their having larger receptive fields.

While the limitations of our recording methods make us concentrate on the activity of individual ganglion cells, the visual information is really transmitted from the eye to the brain by the activity of arrays of these cells. In Fig. 8 I have drawn idealised arrays of ganglion-cell receptive field centres. They are idealised in the sense that the receptive field centres are regularly arranged and their diameters show neither scatter nor

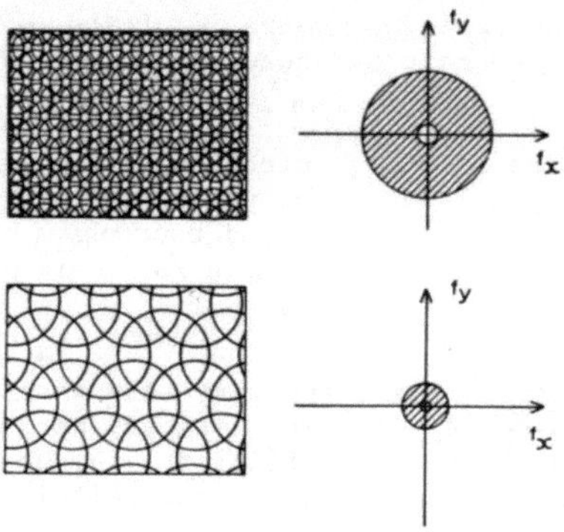

Fig. 8. Idealised arrays of retinal ganglion-cell receptive fields to indicate the approximate relationship between the diameters of receptive field centres and their separation. The contours are loci of half-maximum responsivity. The upper array represents X-cells and the lower the linear cells with larger receptive fields. The right-hand diagrams indicate the half-maximum contours for the sine-wave responsivity functions of these cells plotted in the two-dimensional frequency plane

the increase associated in the real retina with increasing eccentricity. However we may accept these sketches as adequately representing a fairly small area of the array. Sampling theory shows that a signal that is band-limited, that is one which contains frequency components only up to some limit, can be exactly represented by a finite set of discrete samples. If we allow that the image represented by the activity of a ganglion-cell array is band limited by the integration that occurs over each cell's receptive field centre (that is if we consider the spatial frequency response function of the individual cells as the characteristic of a spatial filter which has as its input the visual image) then we can ask whether the cell array is dense enough for its set of discrete ouput signals to be a complete representation of the image. HUGHES [13] has recently examined this question for X-cells using anatomical information about the cell densities. It seems that the representation will indeed be complete if the on- and off-centre cells co-operate in the sampling of the image. We do not have such precise information about the linear cells with larger receptive fields but it seems likely that their representation of the more frequency-limited image which they transmit is equally complete.

Whether these are further arrays of retinal ganglion cells with even larger receptive fields is unclear. Unfortunately one would expect there to be so few of these as to make it difficult to find them with the random sampling inherent in neurophysiological studies.

Anyway we may certainly visualise the retinal output as two (or possibly more) sampled pictures, each band-limited sufficiently to be completely represented by the available array of samples. Each picture contains relatively less energy than the input at spatial frequencies low compared to the band limit, but it is pretty clear that the spatial frequency selectivity of the different cell types is much too low to make the information in the outputs of the different arrays even roughly orthogonal and that we have not yet found the physiological substrate of the spatial frequency selectivity demonstrated psychophysically.

Parenthetically it may be worth adding that contrary to the suggestion that one function of the retinal processing is to generate a signal which is the second spatial derivative of the visual image (more specifically contrary to MARR and HILDRETH's suggestion [14] that a $\nabla^2 G$ function is computed), it seems to me that it is more appropriate to consider the images transmitted to the brain as being fairly faithful reproductions of the visual input. Although it is indeed entirely appropriate to describe the spatial weighting functions of X-cells (and also the linear cells with larger receptive fields) as "differences of Gaussians", the ratio of the radii of the centre and surround functions is too large to generate an acceptable approximation to the $\nabla^2 G$ function.

Enough of the retina, for though it serves as a simple introduction to the visual nervous system it is clearly essentially only an image pre-processor and not in any real sense an image-analysing structure. To find neurophysiological mechanisms which may more reasonably be considered as stages in image analysis we must look at the visual cortex.

Here things are really quite different. Firstly there are very many more cortical cells than there are retinal ganglion cells. In man there may be something like 500 times as many, a fact which may not be sufficiently emphasized. But rather than speculate on the significance of such an anatomical fact let us look to see how the behaviour of the individual cells compares with that of the retina. Fortunately, as we have mostly looked at the behaviour of retinal ganglion cells with grating stimuli we can make the comparison directly. Mapping the receptive fields of ganglion cells with spots may provide the same kind of information as measuring a spatial frequency response function but the use of spots of light as stimuli for cortical cells tends to be unproductive. They often will not respond at all to such stimuli. Drifting gratings are, however, generally quite adequate stimuli so long as their direction, orientation, velocity, spatial frequency and possibly the length of their bars have been adjusted to be within the appropriate range. For all these stimuli variables cortical cells are often significantly more selective than retinal ganglion cells (or, for that matter, the cells of the lateral geniculate nucleus).

Before considering the dependence of cortical cell response upon the stimulus parameters, it is probably worth noting one or two things about the nature of the response which is rather different from that of retinal ganglion cells. Firstly cortical cells are generally much less active than ganglion cells in the absence of specific stimulation. With a steady uniform field in front of the animal most visual cortical cells are silent or fire just a few times every second. This background activity, and also that evoked by a stimulus, is very irregular. The variability of the response can also be very marked [15]. It is not unusual to find cells failing on occasion to respond to a discretely presented stimulus which on other occasions produces a large response. This fluctuation in responsiveness has not been much studied and is usually obscured by the reporting of cell behaviour in terms of the average responses to repeated stimulus presentations. It is obviously of considerable importance to know whether variations in responsiveness of cortical cells are correlated between different cells or classes but as yet we just have no useful information about this.

Another aspect of cortical cell responses about which we have rather little information relates to the direct correlation of discharges in different cortical cells (a similar problem in fact arises in the case of ganglion cells). Although it is universally assumed to be appropriate to consider the discharge train from each individual cell on its own, it is in fact far from certain that significant signals may not be encoded in some way which involves activity in two or more fibres. Such a possibility has, as far as I know, never been experimentally examined. So we must assume, *faute de mieux*, that we are not missing anything important by examining cortical cells individually.

With these provisos we can look at the responses of a typical cell in the visual cortex to a drifting grating. In so far as it is possible to identify a typical cortical cell this will be one of the type called by HUBEL and WIESEL [16] a "simple cell". This will have little or no spontaneous discharge in the absence of stimulation but if it responds to the grating it will do so with a burst of action potentials as each bar of the grating passes across the screen. The average discharge rate of the cell will look very much like a half-wave rectified sine wave. The amplitude of this wave can be used as a measure of cell response.

Figure 9 is from the extensive study of DEVALOIS, YUND and HEPLER [17] and shows for a monkey cortical cell the way in which the response amplitude can vary as a function of orientation. While retinal ganglion cells show little or no orientation selectivity cortical cells are mostly markedly orientation selective, though the orientation bandwidth may vary by a factor of around 4. This will be clear from Fig. 10 which shows how the orientation was distributed in DEVALOIS' et al. sample of cells.

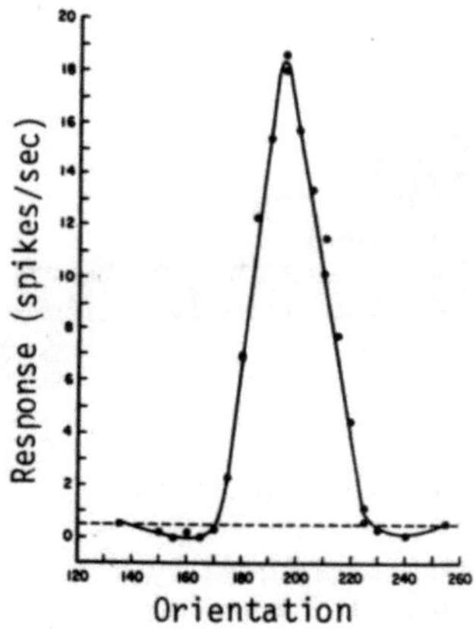

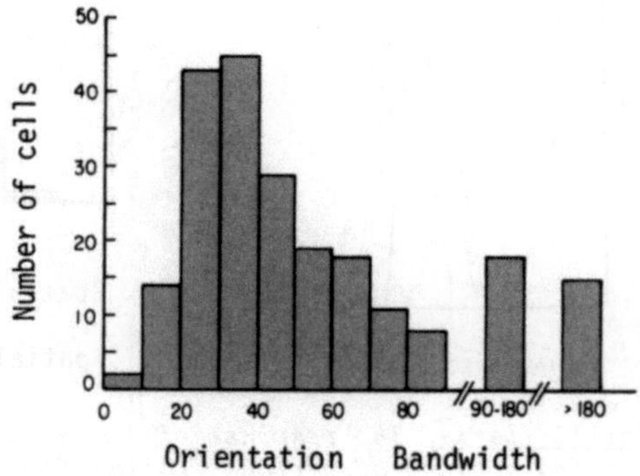

Fig. 9 (left). The average rate of discharge of a cell in monkey striate cortex responding to a drifting grating at various orientations. The dashed line indicates the discharge rate in the absence of a stimulus pattern. The orientation bandwidth *of such a cell is the full width of the curve at a level corresponding to half the response at the optimal orientation. (From ref. [16])*

Fig. 10 (right). The distribution of orientation bandwidths of all cells recorded in monkey striate cortex. (From ref. [16])

This is may be a good place to say that one should probably place rather little trust in the exact form of distribution data such as these. Recording with a microelectrode from a complex neural structure is as much an art as a science and subject to unexplained, but none the less real, sampling biases. However, even if the form of an experimentally determined distribution should be treated with caution, its range can probably be accepted as a fair reflection of reality. Before leaving orientation selectivity we should note that the preferred orientations of cortical cells appear to be more or less uniformly distributed.

If we now fix the orientation of the grating stimulus at its optimum and examine the effect of changing the spatial frequency we may expect to get a result something like Fig. 11 (from ALBRECHT and HAMILTON [18]). This shows the sine-wave response function for a cortical cell measured at three different contrast levels. Interestingly the form of the function does not seem to depend upon the contrast at which it was measured even though the response versus contrast relation for cortical cells often shows marked saturation-type non-linearity [18].

The spatial frequency selectivity of this cell is quite high, the ratio of the frequencies at half amplitude being about 1.6 (often referred to as a bandwidth of 0.7 octaves). This degree of selectivity is much higher than that of any ganglion cell. We may also note that the attenuation at low spatial frequencies is not only much greater than for a ganglion cell, it is also much more rapid. The slope of the low frequency limb of such curves is usually considerably greater than 2, the value appropriate for a second-differential operator.

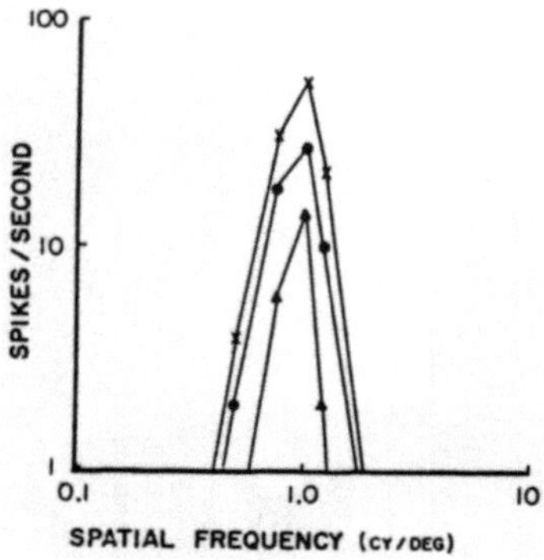

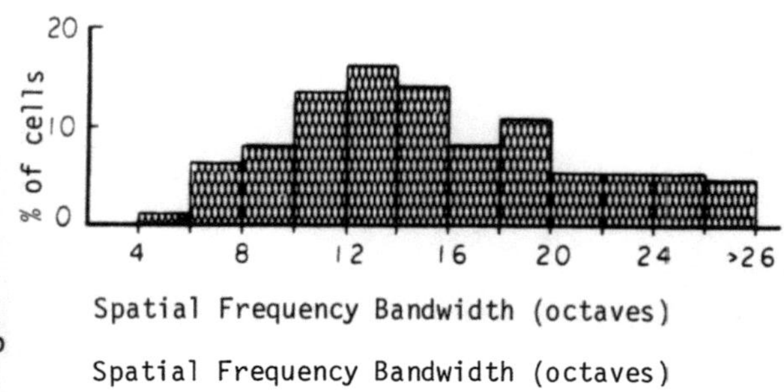

Fig. 11 (left). *The response of a monkey cortical cell stimuluated by a drifting grating of various spatial frequencies measured at three levels of contrast (crosses: 0.33; circles: 0.066; triangles: 0.040)*

Fig. 12 (right). *The distribution of spatial frequency bandwidths of simple cells in the foveal projection area of monkey striate cortex. (From ref. [18])*

Like the orientation bandwidths of cortical cells, their spatial frequency bandwidths also vary considerably from cell to cell [19]. Figure 12 shows the distributions for simple cells with foveal receptive fields.

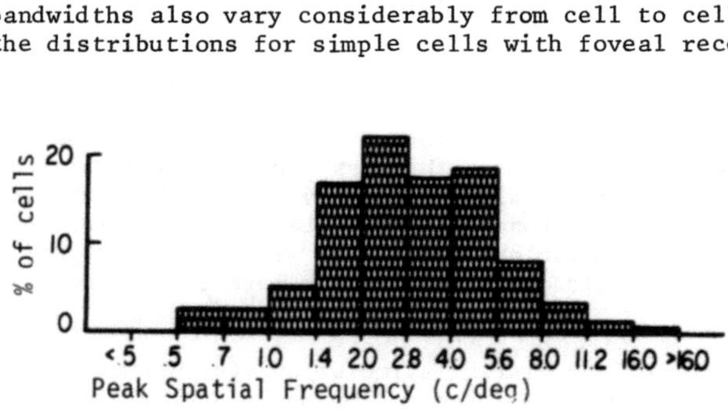

Fig. 13. Distribution of optimal spatial frequencies of simple cells in the foveal projection area of monkey striate cortex. (From ref. [18])

The other obvious parameter of cortical cell spatial frequency response functions is their position in the frequency scale. Figure 13 [19] shows that this again is quite widely distributed even if we look at cells with receptive fields in a relatively restricted part of the visual field.

It would simplify our task of codifying these observations if there were strong correlations between the various parameters of orientation and spatial frequency tuning functions. Unfortunately the correlations are mostly non-existent or weak, the strongest being between orientation and spatial frequency bandwidth [19]. Cells that are particularly orientation selective tend to have narrow spatial frequency bandwidths.

So can we make any sense of the behaviour of cortical cells as revealed by those measurements made with gratings?

By analogy with the retina we might suppose that we are looking at the individual cells in various neuronal arrays each of which carries a differently filtered version of the visual image. On this basis we would suppose that although the cells' parameters we have been considering appear to be continuously distributed rather than grouped, each set of cells which have the same characteristics should be considered as a separate array. The cells in each such array would share the same preferred orientation, the same orientation bandwidth, the same optimum spatial frequency and the same spatial frequency bandwidth. They would differ in the positions of their receptive fields and it would be presumed that their density would be sufficient to provide a complete representation. There would probably be hundreds of such arrays and any real scene would no doubt give rise to significant activity in a large number of them. Many of the differently filtered images, which is what the activity of each different array would represent, could be expected to contain significant information about some aspect of some parts of the visual scene. But how such dispersion of the available information might be helpful in analysing the scene is unclear.

Another view of the cortical representation can be derived from considering somewhat more explicitly the receptive fields of the cortical cells whose

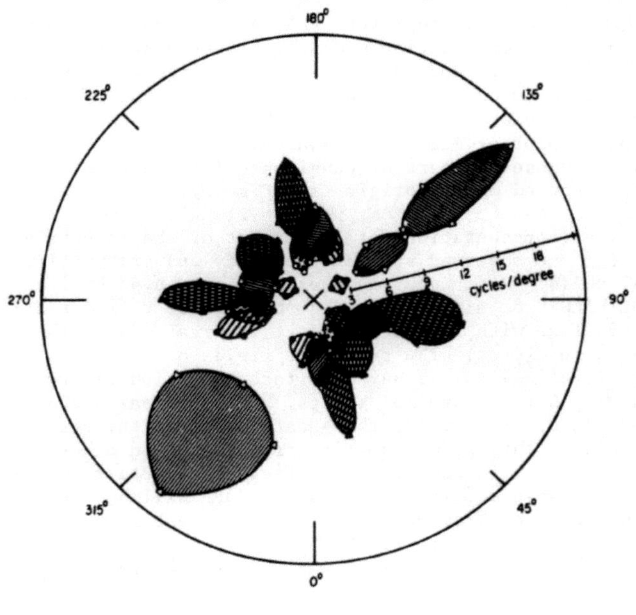

Fig. 14. Half-maximum response contours of the most frequency-selective cells in the parafoveal projection of monkey striate cortex stimulated by drifting gratings of various spatial frequencies and orientations. (From ref. [18])

characteristics we have so far considered only in terms of their orientation and spatial frequency selectivity. But first let us look at a way of representing these two things together. Figure 14 shows data relating to a sample of the more spatial frequency selective striate cortex cells examined by DEVALOIS, ALBRECHT and THORELL [19]. In this plot we can see the half-maximum response contours of a number of cells. At the centre of each shaded patch we can imagine a point representing the optimum spatial frequency and orientation for a given cell. The distance of this point from the origin represents the spatial frequency and the angle of the line joining the point to the origin represents the orientation. The contour around each patch now represents those combinations of orientation and spatial frequency which produce a response one half as large as that with the optimum combination. Although the contours in this picture are somewhat radially elongated, those for less frequency-selective cells are often more nearly circular and I shall idealise them all as circles for simplicity.

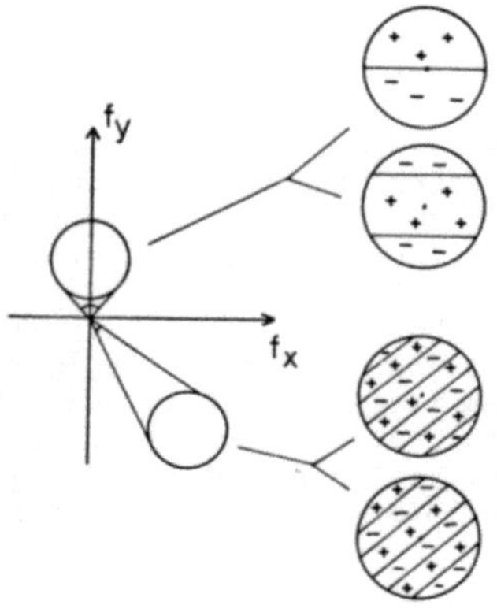

Fig. 15. Idealised two-dimensional frequency response contours of simple cortical cells (left) showing a relatively selective cell (lower) and a less selective cell (upper). On the right are corresponding schematic receptive field structures. The symmetry of the receptive field layout is not determined by the magnitude *of the sine-wave response function alone.*

In Fig. 15 I have drawn two circles to represent the half-maximum response contours for a cell with rather low orientation and spatial frequency selectivity (the one at the top on the left) and one with about as high an orientation and frequency selectivity as usually observed (the lower of the circles on the left). The axes are appropriately labelled for the 2-dimensional frequency space to which the polar plot of Fig. 14 is equivalent. It may be noted that the half-maximum response contours of ganglion cells represented in this way are broad annuli centred on the origin (see Fig. 8).

On the right of Fig. 15 are representations of the layout of the receptive fields which could account for the observed spatial frequency and orientation characteristics of cortical cells. Unfortunately these are only hypothetical receptive fields because DE VALOIS et al. [17, 19] did not examine them directly, but they are certainly in line with what more direct observations, both in monkey and in cat, lead us to expect. The receptive fields of cortical simple cells, unlike those of retinal cells, have excitatory and inhibitory regions arranged not circularly but in parallel strips. The more pairs of strips within a receptive field of a given size the greater will be the orientation selectivity and the higher the optimum spatial frequency. In other words the more strips that there are in the receptive field the further the cells' response contour in the two-dimensional frequency plane will be from the origin. In fact the two cells diagrammed in Fig. 15 represent more or less the extremes that probably actually occur.

There are two other points to be made. Firstly, as indicated on the right of Fig. 15, the same spatial frequency and orientation selectivity can be obtained having the alternating excitatory and inhibitory strips arranged with either odd or even symmetry with respect to the centre of the receptive field. POLLEN and RONNER [20] have recently produced some evidence that in the cat's visual cortex, cells having receptive fields with such odd and even symmetry may in fact occur associated in pairs. Secondly I should point out that all cells giving similar sized response contours in the 2-dimensional frequency plane will have the same size receptive fields (the same overall size, that is). In this context it should be recalled that it has been known since HUBEL and WIESEL's investigations [21] that the receptive fields of visual cortical cells are found to come in a range of sizes.

We can now come to the second view of the cortical representation towards which we were progressing. This is obtained by considering in frequency space the behaviour of all those cortical cells which have the same size receptive fields *located in the same place*. Figure 16 shows for three such cells their response contours in the frequency plane on the left and their receptive fields in the 2-dimensional space plane on the right. This is much like Fig. 15 but emphasises that the receptive fields are coincident in space.

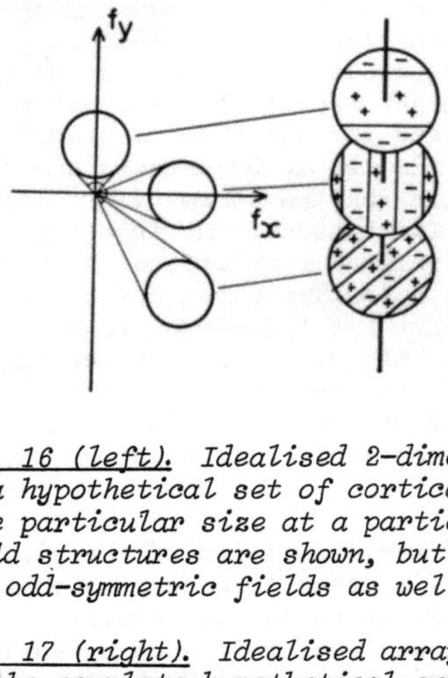

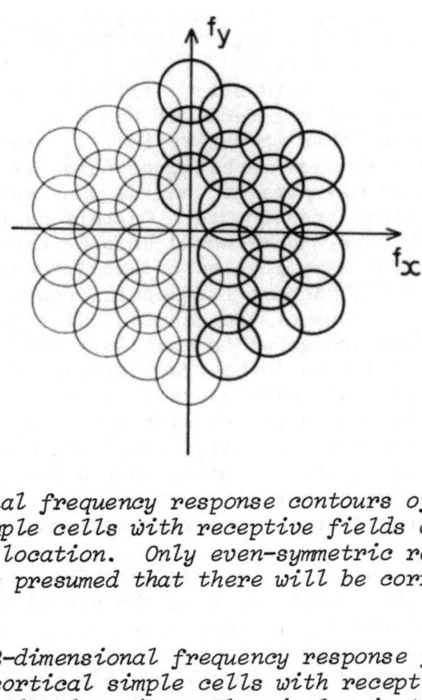

Fig. 16 (left). Idealised 2-dimensional frequency response contours of three of a hypothetical set of cortical simple cells with receptive fields of the same particular size at a particular location. Only even-symmetric receptive field structures are shown, but it is presumed that there will be corresponding odd-symmetric fields as well

Fig. 17 (right). Idealised array of 2-dimensional frequency response functions of the complete hypothetical set of cortical simple cells with receptive fields of particular size at a particular location. The circles in the left and right half planes may be thought of as corresponding to cells with odd- and even-symmetric receptive fields. The density and distribution in this plane have been chosen to provide roughly the same sampling as occurs in 2-dimensional space at the retinal level.

Now, in Fig. 17, we have what I believe may be a representation in the spatial frequency plane of a set of cells whose activity could completely represent one patch of the band-limited visual image. With reasonable (if not very precise) assumptions about the range of spatial frequency characteristics which can actually be found in real cortical cells, as well as assumptions about the density of the sampling in the frequency plane, and allowing for the necessity of having odd and even symmetric pairs, as well as on and off pairs (to signal both positive and negative values), we can guess that about *seventy* cells might be involved in representing each patch of visual scene. The activity of each such set of cortical cells can be considered to be a *discrete transform* of the light distribution in one patch of the visual scene effectively band-limited and sampled by the retinal ganglion cells.

The nature of the transform depends upon the particular spatial weighting functions of the set of cortical cells. While sufficiently precise and detailed information about cortical cell characteristics remains unavailable we cannot be certain of the exact nature of the transform. However, if Fig. 17 is a reasonably correct guess as to how the responsiveness of each of a set of cortical cells is localised in 2-dimensional frequency space, then it is certainly appropriate to think of the cells' responses as a *spatial frequency transform*.

While it would presumably be possible for the whole visual scene to be completely represented by a relatively small number of these cortical local transform sets (all the sets together might require not many more cells than there are retinal ganglion cells), this would imply that the patches of the visual scene represented by the different transform sets would be spaced to cover the whole field with relatively little overlap. However, if we take into account the existence of the very much larger number of cells in the visual cortex than in the retina, and the likelihood that the visual system, being a rather slow machine, performs as many operations as possible in parallel, we may presume that as many patches as could usefully be analysed are separately transformed. On this basis we might expect that for patches of a given size, there would be one patch analysed for every location appropriate for sampling the original scene band-limited to an extent corresponding to the highest frequencies represented in the transform. This would mean that as many of the smallest size patches (with transforms involving spatial frequencies up to the limit set by the retinal X-cells) as there are retinal X-cells would be transformed. This would involve about seventy times as many cortical cells as retinal cells. Allowing for transforms to be computed on as many other smaller scales as useful (say each scale is one eighth smaller than the next larger one) would increase the number by a factor approaching eight. Assuming that about one half of retinal cells are X-cells this would mean that altogether about half the cortical cells might be involved in local transform representations.

Even if this is a correct description of the representation of the visual scene by the simple cells of the visual cortex (it seems to encompass most of the information we currently possess about simple cells) it remains to be determined whether it is really appropriate or just what role such a representation could play in the brain's image-analysis function. Maybe the usefulness of such transformations will be obvious to those who endeavour to mimic with machines the visual system's still almost incredible powers of analysis. If it is, I hope we shall get some helpful leads about what to expect at higher levels of the visual system, for without such help we must rely on serendipity alone.

References

1 E.W.H. Selwyn, Photographic Journal, Sect. B, Vol. 88B, 6 (1948).
2 O.H. Schade, J. Opt. Soc. Am. 46, 721 (1956).
3 J.G. Robson, J. Opt. Soc. Am. 56, 1141 (1966).
4 F.W. Campbell and J.G. Robson, J. Physiol. 197, 551 (1968).
5 C. Blakemore and F.W. Campbell, J. Physiol. 203, 237 (1969).
6 G.F. Cooper and J.G. Robson, IEE-NPL Conference on Pattern Recognition, I.E.E. Conference Publication 42, London (1968).
7 C. Enroth-Cugell, J.G. Robson, D. Schweitzer-Tong and A.B. Watson, Spatio-temporal interactions in cat retinal ganglion cells showing linear spatial summation. Submitted for publication.
8 W.R. Levick and L.N. Thibos, J. Physiol. 329, 243 (1982).
9 S.W. Kuffler, J. Neurophysiol. 16, 37 (1953).
10 R.W. Rodieck and J. Stone, J. Neurophysiol. 28, 831 (1965).
11 C. Enroth-Cugell and J.G. Robson, J. Physiol. 187, 517 (1966).
12 B.G. Cleland and W.R. Levick, J. Physiol. 240, 421 (1974).
13 A. Hughes, Expl Brain Res. 42, 196 (1981).
14 D. Marr and E. Hildreth, Proc. R. Soc. Lond. B. 207, 187 (1980).
15 D.J. Tolhurst, J.A. Movshon and I.D. Thompson, Expl Brain Res. 41, 414 (1981).
16 D.H. Hubel and T.N. Wiesel, J. Physiol. 195, 215 (1968).
17 R.L. De Valois, E.W. Yund and N. Hepler, Vision Res. 22, 531 (1982).
18 D.G. Albrecht and D.B. Hamilton, J. Neurophysiol. 48, 217 (1982).
19 R.L. De Valois, D.G. Albrecht and L. Thorell, Vision Res. 22, 545 (1982).
20 D.A. Pollen and S.F. Ronner, Science 212, 1409 (1981).
21 D.H. Hubel and T.N. Wiesel, J. comp. Neurol. 158, 295 (1974).

Psychophysical Evidence for Spatial Channels

H.R. Wilson

Department of Biophysics and Theoretical Biology, The University of Chicago, Chicago, IL 60637, USA

1. Introduction

More than a decade of research spawned by the work of CAMPBELL and ROBSON [1] and GRAHAM and NACHMIAS [2] has led to a general consensus that the human visual system contains several different classes of spatial mechanisms or channels that process the retinal image in parallel. Thus, elucidation of the spatiotemporal filtering characteristics of these mechanisms has become one of the major tasks of psychophysics. In general, of course, it is to be expected that complex interactions will occur between mechanisms responding to neighboring portions of the retinal image; and we know this to be true for the processing of movement and depth information. Fortunately, experimental paradigms have been developed which have enabled us to estimate mechanism characteristics at a lower level than that at which these interactions occur. This article will focus on two of these techniques and the resulting characterization of human spatiotemporal mechanisms.

Before proceeding, a comment about the physiological nature of a visual mechanism is in order. As visual information passes successively through the retina, lateral geniculate nucleus, and several areas of the cerebral cortex; it is difficult to determine just which level of processing is reflected in psychophysical data. In general, therefore, a psychophysically defined visual mechanism is probably best thought of as a small network of cells perhaps spanning several successive anatomical levels. However, the results discussed below show such close agreement with the neurophysiology of single cells in the primate visual cortex [3] that it is tempting to assume that the psychophysical results reflect the responses of cells in human visual cortex.

2. Visual Line Spread Functions

My earliest measurements of the spatial sensitivity of visual units were made with the use of a stimulus consisting of three equally spaced thin vertical lines superimposed on a CRT screen of fixed mean luminance (8.6 cd/meter^2). The object was to measure the threshold for these patterns (i.e., the lowest contrast at which they just became visible) as a function of the distance between the lines. As the center line was of higher contrast than the flanking lines, it alone was visible at threshold, yet the presence of the flanking lines had a pronounced effect. Figure 1 shows data obtained at fixation and at eccentricities of 2.0 and 4.0 degrees [4]. The ordinate plots sensitivity, which is defined as the reciprocal of the threshold contrast. For reference, the data point on the far right in each panel indicates the sensitivity to the center line alone. Thus, the presence of the flanking lines makes one more sensitive when the three lines are close together. As the inter-line separation increases, however, there is a region in which the flanking lines inhibit the visibility of the central line. Thus, this technique provides a quantitative estimate of the spatial extent of central excitatory and flanking inhibitory regions of human visual mechanisms. In analogy to optics, this spatial sensitivity profile will be termed a line spread function or LSF. Note, however, that the presence of inhibition means that the LSF primarily reflects neural rather than optical properties. The solid curves in Fig.1 show the fit of a difference of two spatial Gaussian functions to the data. In symbols:

$$\mathrm{LSF}(x) = A \exp[-(x/s_e)^2] - B \exp[-(x/s_i)^2] \quad , \qquad (1)$$

where s_e and s_i are the excitatory and inhibitory space constants respectively; and $s_e < s_i$ to reflect the greater range of the inhibition. The spatial variable x is measured in degrees of visual angle.

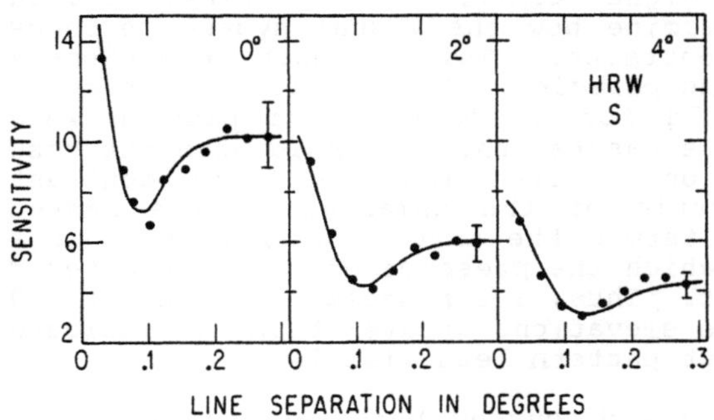

Fig. 1. Sensitivity to a three line pattern as a function of line spacing. Data from three eccentricities are shown

Comparison of the LSFs at 2.0 and 4.0 deg. in Fig.1 with that at fixation immediately shows that the visual system is spatially inhomogeneous, as has been known at least since 1894 [5]. This inhomogeneity appears to consist both of a decrease in absolute sensitivity and an increase in the width of the LSF at progressively greater eccentricities. More detailed examination of the data suggests that this width increase is approximately linear with eccentricity, which agrees well with both retinal [6] and cortical [7] physiology. This eccentricity dependence can be represented mathematically by making s_e and s_i in (1) linear functions of eccentricity. It has also been shown that the appropriate form for A and B is a reciprocal linear function of eccentricity. The possible functional significance of this inhomogeneity will be discussed subsequently.

Further measurements similar to those just described led to the conclusion that there must be several different visual mechanisms responding in parallel to the retinal image [4,8]. WILSON and BERGEN [8] summarized these results in a quantitative model having considerable predictive power. This model was composed of four mechanisms at each eccentricity, all of which had LSFs with the mathematical form in (1), and all of which increased in width at the same linear rate as a function of eccentricity. The predicted response R_i of the ith mechanism to a one-dimensional luminance pattern $P(x')$ was calculated from the formula:

$$R_i = \int_{-\infty}^{\infty} LSF_i(x,x')\, P(x')\, dx' \qquad (2)$$

Note that this is not identical to convolution due to the fact that the parameters of LSF_i are a function of eccentricity x'. The responses of the different mechanisms are combined probabilistically to reflect the stochastic nature of the detection process [4,8,9,10]. Thresholds predicted in this manner show excellent agreement with data for cosine gratings, square-wave gratings, and a variety of aperiodic spatial patterns [8,11].

3. Masking and Suprathreshold Studies

The model just outlined was designed to deal only with threshold measurements. As most normal vision operates at suprathreshold levels, however, it is clearly appropriate to extend this type of modeling to higher contrasts. The first step in this direction is to determine how the visual system responds to a range of pattern contrasts. This is most conveniently done using a visual masking paradigm. In a masking experiment, one measures the threshold for a spatial test pattern when presented alone and then remeasures the threshold when the test pattern is superimposed on a fixed (usually high) contrast masking pattern. The ratio of the unmasked to the masked threshold, which will be termed the threshold elevation, is a measure of the degree to which the presence of the mask altered the visibility of the test. Thus, a threshold elevation of 1.0 indicates no effect, while elevations greater than 1.0 indicate that the mask made the test pattern less visible.

In all of the experiments discussed here test stimuli were vertical, one-dimensional patterns with a horizontal luminance profile defined by the sixth derivative of a spatial Gaussian function. These can readily be shown to be the product of a Hermite polynomial and a Gaussian function [12]. The advantages of these patterns are that they are localized in both space and spatial frequency content, having a Fourier transform with full bandwidth at half height of 1.0 octaves. In addition, they have even symmetry and approximate the center-surround structure of a typical receptive field. Figure 2 shows the luminance profile of such a stimulus.

Masking patterns consisted of cosine gratings (i.e. one-dimensional gratings with a cosine luminance profile perpendicular to the bars) of a range of spatial frequencies. In the main series of experiments the cosine mask was oriented at 14.5

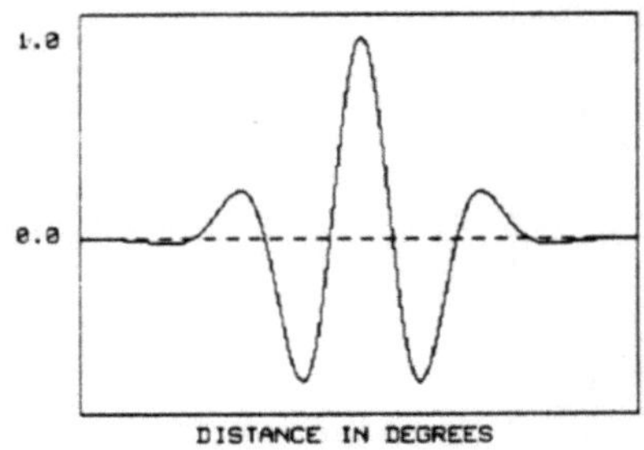

Fig.2. Horizontal luminance profile of the test stimulus used in masking experiments

deg. relative to the vertical test pattern. This was done to average out any phase effects and to keep the contribution of probability summation constant [13].

The first goal was to determine how suprathreshold contrasts are processed by the visual system. This was done by masking the test stimulus with cosine gratings of a spatial frequency identical to the peak frequency of the test. Figure 3 shows data for four different spatial frequencies and two subjects as a function of the mask contrast. The fact that the data are well fit by a straight line on double logarithmic coordinates indicates that threshold elevation is a power function of mask contrast. The exponent of this function is about 0.55 at low and intermediate frequencies and declines to about 0.25 at high frequencies. Several other laboratories have obtained similar data and have interpreted this to mean that above threshold the visual system contains a compressive nonlinearity for contrast [14-17]. Thus, the picture emerges of visual mechanisms containing a linear filter followed by a contrast transfer function consisting of a threshold followed by a suprathreshold compression. A schematic of such a visual system is shown in Fig.4.

4. Masking and Spatial Filter Characterization

Given the experimental evidence for nonlinear contrast processing in Fig.3, it is now possible to use masking to determine the characteristics of the filters in Fig.4. This was done by masking a single vertical test stimulus with cosine grating masks spanning a range of spatial frequencies centered on the peak frequency of the test. To obtain the largest effects, high contrast (40%) masks were used. As before, these masks were oriented at 14.5 deg. relative to the vertical test pattern so that all relative phases between the two would be present.

In separate experiments threshold elevations were measured as a function of mask frequency for each of fourteen different test spatial frequencies spanning the range from 0.25 to 22.0 cycles/deg. in 0.5 octave steps. Threshold elevation curves for test frequencies of 0.35 (open circles), 0.50 (solid circles), and 0.71 (open squares) cycle/deg. are plotted in Fig.5

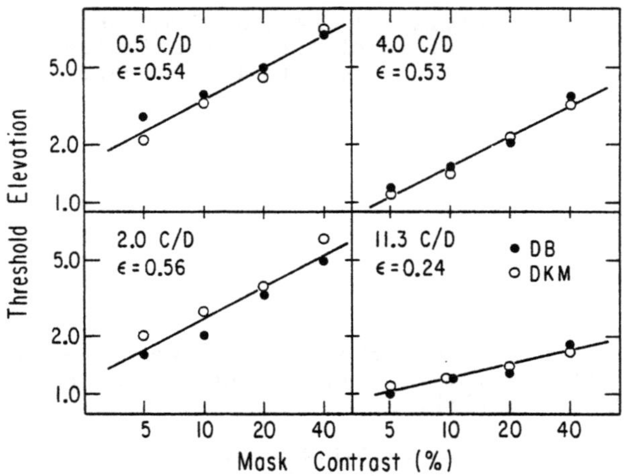

Fig.3. Threshold elevation as a function of mask contrast for two subjects and four spatial frequencies. Solid lines are least mean squares fits of power functions with the indicated exponents

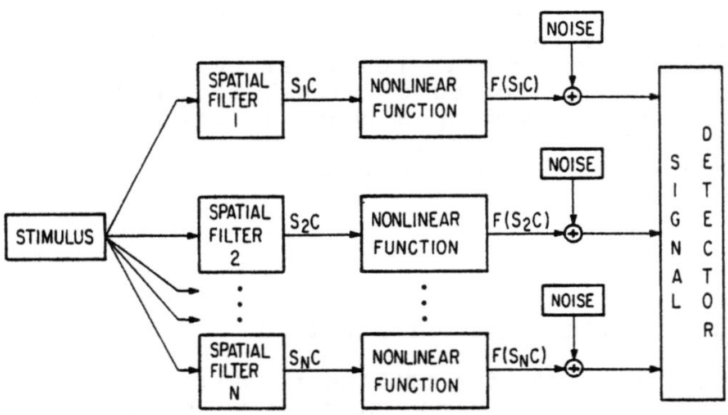

Fig.4. Schematic of the visual system model obtained from masking experiments. S_i is filter sensitivity, while C is pattern contrast

for two different subjects. In every case the data peak at an intermediate frequency and fall off at both higher and lower frequencies, thus indicating a bandpass characteristic for the spatial frequency dependence of the underlying filter. Two other aspects of the data are particularly striking. First, notice that the most effective mask does not coincide with the test frequency in any case. Rather, it is displaced toward higher frequencies by 0.5 to 1.0 octave. Second, the three

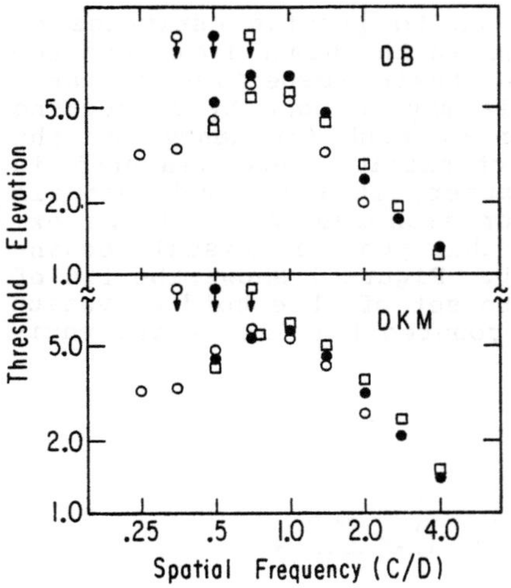

<u>Fig. 5.</u> Masking as a function of mask spatial frequency for three different test frequencies and two subjects. Open circles: 0.35 cycles/deg., closed circles: 0.50 cycles/deg., open squares: 0.71 cycles/deg.

threshold elevation curves for each subject in Fig.5 are identical to within experimental error. These two points strongly suggest that all three test patterns are being detected by the same underlying visual mechanism.

Similar conclusions emerge from consideration of the data in Fig.6. Here the test frequencies are 8.0 and 11.3 cycles/deg. In this figure the most effective mask was about 8.0 cycles/deg. for both subjects and both test frequencies, and the two data sets are not significantly different. Thus, it appears that these two test patterns are also being detected by a single bandpass visual mechanism that is sensitive to much higher frequencies than the one in Fig.5. Similar patterns appear in the data for intermediate test frequencies [13].

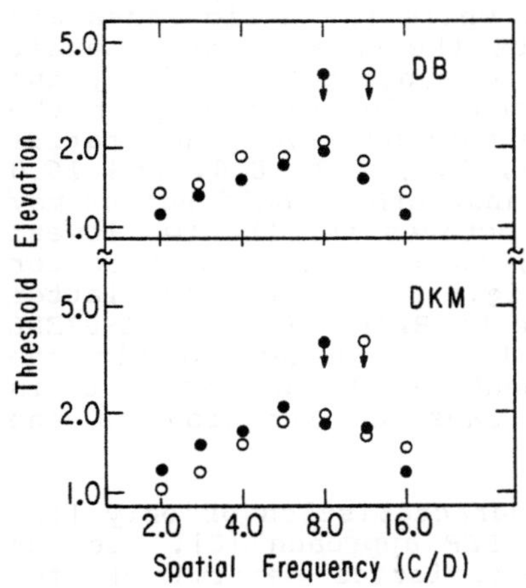

<u>Fig. 6.</u> Masking as a function of mask spatial frequency for test frequencies of 8.0 and 11.3 cycles/deg. (solid and open symbols respectively)

The goal of these experiments was to provide estimates of both the number of visual mechanisms in the human fovea and the spatial frequency sensitivities of their respective filters. From Fig.4 it is apparent that this may be done by correcting the data on masking as a function of mask frequency for the contrast nonlinearity whose characteristics were measured in Fig.3. When this was done by computer, it was found that all of the threshold elevation data for each subject could be extremely well fit on the assumption that just six visual mechanisms were present in the fovea [13]. Figure 7 shows the fit of these six mechanisms to the complete set of threshold elevation data for one subject. This fit accounted for 93% of the variance of the data.

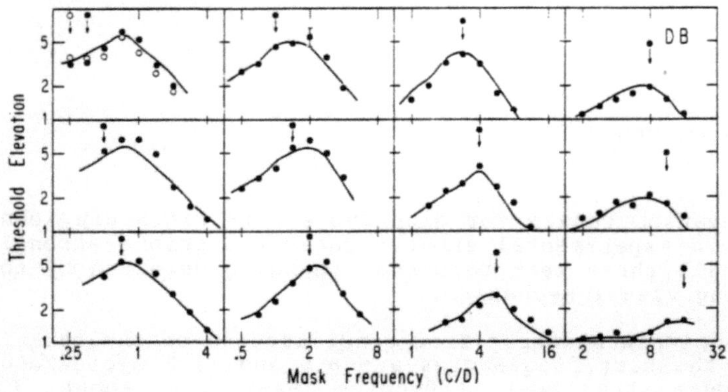

Fig.7 Masking functions for one subject and all test frequencies from 0.25 to 16.0 cycles/deg. Solid curves show fits of six mechanisms to the data.

The spatial frequency sensitivities of each of the six mechanism filters are plotted in Fig.8. A different symbol shows results for each of three subjects, while the solid lines are drawn through the geometric means of the data in each panel. Although there is some variation from subject to subject, the overall similarities are quite strong. Considering just the averages (solid curves), the six mechanisms have peak spatial frequency sensitivities at 0.8, 1.7, 2.8, 4.0, 8.0, and 16.0 cycles/deg. In addition, the full bandwidth at half height may be seen to decrease from about 2.0 octaves for the two lowest frequency mechanisms (Fig.8 A and B) to about 1.33 octaves for the highest frequency mechanisms (Fig.8 E and F). The latter number is identical to that obtained by BLAKEMORE and CAMPBELL [28] and STROMEYER and JULESZ [29]. In order to fit the filters with bandwidths below about 1.70 octaves, it is necessary to add a third spatial Gaussian function to the expression for LSF(x) in (1).

It will be remembered that the characteristics of only four mechanisms were estimated using the LSF approach [8]. Let us therefore compare the filter characteristics determined from

LSF data with those obtained from masking. This is easily done by taking the Fourier transform of the analytic expression for the LSF sensitivity profiles from (1), as this gives the response to cosine gratings. The resulting four filters are plotted with dashed lines in Fig.8 A, B, D, and E. The agreement in filter characteristics estimated in these two very different manners is obviously quite good. The major difference is that the filters estimated from LSF data tend to be of slightly broader bandwidth than those determined by masking. This is probably due to the fact that the masking technique produces bigger effects and thus has better resolution than the LSF approach. Of the two mechanisms that were not revealed by the LSF approach, the high frequency one (Fig.8 F) is beyond the range in which LSF measurements could be made reliably. Evidence for this very high frequency mechanism has been provided independently by others [18, 19]. The remaining mechanism in Fig.8 C was probably missed because its characteristics are similar to those in Fig.8 D. It may be concluded that the masking approach gives a more precise description of the mechanisms operating in the human fovea.

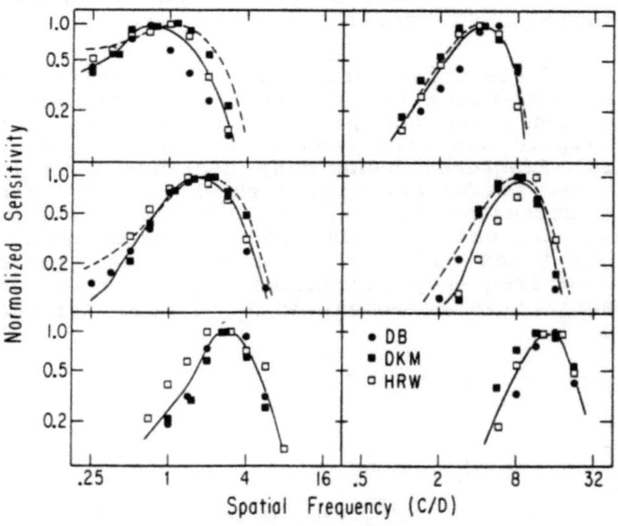

Fig.8. Sensitivities of the six mechanisms estimated from masking for each of three subjects. Solid curves are geometric means of the data, while dashed curves are Fourier transforms of LSF data

5. Orientation and Temporal Tuning

As the masking studies discussed above utilized stimuli oriented at or near the vertical, no information has yet been presented that would enable one to determine whether these mechanisms are orientation selective. However, such information may be easily obtained using a variation of the masking technique. Instead of keeping mask orientation constant and varying mask spatial frequency, mask frequency was now held constant, and mask orientation was varied. The vertical test stimulus had the same peak frequency as the mask, whose contrast was maintained at 40%. The data in Fig.9, which were obtained with 4.0 cycle/deg. test and mask, demonstrate that visual mechanisms clearly are orientation dependent. The arrow marks the 14.5 deg. mask orientation used in the earlier masking studies and shows that this is well within the half amplitude half bandwidth of this mechanism.

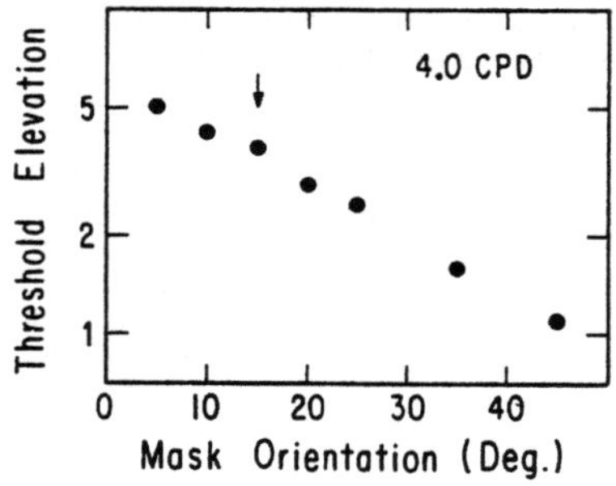

Fig. 9. Threshold elevation as a function of mask orientation for 4.0 cycle/deg. mask and test. Arrow shows orientation used in other masking experiments

The data on threshold elevation versus mask orientation must again be analyzed to compensate for the contrast nonlinearity in Fig.3. This was done, and the orientation half amplitude half bandwidths in Fig.10 were obtained [20]. Just as Fig.8 revealed that spatial frequency bandwidths decrease with increasing mechanism peak frequency, Fig.10 shows that orientation half bandwidth decreases from about 30.0 deg. to about 15.0 deg. as frequency increases. Thus, high-frequency mechanisms are more sharply tuned for both orientation and spatial frequency than are low-frequency mechanisms, which is in both qualitative and quantitative agreement with primate cortical neurophysiology [3].

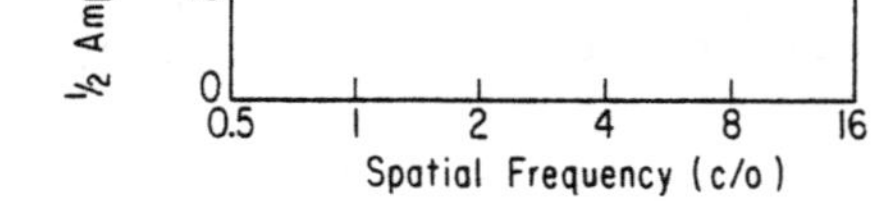

Fig. 10. Orientation half-amplitude half-bandwidth as a function of test spatial frequency. Solid and open circles are data for sustained and transient temporal presentations

Thus far little mention has been made of temporal variables. In fact, all of the masking data reported thus far were obtained using low-frequency temporal modulation (i.e., a temporal Gaussian with a 0.25 second time constant). To determine the effects of temporal modulation, most of the masking experiments have been repeated with both test and mask modulated at 8.0 Hz. Figure 10 compares low temporal frequency (solid circles) with high temporal frequency (open circles) orientation bandwidth estimates, showing that there is no measurable difference between the two [20]. We are currently completing masking experiments to determine the temporal frequency tuning of each of the six mechanisms. The preliminary indication is that the three lowest frequency mechanisms (Fig.8 A-C) are bandpass or transient, while the three highest frequency ones are low pass or sustained. Similar results have been obtained in other studies [4, 21, 22].

6. Discussion

The picture that emerges from these studies is of six discrete classes of vertically oriented visual mechanisms processing spatiotemporal patterns imaged on the human fovea. Both orientation and spatial frequency bandwidths are broadest for the lowest spatial frequency mechanism and become progressively narrower for mechanisms tuned to successively higher frequencies. The orientation tuning of these mechanisms suggests that they are located in the visual cortex, as this is the first site at which orientation tuning has been physiologically observed in primates. Orientation and spatial frequency bandwidth estimates as a function of mechanism peak frequency are in quantitative agreement with the most recent physiological measurements of these properties for single cells in the primate visual cortex [3]. Other mechanisms doubtless exist with different preferred orientations.

None of this research has yet dealt either with the spatial distribution of these mechanisms (except spatial inhomogeneity) or with the processing of information concerning motion, disparity, or color. However, an interesting conjecture may be made here concerning motion and disparity. Suppose that neighboring cells of the same mechanism type were located about one-half receptive field center diameter apart. This would be about optimal on the basis of the sampling theorem [23]. Suppose further that the six mechanism types characterized here are monocular (all of our data were obtained monocularly) and provide the inputs for the subsequent processing of motion and disparity. Then appropriate spatiotemporal comparisons of the activity of nearest neighbor cells of the same mechanism type receiving input from the same eye would produce direction selectivity and motion information. Similarly, comparison of the response of a cell stimulated through one retina with that of a nearest neighbor to the corresponding point on the other retina would provide disparity information. This approach to the extraction of disparity has already been developed into models of stereopsis by MARR and POGGIO [24] and MAYHEW and FRISBY [25]. Certainly, the nearest neighbor comparison approach to the processing of motion and disparity would be both parsimonious and easy to implement anatomically.

Finally, let us consider the functional significance of the spatial inhomogeneity of the visual system. Psychophysical data presented above [4, 8] plus primate cortical neurophysiology [7] both indicate a linear increase in receptive field size with increasing distance from the center of the fovea. This is accompanied by a roughly linear increase in the dimensions of Panum's fusional area (the range of disparities permitting single vision) [26] and by an increase in the maximum velocity at which motion can still be perceived [27]. Thus, in progressing outward from the center of gaze, the visual system becomes optimally sensitive to larger objects moving at greater velocities and at greater depths relative to the point of convergence of the two eyes. I would suggest that this may represent an evolutionary adaptation of the visual system to permit man to efficiently run through a forest chasing his prey (or mate) while avoiding the trees. As one approaches a tree or other obstacle and assuming fixation on the prey, its image will expand and move outward on the retina at ever increasing velocities and at ever greater depths relative to the prey. Although this is currently only a suggestion, the smooth and approximately linear variation of most visual information processing capabilities as a function of eccentricity certainly deserves greater theoretical attention than it is usually given.

7. References

1. F. W. Campbell and J. G. Robson: J. Physiol. 197, 551 (1968).
2. N. Graham and J. Nachmias: Vision Res. 11, 251 (1971).
3. R. L. DeValois, D. G. Albrecht, and L. G. Thorell: Vision Res. 22, 545 (1982).
4. H. R. Wilson: Vision Res. 18, 971 (1978).
5. T. Wertheim: Z. Psychol. Physiol. Sinnesorg. 7, 172 (1894).
6. B. Fischer: Vision Res. 13, 2113 (1973).
7. D. H. Hubel and T. N. Wiesel: J. Comp. Neurol. 158, 295 (1974).
8. H. R. Wilson and J. R. Bergen: Vision Res. 19, 19 (1979).
9. R. F. Quick: Kybernetik, 16, 65 (1974).
10. J. G. Robson and N. Graham: Vision Res. 21, 409 (1981).
11. J. R. Bergen, H. R. Wilson, and J. D. Cowan: J. Opt. Soc. Am. 69, 1580 (1979).
12. W. H. Swanson, H. R. Wilson, and S. C. Giese: Submitted for publication (1982).
13. H. R. Wilson, D. K. McFarlane, and G. C. Phillips: Submitted for publication (1982).
14. H. R. Wilson: Biol. Cybernetics, 38, 171 (1980).
15. C. R. Carlson: Photo. Science and Eng. 22, 69 (1978).
16. G. E. Legge and J. M. Foley: J. Opt. Soc. Am. 70, 1458 (1980).
17. J. Nachmias and R. V. Sansbury: Vision Res. 14, 1039 (1974).
18. D. Marr, T. Poggio, and E. Hildreth: J. Opt. Soc. Am. 70, 868 (1980).
19. A. B. Watson: Vision Res. 22, 17 (1982).
20. G. Phillips and H. R. Wilson: Submitted for publication (1982).
21. A. B. Watson and J. Nachmias: Vision Res. 17, 893 (1977).

22. H. R. Wilson: Vision Res. 20, 443 (1980).
23. B. Sakitt and H. B. Barlow: Biol. Cybernetics, 43, 97 (1982).
24. D. Marr and T. Poggio: Proc. Roy. Soc. B, 204, 301 (1979).
25. J. E. W. Mayhew and J. P. Frisby: Perception, 9, 69 (1980).
26. K. N. Ogle: Researches in Binocular Vision, Hafner, New York (1964).
27. R. Sekuler: In Handbook of Perception, 5, ed. by E. C. Carterette and M. P. Friedman, Academic Press, New York (1975).
28. C. Blakemore and F. W. Campbell: J. Physiol. 203, 237 (1969).
29. C. F. Stromeyer and B. Julesz: J. Opt. Soc. Am. 62, 1221 (1972).

Detection and Recognition of Simple Spatial Forms

A.B. Watson

NASA Ames Research Center, Moffett Field, CA 94035, USA

1. Models in Vision

In spite of the large number of intelligent, energetic people engaged in the study of vision, it often seems that we advance our understanding of this process at an excruciatingly slow pace. Why is this so? I believe that it is the fault of our models. We have been seduced by the simplicity of explanation available in the physical sciences, and try to describe the vast complexity of vision with models that would not do justice to a sphere rolling down an inclined plane.

Today I will describe a model which in a small way attempts to remedy this situation. The principles that underly it are simple, but it attempts to represent more adequately the inherent complexity of vision. Before describing this model in detail, I will emphasize its unusual aspects, and some of the benefits we may gain from this departure.

First, the model is explicit. No "channels", "mechanisms" or other ill-defined entities appear. Put another way, the model is computable. It is well enough defined to permit numerical predictions to be made for a given experimental situation. Second, the model adequately represents all the dimensions that are pertinent to the experimental domain. In the case I consider, these are the two spatial dimensions of a monochromatic, stationary image. Although this is an inherently two-dimensional (2D) situation, visual models in this context have almost invariably been one-dimensional (1D). Third, the model is general within its specified domain. The same model can be used to predict performance in a wide variety of different visual tasks on a wide variety of images. This contrasts with most visual models, which only attempt to explain the data from a very restricted domain. Fourth, the model draws many of its parameters and assumptions directly from physiological results. While this has long been an ambition of psychophysical theories, it has rarely been realized. Finally, the model attributes intelligence to the observer. It is perhaps understandable that this has not been done in the past, since psychophysicists are most often their own observers. The model assumes that the observer will make optimal use of the available information. This assumption is of course a commonplace in signal detection theory [10], but rarely appears in explicitly visual models.

The price of these amendments is a model that is somewhat more complex, requiring more assumptions and parameters than average. The benefits, however, are worth it. It provides a mechanism with which to integrate information within the field. Since the model can be applied in diverse contexts, it provides a common repository for results from many different sources. The model also insures the consistency of interpretation from one experiment to the next. Too often a model is constructed in one context that is quite obviously incompatible with data from another context. The model also provides a natural path along which to specify in ever greater detail the relation between our visual experience and the physiological mechanisms of the visual brain. Finally, to the extent that the model is successful, it will permit us to turn our attention from the early, image-driven stages of vision to the more complex, cognitive processing that must subsequently occur.

It should be clear that the argument I have advanced is on behalf of any model that meets the criteria of explicitness, adequacy, and generality noted above. The model I will describe is just a candidate, and a green one at that.

2. Domain of the model

What is the domain of my model? It is intended to account for psychophysical responses of an accommodated, fixating human observer viewing binocularly a 2D stationary achromatic image at a fixed adapting level of 340 cd m^{-2}. Specifically, I have excluded both color, stereo, and dynamic imagery since it seems likely that they are analyzed independently of spatial luminance contrast.

3. General properties of the model

Since this model is a first approximation whose parameters and structural details are quite open to improvement, it is worth giving a brief overview of its essential structure. The heart of the model is a set of *feature sensors*, which perform different measurements upon the input image. The set of measurements taken from a particular image make up a *feature vector*. The sensors are perturbed by noise, so that from presentation to presentation, the same image will give rise to somewhat different feature vectors. Following each trial of an experiment, the observer processes the feature vector in an optimal way to arrive at a psychophysical decision.

The model thus divides naturally into two parts: generating the feature vector, and processing the feature vector.

4. Generating the Feature Vector

Each feature sensor is defined by a *spatial weighting function,* which is a model or template of the feature to be sensed. The measurement is performed by cross-correlating the contrast image and the weighting function. The features used in the model are typified by the pattern in Fig.1(top). It is the product of a 2D sinusoid and a 2D Gaussian. Its various parameters, which may differ from sensor to sensor, are: *spatial frequency* (the frequency orthogonal to the bars), the *orientation* (the angle of the orthogonal to the bars), the *width* (the size at half height of the Gaussian defined orthogonal to the bars), the *height* (the comparable measure parallel to the bars), and the *phase* of the sinusoid (defined relative to the center of the Gaussian). Finally, each function is located at a particular point in the visual field. Of these seven parameters, we shall see that two can be defined in terms of the others, leaving five feature dimensions that may vary from sensor to sensor.

This general form of weighting function closely resembles the receptive field profiles of simple cells, the most numerous class of visual neurons in the striate cortex of cat and monkey [3,13]. Most cortical cells respond only over a modest region of space, a modest band of spatial frequencies, and a modest range of orientations [3,4,15,24]. Psychophysical data are also consistent with a moderate selectivity in space, spatial frequency and orientation [1,29,32]. The Fourier transform of the feature is shown in Fig.1(bottom). where it can be seen that the sensor also responds only over a small band of spatial frequencies and orientations.

This sort of function is often named after GABOR, who showed that (in the 1D case) it minimizes the width in both space and frequency [8]. DAUGMAN has noted the virtues of the 2D version of the Gabor function [2].

4.1. Width and Height

The data of DEVALOIS *et al.* [3] indicate simple cortical cell receptive fields tend to be slightly taller than they are wide, but including this subtlety did not seem worth the extra computational effort. Accordingly, I have equated height and width, so the supports of both pattern and transform are circular. This also allows us to specify the width of a pattern, or of its transform, by a single number: the diameter at half height. With this amendment, the weighting function for the sensor can be written

$$w(x,y) = e^{-4\ln 2(x^2+y^2)/w^2}\cos[2\pi f(x\cos\theta + y\sin\theta) + \varphi] \tag{1}$$

where f is the spatial frequency, θ is the orientation, w is the width, and φ is the phase.

The width of a feature and the bandwidth of its transform are inversely related (bandwidth $= 4\ln 2/(\pi\,\text{width})$). I have set the width of each sensor to about 1.324 cycles of the sinusoid. This means that sensor bandwidth is proportional to sensor spatial frequency (bandwidth

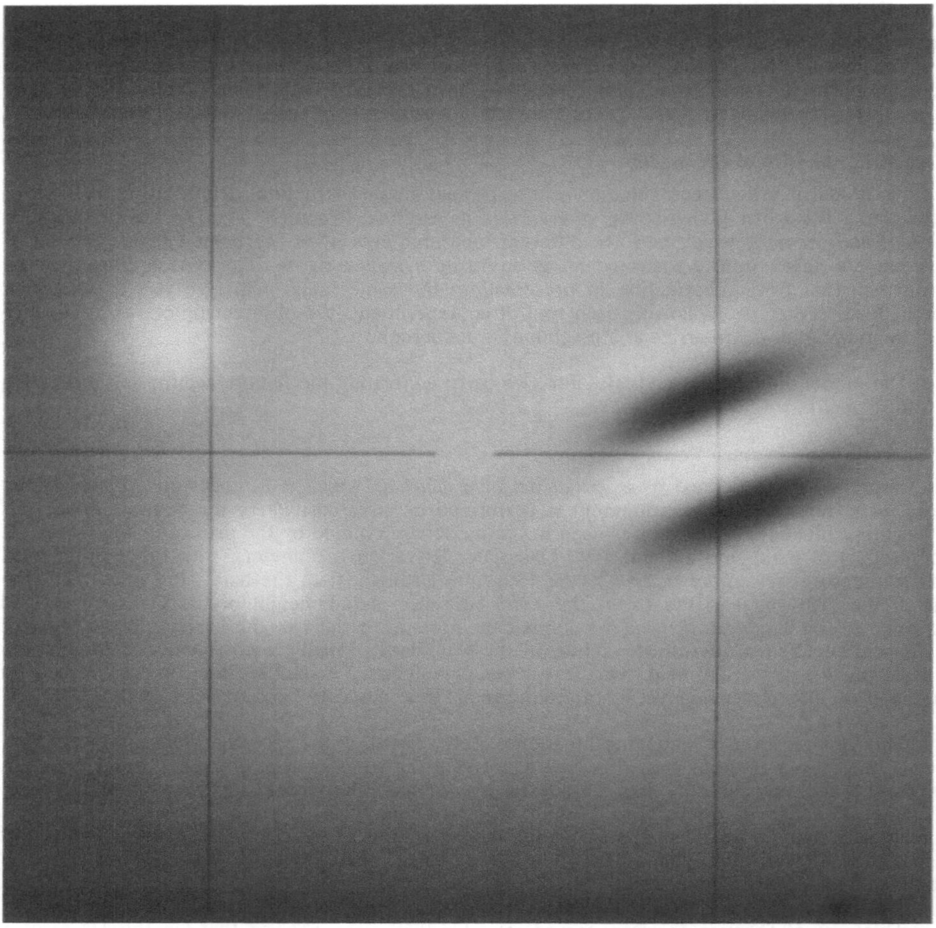

Fig. 1. Top: an example of a spatial feature consisting of the product of a sinusoid and a Gaussian. The two axes are horizontal and vertical visual field position. Width of the feature is 1.324 cycles of the sinusoid. The orientation is 22.5 degrees, and the phase is 0. Bottom: the Fourier transform of the feature. The axes are horizontal and vertical spatial frequency. Top and bottom figures are to scale when 1 degree on the top scale equals 1 cycle/degree on the bottom scale

= 2/3 frequency), or, in logarithmic terms, that each sensor has a bandwidth of one octave. It also means that a low frequency sensor will be large, and a high frequency sensor will be small.

Spatial frequency bandwidths for simple striate cells rise approximately in proportion to frequency. Bandwidths appear to range from one half to 2.5 octaves, but one octave bandwidths comprise a large fraction of those recorded [3,15]. Psychophysical data do not lead so directly to estimates of bandwidth, but they are consistent with a proportionality between frequency and bandwidth, and with log bandwidths of about one octave [29,30,31].

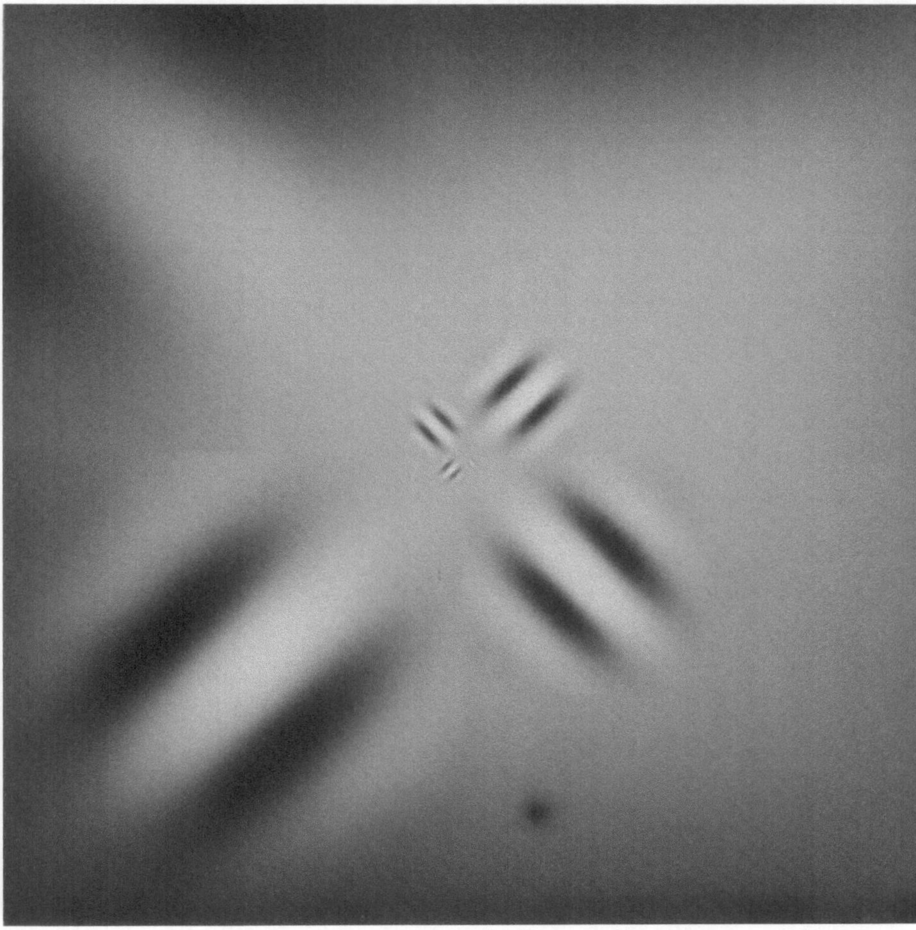

Fig.2. The basic set of sensors. The sensor frequencies range from 0.25 to 32 cycles/degree in octave steps. Each sensor has a bandwidth of one octave (width = 1.324 cycles)

4.2. Frequency

Since each sensor covers only one octave of frequency, we need a number of sensors to cover the full range of frequency which the human can sense (about 0 to 60 cycles/degree). The number required will also depend upon the density of the sensors in the frequency domain. This cannot be less than about one sensor per bandwidth, or information in the image will be lost, and sensitivity will show dips between sensors [29]. In the model I have tried to include as few sensors as possible, so just one sensor per bandwidth has been used. This yields a set of eight basic sensors, whose frequencies range from 0.25 to 32 cycles/degree in octave steps. This basic set is shown in Fig.2. The largest sensor, a small part of which is barely visible in the upper left corner, is 128 times larger than the smallest. The sensors have been arranged in a spiral so as to fit in a square, but should all be considered to lie at the center of the visual field. This basic set will be present only at this point; the sensors at other locations will be generated by a rule described below.

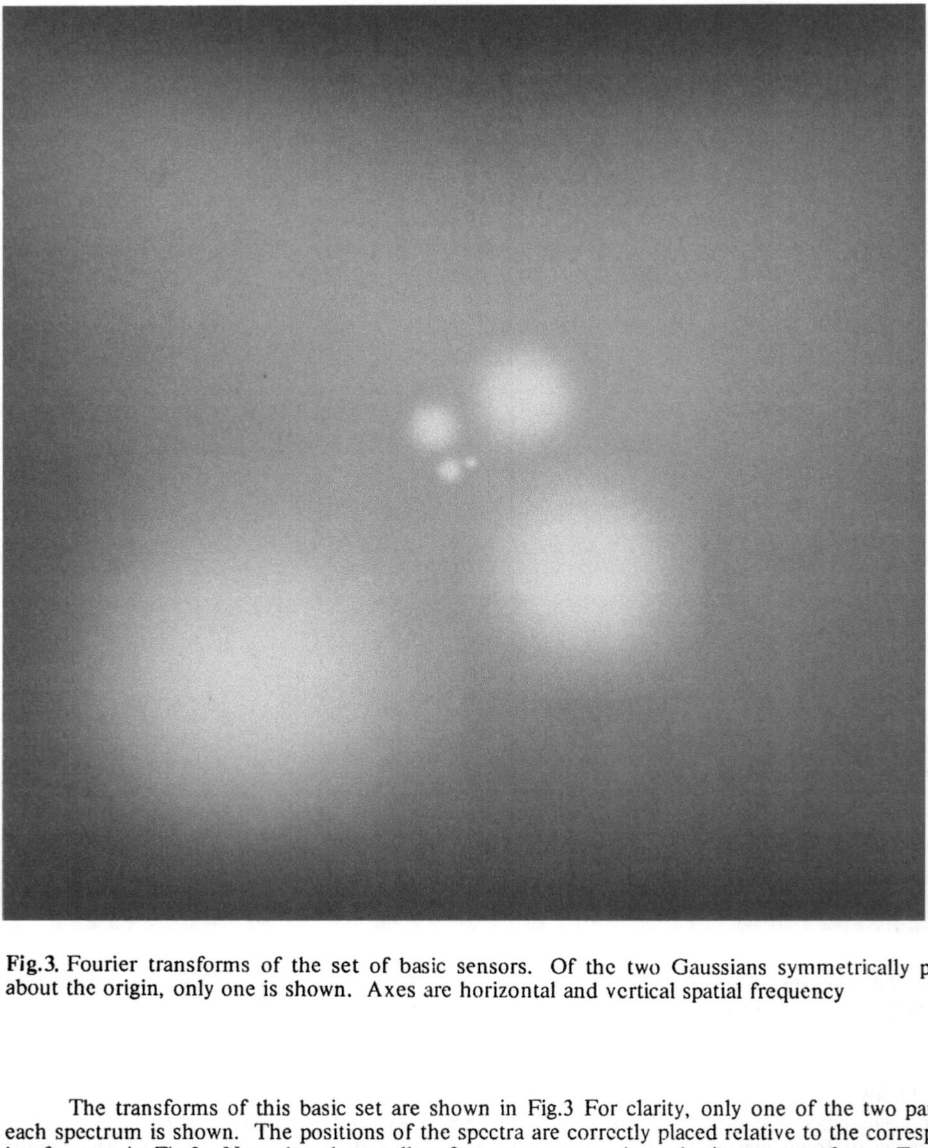

Fig.3. Fourier transforms of the set of basic sensors. Of the two Gaussians symmetrically placed about the origin, only one is shown. Axes are horizontal and vertical spatial frequency

The transforms of this basic set are shown in Fig.3 For clarity, only one of the two parts of each spectrum is shown. The positions of the spectra are correctly placed relative to the corresponding features in Fig.2. Note that the smallest feature corresponds to the largest transform. For clarity I have shown features with different orientations. If all the features were similarly oriented, all the transforms would lie at a single angle, and would overlap considerably. This illustrates that all of frequency space is covered.

4.3. Spatial Sampling

To preserve the information in the image the 1D spatial density of the sensors of a particular frequency must be at least twice the frequency extent of the sensor. This extent is not well defined for

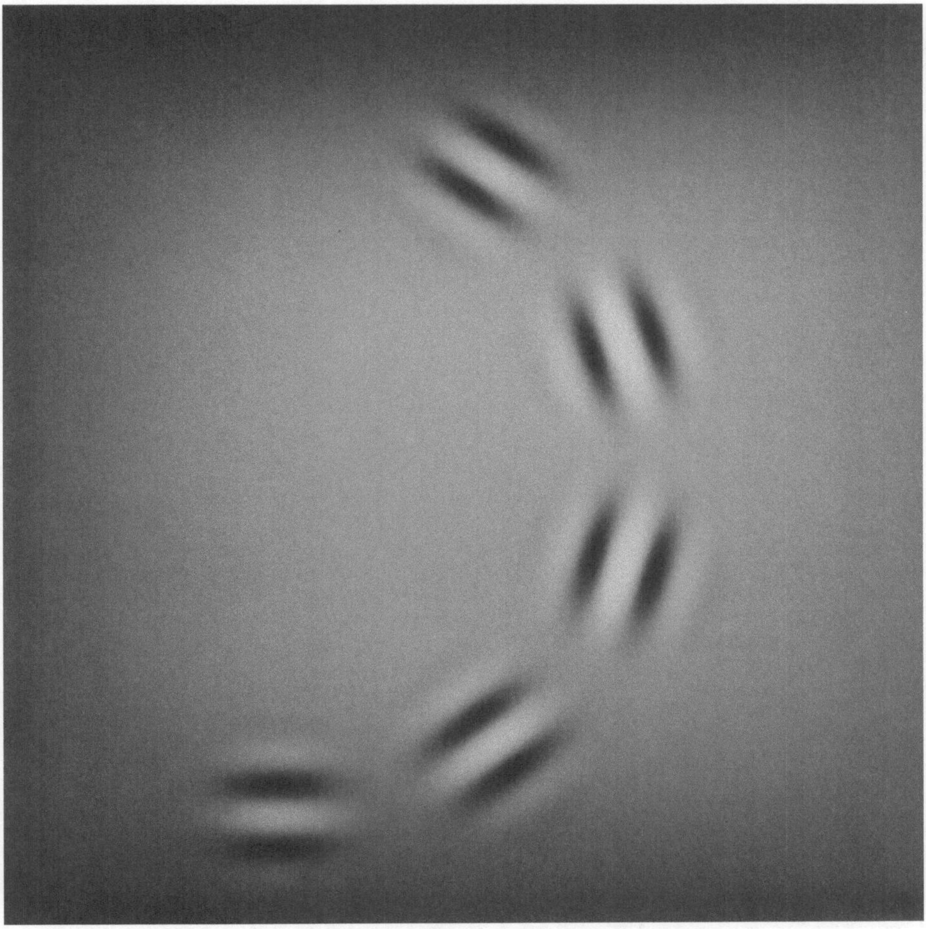

Fig.4. The five sensor orientations. They range from 0 to 144 degrees in steps of 36 degrees

a Gaussian spectrum, but if we used the bandwidth, the required sample frequency would be 4/3 the sensor frequency. I have used the more conservative factor of 1.8. Thus for the basic set of eight frequencies at the fovea, sampling intervals range from 0.0174 degrees for the highest frequency to 2.22 degrees for the lowest.

4.4. Orientation

The orientation bandwidth of a sensor (as measured with the spatial frequency to which the sensor is tuned) is completely determined by the spatial frequency bandwidth. For our one octave spatial frequency bandwidth, the orientation bandwidth is about 38 degrees. This can be seen in Fig.3, where the orientation bandwidth will be given by the range of angles that intersect the spectrum at half-height or above. This 38 degree figure agrees well with the 40 degrees reported by DEVALOIS *et al.* [4] as the median orientation bandwidth for simple cells in monkey striate cortex. There is a great deal of psychophysical data related to orientation tuning, but as yet there has been no convincing route from the data to an estimate of the orientation bandwidths of underlying sensors.

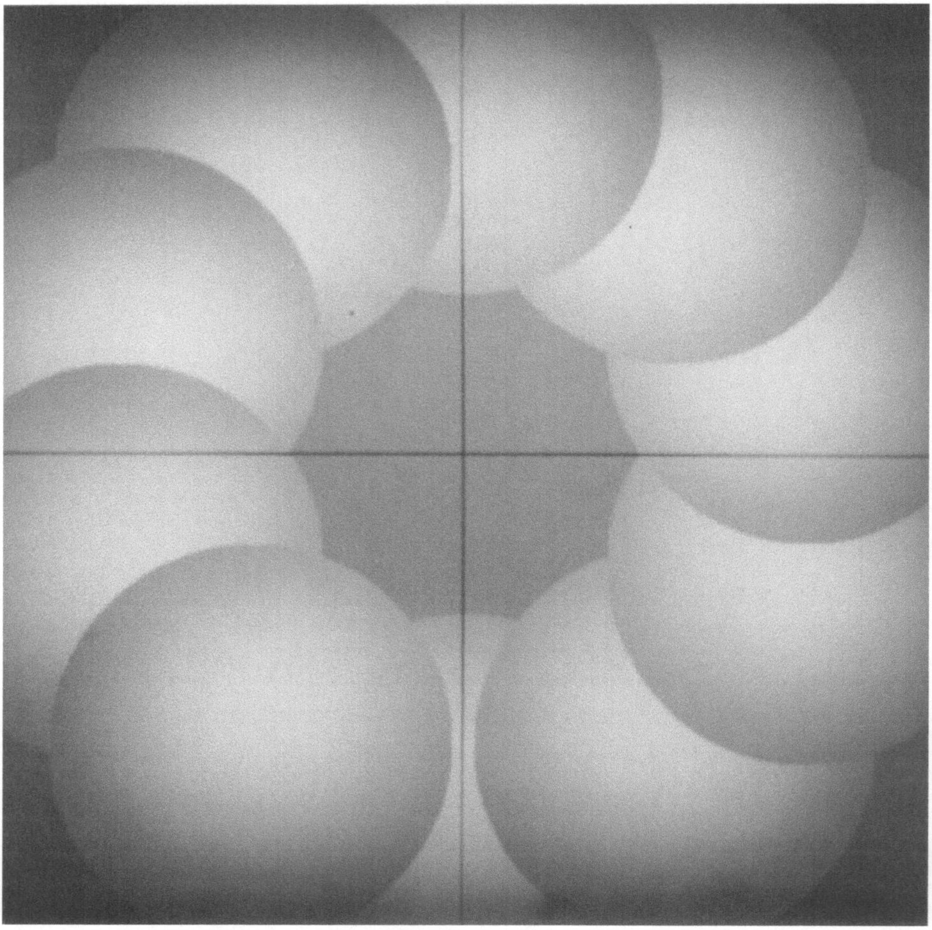

Fig.5. Transforms of the five sensor orientations. Each Gaussian has been truncated at one width for clarity. The orientation bandwidth of each sensor is about 38 degrees

In specifying sensor density in the orientation domain I have again followed the rule of about one sensor per bandwidth. This yields the set of five orientations illustrated in Fig.4. The transforms of these features are shown in Fig.5. Successive transforms are overlaid, and each is truncated at about one width, illustrating the degree of sensor overlap in the domain of orientation.

4.5. Phase

I include sensors of two phases, 0 and 90 degrees. At least two phases are required to preserve the odd and even information in the image. Beyond this, POLLEN and RONNER [19] have found that adjacent simple cells in cat striate cortex usually share a common spatial frequency and orientation, but differ in phase by 90 degrees. Existing psychophysical data do not lead directly to any strong hypothesis regarding sensor phase.

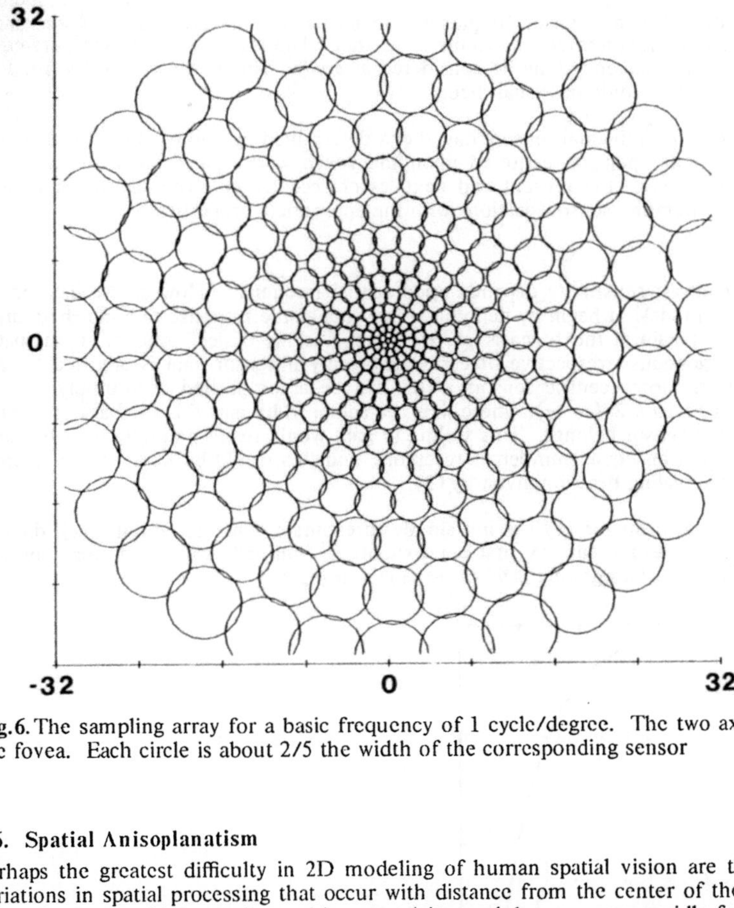

Fig.6. The sampling array for a basic frequency of 1 cycle/degree. The two axes show degrees from the fovea. Each circle is about 2/5 the width of the corresponding sensor

4.6. Spatial Anisoplanatism

Perhaps the greatest difficulty in 2D modeling of human spatial vision are the large and complex variations in spatial processing that occur with distance from the center of the visual field. In general, contrast sensitivity declines with eccentricity, and does so more rapidly for high spatial frequencies than for low [11,20,27]. However it has recently been suggested that spatial processing in all portions of the visual field is identical save for a change in the 2D spatial scale [5,11,22]. This would be the case if ganglion cells changed in size with eccentricity, but all subsequent processing were homogeneous with respect to eccentricity. Though the evidence for this hypothesis is not overwhelming, it is too great a simplification to be resisted. Accordingly I have taken the spatial scale s to be given by the *scaling function*

$$s = 1 + ke \tag{2}$$

where k is a constant and e is eccentricity in degrees. For my own eyes, k is about 0.4. This agrees reasonably well with the estimates of other authors, and with the changes in ganglion cell density and cortical magnification factor with eccentricity [5,21]. It should be emphasized that this function is intended only as a first approximation.

This scaling function has been used to scale the size, frequency, and spatial density of the sensors at each eccentricity. Thus if a sensor at the fovea has a basic frequency of f width w, and 1D density d, the related sensor at eccentricity e will have frequency f/s, width ws, and density d/s. These relationships are shown in Fig.6, where I have drawn a sampling array for a basic frequency of 1 cycle/degree. Each circle in the figure is about 2/5 the width of the corresponding sensor, showing the considerable degree of spatial overlap among sensors of a common frequency. Note that as eccentricity increases, sensor frequency and spatial density are scaled by the same amount, so that the sampling requirements continue to be met. The set of all sensors generated in this way from a given basic frequency may be considered a *family.* This is the closest we get to a "spatial frequency channel" of traditional visual theory.

The actual positions of individual sensors have been determined by finding, for each family at each eccentricity, that regular polygon centered upon the fovea whose sides are about equal in length to the difference between the present and next lesser eccentricity. This algorithm is used only as a simple way of generating sensor locations with the appropriate properties.

4.7. Sensor Gain

It is well known that contrast sensitivity depends upon spatial frequency. How do we incorporate this dependence into the model? I begin by defining *sensor gain* as the response to a matched target at unit contrast. If sensor gain at the fovea is $\chi(f)$, where f is the basic frequency, then spatial processing will be homogeneous irrespective of eccentricity only if sensor gain is adjusted to take into account the larger size of eccentric sensors. This can be accomplished by multiplying each weighting function by $\chi(f)\ 16\ ln2/(w^2\ \pi)$, where w is the actual width and f is the basic frequency of the family to which the sensor belongs. This scaling of gain would result naturally if each sensor received contributions from an equal number of receptors, and this would be likely if the receptor density followed the same scaling function given by (2).

The actual functional form for $\chi(f)$ is not simple to estimate, since it is not given directly by any simple empirical measurement. A first approximation, indirectly estimated from contrast sensitivity to one octave bandwidth grating patches, is shown in Fig.7.

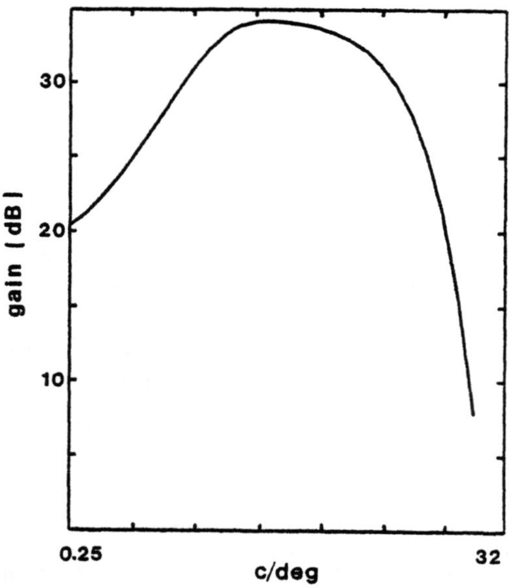

Fig.7. Sensor gain, $\chi(f)$

4.8. Computing the Feature Vector

The preceding assumptions permit us to compute a feature vector, z, each entry of which is the response of a single feature sensor. The many cross-correlations involved are generally more easily computed by way of well-known detours through the frequency domain. Further savings can be gained by disregarding sensors that are well outside the space or frequency support of the image. Even so, the amount of computation required to obtain the feature vector may be formidable, especially for large stimuli with high frequency content. Fortunately the sensors and all the images considered here have Fourier transforms that can be obtained analytically, so that finite transforms need not be resorted to.

5. Processing the Feature Vector

The psychological literature is filled with suggestions as to how an observer might make use of the feature vector to detect and discriminate among patterns. One might look for the largest single entry, look for the centroid of some distribution of vector entries, check whether at least one entry is larger than a criterion value, and so on. But these procedures are largely *ad hoc*, and rarely generalize much beyond the experimental context they are designed to explain. For example, looking for the largest single entry might work at threshold but is not a reasonable rule above threshold. Also, these procedures are not derived from basic principles but rather appeal to the intuition, sometimes not very forcefully.

In contrast, the statistical theory of pattern classification provides a logical basis for constructing optimal procedures with which to analyze the feature vector [6,14]. Furthermore, pattern classification theory is general, and could in principle be applied to most tasks the human observer confronts in the psychophysical lab, if not in the world at large.

5.1. The Optimal Bayesian Classifier

Let me illustrate how I apply this theory to the feature vector constructed a moment ago. First I assume that each sensor is perturbed by zero mean, unit variance Gaussian noise. The feature vector then has a multivariate normal density. I further assume that the covariance matrix of this density is the identity matrix, which means that the sensors are statistically independent.

The job of the classifier is to examine the feature vector and decide which of several possible images was in fact presented. Each of the possible images is associated with a *mean vector* $\mathbf{m}_k$. Presentation of one of the images gives rise to a feature vector $\mathbf{z}$. If we also know the prior probability of each alternative, Bayes' rule allows us to use the feature vector to calculate the posterior probability of each of the alternatives. In the cases I will consider, all alternatives have equal priors. A reasonable, and in fact optimal rule is to choose the image with highest posterior probability. Since we care only which has highest probability, it is sufficient to calculate for each alternative any quantity that is monotonic with posterior probability. Such functions on the feature vector are called *discriminants*.

In the case I have developed so far, a set of optimal discriminants are

$$d_k(\mathbf{z}) = -(\mathbf{z}-\mathbf{m}_k)^T(\mathbf{z}-\mathbf{m}_k) = -\sum_{i=1}^{n}(z_i-m_{ki})^2 \tag{3}$$

where the superscript T denotes the transpose of a matrix. Note that if $\mathbf{z}$ has n elements, then it can be considered a point in n-dimensional space. Likewise for each of the mean vectors $\mathbf{m}_k$. This form of discriminant tells us to pick the alternative whose mean vector is closest to $\mathbf{z}$ in this space. For this reason this is often called a *minimum distance* classifier.

5.2. Prior Information

One objection to this scheme is that it assumes that the observer has perfect prior information regarding each mean vector $\mathbf{m}_k$. Since this vector is obtained through experience, and each experience is subject to variability, this assumption is unrealistic. Empirical data also clearly show that the observer is less than optimal, as though he were uncertain regarding $\mathbf{m}_k$ [7,16,25]. Fortunately, for the detection task I will consider, an approximation is available for a more realistic uncertain observer [18]. In the case of discrimination, we must be content for the moment to examine the behavior of the ideal. This is unfortunate, since to the extent that the sensors preserve the informa-

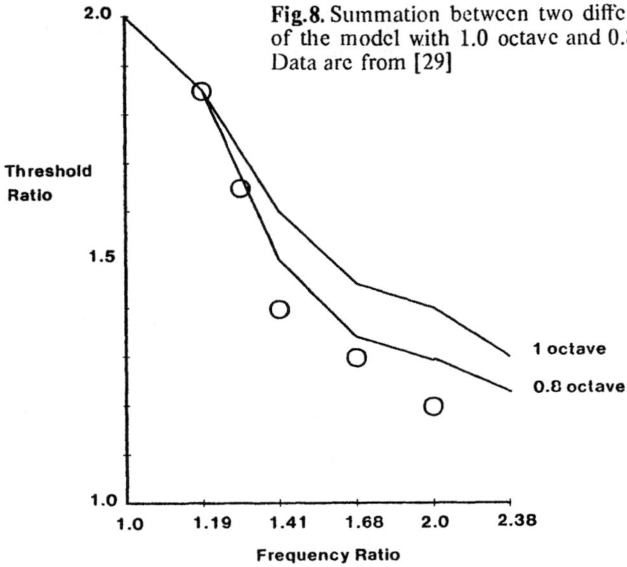

Fig.8. Summation between two different spatial frequencies. Predictions of the model with 1.0 octave and 0.8 octave bandwidths are shown. Data are from [29]

tion in the image, discrimination performance tells us more about the orthogonality of the alternative images than about the properties of the sensors.

5.3. Detection

The ability of an observer to detect a pattern is often measured by a two-interval forced-choice (2IFC) method, in which the observer must judge which of two time intervals contained a signal. This may be viewed as a discrimination between two images: the test image and a null image. From a number of trials we can determine the proportion correct, or d', defined here as $\sqrt{2}$ times the normal deviate of the proportion correct. Since there are only two alternatives (image in first interval or in second interval), we can difference the two discriminants and use the sign of the result to choose an alternative. This quantity is normally distributed with a mean equal to the squared length of the mean vector, and a variance of twice the squared length. From this it is easy to show that for any given pattern,

$$d' = \sqrt{\mathbf{m}^T \mathbf{m}} \qquad (4)$$

$$= \left[\sum^{n} m_i^2 \right]^{1/2}$$

As noted above, this prediction assumes an observer with perfect prior information. For an uncertain observer, performance can be approximated by specifying that for all stimuli at some fixed d',

$$1 = \left[\sum^{n} m_i^{\beta} \right]^{1/\beta}$$

where β is a constant of about 3.5 [18]. This approximation has been used in the following predictions of detection performance.

The model gives a good account of sensitivity to a wide variety of different spatial targets. For example, the model does a reasonable job of predicting the contrast sensitivity function at various eccentricities, and the effects of size of a grating pattern upon sensitivity. Among the earliest experiments to suggest that visual mechanisms were selective for spatial frequency were those which measured sensitivity to mixtures of two frequencies [9,23]. Data from a more recent version of this experiment are shown in Fig.8 [29]. Each point shows the sensitivity to a mixture of two frequen-

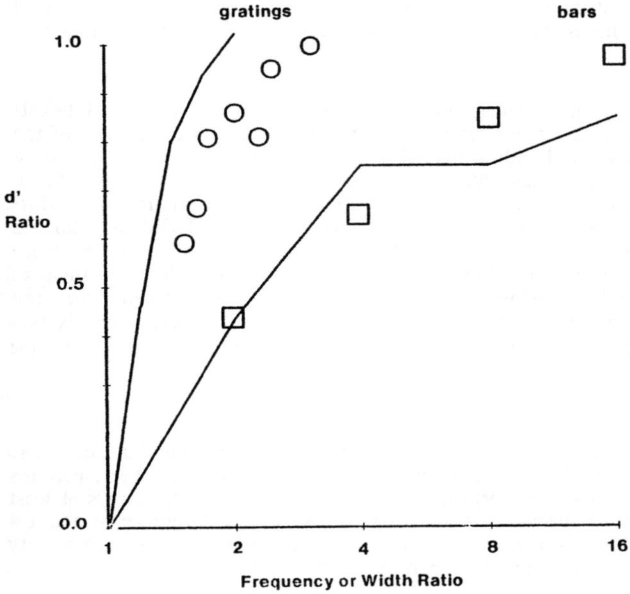

Fig.9. Discrimination between grating patches of different frequencies (circles) and Gaussian bars of different widths (squares). Curves are predictions of the model. Data are from [31] and [28]

cies, relative to the sensitivity to either frequency alone. If the two frequencies add linearly, we get a ratio of two, if they do not add at all, we get a ratio of 1. The data show a decline in summation as the two frequencies move farther apart. The prediction of the model is shown by the upper curve. It does not agree precisely with the data, but given that most sources of error in the data will tend to lower the ratio it does not do a bad job. I have also shown the prediction in the case of 0.8 octave bandwidth sensors. The fit is better, and this amendment may have to be resorted to in the future.

5.4. Discrimination

A particularly sensitive way of measuring discrimination performance is by a two-by-two forced choice (2X2FC) method [17,26,31]. As in the 2IFC method, on each trial the observer is presented with an image in just one of two time intervals. But in this case the image is selected at random from a set of two. The observer must choose both the interval containing the image, and which image it was. The advantage of this technique is that it measures detection and discrimination concurrently, so the two sorts of performance can be compared.

In the case of 2X2FC the observer must choose among four alternatives, so we must determine four discriminants. If we call the feature vector from the first interval $\mathbf{z}_1$ and that from the second interval $\mathbf{z}_2$, and the mean feature vectors for the two images $\mathbf{a}$ and $\mathbf{b}$, then these discriminants are

$$d_{1a} = \mathbf{z}_1^T \mathbf{a} \qquad d_{2a} = \mathbf{z}_2^T \mathbf{a} \tag{5}$$

$$d_{1b} = \mathbf{z}_1^T \mathbf{b} \qquad d_{2b} = \mathbf{z}_2^T \mathbf{b}$$

where d_{2b} for example is the discriminant for image b in the second interval. Here we cannot reduce these to a single function of $\mathbf{z}$ as was possible in the case of 2IFC. Furthermore, although each discriminant is again normally distributed, the four are not independent, so we cannot easily derive the probability with which any one is the largest and must resort to Monte-Carlo techniques.

5.4.1. Spatial Frequency

Since our model observer discriminates among images on the basis of certain features, it is interesting to ask how well the human observer discriminates between two images that are members of the feature set. Fig.9 shows some data collected with a 2X2FC method in which the two images to be discriminated were Gaussian-windowed sinusoids, like the features of the model [31]. The ordinate plots the ratio of d' for discrimination and detection, which is a measure of discrimination performance relative to detection. This method of plotting 2X2FC data is due to THOMAS, who has done much of the pioneering work in this area. The ratio rises rapidly as the two frequencies are moved farther apart, so that when they differ by about an octave, we discriminate between them as well as we detect either one. The model's performance, shown by the solid line, is better than the human observer. Introduction of uncertainty into the model would reduce this discrepancy. Notice that, as a rule of thumb, the model predicts a ratio of one when the two patterns differ by about one bandwidth.

5.4.2. Bar Width

What if we repeat the experiment with images which do not resemble the sensor features? The square symbols in Fig.9 are comparable data for discriminations between Gaussian bars: patterns with a Gaussian profile in both horizontal and vertical dimensions and with height always at least twice the width [28]. Discrimination performance is vastly poorer, so that a difference of about 4 octaves is required before discrimination is as good as detection. This poorer performance is very nicely described by the model, whose predictions are shown by the rightmost line.

5.4.3. Orientation

The last set of data I will show you were collected by THOMAS and GILLE [26]. They used a 2X2FC method to discriminate between gratings of different orientations. The gratings were large, extending 15 cycles in each direction. Their data are shown in Fig.10. Discrimination improves rapidly as the difference in orientation increases, and is almost as good as detection when the difference is 10 degrees. The model's simulated data are shown by the leftmost solid line. The agreement is quite good; perhaps too good considering uncertainty has not been included.

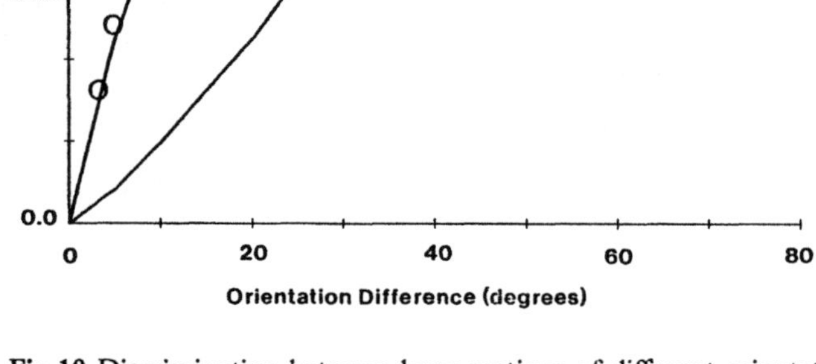

Fig.10. Discrimination between large gratings of different orientations (circles). Data are from [26]. Curves show the predictions of the model for large and small gratings

On the basis of a 1D model, THOMAS and GILLE estimated mechanism orientation bandwidths of 10.5 to 20.5 degrees, depending on observer. These are 2 to 4 times narrower than the sensor bandwidths used in the prediction in Fig.10. These authors acknowledged that their estimates were much narrower than bandwidths of cortical cells, and speculated that "psychophysically defined channels represent physiological mechanisms other than single cortical cells".

How can the present model with 38 degree bandwidths perform as well as the THOMAS and GILLE model with bandwidths about 3 times narrower? The most likely answer is that the present model is 2D, and can make intelligent use of the pattern of activity across a large number of spatially distributed sensors. When the stimulus is confined to a small area, performance deteriorates markedly, as shown by the rightmost curve in Fig.10 (though much of this is due to the broadened orientation bandwidth of the stimulus itself).

These observations illustrate an important point: sensor properties cannot be estimated without an explicit, computable, 2D model of the sensors and of their distribution over the visual field. The model must also allow the observer to make intelligent use of the sensor outputs. This argument applies to any effort to derive sensor parameters from psychophysical data, that is, to almost all psychophysical research.

Acknowledgements

I thank Albert Ahumada for useful discussions and help in the construction of Figs.1-5.

References

1. C. Blakemore, F. W. Campbell *J. Physiol.*, **203**, 237-260 (1969).
2. J. D. Daugman *Vision Res.* **20**, 847-856 (1980).
3. R. L. De Valois, D. G. Albrecht, L. G. Thorell *Vision Res.* **22**, 545-559 (1982).
4. R. L. De Valois, E. W. Yund, N. Helper *Vision Res.* **22**, 531-544 (1982).
5. N. Drasdo *Nature* **266**, 554-556 (1977).
6. R. O. Duda, P. E. Hart *Pattern classification and scene analysis.* (John Wiley, New York, 1973)
7. J. M. Foley, G. E. Legge *Vision Res.* **21**, 1041-1053 (1982).
8. D. Gabor *J. IEE London*, **93**(III), 429-457 (1946).
9. N. Graham, J. Nachmias *Vision Res.*, **11**, 251-259 (1971).
10. D. M. Green, J. A. Swets *Signal Detection Theory and Psychophysics.* (Wiley, New York, 1966).
11. J. J. Koenderink, M. A. Bouman, A. E. Bueno de Mesquita, S. Slappendel *J. opt. Soc. Amer.* **68**, 845-849 (1978).
12. J. J. Koenderink, M. A. Bouman, A. E. Bueno de Mesquita, S. Slappendel *J. opt. Soc. Amer.* **68**, 854-860 (1978).
13. S. Marcelja *J. Opt. Soc. Am.* **70**, 1297-1300 (1980).
14. J. G. Moik *Digital processing of remotely sensed images.* (NASA SP-431, Washington, 1980).
15. J. A. Movshon, I. D. Thompson, D. J. Tolhurst *J. Physiol.*, **283**, 53-77 (1978).
16. J. Nachmias, E. C. Kocher *J. Opt. Soc. Am.*, **60**, 382-389 (1970).
17. J. Nachmias, A. Weber *Vision Res.*, **15**, 217-233 (1975).
18. D. G. Pelli *Effects of visual noise.* (Ph.D. Thesis, University of Cambridge, Cambridge, England, 1981).
19. D. A. Pollen, S. F. Ronner *Science* **212**, 1409-1411 (1981).
20. J. G. Robson, N. Graham *Vision Res.* **21**, 409-418 (1981).
21. J. Rovamo, V. Virsu *Exp. Brain Res.* **37**, 495-510 (1979).

22. J. Rovamo, V. Virsu, R. Nasaren *Nature* **271**, 54-56 (1978).
23. M. B. Sachs, J. Nachmias, J. G. Robson *J. opt. Soc. Am.* **61**, 1176-1186 (1971).
24. P. H. Schiller, B. L. Finlay, S. F. Volman *J. Neurophysiol.* **39**, 1334-1351 (1976).
25. W. P. Tanner Jr., J. A. Swets *Transactions of the IRE*, **PGIT-4**, 213-221 (1954).
26. J. P. Thomas, J. Gille *J. Opt. Soc. Am.* **69**, 652-660 (1979).
27. A. J. van Doorn, J. J. Koenderink, M. A. Bouman *Kybernetik* **10**, 223-230 (1972).
28. A. B. Watson *Optical Society of America, Technical Digest of Topical Meeting on Recent Advances in Vision*, SA4 (1980).
29. A. B. Watson *Vision Res.* **22**, 17-25 (1982).
30. A. B. Watson, H. B. Barlow, J. G. Robson *Invest. Ophthlmol. and Visual Science* Suppl. to **20**, 178 (1981).
31. A. B. Watson, J. G. Robson *Vision Res.*, **21**, 1115-1122 (1981).
32. D. W. Williams, H. R. Wilson, J. D. Cowan *J. opt. Soc. Amer.* **72**, 878-887 (1982).

Perceptual Filters

G.D. Sullivan

Laboratory of Experimental Psychology, University of Sussex,
Brighton, Sussex, United Kingdom

1.1. Introduction

Research into visual perception is currently dominated by three types of investigation: neurophysiology, psychophysics and computational studies. A complete understanding of a visual task requires a synthesis of all three. At present, our conception of the mechanisms of early vision is mainly derived from psychophysics and neurophysiology, and from these has emerged in recent years a fairly robust theory of the initial stages of vision. It is now widely accepted that early visual processes are served by multiple, independent, band-pass filters, each acting approximately linearly. The theory explains many phenomena in the vision of low-contrast repetitive patterns, but little progress has been made in understanding its implications in the more "natural" case of high-contrast, non-repetitive patterns. We have good reason to believe that the same mechanisms operate. Neurophysiological studies have shown that the single neural units which are assumed to underlie the psychophysically determined channels do not greatly change their properties in high-contrast conditions [1]. Furthermore, some aspects of high-contrast vision, such as the spatial frequency shift [2], and the frequency specific disruption of contrast matching after adaptation [3], make it likely that the band-pass channels at work in low-contrast conditions also serve high-contrast vision.

But knowledge of mechanism is not knowledge of function. What is missing from the psychophysical theory is a coherent view of how multiple band-pass channels are used in the performance of perceptual tasks. Computational studies provide a means to study the link between visual mechanisms and visual behaviour, and MARR and his colleagues [4,5,6] have contributed a great deal to our understanding of the role of band-pass filters in stereopsis and movement analysis. Little has been said though about the analysis of geometrical form. How does the visual system detect and discriminate different 2-dimensional shapes? A key issue is whether the visual system is dedicated to a fixed set of perceptual filters, which provide the "primitives" for the perception of shape. If so, then we might expect to find psychophysical evidence of the filters in the detection of patterned stimuli.

1.2 The detection of isolated patterns

Early experiments on the perception of patterns of light on a bright background (e.g. [7,8]) found that the contrast threshold for seeing very small targets depended mainly on the energy (the intensity times the area) of the target, whereas for targets subtending more than about 5 mins of arc the length of the circumference appeared to dominate sensitivity. Unfortunately these experiments say little about the role played by

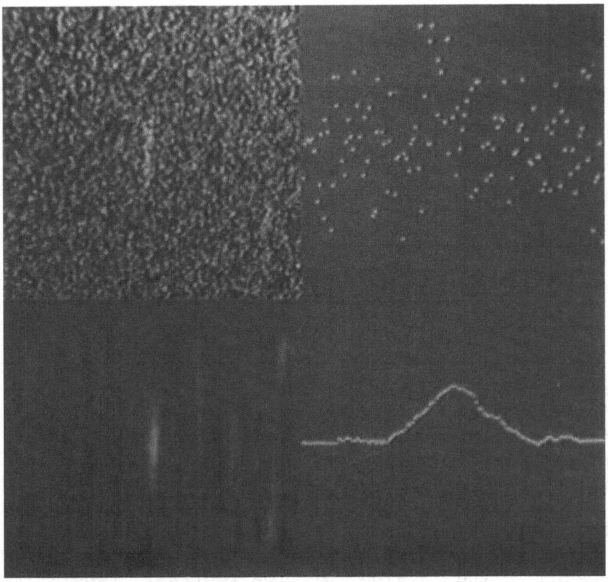

Fig.1. Top: Bar target in noise, and (right) vertical luminance profile. Bottom: Output of filter matched to the bar, with luminance profile

multiple channels in vision - indeed they may be explained (qualitatively at least) by a single channel model. We know that at contrast-threshold the overall frequency response of the visual system is strongly band-pass in nature. The high frequency cut beyond about 5c/deg accounts for the energy summation which occurs with small targets, and the low-frequency cut explains why the edges of large targets are more important than the body of the target.

Contrast-thresholds for isolated targets are therefore primarily determined by the dominance of the most sensitive channels in human vision (this observation leads to predictions which have been confirmed in a study of adaptation to single lines [9]). We do not know whether the same also holds in suprathreshold perception. The low-contrast response of the visual system can be avoided by using random noise to mask the targets, and psychophysical thresholds can then be measured for patterns which are well above the absolute contrast threshold. As will be seen, masking the target patterns with noise also allows direct comparisons to be made between psychophysical performance and computing algorithms.

1.3 The detection of patterns in noise

Figure 1 shows an example of our perceptual task: in the top left quadrant a rectangular bar (the target) is masked by random noise. The top right quadrant of Fig.1 shows the luminance distribution along a vertical line through the middle of the bar. What computational strategies do we use to detect the target? The contrast of the bar is about 10% (approximately 1 log unit above its contrast threshold), so the sensitivity of early visual processes plays little part in determining our ability to perceive the target. Instead, performance depends on the decision-making algorithms used by the observers. To detect the target the visual system must analyse the significance of a cacophony of signals - not the existence of any single reliable signal. Ultimately, the same may be true of the normal psychophysical threshold, but in this case the disruptive noise arises internal to the perceiver, being due to physical and physiological factors (such as quantum effects in the transducers, or the limitations of the sampling density and dynamic range of cells). In Fig.1 the noise is readily visible, so it must considerably exceed any internal noise present, which (to a first approximation) can be neglected.

The noise level in the task is known, so the theoretical limit of the detection task can easily be stated, and this gives a measure of the ideal performance (see below). BARLOW [10] pointed out the value of expressing human sensitivity in terms of its efficiency relative to an ideal detector. He argued that if the visual system were specialised to a fixed set of mechanisms (as seems to be implied by its physiology), then perceptual efficiency would be best when the task matched the mechanisms. BARLOW's experiments measured the detection of differences in dot density in stimuli consisting of points of light at random positions. One problem he encountered was that the size of the target regions was seriously constrained, since for small targets, the increase in dot density led to locally discriminable sub-patterns. The notionally independent dots fused into miniature lines and blobs, which provided an independent clue for discriminating the target region. Our task avoids this problem since the statistical noise affects the intensity of the pixels, not the density of brightened points. Targets as small as one pixel (here, 1 min of arc) are meaningful.

2. Human efficiency for detecting targets in noise

2.1 The psychophysical method

Except where otherwise noted, the psychophysical task remained the same in all the experiments. Target patterns, of either positive or negative polarity were presented at the centre of the 256 min square display. They were added to fields of noise drawn randomly from a gaussian distribution of intensities, whose standard deviation was equivalent to 16% contrast, and whose mean luminance was 25 lux. The overall contrast of the display was ramped on gradually (0.7 secs) to avoid transients in the stimulus, and remained present until terminated by the observer. On each of 100 trials the observer signaled by means of 6 switches whether he thought the target was positive or negative, together with his degree of confidence in the judgement - strong, moderate or unsure. This provided a 6 category confidence rating, which was analysed by conventional signal detection techniques [11]. Pressure of time precluded a probit analysis of the results, and as the measure of d' (the detection performance), we have used the mean of the estimates corresponding to the 5 confidence boundaries.

2.2 The ideal detector

On each trial the target shape, size and position on the screen was known, the only uncertainty about the target being its polarity. Furthermore, the targets had homogeneous luminance and the mean background luminance was constant (excepting the noise). There is therefore a well-defined optimal strategy for the task: sum all the pixel values falling within the target area, and respond positive or negative according to whether the sum is greater or less than an equivalent sum taken over the mean luminance of the background. Different confidence ratings can be generated by a further partition of the summed value. This strategy is equivalent to using a matched filter to evaluate the correlation between the stimulus and the target pattern.

The bottom left quadrant in Fig.1 shows the output of a filter matched to the target bar, and the vertical cross-section through the centre of the target is shown on the right. Familiar statistical arguments show that the detectability (d') for the matched filter is given by:

$$d' = \frac{2I\sqrt{A}}{\sigma}$$

where I is the (mod) intensity, and A is the area of the target; and σ is the standard deviation of the noise distribution (the factor 2 arises because positive and negative targets differ by 2I).

2.3 The detection of square targets in noise

Figure 2 shows the results obtained for two observers, for square targets of side ranging from 2 to 32 mins of arc. Pilot studies were first carried out to determine an intensity of the 8x8min square which yielded moderate values of d', and the intensities of the other squares were made inversely proportional to their (linear) size. The ideal performance is therefore the same for all targets; it corresponds to a d' of 6. [In practice, because of limited spatial and grey-level resolution, we had to accept a small error for some targets; it never exceeded 2%.]

Despite the large variance inherent in the d' estimates, it can be seen that there is very little, if any, difference in the measured performance. The lower abscissa in Fig.2 shows the intensities used, expressed as contrast (dI/I). Human performance at this task is evidently well-described by the statistical constraints, and there is no evidence of selective preference by the visual system for squares of a particular size. The right-hand ordinate shows the data expressed as absolute efficiency (the square of the ratio between experimental and ideal d' [10]). Our measured d' lay between 1 and 2, corresponding to efficiencies of between 2 and 11%. These results are a great deal lower than those obtained by previous studies of human perceptual efficiency, and probably reflect the unrealistic specification of the ideal detection algorithm (see below).

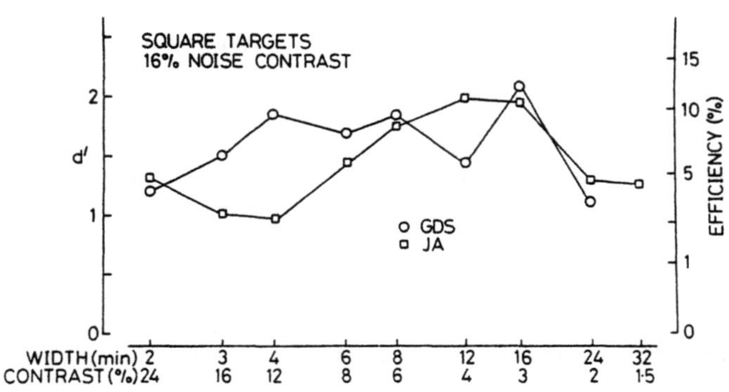

Fig.2. Detection of square targets in random noise. Contrast (lower abscissa) was weighted to preserve constant ideal performance

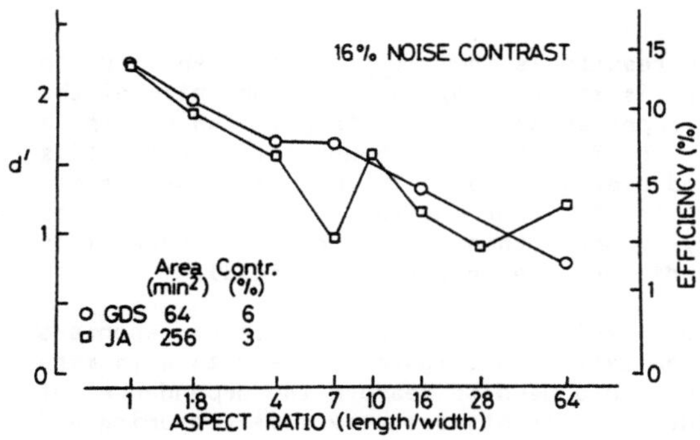

Fig.3. Detection of vertical bars, constant area, variable aspect ratio

2.4 The detection of rectangular bars in noise

Figure 3 shows the results of a similar study, in which the aspect ratio of (vertical) rectangular bars was varied, keeping the area constant (64 square pixels, at 6% contrast for observer GDS; 256 square pixels, 3% contrast, for observer JA). Since the area and the contrast of each target was the same, the ideal performance was also constant (d'=6). These results show a systematic decline in visual performance with increasing aspect ratio, the square targets being most detectable. The decline in performance seems to set in at very low aspect ratios, and there is no sign of enhanced performance at intermediate values, such as might arise from elongated visual filters.

2.5 The detection of rectangular bars, as a function of length

Figure 4 shows the results obtained in a study in which the width of the targets was kept constant (4 mins), and the length was varied from 4 mins to 64 mins. As in the first experiment, the intensity of the targets was inversely proportional to the square root of the area, so that the ideal detector again performs with a d' of 6. The contrast of the targets is shown on the lower abscissa. Once again there is no sign of selective preference for particular patterns.

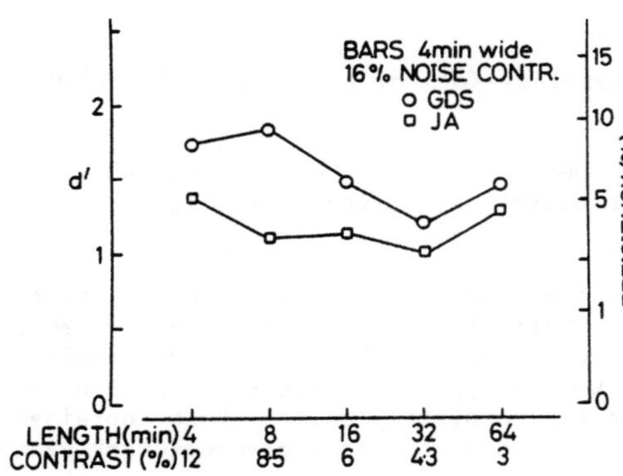

Fig.4. The detection of vertical bars of variable length. Contrast was weighted to preserve constant ideal performance

2.6 Discussion.

The most striking aspect of these results is how little the ability to detect targets in noise fields is affected by the shape and size of the target pattern. The experiments reported here used targets which in a statistical sense were equally difficult, even though their intensities (and hence their contrasts) varied over a range of 16 to 1, and their areas varied over a range of 256 to 1. For neither observer did the detectability of the targets vary by more than a factor of three (the greatest variation occurring for different aspect ratios, Fig.3).

It is important to note that even this range of performance represents a very small change in perceptual ability. To compare our results with more conventional threshold measures we would need to measure the dependence of d' on the stimulus intensity. Such experiments are very time consuming and our data on the matter are limited. Fig.5 shows measurements for three different lengths of a 2 min wide bar. The abscissa is scaled according to $I*\sqrt{area}$, so the three sets of points may be compared directly. All three show clear evidence of a threshold non-linearity (d'=0 at finite intensities), and are quite different from the linear expectation of the ideal detector (dashed line through the origin). Reasons for this will be discussed later; for the present purpose it is sufficient to point out that if these results are representative of the psychometric functions in all of our experiments, the small range of d' we have observed reflects an even smaller range of equivalent contrast. A two fold change of target contrast is sufficient to account for our most extreme results.

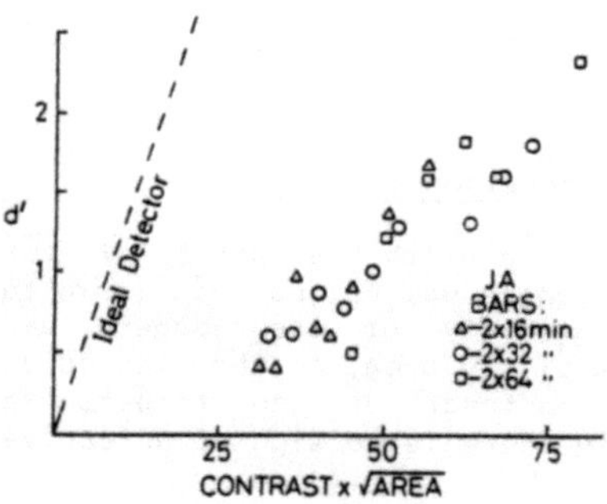

Fig.5

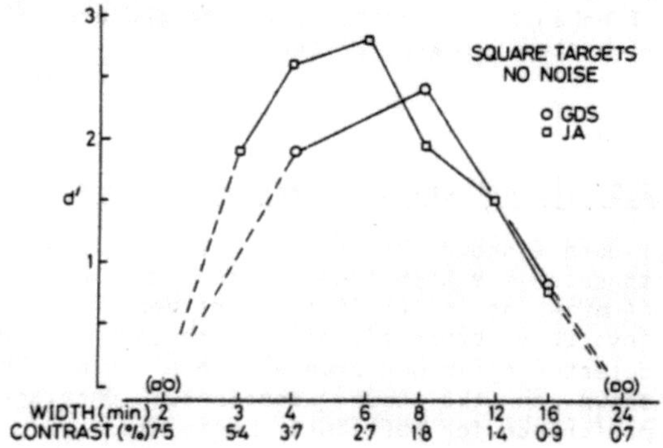

Fig.6

Fig.5. Psychometric functions: detectability (d') of 3 bars as a function of scaled intensity (see text)

Fig.6. The detectability of square targets on a homogeneous background. The contrast of the targets (lower abscissa) was scaled as in Fig.2

Another way of demonstrating how slight the observed variation is, is to measure the conventional contrast threshold in a comparable experiment. Fig.6 shows the effect of repeating Experiment 1 without the added noise. As before, the stimulus intensities were weighted in inverse proportion to the square root of area. The results are completely unlike those obtained with noise fields: within the range of sizes used, the perceptual task went from impossible through facile and back to impossible again. These findings may be attributed to the strong band-pass tuning of the visual system at contrast threshold, which clearly did not affect our suprathreshold results.

The broad conclusion then is that in detecting targets embedded in noise, the human visual system is remarkably insensitive to changes in stimulus configuration, and operates at approximately constant (though very low) efficiency. BARLOW [10] came to a similar conclusion, using an entirely different perceptual task. He pointed out that the lack of specificity to particular stimulus patterns is difficult to reconcile with the neurophysiological discoveries, if they are taken to suggest that the visual system is pre-disposed to analysing specific patterns. If the function of a neuron which responds best to (say) a bar of a fixed length and width is to detect such a bar, then the system as a whole will display preferential efficiency for target patterns which match its set of neural analysers. The argument is not conclusive, since there may be a host of neural analysers in the visual system which between them match all possible patterns. In the case of square targets, this possibility is not too far-fetched, since single units do show a wide range of preferred sizes, but as far as I am aware, receptive fields of such extreme aspect-ratios have not been reported.

3. Computing algorithms for detecting targets in noise

3.1 Against matched filters

The ideal detector, against which human performance was compared when calculating perceptual efficiency, relies on matched filters, selected with no uncertainty for each target pattern. Few theorists are prepared to consider this as a model of human vision: the number of filters needed is unlimited, and it is unreasonable to suppose that in normal visual tasks, the perceiver knows exactly what it is he is looking at (and where it is), before he sees it. In any case, how could such a scheme be reconciled with the multiple band-pass channels revealed by psychophysics and neurophysiology? The patterns matched by such filters are hardly common.

This approach seems fundamentally misguided. Early visual processes should not be expected to solve complex visual problems. Rather, their function must be to provide a general-purpose data-base which facilitates higher level decisions without excessive loss of information. A good data-base should exploit consistencies and constraints of the visual process, and make explicit the information relevant to a wide range of perceptual problems.

The band-pass filters of vision can be simulated by "difference-of-gaussian" (DOG) filters (2-D linear filters whose spatial weighting functions consist of a gaussian distribution subtracted from another having a smaller standard deviation). I have previously shown that a suitably chosen set of DOG-filters provides a complete decomposition of an input image, i.e. all the information in the image is preserved [12]. MARCELA [13] made an equivalent observation by pointing out the similarity between the physiological properties of cells and Gabor functions, which also provide a complete description. In consequence, DOG-functions are capable of mimicking any pattern, to any desired accuracy. Our previous work demonstrated that multiple DOG-filters can be used in image-enhancement to remove noise and blur from an image, without prior knowledge of the signal distribution. This suggests a means whereby the visual system might achieve matched-filter like behaviour in detecting targets in noise.

3.2 The efficiency of DOG-filters

The ideal detector algorithm can readily be adapted to DOG-functions. Instead of classifying the targets according to the response in a filter matched to the target, we can apply the same decision algorithm to the response in DOG-functions. The efficiency of the strategy will once again

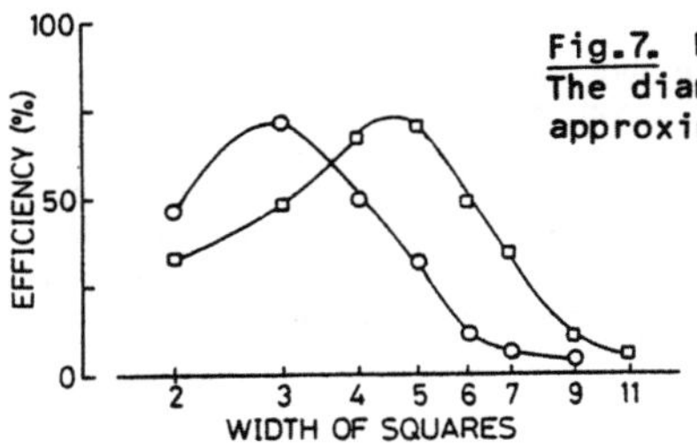

Fig.7. Efficiency of DOG-filters - simulated data. The diameters of the central summing areas were approximately 3 and 5 units

be given by the signal-to-noise ratio in the filter. The mathematics is harder, so it is convenient to simulate the algorithm; Fig.7 shows the d' values obtained for 2 different DOG-filters, whose component gaussians had space constants in the ratio of 1:1.4, for square targets. In each case the peak response in the filter (no matter where in relation to the target it fell) was used to compute the signal strength, and the noise level was estimated by the standard deviation of the filter's response across a sample of pure noise.

DOG-functions achieve a maximal efficiency of about 70-75% in discriminating square targets from noise. Notice how rapidly the efficiency falls off as the target size fails to match the central summing region of the filter. Clearly the psychophysical efficiency for squares (2%-11%) can easily be bettered by a set of DOG-functions provided that they span a sufficiently large range of sizes. However, the same cannot be said for elongated targets. The best response to a bar occurs in a DOG-function which matches the width of the bar; the peak responses are at the ends of the bar, and are independent of the bar's length. The performance of the algorithm therefore falls off as the square root of the length.

However, this detection algorithm pays no attention to the fact that the response to the bar target spreads out in the filtered image. In the case of matched filters this is unimportant, since the "best" value to use is that which occurs when the filter and target are exactly in register (Fig.1), and a filter so positioned uses all the available information. But the response in (say) a DOG-filter to an elongated target is itself elongated. The most reliable single point is at the end of the bar, but responses along the bar convey additional, independent, information. In effect the algorithm so far considered is equivalent to a neural model in which only one neuron is allowed to influence the decision.

Figure 8 illustrates the problem. The input image (top left) is the same as in Fig.1, and the result of filtering by a DOG-function is shown on the top right. The bottom left image shows the effect of applying a

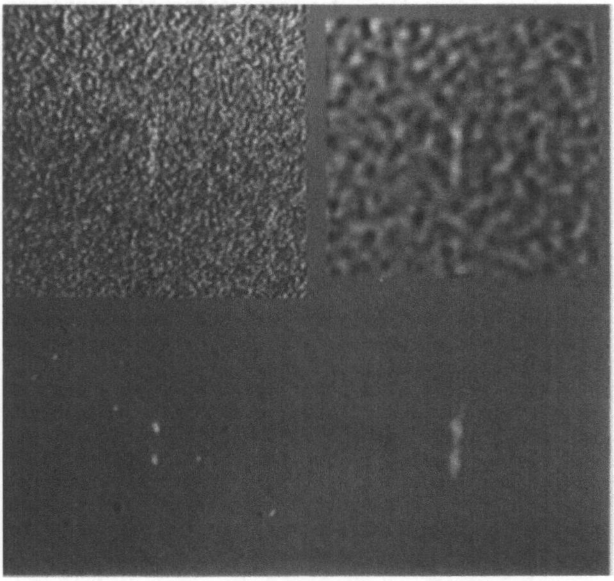

Fig.8. Top left: bar target in noise field (as Fig.1).
Top right: Output of a DOG-filter (central summing width = 4 pixels).
Bottom left: Thresholded image of top right.
Bottom right: Output of region-testing algorithm (see text)

threshold to the filtered image. Notice that at this level of threshold, which suffers several "false alarms", the bar has been fragmented, and only its ends are passed. However, the presence of the target is clearly indicated by the elongated blob in the filtered image. One very simple means of aggregating the information is to sum along the blob, but this is equivalent to using a filter matched to the length of the target line, and suffers the disadvantages of matched filters. A more interesting possibility is to segment the DOG-filtered image into its constituent blobs and then test each blob for statistical significance as a whole. Such an algorithm can reap the statistical benefit of large samples, without prior knowledge of the shape, size and position of the target.

3.3 Segmentation of DOG-images

In his computational theory of stereo-matching, MARR [6] pointed out that zero-crossings in DOG-filtered images of Julesz stereograms occur with predictable regularity. They are usually spaced at intervals equal to the width of the central summing region of the DOG-filter. This effect can be seen in Fig.8: the noise background generally gives rise to small blobs, localised in 2 dimensions. This is not the case in the presence of the target, and we can use this fact to lower the statistical criterion applied to large blobs. The effect of the algorithm depends on the segmentation method. MARR and HILDRETH [14] favoured the use of zero-crossings in the DOG-filtered images to locate edges. Edge segments could be used to partition the image, but this seems an inefficient strategy. We have found it easier to grow the regions directly, by aggregating all connected positive (or negative) points.

The bottom right image in Fig.8 shows the effect of the algorithm. The line target has been distinguished from the noise background, by using the redundancy of the pattern, <u>without</u> prior knowledge of the target shape. The technique provides some of the advantages of matched filters, without the crippling disadvantage of either having to know precisely what it is that is being sought, or having to use an unlimited number of filters simultaneously.

3.4 Context-sensitive filters

Only one DOG-filter was used in Fig.8, but because of the reconstruction properties of a suitably chosen set of DOG-filters [12], the blob-growing algorithm can be applied to targets of arbitrary shape and size. Fig.9 shows the results of applying a system of 6 DOG-filters to a more complex image. The top left quadrant shows letters of two different intensities degraded by noise. A thresholded DOG-image is shown without (top right) and with (bottom left) the region-growing algorithm. The bottom right image shows a reconstruction of the input, generated by summing the outputs of 6 DOG-functions whose scales differ by 1/2octave. It seems to capture remarkably well the information our own eyes can extract from the original input.

In effect the algorithm approximates a self-tuning matched filter, and accounts satisfactorily for the unselective behaviour we observed psychophysically, but there remains one major problem. Our psychophysical experiments show that the absolute efficiency of human observers at the task of detecting targets in noise is very low (2-11% in the cases studied here). But perhaps it is unreasonable to compare human performance with the ideal detector, which needs exact knowledge not only of the shape and size of the target, but also its position on the screen and the expected response to the background intensity. If we simply relax the positional constraint, and allow the target to appear anywhere on the screen, then the decision criterion of the ideal detector becomes more complex, since before judging the sign of the target we must first find it. Uncertainty about the signal generates threshold-like behaviour [11], essentially as in Fig.5. Unfortunately it is difficult to incorporate this into the mathematical specification of the ideal detector, since the magnitude of the effect depends on the precise size of the search area (and hence the number of effectively independent samples of the noise field which must be eliminated). As it stands, the region-growing algorithm suffers from the same problem; it too has poor absolute efficiency, since it does not use *a priori* knowledge of the target position. Perhaps the visual system is forced to adopt a similar strategy - information must first be identified as being different from its surroundings before it can be acted on.

This accords with our introspections. Even though, as observers in our experiments, we knew that the target would always appear at the centre of the screen, we felt that we were unable to use this information adequately. Our judgements seemed to be based on the most prominent feature which fell approximately in the right place. It was not uncommon to experience trials in which the random noise field caused a bright patch to be adjacent to a dark patch at the centre of the screen, and we were uncertain which one to pay attention to. An ideal detector has no such problem, but in normal vision, it must be extremely rare to encounter such well-defined targets, so the apparent inefficiency of human vision may not be surprising.

Fig.9. Top-left: Letters of two intensities in noise.
Top right: Thresholded DOG-image (as Fig.8, bottom left).
Bottom left: Thresholded after region-testing (as Fig.8, bottom right).
Bottom right: Sum of 6 different DOG-filters after region-testing

4.1. Conclusions

The most important outcome of our psychophysical experiments is that the ability of humans to detect targets in noise fields shows little evidence of the selectivity for shape manifested by single neurons in the early visual system. Our computing experiments show that this is not incompatible with single unit recordings, provided that perceptual decisions can make use of aggregated responses in banks of neurons. This is probably an unexceptionable hypothesis, since few would argue in favour of discrete "perceptual units", separately responsible for each perceptual experience. The implication though is that single neurons do not on their own analyse patterns, rather their function is to encode the spatial arrangement of the stimulus for subsequent stages of the visual process to interpret. Perceptual filters must be viewed as a data-base for vision, which is quite distinct from the decision rules.

5. Acknowledgement

James Anderson provided invaluable assistance with programming the psychophysical experiments, and acting as observer.

6. References

1. Movshon, J.A., I.D. Thompson, and D.J. Tolhurst. (1978) Spatial and temporal contrast sensitivity of neurones in areas 17 and 18 of the cat's visual cortex. J. Physiol. (Lond) 283, 101-120.

2. Blakemore, C., J. Nachmias, and P. Sutton. (1970) The perceived spatial frequency shift: evidence for frequency selective neurones in the human brain. J. Physiol., Lond. 210, 727-750.

3. Blakemore, C., J.P.J. Muncy and R.M. Ridley. (1973) Stimulus specificity in the human visual system. Vision Res. 13, 1915-1931.

4. Marr, D. (1982) "Vision", Freeman.

5. Marr, D. and T. Poggio. (1979) A computational theory of human stereo vision. Proc. Roy. Soc. Lond., B 204, 301-328.

6. Ullman, S. (1979) "The Interpretation of Visual Motion". MIT Press.

7. Lamar, E.S., S. Hecht, S. Schlaer and C.D. Hendley (1947) Size, shape, and contrast detection of targets by daylight vision. J. Opt. Soc. Am. 37, 531-545.

8. Blackwell, H.R. (1946) Contrast thresholds of the human eye. J. Opt. Soc. Am. 36, 624-643.

9. Sullivan, G.D., M.A. Georgeson and K.G. Oatley. (1972) Channels for spatial frequency selection and the detection of single bars by the human visual system. Vision Res. 12. 383-394.

10. Barlow, H.B. (1978) The efficiency of detecting changes in random dot patterns. Vision Res. 18, 637-650.

11. Green, D.M. and J.A. Swets, (1966) "Signal Detection Theory and Psychophysics". Wiley and Sons.

12. Baker, K.D., and G.D. Sullivan, (1980) Multiple bandpass filters in image processing. IEE Proc., 127 E(5).

13. Marcela, S. (1980) Mathematical description of the responses of simple cortical cells, J. Opt. Soc. Am. 70, 1297-1300.

14. Marr, D. and E. Hildreth (1980) Theory of edge detection. Proc. Roy. Soc. Lond. B 207, 187-217.

Part III

Early Stages of Image Interpretation

Visual Algorithms

T. Poggio

M.I.T., Artificial Intelligence Laboratory and
Department of Psychology, Cambridge, MA 02139, USA

1. Early Visual Algorithms

1.1. Algorithms Depend on Computation and Hardware

One can distinguish (or at least I did so with David Marr:see Marr and Poggio, 1976; Marr, 1982) at least three levels at which a visual processor must be understood. At the top level is the computational theory of the device in which the problem to be solved is characterized, and the natural constraints are made explicit. At the bottom is the level of the detailed neuronal "hardware" - neural circuits, synapses and so forth - that perform the computation. In the middle is a study of the algorithms used to compute the solution. This second level is the hardest to define precisely since it represents a bridge between the computational level and the hardware level. Thus, while the circuitry is determined by the available mechanisms and the computation by the nature of the problem, the algorithm itself is determined by the computation *and* by the available hardware.

David Marr has especially stressed the computational level of analysis since it is a level of explanation which is still new to neurobiology. Together we have stressed that the relationships between these levels are rather loose. In this meeting I find it especially appropriate to emphasize that it is hopeless to understand the algorithms used by a biological or artificial processor without knowing which computational problem is solved and what are the properties and the limitations of the hardware.

Both the mechanisms and the problem provide powerful constraints to the possible algorithms. Horace Barlow made this point very clearly when, in his Ferrier lecture (1981), he spoke about the "limiting requirements" imposed by the physics of light, i.e., the nature of the visual world, and the properties of the nervous mechanisms, for instance the limited precision of the connections and the noise of nervous transduction. For instance, the von Neuman architecture of classical computers depends almost entirely on the type of available processing elements which made concurrency cumbersome to implement. In the nervous system the processing elements - neurons and synapses, as I will discuss later - are abundant and flexible. VLSI is now bringing similar advantages to circuits. Connections, however, are still vastly more numerous and more flexible in the brain than in solid state electronics, where they are restricted to 2-D surfaces. The costs of internal communication are still exorbitant in today's computers.

It is therefore not surprising that algorithms strongly depend on the constraints imposed by the hardware. I would argue that the main reason for the large gap presently existing between computational theories and computer scientists on one end and physiology on the other end is our ignorance of the nature and the properties of the biological hardware performing the elementary steps of information processing. Biophysics of information processing, which I will discuss later, is as necessary for analyzing and understanding the algorithms used by the brain as the computational analysis of the specific tasks.

1.2. Systems and Algorithms

Visual information processing begins with a large array of photoreceptors that transduce local light intensities into time dependent signals. The information about the outside world and how it changes is implicit in this retinotopic array of signals and must be decoded by a variety of processes or algorithms. Formally these algorithms, considered as "black boxes", have many inputs – the photoreceptors – and are in general nonlinear.

Since I restrict my discussion to the first steps of visual information processing, I will consider algorithms that operate almost directly on the photoreceptor signals, i.e., on the primary intensity representation (Nishihara, 1981). At this level, and especially for biological systems, it is natural to treat algorithms as systems or operators mapping a set of inputs into a set of outputs. From this point of view, two simple dichotomies can characterize, albeit rather superficially, early visual algorithms: 1) linear vs. nonlinear; and 2) parallel vs. serial.

Systems have inputs and outputs. Mathematically a system is equivalent to an operator which maps functions into functions (in a suitable space). An operator can be defined in at least two ways: (a) as a catalogue of all the inputs and the corresponding outputs; and (b) as an algorithm, i.e., an explicit law or set of rules that enables one to compute the output for any given input.

These two descriptions are met under different forms in various contexts. In information theory a view close to (a) leads to the classical combinatorial definition of information measure, while an algorithmic view leads to the Kolmogorov's definition of information entropy. In computer technology logical operations are often performed through a look-up table, i.e., a catalogue.

1.3. Linear vs. Nonlinear

Nonlinear systems represent in the space of all systems a much larger class than linear systems. The restriction of linearity is very strong and sets powerful general constraints on the system's behavior and properties. A linear system is a map L satisfying

$$L(ax) = aL(x)$$

$$L(x_1 + x_2) = L(x_1) + L(x_2)$$

From the information processing point of view, the limitations of linear systems are clear. Linear operations cannot perform conjunctions and discriminate the intersection of events. In a sense which can be made more precise multiplications or divisions are necessary to provide a sufficiently powerful set of basic operations. The crucial processes of an information processing device require logical operations that are essentially nonlinear and more like multiplications than addition or subtraction.

For a computer scientist this question of linear vs. nonlinear algorithms may indeed seem a straw problem: after all every computing machine is intrinsically nonlinear, is full of nonlinearities. For neurobiologists, however, the question of linearity vs. nonlinearity of some nervous subsystem – and of the operation thereby implemented – is non trivial. As we will see later, the classical view of the neuron and the concept of integrative action may be flawed by the failure to recognize the necessity of nonlinear operation on graded synaptic inputs.

1.4. Parallel vs. Serial

The distinction between parallel vs. serial processing is almost a commonplace in a variety of different areas (of course the concepts of parallel and serial processing have a range of meanings). It is, for instance, the most frequently cited difference between present computers and brains. The amount and the nature of wiring are vastly different: it is easy to create many very small transistors with present solid state technologies but more difficult to produce extensive connections among them. In a brain each nervous cell receives thousands of inputs. Nervous wiring is not restricted to two-dimensional surfaces. Especially in the first parts of the visual pathway nervous processing is indeed undoubtedly spatially parallel and preserving the topography of the image (at least up to area 17 and other visual areas, modulo a conformal mapping that preserves local geometry).

Algorithms with a more serial flavor are certainly used at later stages in the nervous system. Even in vision, however, serial processing is likely to play an important role quite early on, certainly earlier than most neurobiologists accustomed to the idea of topographic maps and inner screen would be ready to admit.

From a computational point of view the most interesting issues about parallel algorithms revolve around the notion of "local and global" or "parts and wholes". Loosely speaking, a computational problem is inherently local if it can be divided into small, non interacting modules. It is inherently global if any way of dividing it into subcomponents must entail substantial interaction among the modules.

1.5. Plan of the paper

It is, however, very difficult to characterize the locality or globality, as well as the linearity or nonlinearity, of a computational task without any reference to a specific and possibly abstract class of machines on which the computation will run. I will thus consider a class of algorithms and related abstract machines that are a natural extension of Perceptrons and briefly formalize in this framework the issue of *local and global* as well as of *linear and nonlinear*. The motivations for considering this particular class of algorithms is more fully discussed by Poggio and Reichardt (1980). The main attraction of polynomial algorithms is their generality: they approximate, under rather weak conditions, all *smooth* input-output transductions. Heuristically, several early visual computations seem indeed to have this smooth character of transducing input signals continuously into output functions, without major discontinuities or "decisions". The main results on polynomial algorithms are summarized in section 2 , mainly from Poggio and Reichardt (1980), to which we refer the reader for detailed definitions, proofs and references. Some of the results on coding of the input set (section 2.8) and nonlinear associative mappings (section 2.9 and 2.10) are new. In the context of this paper the main thrust of the next section is to show formally that several simple but important computations require local (i.e., highly parallel) algorithms with nonlinearities of a simple kind. The question is then how could these interactions be implemented in neuronal hardware? Section 3 suggests that a specific biophysical mechanism may perform the key operation in many local, nonlinear visual algorithms. The section is a brief and incomplete summary of several recent papers. Three examples of specific algorithms based on this mechanism are proposed in the last section for (a) detecting directional motion, (b) separating figure from ground and (c) localizing zero-crossings . It is conjectured that the corresponding circuitries may indeed be used by some biological visual systems. The first two example are summarized from Torre and Poggio (1978) and Poggio et al. (1981a), respectively. The neuronal circuitry for the detection of zero-crossings is original. It may be useful to point out that this paper is not fully selfcontained. Its main purpose is rather to outline several recent developments and to connect them in a new and coherent framework: the many gaps and missing details should be filled in from the original papers.

2. Polynomial Algorithms

As I mentioned, the input space relevant for us is a 2-D array of time-dependent signals. We formalize this, defining as *retina* the collection of N photoreceptors arranged on a 2-D lattice, and by a pattern on the retina a set of input functions, seen by the photoreceptors. Then a polynomial algorithm on the pattern X is a mapping to an output function with the form

$$S[x] = \sum_{i=1}^{N} \sum_{j_1 \dots j_i} L_{j_1 \dots j_i}[x(t)] \tag{1}$$

where $L_{j_1 \dots j_i}$ is an i-linear form in the i components of the input array $[x_1(t) \dots x_N(t)]$. Eq. 1 is a natural extension of the usual algebraic polynomials: inputs and outputs are here *functions* instead of real (or complex) numbers.

2.1. Inputs

When $x_i(t) = x_i$, the input pattern is a grey-level "figure" on the retina. If the x_i take only 0 and 1 values the pattern is essentially a "geometric figure" as considered in "Perceptrons" by Minsky and Papert (1969). Since a linear transformation of the pattern does not change the type of representation (eq. 1), it is often useful to think of the pattern as a (linearly!) filtered version of the brightness array (for instance through a difference of Gaussians operator, a DOG).

2.2. Graphs

Eq. 1 is equivalent to the decomposition of the operator S into the sum of interactions of different orders between the input functions. Each term can be represented by a certain graph and thus an N-input system can be decomposed into a sequence of graphs (see Fig. 1). The graphs are actually another notation for the polynomial operator itself. In particular, composition of systems can be computed directly in terms of the graphical notation.

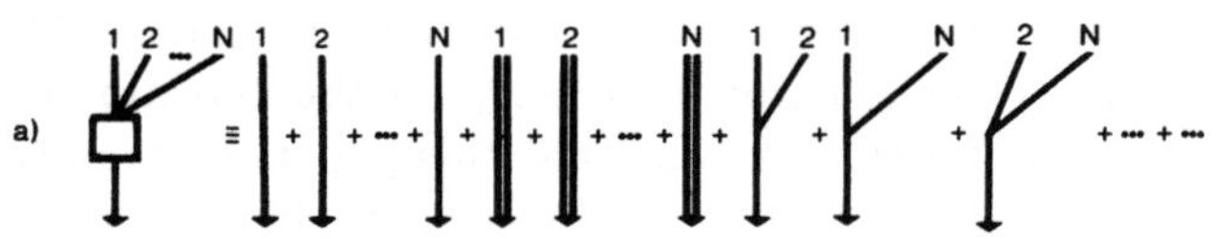

Figure 1.
The graphic representation for equation 1

2.3. Three Questions

Three interesting questions can be asked about these operators. The first one concerns the existence of an explicit representation. The second problem is how wide is the class of algorithms that can be approximated by polynomial systems? Finally, we would like to characterize the computational properties and limitations of polynomial algorithms, especially in the framework of the local-global and linear-nonlinear dichotomies.

2.4. Representation and Approximation

(a) The answer to the first question (see Poggio and Reichardt 1980; Palm 1978, 1979) can be stated in terms of the following

PROPOSITION 1 (see for precise versions of this theorem Palm and Poggio 1977; Palm 1978).

All polynomial systems can be represented by symbolic integrals. Time invariant polynomial systems can be represented by Volterra series (with the kernels being distributions).

Thus the class of Volterra series coincides with the class of polynomial functionals and a polynomial algorithm (or its graph) can be written in terms of integrals and associated kernels. For instance, a linear time-invariant system can be written as a convolution

$$L_1[x(t)] = \int K(t-\tau)x(\tau)d\tau \tag{2}$$

and a second order interaction as

$$L_{12}[x_1(t), x_2(t)] = \int\int K_{12}(t-\tau_1, t-\tau_2)x_1(\tau_1)x_2(\tau_2)d\tau_1 d\tau_2 \tag{3}$$

(b) The answer to the second question depends on the type of approximation, i.e., topology, that is used (see Palm and Poggio, 1978;Palm 1978, 1979; Poggio and Reichardt, 1980). For instance, under the stochastic approximation and the pointwise approximation – rather weak approximations – polynomial systems like equation 1 approximate essentially every mapping between L^2 and R, but only a subclass of continuous mappings under the uniform approximation. Stronger results clearly hold for discrete-time systems and for discrete time, finite value input spaces. For discrete time systems all different topologies are equivalent and thus all continuous systems can be approximated by polynomial systems. If the input set is finite, then any mapping can be written as a polynomial.

2.5. Computational Properties: degree and p-order

A representation like eq. 1 is essentially a canonical decomposition of the system into the sum of simpler, standard components. Computational properties of the mapping are then "additively" determined by the computational properties of the standard components, the graphs of Fig. 1. Typically one would like to know what is the "simplest" set of graphs or interactions that can perform a given computation. Intuitively, the terms in the representation of Fig. 1b (or eq. 1) become increasingly complex going from left to right. It turns out that the concept of simplicity of a graph can be formalized in terms of the notion of degree, which measures its "nonlinearity", and p order, which measures its "locality".

DEFINITION

A canonical graph is of degree k, if it corresponds to a k-linear form.

The degree of a graph is the total number of incoming lines. Linear graphs have degree 1, quadratic ones (bilinear forms) have degree 2.

DEFINITION

A canonical graph has p order h, if it has inputs from h distinct "photoreceptors".

DEFINITION

The degree and p order of a polynomial system are the maximum degree and p order in the graphs of its canonical decomposition.

Although degree, p order, and rank (see Poggio and Reichardt, 1980) all characterize a polynomial algorithm, the p order is probably the single most important measure of the simplicity of a graph [linear graphs have p order 1 (and degree 1)]. The main reason is that the notion of p order formalizes the issue of "local vs. global" for polynomial algorithms. A lower p-order system is local: the canonical subsystems make independent, nonlinear computations based on small patches of the retina. A high p-order algorithm is global: individual graphs receive inputs from many photoreceptors in the retina. One can ask, similar to "Perceptrons", whether a certain computation is of finite p order, i.e. if it can be computed by a polynomial mapping of some fixed p order, regardless of the size of the retina. Notice that a linear algorithm may use inputs from many photoreceptors, but its canonical representation consists of p-order 1 graphs. Linear operations on the input patterns do not change p order (or degree) of an algorithm (see later).

2.6. Computational Properties: Standard Machines

Several abstract computational machines could be considered for a comparison, like finite state machines, McCulloch-Pitts networks, difference equations, perceptrons, etc. The comparison clearly depends on the input set (most machines are defined only for discrete inputs with a binary number of values). If we restrict ourselves to a discrete, finite set the previous discussion implies that it is always possible to synthesize a polynomial mapping that simulates exactly the behavior of any specific machine. If we relax the constraint of a finite set, polynomial mappings are less powerful than systems with infinite memory like finite state machines, difference equations, and McCulloch-Pitts networks with loops. In practice, time is always finite and thus polynomial algorithms are equivalent to these other machines. The situation, however, clearly indicates that a polynomial description, for instance, of a finite state machine may almost always be too cumbersome to be useful. These limitations of polynomial algorithms are also indicated by their equivalence with standard perceptrons (on appropriate patterns).

Computation	p-order	Computation	p-order
⌈X is a circle⌉	4	Translation invariant recognition of figures	3
⌈$\|X\| \geqq M$⌉	1	⌈X is odd⌉	$\propto(\|R\|)$
⌈$\|X\| \leqq M$⌉	1	$\exists\, \psi_1, \psi_2$ p-order $(\psi_1) = 1$ p-order $\psi_2 = 2$: $\psi_1 \wedge \psi_2$	$\propto$
⌈$\|X\| = M$⌉	2	⌈X is connected⌉	$\propto(> c\|R\|^{1\ 2}$
⌈The n-th central moment of X about the origin is $> \vartheta$⌉	1	Translation invariant geometric spectra	2
All Boolean functions of 2 variables, except $x \oplus y$ and $x \equiv y$	1	ψ in context	$\propto$
$x \oplus y$	2	⌈X is a hollow square⌉	3
$x \equiv y$	2	⌈One component of X is a hollow square⌉	$\propto$
⌈X is convex⌉	3	⌈X has symmetry under reflection⌉	$\leqq 4$
		⌈Pattern A is a translate of pattern B⌉	$\leqq 5$

Figure 2. *P-order of some computations on geometrical figures*

2.7. Polynomial Algorithms are equivalent to Perceptrons for Geometrical Figures

The input set is restricted to "geometrical figures", i.e. to the set $[0,1]^N$ where N is the number of photoreceptors. A predicate on R is a function Ψ from a figure of R to $[0,1]$. A perceptron is a predicate on the retina R of the form (Minsky and Papert, 1969)

$$\Psi(R) = [\sum a_i \phi_i(R) > \theta]$$

where [some condition] if the is 1 condition is true and 0 if it is false. The support of ϕ is the set of all photoreceptors which affect the value of ϕ, and the order of ϕ is the size of the support of ϕ. It is straightforward to prove

Lemma

Any (perceptron) ϕ function of support n can be represented exactly for all figures by a set of polynomial graphs of p order n and degree $(2^n - 1)$.

THEOREM 2

For geometrical figures on $[0,1]^N$ *perceptrons of order* $\leq n$ *and polynomials of p order n and degree* $(2^n - 1)$ *are equivalent.*

Thus the results on the order of perceptrons which compute various geometrical predicates also apply to the p order of the corresponding polynomial mappings. Fig. 2 lists some of these results.

For both perceptrons and polynomial algorithms many simple computations turn out to be nonlocal. The limitations of perceptrons carry over to polynomial algorithms, and probably hold also for more general time dependent input patterns. This is not a surprise but is especially interesting from the point of view of the globality of given computational problems in vision. The implication is that simple, parallel algorithms, not just restricted to the Perceptron machine, can easily, i.e.,

Computation	p-order	Degree
Directional movement (translation invariant)	2	2
Directional movement (translation- and rotation invariant)	1 or 2	2
Relative movement'	2 or better 4	4
Absolute position of a stimulus on the retina (not invariant to any geometric group)	1	
Expansion or contraction of a texture (translation invariant)	≧4	
First order statistics of a (timeless) texture	1	High
Second order statistics of a (timeless) texture	2	High
Expansion-concentration of a texture (around the coordinate origin)	2	2

Figure 3.
P order of some computations on time dependent patterns

locally, compute a range of important predicates, but cannot compute all of them. Apparently simple computations – like connectedness, straightness – probably require different types of algorithms, perhaps more "serial". Other computations, however, such as the computation of the direction of motion and of discontinuities in the motion field, can be performed by simple parallel algorithms (see fig.3). It may be interesting to consider the various processes involved in the first stages of visual perception from this point of view. In particular, Treismann's notion of separable and not separable features, Julesz' concept of instantaneous perception and Barlow's idea of topographic maps may be connected to the intrinsic "locality vs. globality" (and nonlinearity) of algorithms.

2.8. Coding of The Input Set

I mentioned earlier that linear transformations of the input patterns do not change the main characteristics of a polynomial algorithm (they simply change the kernel's values). More generally a transformation of the retina R_1 into R_2 by the mapping f provides for every pattern

$$X = [x_1(t) ... x_n(t)]$$

another pattern

$$Y = [y_1(t), ... y_n(t)] = f[X] = [f_1[X], ..., f_m[X]]$$

Given a polynomial mapping S_2 on R_2 we define a mapping S_2 on R_2 through

$$S_1(X) = S_2(Y) = S_2(f(X))$$

The following property is self-evident (Poggio and Reichardt, 1980)

PROPERTY

If for $i = 1...m$ the support of f is at most equal to 1, then p order$(S_1) \leq$ p order(S_2).

Thus nonlinear scaling of each input separately does not increase the p order of a polynomial algorithm. Coding of this type will, however, change the order and possibly decrease it. In other words, this simple input coding may make a polynomial algorithm simpler without changing its essential properties. We can then consider the class of polynomial algorithms defined by transformation of the retina of support 1. Two principles can be used to guide the choice of appropriate transformations (Resnikoff, 1975).

Principle 1

The domain and the range of the induced polynomial algorithm must coincide with the domain and range of the input-output operation (considered as function on $\mathbf{R}$).

Principle 2

The transformations of the retina shall be birational functions, the exponential functions and its inverse and the compositions of these functions.

Thus the transformations f_i of the retina affect the necessary conversion of domain and range with a minimal disruption of the algebraic structure of the input space . The extension from functions f_i to operators seems quite natural, when the inputs are time-dependent.

Example

The operation $[x_2$ vetoes $x_1]$ can be represented as a polynomial algorithm, with $0 \leq x_1 < \infty$, $0 \leq x_2 < \infty$. For fixed x_1, the output y is $0 < y \leq M$ for $\infty < x \leq 0$; for fixed x_2, the output y is $0 < y < \infty$ for $0 < x < \infty$. Thus $f_1(x_1) = x_1$ and $f_2(x_2) = \frac{1}{(a+x_2)}$. The polynomial algorithm has p order 2 and degree 2 on the transformed input; it would have a much higher degree without the transformation of the retina.

Input coding of this type simply tries to "linearize" the computation as much as possible. Input coding with support greater than 1 changes the p order of the polynomial algorithm. Output coding also changes the properties of the algorithm. This can easily be seen in Fig.4 which shows Kolmogorov's (1963) solution of Hilbert's 13th problem: a continuous function can always be represented as the superposition of functions of 1 variable, as shown (for 3 variables) in the figure. Thus with appropriate output and input coding a p order 1 polynomial can represent all continuous mappings. Interestingly, this result does not hold for continuous time functionals (Palm, 1978).

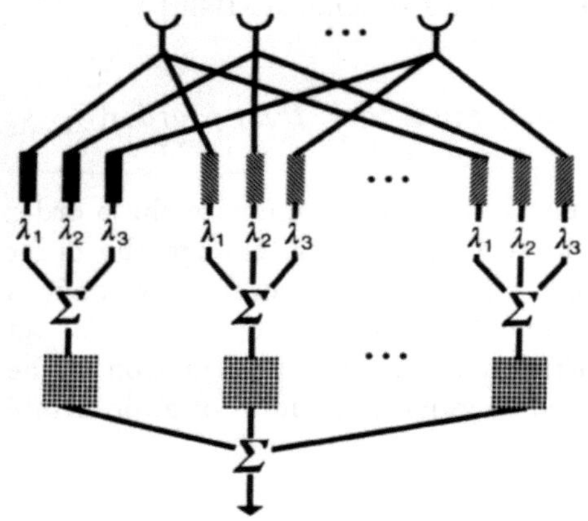

Figure 4. *Kolmogorov's and Arnold's decomposition of a function of three variables*

2.9. Associative Memory and Synthesis of a Polynomial Algorithm

How can one synthesize (by 'learning') a non linear polynomial algorithm from a set of given inputs and desired outputs? In the case of a finite, discrete space of inputs the problem is well known. It is equivalent to the problem of optimal estimation of a system and to the problem of synthesizing an associative memory (Poggio, 1975 a,b). The usual problem considered in the literature concerns the synthesis of an optimal *linear* mapping M from a set of input vectors X and a set of desired output vectors Y. The best approximate mapping is given by $M = YX^+$ where X^+ is the pseudouinverse of X (see Kohonen, 1977). This technique can be extended to find the optimal polynomial mapping (Poggio, 1975 a,b). In particular, then, any mapping can be constructed in this way, at least in principle, since any mapping between two finite sets of vectors can be written as a polynomial (see earlier). This can also be derived from another interesting result, that I discuss next.

2.10. Associative Memories and Nonlinearities

The role of nonlinearities in an associative memory scheme has been long recognized as critically important. The following simple theorem provides a general connection between linear associative schemes and nonlinearities.

PROPOSITION 3

Any nonlinear associative mapping between two finite sets of vectors is equivalent to a linear associative mapping preceded by (nonlinear) input coding.

PROOF

The proof is based on the following two obvious lemmas (given, implicitly, in Palm 1978; Poggio, 1975b).

LEMMA 1 Any (nonlinear) mapping between two finite sets of vectors can be written as a polynomial.

LEMMA 2 Any polynomial mapping between two finite sets of vectors

$$Y = L_0 + L_1[X] + L_2[X,X]...$$

is a linear mapping on appropriate crossproducts of the elements of X (i.e., on the tensor products X, $X \otimes X$,...).

The synthesis of these crossproducts can be seen as nonlinear input coding (of support >1). [Output coding instead of input coding has similar properties (see Poggio 1975b)]. A simple but striking result follows again:

COROLLARY

Any nonlinear mapping between two finite sets of vectorscan be synthesized associatively with the pseudoinverse technique.

For more general input sets the problem cannot be answered exactly but only in terms of approximate associative mappings. The idea of coding an input set before performing the bulk of the computation or association is clearly powerful and can be found in a variety of contexts. In the framework of associative memory it is again interesting to notice the connection with the Kolmogorov result: input coding here does *not* depend on the mapping to be synthesized but the final "output" function g does. The result shows that for a continuous function an n-dimensional table can be replaced by a 1-dimensional table representing g and some input coding. This does not necessarily reduce the memory requirements in all cases. However, it may be conjectured that an appropriate choice of the input coding for a specific class of input-output operations may allow significant reductions in memory size (see Poggio and Rosser, 1982).

In summary, polynomial algorithms are powerful parallel computational devices and it is important to stress not only their limitations but also their locality in a number of important computations. They may be useful for characterizing simple parallel processing operations in a visual system. The detection of motion and relative motion can be characterized in terms of simple polynomial algorithms and general properties can be proved for a whole range of specific models. As discussed earlier, however, the algorithms used by a system are strongly constrained by the available hardware. In the next section I will briefly discuss a class of mechanisms – local interactions between synaptic inputs – possibly used by the nervous systems. I will also show that these interactions compute in fact specific polynomial functionals of the inputs. Thus the link between polynomial algorithms and nervous hardware may turn out to be, at least in some instances, rather direct.

3. Biophysics of Information Processing

Many different types of algorithms could be used in early vision. The computational problem does not provide sufficient constraints to uniquely define the algorithm. Even properties like local vs. global may vary for the same computation between different classes of algorithms. Ultimately the hardware of the computer or of the brain imposes critical limiting factors that constrain the class of algorithms. What is then the hardware of the brain? Where are the elementary operations performed and what do they look like?

The traditional view is that the threshold mechanism associated with spike generation performs the elementary logical operations: a neuron fires if the sum of its inputs exceeds a certain threshold and is otherwise silent. All logical operations can be implemented in this way, via McCullogh-Pitts networks.

It is, however, clear by now that there are probably several other mechanisms as important or more important than the McCullogh-Pitts neuron. For instance, it is now well recognized that much processing takes place without somatic spikes, simply in terms of graded potentials. If graded signals play an important processing role, there must be *nonlinear* interactions between synaptic signals. The need for nonlinear operations that are more like multiplication than addition or subtraction has been customarily neglected by most neurophysiologists but is clearly critical for even the first stages of visual information processing.

A simple biophysical mechanism that could underlie nonlinear interactions between graded signals is already known. Since synaptic inputs are not current inputs but conductance changes to specific ions, synapses which are electrially close to each other on a cell's dendrite will mutually influence each other and result in a potential change at the soma which depends nonlinearly on the input signals. Probably the simplest and most common interaction of this type involves two synapses (or sets of synapses), one excitatory and the other inhibitory, increasing conductance to an ion with a battery close to the cell's resting potential. Activation of the inhibitory channel by itself will contribute nothing to the potential, but it may have a very powerful effect in shunting the potential towards the resting state when a neighbouring excitatory synapse becomes active. This shunting effect can be powerful and local. It can also be shown from the membrane's equations (Torre and Poggio, 1978; Poggio and Torre, 1981) that the interaction implemented is multiplication-like, of the type $g_1 - \alpha g_1 g_2$. This is in turn formally equivalent to an 'analogue' AND-NOT operation, one input (g_2) *vetoing* the other (g_1).

3.1. Synaptic interactions are polynomial functionals

The multiplication-like character of these synaptic interactions can be indeed demonstrated rigorously. An extension of cable theory shows that the voltage potential in an arbitrary dendritic structure is given by a specific Volterra series of the conductance inputs.

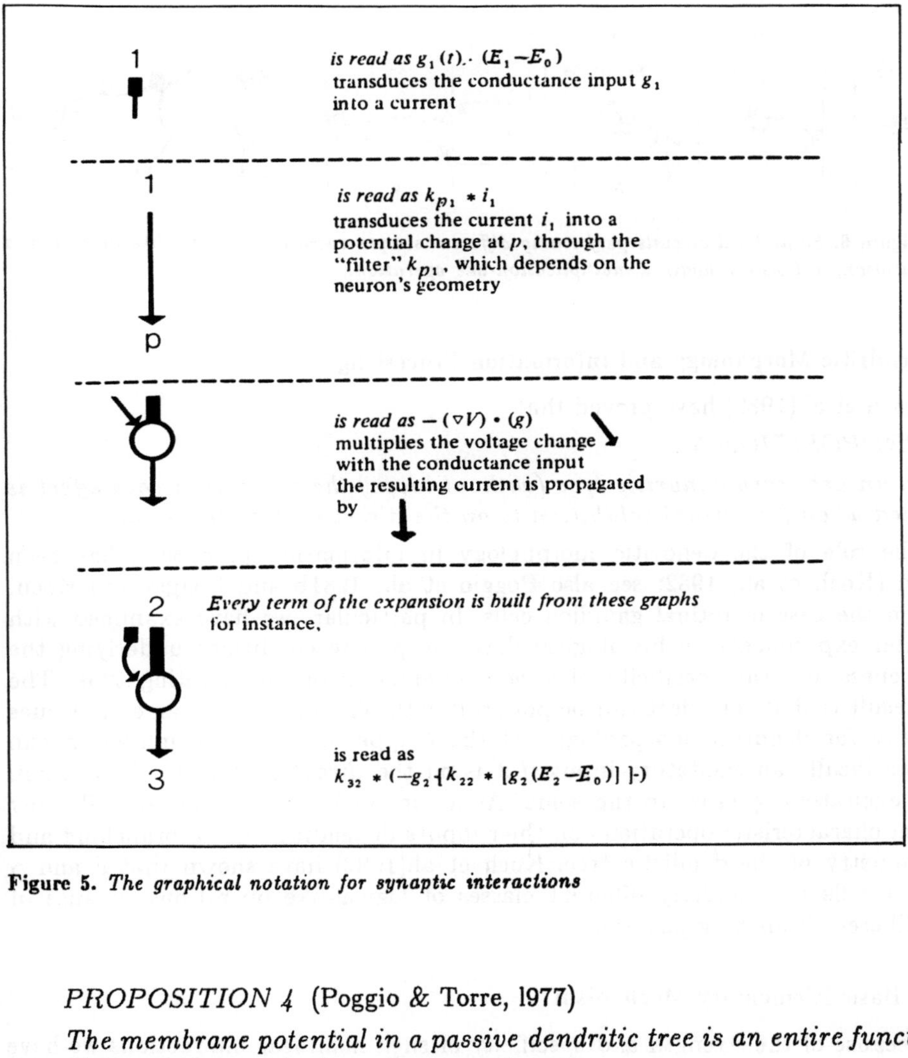

Figure 5. *The graphical notation for synaptic interactions*

PROPOSITION 4 (Poggio & Torre, 1977)

The membrane potential in a passive dendritic tree is an entire functional for all bounded, transient conductance inputs.

Thus synaptic interactions of this type implement a specific type of the polynomial algorithms described earlier (their Volterra kernels have a specific structure). The polynomial structure of the interactions can be represented again by a graphic notation, where the various graphs give multiplication-like interactions of increasing degree and p order between conductance changes (see fig.5).

This mathematical analysis begins to connect the morphology of the dendritic tree and the geometry of the synapses with the algorithms thereby implemented. Fig.6 shows several local circuits implementing simple nonlinear operations.

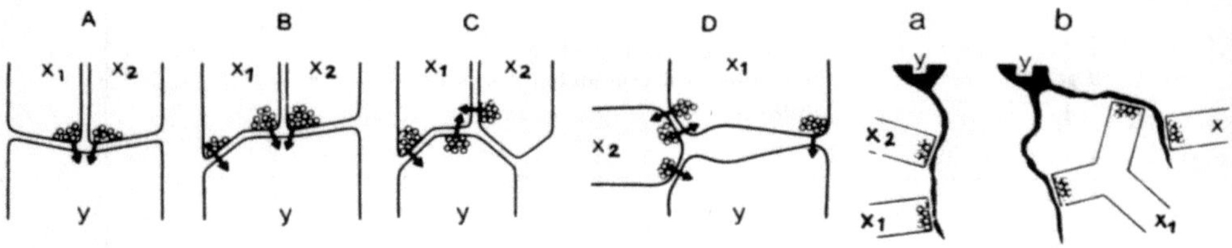

Figure 6. *Some local circuits performing different simple operations: a veto-like operation, a multiplication, a division, again a multiplication-like operation.*

3.2. Dendritic Morphology and Information Processing

Koch et al.(1982) have proved that

PROPOSITION 5

In an arbitrary dendritic tree (without loops) the maximum veto effect is obtained when (shunting) inhibition is on the direct path to the soma.

The role of the dendritic morphology in information processing has been studied (Koch et al., 1982; see also Poggio et al., 1981b and Poggio and Koch, 1981) in the case of retinal ganglion cells. In particular, we have examined with computer experiments on histological data the precise conditions underlying the effectiveness and the specificity of a *veto* interaction of the shunting type. The main result is that the effect can be powerful with physiological parameters values especially for dendritic morphologies of the δ type. In this case inhibition can veto specifically an excitatory input *if* it is on the *direct path* from the location of the exçitatory synapse to the soma. As a consequence each class of cells may perform characteristic operations on their inputs depending on the branching and the geometry of the dendritic tree. Koch et al.(1982) have shown that δ and γ ganglion cells may underly different classes of logical-like operations because of their different branching patterns.

3.3. A Basic Elementary Mechanism

Because of the strength and specificity of such nonlinear interactions we have proposed that they may perform characteristic information processing operations in passive dendritic trees. Since inhibition *vetoes* effectively more distal excitatory inputs only when it is *on-path* to the source a variety of local operations can be performed, exploiting the branching geometry of a dendritic tree with a suitable localization of excitatory and inhibitory inputs. If this is true, a neuron would probably resemble an analogue LSI circuit with thousands of elementary processing units - the synapses - rather than a single logical gate. The idea that a veto- like operation plays an important role in visual information processing in the brain is not new, though its specific synaptic nature and properties probably are: Barlow has stressed many times, since his classical study of direction sensitive ganglion cells,

(Barlow and Levick, 1965), that a veto-like operation is an important physiological mechanism in the retinal and cortical processes that underly perception. In the next section I will propose several simple visual algorithms—and corresponding neuronal circuits—that use this elementary veto mechanism.

4. Three Examples of Nonlinear, Local Algorithms

4.1. Direction selective Motion Detection

The computation of local motion – in the simple sense of detecting the direction of motion – is a simple but fundamental step in machine and biological vision systems. It is straightforward to show that it is a nonlinear computation – i.e., linear operations alone could never be said to be truly direction selective. In terms of polynomial algorithms direction selectivity requires at least p order 2 , i.e., a multiplication-like operation between 2 photoreceptors (or second-order 'cells' linearly filtering the photoreceptor array). A linear system cannot give a time-averaged output that inverts sign for inversion of direction of motion, since $< L[x_1, x_2] >= L[< x_1 >, < x_2 >]$, if L is a linear mapping. A p-order 2 polynomial algorithm with a non zero antisymmetric kernel component has the correct property. Thus:

PROPOSITION 6 (Poggio & Reichardt, 1976, 1981)

Direction selective motion detection – the average output must reverse sign for inversion of direction of motion – is p order 2 (and degree 2).

In the specific case of the visual system of the fly there is convincing evidence that the p order of the algorithm used is indeed 2 and not higher. The evidence rests on a variety of experiments briefly reviewed in (Poggio and Reichardt, 1976). The main ones are:

(a) the optomotor response depends in a characteristic way on the phase of 2 adjacent, flickering bars of light

(b) the average optomotor response is blind to the phase of the spatial Fourier components of a pattern moving at constant speed.

(c) the response to a moving grating does not depend on velocity *w per se* but can be factorized into a term depending on the wavelength λ and a term depending on the contrast frequency $\frac{w}{\lambda}$.

(d) the resolution limit of the motion detection mechanism.

(e) the dynamics of the response to oscillating patterns.

Thus the basic algorithm for direction selective motion detection is based on a multiplication-like interaction between pairs of inputs after asymmetric filtering

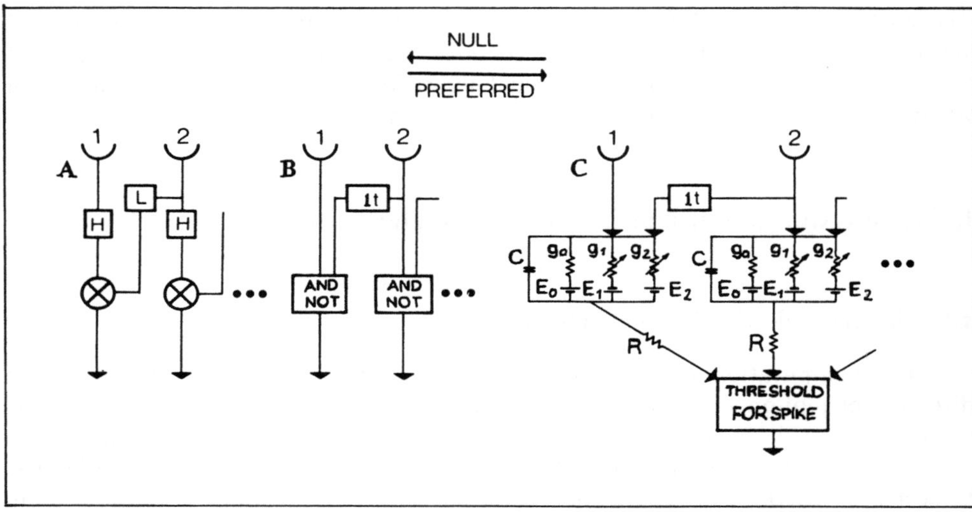

Figure 7. *Reichardt and Hassenstein's movement detector model, Barlow and Levick's scheme and Torre and Poggio's synaptic mechanism.*

(low-pass but also high-pass filters are possible). A particularly simple filtering operation is an asymmetric delay (see fig. 7). It is easy to prove that correlation models are equivalent to p-order 2 polynomial algorithms:

PROPOSITION 7 (Poggio & Reichardt, 1976)

Correlation models of motion detectors (in the sense of Reichardt) are a subclass of antisymmetric, p-order 2 polynomial algorithms.

How can this algorithm be implemented in neural hardware? If we follow the ideas outlined earlier, the obvious choice would be to use a synaptic mechanism of the veto type at the level of a cell's dendrite. As shown in fig. 7, Barlow and Levick had in fact proposed from their physiological experiments on directional selective ganglion cells, an AND-NOT operation as the basis for directional selectivity to motion. Torre and Poggio (1978) conjectured that the synaptic veto effect described earlier may be the mechanism whereby directional selectivity is achieved. Provided that suitable conditions on the ionic channels, the geometry of the dendritic tree and the localization of synapses are satisfied, their conjecture certainly fits two of the main experimental properties of direction selective cells, namely that the interactions responsible are between local subunits of the receptive field, and that they are inhibitory. Thus the basic algorithm used for simple motion detection in various biological systems may indeed by based on the synaptic veto mechanism. In particular, Koch et al. (1982) have recently proposed that a δ-cell-like morphology is the substratum of direction selectivity in the retina of the cat.

4.2. Detection of Relative Movement

Discontinuities in the optical flow field—the distribution of apparent velocities on the eyes—are a good indication of object boundaries and can be used to segment images into regions that correspond to different objects. In particular, the relative motion of an object against a textured background can be used to reveal its presence and to delineate its boundaries. The human visual system is very efficient at this task. Quite similarly a fly is able to detect and discriminate an object that moves relative to a ground texture.

In terms of polynomial algorithms this computation is of p order 4 (although p-order 2, degree 4 may also be sufficient in specific cases, see Reichardt and Poggio, 1979). Many experiments have established that the fly indeed uses an algorithm which is mainly p order 4 (it has also higher terms). More precisely, the behavioral data which measure the fixation response of the fly to a textured small figure oscillating sinusoidally with various phases in front of an oscillating ground texture shows that the basic algorithm relies on an inhibitory multiplication - like operation between motion detector units (Reichardt and Poggio, 1979).

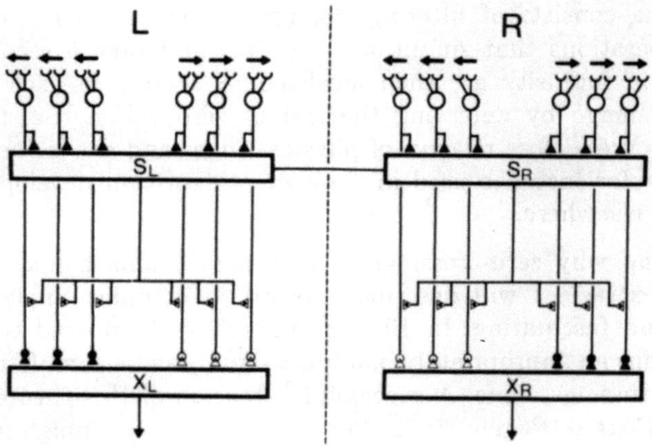

Figure 8. *The circuit (and algorithm thereby implemented) possibly used by the visual system of the fly to perform the detection of discontinuities in the optical flow. Redrawn from Poggio et al., 1981*

Again the synaptic veto mechanism of shunting inhibition seems an ideal candidate for implementing this operation. The overall circuitry is shown in fig. 8 (Poggio et al., 1981a). Large field cells summate the output of many elementary motion detectors and inhibit via presynaptic shunting inhibition the single elementary motion detectors. This circuitry accounts well for a large body of existing behavioral experiments; many more predictions have been successfully tested. In particular, the dynamics of the fly's behavioral response is quantitatively predicted by this algorithm for a variety of figure-ground descrimination tasks. Presynaptic inhibition with an equilibrium potentical near the resting potential is conjectured to implement the key operation in the algorithm, which amounts to a comparison of large field motion with local measurements. This circuitry, considered as an algorithm for detecting discontinuities in the optical flow (it has p order 4 and higher), is efficient and reliable, as shown by several computer experiments on textured patterns.

4.3. The Detection of Zero-crossings

Over the past twenty years researchers in computer vision have proposed several algorithms to detect and represent various kinds of intensity changes. I will focus here on one of them because of its potential implications for cortical processing. The basic ideas were suggested to D. Marr and myself (Marr and Poggio, 1977, 1979), while working on the problem of human stereo, from a combination of psychophysical data (by H. Wilson) and of recent results in the field of complex analysis . Briefly the scheme consists of filtering the image through a number of independent bandpass operations that simultaneously blur and take a second spatial derivative. Changes in intensity are then localized separately in each of the filtered versions of the image by detecting the loci of zero values, i.e., the zero-crossings. Zero-crossings are a close relative of physical edges and can be used for later processing; they are for instance used in the stereo algorithm developed at MIT (Grimson, 1981) and elsewhere.

To help in understanding why zero-crossings in bandpass channels may be useful discrete symbols to extract, I will describe a result in complex analysis that I still find intriguing and fascinating. In 1977 B. Logan (1977) proved that under some technical conditions an appropriately bandpass signal can be completely reconstructed from its zero-crossings alone. A successful extension of this theorem to images by Nishihara and Poggio (Poggio, 1982; Poggio et al., 1982), though not strictly valid for the channels described by H. Wilson, suggests that zero-crossings are very rich in information about the filtered image. Ideas based on Logan's type of results are attractive especially from the point of view of visual psychophysics and physiology, since they seem to provide a theoretical basis for the existence of edge detectors in the output of bandpass channels in the visual system, thus providing a potential synthesis of the edge detectors ideas with the frequency channels evidence. Marr and Hildreth (1980) have provided a number of attractive heuristic arguments for justifying a slight variation of the original scheme (Marr and Poggio, 1977). In particular they proposed that the initial filtering of the image was performed by nondirectional (as opposed to oriented) receptive fields, again described as differences of gaussians (DOG). Since X retinal ganglion cells have a DOG receptive field and are usually described as linear filters, it is not too unreasonable to propose that the filtering operation is indeed performed in the retina and represented by the activity in the ON and OFF layers of ganglion cells, positive values being represented by ON center X cells and negative values by OFF center X cells. Thus the binary map of the convolved image shown in fig. 9 would represent the combined map of activity in the OFF and ON layers of ganglion cells in the retina.

How can the zero-crossings - the transition of activity between ON and OFF cells – be detected?

Fig. 10 shows that a mechanism connecting neighboring ON and OFF cells with an AND gate, possibly implemented via synaptic mechanisms of the Poggio-Torre type (with a shunting conductance decreasing input, see Koch et al., 1982), could detect zero-crossing lying between the two rows of cells. This scheme, proposed by Marr and Hildreth (1980) does not require the inhibition which seems to be involved in the main properties of cortical cells, like orientation and direction selectivity. An alternative scheme can, however, be based on the synaptic veto mechanism.

The critical observation is that a zero-crossing is also defined by activity in the ON layer and *absence* of activity in neighboring ON cells (and conversely for the OFF layer). Thus a zero-crossing can be detected by avoidance of inhibition, logically equivalent to an AND-NOT operation. It is a simple matter to adapt this idea to create an oriented zero-crossing segment detector as shown in fig. 10. Since the veto operation can be performed by distal excitation (on spines ?) and inhibition of the shunting type on the proximal part of a single dendrite, the same cell may perform independently this operation on the OFF and on the ON layer on different dendrites, adding the two results for increasing reliability. Interestingly, however, either the ON or the OFF layer alone are sufficient. Notice that in a standard map of the receptive field inhibition may be invisible and only excitatory inputs from ON and OFF cells (on different dendrites) may be measureable (and linear). In this scheme unbalanced receptive fields (loosely corresponding to sustained properties) are not only advantageous but probably (as suggested by K. Richter) necessary for a robust physiological implementation. A trivial property of the circuitry follows from the preceeding sections:

COROLLARY

A veto-like zero-crossing detector can be regarded as a p-order 2 polynomial algorithm (on the array of ganglion cells).

With an appropriate transformation of the input array its degree can be as low as degree 2 (see earlier example). Thus this way of detecting zero-crossings is equivalent to taking measurements on the ganglion cells signals that are degree 2, p-order 2 (nonlinear) functionals.

This idea can be easily extended to account for directional selective properties, of some cortical cells. The first possibility is to gate the schemes of fig.10 (left) with an hypothetically transient Y-cell input. The resulting algorithm would be similar to

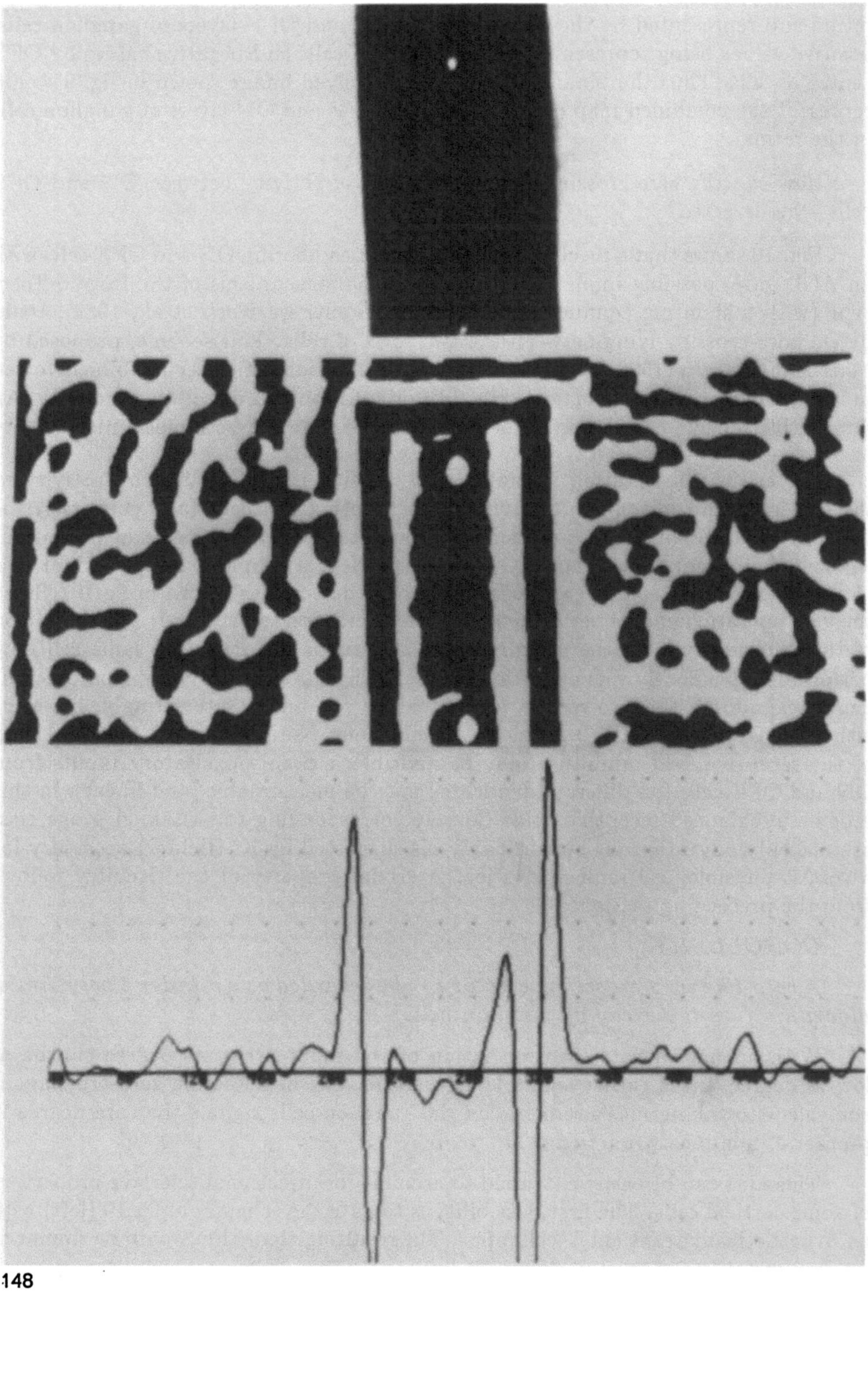

the scheme proposed by Marr and Ullman (1981), where all AND operations would be substituted with AND-NOT operations in the way suggested earlier. Another possibility is a scheme similar to the models of fig.7. As in the fly motion detector scheme, a low-pass operation on one of the two channels (or high-pass on the other) endows the zero-crossing detector scheme with direction selective properties (see fig. 11). Since precision is needed in the detection of the zero-crossing, the low-pass element must operate on the inhibitory input. It follows then that light edges can be detected only by the ON system and dark edges by the OFF system (if directional selectivity is required; otherwise there is no such restriction). This prediction may be supported by recent pharmacological experiments of P. Schiller. Interestingly, this algorithm is again very similar to the fly's (and Barlow's) movement detector,

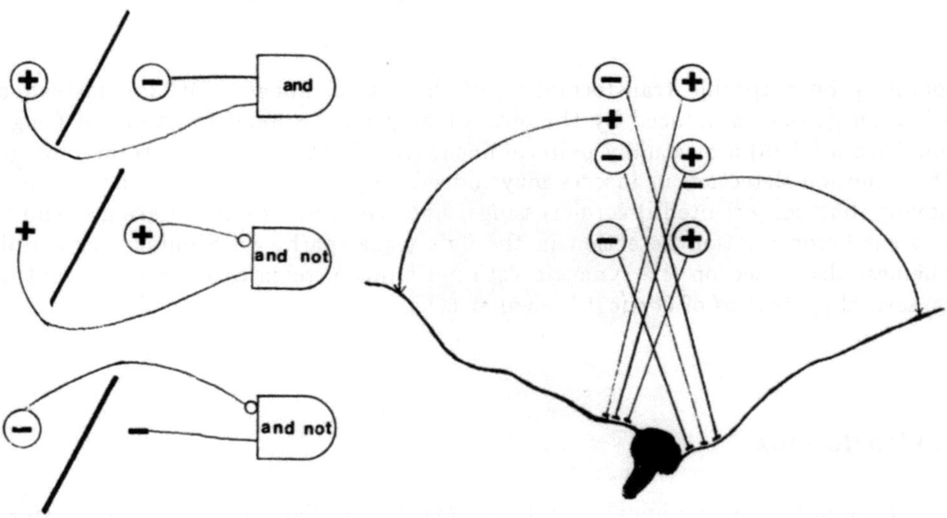

Figure 10. *Various models of zero-crossing detectors. In (a) a zero-crossing corresponding to transition of activity between the ON layer and the OFF layer (like the zero-crossing corresponding to the first edge from the right in Fig. 9) can be detected by an AND mechanism between adjacent regions of the ON and the OFF layer, as proposed by Marr and Hildreth (1980): either input alone is "invisible". In (b) and (c) the same zero-crossing is detected by an AND-NOT operation on either the ON or the OFF layer (circled inputs are "invisible", unless the other excitatory input is simultaneously active, as it is the case for so-called "silent" inhibition). All these operations may be performed by nonlinear, p-order 2 interactions on the ganglion cells arrays. A cell summating linearly these last two operations, performed on different dendrites, is sketched on the right side of the figure.*

← **Figure 9.** *The image of a dark piece of metal on a whitish background (top). The middle represent the sign of the convolution of the image with a center-surround type of receptive field (DOG). The filtering operation was performed by the M.I.T. convolver, developed by K. Nishihara and N. Larson (1981). The bottom graph shows a horizontal scan through the convolution array. Black would correspond to activity in the OFF ganglion cell layer in the retina and no activity in the ON ganglion cells, while white would correspond to the complementary activity pattern.*

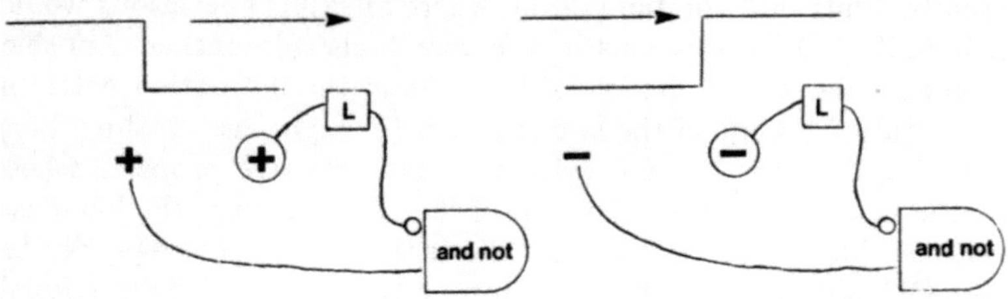

Figure 11. *A prediction: moving zero-crossings are detected – with directional selective properties – by the schemes sketched here: light edges by the ON mechanism, dark edges by the OFF mechanism. If a cell has to detect the same physical edge moving in both directions, then it may use both the ON and the OFF mechanism – on different dendrites*

operating on a specific transformation of the retinal image (DOGs), instead of the usual gaussians induced by the optics (the ρ point-spread functions of Geiger and Poggio (1975) are actually *general* linear transformations of the retinal image) Thus, motion detection in insects may indeed be very similar to the detection of moving (but non oriented !) zero-crossings, since center-surround filtering is known to occur before motion detection in the fly's visual pathway! Similar very simple schemes, also based on the synaptic veto mechanism, seem capable of accounting for several properties of cortical binocular cells.

5. Conclusions

To summarize, nonlinear, local and highly parallel algorithms can perform several simple but important visual computations. Specific classes of algorithms can be considered in an abstract way. I have studied the class of polynomial algorithms to exemplify some of the important issues for visual processing like linear vs. nonlinear and local vs. global.

Algorithms however depend strongly on the available hardware and not only on the computational problems. A study of the hardware of the brain is therefore needed not only for its own sake but also for its usefulness at the higher levels of understanding the nervous system. The synaptic veto mechanism based on inhibition of the shunting type may represent (consistently with Barlow's early suggestions) an important elementary information processing operation in the nervous system, thereby influencing the structure of visual algorithms. As specific examples I have considered three local, nonlinear algorithms based on the synaptic *veto* mechanism:

- direction sensitive motion detection, as (possibly) used in the vertebrate retina and in insect visual systems

- detection of discontinuities in the optical flow, as (possibly) computed in the fly's visual system
- detection and localization of stationary and moving zero-crossings, as (possibly) performed by a subclass of cortical cells.

If ideas of this type are correct, the neuron would be a parallel, asynchronous processor performing hundreds of analog operations – mostly of the veto type – on its inputs. This would not make any computation impossible but it would certainly have strong implications for the computational architecture of a visual processor.

Acknowledgement. This paper describes research done in part at the Artificial Intelligence Laboratory of the Massachusetts Institute of Technology . Support for the Laboratory's artificial intelligence research is provided by the Advanced Research Projects Agency of the Department of Defense under Office of Naval Research contract N00014-75-C-0643; this work was supported in part by National Science Foundation Grant MCS-79-23110 and travel for collaboration purposes was supported under NATO grant number 122.82.

References

Barlow, H.B. Critical limiting factors in the design of the eye and visual cortex. Proc. R. Soc. Lond. B 212, 1-34 (1981)

Barlow, H.B., Levick, W.R. The mechanism of directionally selective units in rabbit's retina. J. Physiol. 178, 477-504 (1965)

Geiger, G., Poggio, T. The Orientation of Flies towards Visual Patterns: On the Search for the Underlying Functional Interactions. *Biol. Cyber.*, **19**, 39-54, (1975).

Grimson, W.E.L.: From images to surfaces. M.I.T. Press (1981)

Koch, C., Poggio, T. and Torre, V., Retinal Ganglion Cells: A Functional Interpretation of Dendritic Morphology, Phil. Trans. R. Soc.B, 298, 227-264, (1982)

Kohonen, T.: Associative memory. A system-theoretical approach. Berlin, Heidelberg, New York: Springer (1977)

Kolmogorov, A.N.: On the representation of continuous functions of several variables by superpositions of continuous functions of one variable and addition (in russian). Dokl. Akad. Nauk SSSR 114, 679-681 (1957); Am. Math. Soc. Transl. 28, 55-59 (1963)

Logan Jr., B. F. Information in the Zero-crossings of Bandpass Signals, The Bell System Technical Journal, Vol. 56, 487-510 (1977)

Marr, D.: Vision. W.H. Freeman, San Francisco (1982)

Marr, D., Hildreth, E. 1980 Theory of Edge Detection Proc. Roy. Soc. Lond. B 207, 187-217 (1980)

Marr, D. C., Poggio, T. From Understanding Computation to Understanding Neural Circuitry. In: MECHANISMS IN VISUAL PERCEPTION, E. Poppel, R. Held and J. E. Dowling, eds. *Neurosciences Res. Prog. Bull.*, **15**, No. 3, pp. 470-488, 1977 .Also available as *A. I. Memo 357*, Massachusetts Institute of Technology Artificial Intelligence, May 1976.

Marr, D., Poggio, T. A Theory of Human Stereo Vision. *A. I. Memo 451*, November 1977, Massachusetts Institute of Technology, Artificial Intelligence Laboratory. Also avalable in a shorter version as A Computational Theory of Human Stereo Vision. *Proc. R. Soc. Lond. B*, **204**, 301-328, (1979.)

Marr, D., Ullman, S.: Directional selectivity and its use in early visual processing. Proc. R. Soc. Lond. B 211, 151-180, (1981)

Minsky, M., Papert, S.: Perceptrons. An introduction to computational geometry. Cambridge. Mass., London: MIT Press (1969)

Nishihara, H.K.: Intensity, Visible-Surface, and Volumetric Representations. Artif. Intell. 17, 265-284 (1981)

Nishihara, H.K., Larson, N.G. Proc. DARPA image Understanding Workshop 114-120 Science Application Inc., Arlington, VA (1981)

Palm, G.: On representation and approximation of nonlinear systems. Biol. Cybernetics 31, 119-124 (1978)

Palm, G.: On representation and approximation of nonlinear systems. Part II. Biol. Cybernetics 34, 49-52 (1979)

Palm, G., Poggio, T. The Volterra Representation and the Wiener Expansion: Validity and Pitfalls. *SIAM J. Appl. Math*, **33**, 195-216, (1977)

Palm, G., Poggio, T. Stochastic Identification Methods for Nonlinear Systems: An Extension of the Wiener Theory. *SIAM J. Appl. Math.*, **34**, 524-534, (1978)

Poggio, T. On Optimal Nonlinear Associative Recall. *Biol. Cyber.*, **19**, 201-209, (1975a)

Poggio, T. On Optimal Discrete Estimation. PROCEEDINGS OF THE FIRST SYMPOSIUM ON TESTING AND IDENTIFICATION OF NONLINEAR SYSTEMS, March 17-20, 1975, California Institute of Technology, Pasadena, California. G. D. McCann and P. Z. Marmarelis, eds., pp. 30-37 (1975b)

Poggio, T., Marr's Computational Approach to Vision, Trends in NeuroSciences - October, Vol. 4, No.10, pp.258-262, (1981)

Poggio, T., Trigger Features or Fourier Analysis in Early Vision: A New Point of View, In: The Recognition of Pattern and Form, ed.D. Albrecht, Springer, (1982)

Poggio, T. and Koch, C. Nonlinear Interactions in a Dendritic Tree: Localization, Timing and Role in Information Processing, A. I. Memo No.657, September 1981, Massachusetts Institute of Technology Artificial Intelligence Laboratory (1981)

Poggio, T., Koch, C. and Torre, V., Microelectronics in Nerve Cells: Dendritic Morphology and Information Processing, A. I. Memo No.650, October 1981, Massachusetts Institute of Technology Artificial Intelligence Laboratory (1981b)

Poggio, T., Nishihara, H.K. and Nielsen, K.R.K. Zero-crossings and Spatiotemporal Interpolation in Vision: Aliasing and electrical coupling between sensors, A.I. Memo No. 675, May 1982, Massachusetts Institute of Technology Artificial Intelligence Laboratory (1982)

Poggio, T., Reichardt, W. Visual Control of Orientation Behaviour in the Fly. Part II. Towards the Underlying Neural Interactions. *Quart. Rev. Biophys.*, **9**, **3**, 377-438, (1976)

Poggio, T., Reichardt, W. On the Representation of Multi-Input Systems: Computational Properties of Polynomial Algorithms. *Biol Cyber.*, **37**, 167-186, (1980)

Poggio, T., Reichardt, W. Characterization of Nonlinear Interactions in the Fly's Visual System. pp. 64-84, Appendix 4: A Polynomial Representation of Algorithms, pp. 197-202. In: THEORETICAL APPROACHES IN NEUROBIOLOGY, W. Reichardt and T. Poggio, eds., MIT Press, Cambridge, Mass., (1981)

Poggio, T., Reichardt, W., Hausen, K. A Neuronal Circuitry for Relative Movement Discrimination by the Visual System of the Fly. *Naturwissenschaften*, **68**, 443-466, (1981a)

Poggio, T. and Rosser, B., The Computational Problem of Motor Control, In: Machine Intelligence 10, eds Hayes, Michie, Pao (1982)

Poggio, T., Torre, V. A new approach to synaptic interaction. In: Lecture Notes in Biomathematics 21. Theoretical Approaches to Complex Systems, Proceedings, Tubingen, June 11-12, 1977. Eds. R. Heim and G. Palm Springer-Verlag, Berlin Heidelberg New York, pp. 89-115 (1977)

Poggio, T., Torre, V. A theory of synaptic interactions. In: Theoretical approaches in neurobiology. Eds. W.Reichardt and T.Poggio. M.I.T. Press (1981)

Reichardt, W., Poggio, T. Figure-Ground Discrimination by Relative Movement in the Visual System of the Fly. Part I: Experimental Results. *Biol. Cyber.*, **35**, 81-100, (1979)

Resnikoff, H.L., On the Psychophysical function, J. Math Biol. 2, 295-276, (1975)

Torre, V., Poggio, T. A Synaptic Mechanism Possibly Underlying Directional Selectivity to Motion. *Proc. R. Soc. Lond. B*, **202**, 409-216, (1978)

The Measurement of Visual Motion

S. Ullman and E.C. Hildreth

M.I.T., Artificial Intelligence Laboratory,
Cambridge, MA 02139, USA

1. Introduction

The analysis of visual motion plays a central role in biological systems. Sophisticated mechanisms for extracting and utilizing motion exist even in simple animals. For example, the frog has efficient "bug detection" mechanisms that respond selectively to small, dark objects moving in its visual field [1]. The ordinary housefly can track moving objects and discover the relative motion between a target and its background, even when the two are indistinguishable in the absence of relative motion [2].

In higher animals, including primates, the analysis of motion is "wired into" the visual system from the earliest processing stages. Some species, such as the pigeon [3] and rabbit [4] (see [5] for other examples) perform rudimentary motion analysis at the retinal level. In other animals, including cats and primates, the first neurons in visual cortex to receive input from the eyes are already involved in the analysis of motion: they respond well to stimuli moving in one direction, but little, or not at all, to motion in the opposite direction [6,7].

In some animals, visual motion is used in the guidance of locomotion and the control of body motion. The plummeting gannet [8] uses visual flow information to stretch back its wings a fraction of a second before it hits the water. Perhaps the most remarkable use of visual motion is the recovery of three-dimensional shape using motion information alone [9-11].

The extensive use of motion by biological systems, and in particular the human visual system, demonstrates the feasibility of carrying out certain information processing tasks and helps to establish specific goals for the analysis of time-varying imagery. This analysis divides naturally into two parts. The first stage is the measurement of motion; for example, the assignment of direction and magnitude of velocity to elements in the image, on the basis of the changing intensity pattern. The second is the use of motion measurements; for example, to separate the scene into distinct objects, and infer their three-dimensional structure.

In this article, we present a computational study of the measurement of visual motion. It is a problem which was found to be surprisingly difficult, both in computer vision, and in modelling biological vision systems. We will present the general problem in Section 2, and discuss methods that have been proposed for its solution. Section 3 presents a specific scheme, proposed by Marr and Ullman [12], for extracting the first motion measurements from the changing image. The initial measurements do not yet specify the true motion of objects in the changing image, and must be combined in some way. This raises the *motion integration problem*, which will be discussed in Sections 3 and 4. Section 5 presents some biological implications for motion measurement.

2. Motion Detection and Measurement

The motion of elements and regions in an image is not given directly, but must be computed from a two-dimensional array of time-dependent light intensity values, $I(x,y,t)$. Motion in the image can be described in terms of a vector field $\mathbf{V}(x,y,t)$ giving the velocity of a point (x,y) at time t. The first problem in analyzing visual motion is the computation of $\mathbf{V}(x,y,t)$ from $I(x,y,t)$. In some cases it may be sufficient to detect only certain properties of the velocity field, rather than measure it completely and precisely. For example, in order to respond quickly to a moving object, motion must be detected, but not necessarily measured. Other tasks, such as the recovery of three-dimensional structure from motion, require a more complete and accurate measurement of the velocity field [13-17].

The measurement of motion may be performed at different stages in the processing of an image. It is useful to draw a distinction between two main schemes. At the lowest level, motion measurements may be based directly on the local changes in light intensity values (*intensity-based* schemes). Alternatively, it is possible to first identify features such as edges and their termination points, corners, blobs, or regions, and then measure motion by matching these features over time (*token-matching* schemes). These two modes of motion detection and measurement give rise to different computational problems, and consequently to different kinds of processes in biological as well as computer vision systems.

2.1 Intensity-based Schemes for Motion Measurement

Two main types of intensity-based schemes have been advanced for biological and computer vision systems: correlation techniques and gradient methods. Cross-correlation of raw intensity values has been used in computer vision applications [18-21], and has been proposed as a model for motion measurement in the human visual system [22-24]. Related to cross-correlation schemes are subtraction schemes, involving simple differencing operations between successive frames. In computer vision, such schemes are primarily used for the detection of motion, and object segmentation [25-28]; together with cross-correlation, they have been utilized for the measurement of motion [26,28]. A fundamental problem of most correlation and subtraction schemes is that they assume the image (or a large portion of it) moves as a whole between the two frames. Images containing independently moving objects and image distortions induced by the unrestricted motion of objects in space pose difficult, perhaps insurmountable, problems for these techniques.

Other intensity-based schemes have been proposed for biological systems. A simple motion detector can be constructed by comparing the outputs of two detectors to light increments at two adjacent positions. The output at position p_1 and time t is compared with that at p_2 at time $t - \delta t$ (a low-pass temporal filter may be used instead of the delay [29]). Two variations of this *delayed comparison* scheme have been proposed as models for biological systems. The first is obtained by multiplying the two values, i.e. $D(p_1, t) \cdot D(p_2, t - \delta t)$, where D denotes the output of the subunits (Fig.1a). If a point of light moves from p_2 to p_1 in time equal to δt, this product will be positive. In an array of such detectors, the average output is essentially equivalent to a cross-correlation of the inputs [29].

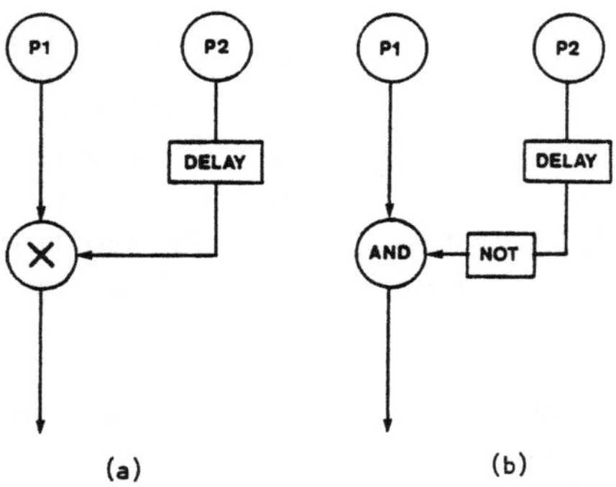

Fig.1 The delayed comparison schemes.
(a) The two inputs are multiplied
(b) The "And-Not" scheme

An alternative method along the same general line is the "And-Not" scheme proposed by Barlow and Levick [30] for the directionally selective units in the rabbit's retina (a similar scheme was suggested for the cat's visual cortex [31]). Evidence for inhibitory interactions within the directionally selective mechanism led to a model in which a motion from p_2 to p_1 is "vetoed" by a delayed response from p_2, whereas motion from p_1 to p_2 produces a positive response (Fig.1b). Poggio and Reichardt [32] have proposed a similar scheme for the visual system of the fly, and an elegant synaptic mechanism that implements these computations was described by Torre and Poggio [33] (see Poggio in this volume).

Some general properties of the delayed comparison schemes are worth noting. First, these detectors respond selectively not only to continuous motion, but also to discrete jumps between p_1 and p_2. Second, the speed of motion must lie within a certain range, determined by the delay (or the low-pass filtering) and the separation between the receptors. A range of velocities can be covered by a family of detectors with different internal delays and interreceptor spacing. Finally, the accurate and reliable measurement of motion will require the combination of the outputs from an array of such elementary detectors.

In gradient schemes, the local motion measurements are derived via a comparison between intensity gradients, and temporal intensity changes. A one-dimensional example is shown in Fig.2. Consider the intensity profile (intensity I as a function of position x), indicated by the solid curve in Fig.2. At the point p, the profile has a positive slope. If the profile moves to the left (dashed curve), the intensity value I at p will be increasing; for a rightward motion (dotted and dashed curve), $I(p)$ will be decreasing. The sign of the temporal change in $I(p)$ thus signals the direction of motion, and from the magnitude of the intensity changes, the speed of motion can be determined. In principle, measurements of motion may be obtained wherever the image intensity gradient is non-zero; however, the measurements are more reliable at the steep intensity gradients around edges.

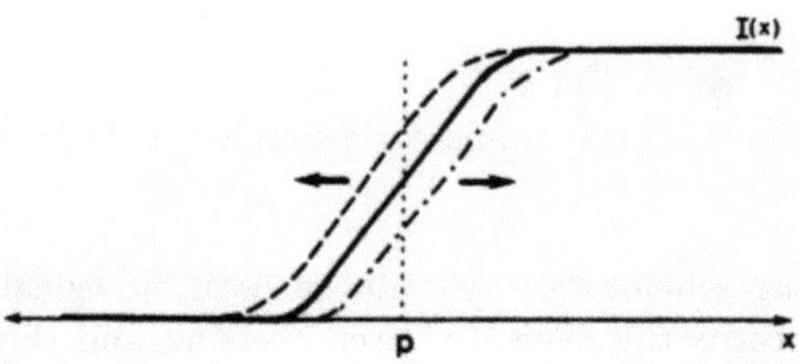

Fig.2

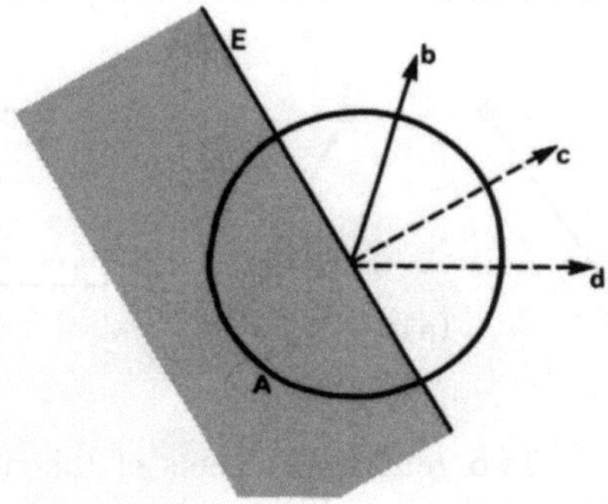

Fig.3

Fig.2 Comparison of the sign of the spatial and temporal derivatives of intensity at the point p yields the sign of direction of motion

Fig.3 The aperture problem. Motion in the directions **b**, **c** and **d** can not be distinguished when viewed through the local aperture **A**.

In two dimensions, the spatial and temporal intensity changes alone are not sufficient to determine the local direction and magnitude of velocity [12, 34-37], because of the *aperture problem*, illustrated in Fig.3. If the motion of the edge E is to be detected by operations which examine an area **A** that is small compared to the overall extent of the edge, the only motion that can be extracted is the component **c** perpendicular to the local orientation of the edge. Such operations cannot distinguish between motion in the directions **b**, **c**, and **d**. To determine the motion completely, a second stage of analysis is required, which integrates the local motion measurements, either over an area of the image, or along contours.

2.2 Token-matching Schemes for Motion Measurement

In token-matching schemes, identifiable elements – tokens – are located and then matched over time. Assuming that the visual input is given as a sequence of discrete frames, a counterpart for each element in one frame must be located in the next. This raises the *correspondence problem* (Fig.4). The filled circles in the figure represent the first frame, and the open circles the second. There are two possible one-to-one pairings between the elements of the two frames, leading to two patterns of perceived motion: diagonal (a) or horizontal (b). In this example, the match is only two-way ambiguous. In general, each frame could contain many elements arranged in complex figures; a correspondence must then be established among them. The rules governing the correspondence process in human vision have been investigated [38-44], but are still far from being completely understood. Token-matching schemes for motion measurement have also been studied for computer vision [45-50].

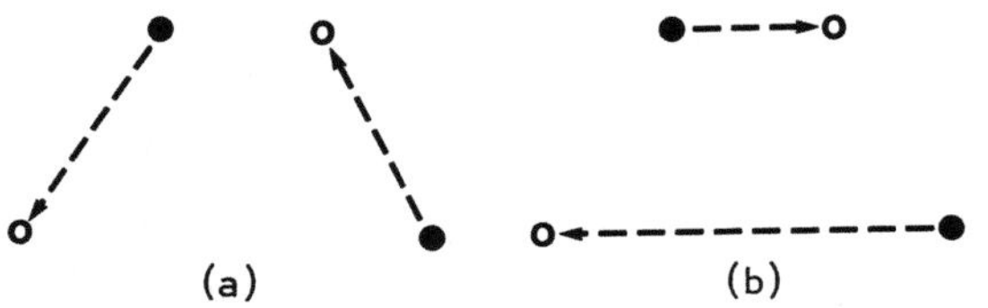

Fig.4
A simple correspondence problem

Two general problems of token-matching schemes are relevant to both biological and machine motion analysis. The first concerns the degree of preprocessing and the complexity of the participating tokens. Matching may be established between simple tokens such as points, blobs, and edge fragments, or complex tokens such as structured forms, or even the images of recognized objects. The use of complex tokens can simplify the correspondence process, since a complex token will usually have a unique counterpart in a subsequent frame. Primitive tokens will usually have many competing possible matches, but their use has two distinct advantages. The first is a reduced preprocessing requirement, which is of special importance in motion perception, where computation time is severely restricted. The second is that a correspondence scheme based on primitive tokens can operate on arbitrary objects engaged in complex shape changes. It seems, therefore, that the correspondence process should operate on the level of rather primitive elements, perhaps at the level of Marr's full primal sketch [51,52].

The second general problem concerns the possible role of intensity-based and token-matching schemes in an integrated vision system. Intensity-based schemes tend to be fast and sensitive, but the ambiguity of the local measurements may make it difficult to recover the velocity field accurately. A token-matching scheme can, in principle, track a sharply localized token (such as a line termination) over long distances, and thereby achieve a high degree of accuracy, at the price of more extensive processing. It is possible that the two motion measurement schemes serve distinct visual tasks. The intensity-based system may be useful as an "early warning" system, and for the separation of moving objects from their background. Token-matching schemes may play an important role in the recovery of structure from motion, where the accurate tracking over considerable distances is useful. A second possiblility is that the two schemes interact to complement each other. For example, a token-matching scheme might be guided by additional constraints supplied by an intensity-based system.

2.3 Two Motion Systems in Human Vision

Psychological studies of motion detection have distinguished two types of visual motion: discrete and continuous. For human observers, a stimulus presented sequentially can produce the impression of smooth, uninterrupted motion (as in motion pictures) [53]. The visual system can "fill-in" the gaps in the discrete presentation even when the stimuli are separated by up to several degrees of visual angle, and by long temporal intervals (400 msec., [54]).

The fact that the visual system can register both types of motion does not imply the existence of two separate mechanisms, since a system that registers discrete motion could in principle register continuous motion as well. Psychophysical evidence supports, however, the view that two different mechanisms are in fact involved in the process of motion detection and measurement [55-60]. The terms "short range" and "long range" processes were suggested by Braddick [58]. The short range mechanism registers continuous motion, or motion presented discretely, with displacements of up to about 15 min. of arc (in the center of the visual field) and temporal intervals of less than about 60–100 msec. The long range mechanism can process larger displacements and temporal intervals.

In the human visual system, it appears that the short range process is an intensity-based scheme, whereas the long range process is a token-matching scheme. Braddick [58] proposed that the directionally-selective units of visual cortex underly the short range process, suggesting that the spatial and temporal limits reflect the spatial and temporal parameters of these neural units. Marr and Ullman [12], present a gradient scheme for the detection and measurement of motion, which includes a model for constructing the directionally-selective units, and an algorithm for combining the local measurements to compute the two-dimensional velocity field. The long range motion phenomena illustrate our ability to derive a correspondence of elements in the changing image, over considerable distances and temporal intervals. Psychophysical studies have shown the long range correspondence to be based on more symbolic primitives, such as edges, bars, blobs, simple groups of primitive elements, and texture edges [13,61].

3. Deriving Velocity Constraints from the Image

In this section, we first present a scheme for extracting initial motion constraints from the image, proposed by Marr and Ullman [12], which was motivated by computational studies of early visual processing, and neurophysiological studies of directionally-selective simple cells in primate visual cortex. Computational studies suggested that the first stage of image analysis should be the detection of intensity changes (see [62] for a review). Marr and Hildreth [63] have proposed that an optimal operator for the initial filtering of the image is the Laplacian of a Gaussian, $\nabla^2 G$, whose shape may be approximated by the difference of two Gaussians. The elements in this convolution output, which correspond to the location of intensity changes, are the zero-crossings [64]. Marr and

Hildreth suggested that the convolution of the image with $\nabla^2 G$ is represented in the output of the retinal ganglion X-cells, and that a class of simple cells in visual cortex assumes the role of zero-crossing detection.

Marr and Ullman [12] have extended this model for simple cells, including a mechanism for their directional selectivity (Fig.5). Fig.5a shows the one-dimensional output of the convolution of a step-edge intensity profile, with the second derivative of a gaussian, (D^2G*I). Fig.5b and Fig.5c illustrate the time derivative, $\frac{\partial}{\partial t}(D^2G*I)$, for motion of the profile to the left and right, respectively. At the location of the zero-crossing Z, the time derivative will be positive for motion to the left, and negative for motion to the right. Similar to the gradient scheme introduced in Section 2, the sign of contrast of the zero-crossing can be compared with the sign of the temporal derivative, to compute the direction of motion of the zero-crossing. By combining the magnitude of the slope of the convolution output as it crosses zero, with the magnitude of the time derivative, rough magnitude of velocity can be computed. In two dimensions, comparison of the spatial and temporal derivatives of $\nabla^2 G*I$ (where I is now a two-dimensional intensity distribution) at the location of zero-crossings, provides only the component of motion in the direction perpendicular to the local orientation of the contour.

Marr and Ullman have proposed that the retinal ganglion Y-cells carry the time derivative of the $\nabla^2 G$ convolution, and that simple cells combine the spatial and temporal derivatives, carried by the X-system and Y-system, to compute the direction of motion of the zero-crossing contours. A possible neural model for the derivation of the spatial and temporal derivatives has been proposed by Richter and Ullman [65]. Recent neurophysiological studies support the role of simple cells in the detection of zero-crossings [Richter, personal communication]. In addition to neurophysiological support, this scheme appears to be consistent with psychophysical studies of the short-range process [12].

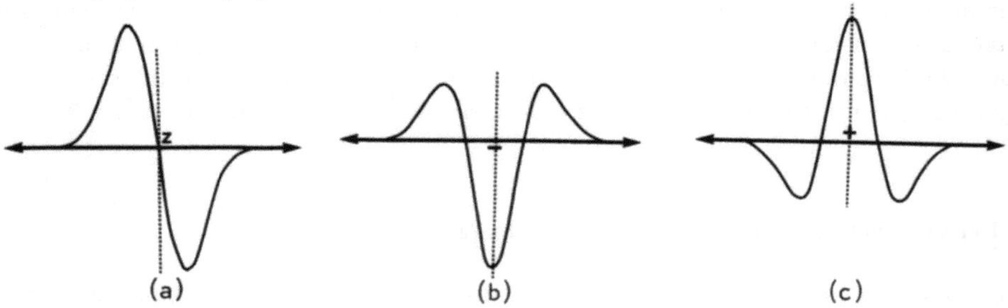

Fig.5 The Marr-Ullman scheme. (a) Convolution of a step intensity change with D^2G (b) and (c) Temporal intensity derivative for motion of the profile to the left and right

From a computational standpoint, restricting the measurement of motion to the location of zero-crossings has two advantages over schemes based only on the raw intensities. First, the zero-crossings of $\nabla^2 G*I$ correspond to points at which the gradient of intensity is locally maximum, yielding the most reliable local velocity measurements. Second, the zero-crossings are tied more closely to physical features. There are many factors that can cause intensity to change locally, such as changing illumination; a change in intensity over time is not necessarily due to the motion of an underlying surface.

The zero-crossing scheme presented above does not yet solve the motion measurement problem. The measurement of the motion of zero-crossings, using a local gradient scheme, provides only the component of motion in the direction perpendicular to the orientation of the contour. More formally, we may express the velocity field along a contour by the function $\mathbf{V}(s)$, where s denotes arclength. $\mathbf{V}(s)$ can be decomposed into components tangent and perpendicular to the contour (Fig.6). $\mathbf{e}_1(s)$ and $\mathbf{e}_2(s)$ are unit vectors in the directions tangent and perpendicular to the curve, and $v^\top(s)$ and $v^\perp(s)$ denote the magnitudes of the two components:

$$\mathbf{V}(s) = v^\top(s)\mathbf{e}_1(s) + v^\perp(s)\mathbf{e}_2(s) \tag{1}$$

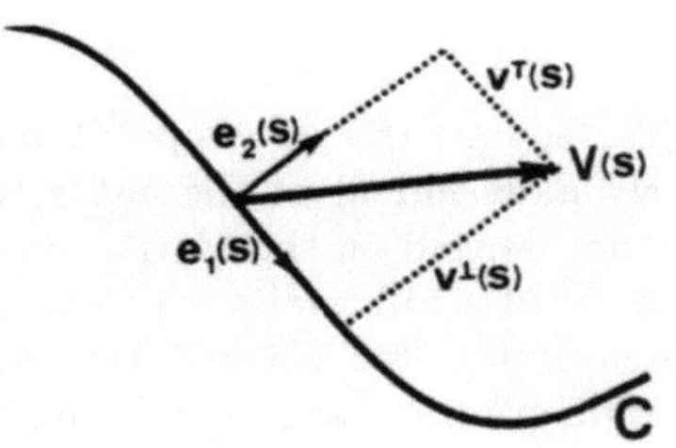

Fig.6 The decomposition of velocity $\mathbf{V}(s)$ into tangential and perpendicular components

The component $v^\perp(s)$ is given directly by the initial measurements; the computation of $\mathbf{V}(s)$ requires the further recovery of $v^\perp(s)$.

At the very least, the computation of $\mathbf{V}(s)$ requires the integration of the constraints provided by $v^\perp(s)$ along the contour. In general however, the solution may still be underdetermined. Additional constraint is required to compute a single velocity field. Fig.7 illustrates two examples, in which the velocity field solution is not unique. In Fig.7a, the solid and dotted lines represent the image of a moving circle, at different instants of time. In the first frame (solid line), the circle lies parallel to the image plane, while in the second frame, the circle is slanted in depth. One velocity field consistent with this sequence is derived from pure rotation of the circle about the central vertical axis, as shown to the left in Fig.7a. (The arrows represent local velocities.) However, there could also be a component of rotation in the plane of the circle, about its center; as shown to the right in Fig.7a. Both velocity fields correspond to valid rigid motions of the circle. This ambiguity is not particular to circles. In Fig.7b, the solid curve C_1 rotates, translates and deforms over time, to yield the dotted curve C_2. The mapping of points from C_1 to C_2 is much less clear (consider, for example, different possible velocities for the point p). The precise computation of the velocity field in this case is important, when one considers the subsequent computation of structure from motion; different choices for the velocity field may yield different three-dimensional structures.

The computation of a unique velocity field requires additional assumptions about physical surfaces, and the velocity field that they may generate under motion.

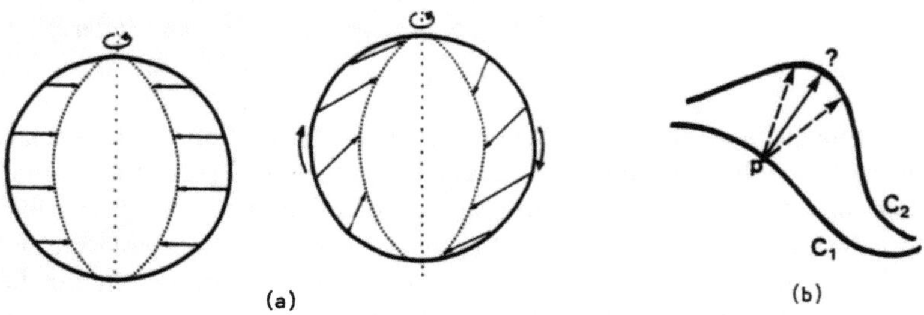

Fig.7 Ambiguity of the velocity field computation. (a) A circle, rotating in depth (b) A deforming curve

4. The Integration of Local Motion Measurements

In this section, we discuss the motion integration problem, strictly from a computational viewpoint. The results that we present here are largely independent of the nature of the initial motion measurements, and in particular, do not depend on the Marr-Ullman scheme discussed previously. We will consider four types of additional constraint on the velocity field: (1) velocity is constant over an area of the image (valid for pure translation); (2) the velocity field is consistent with rigid rotation and translation of objects in the image plane; (3) the velocity field is smooth, and exhibits the least variation among the set of velocity fields consistent with the image constraints; and (4) the velocity field is smooth, exhibits the least variation possible, and is constant over small time intervals. We will discuss methods for combining local measurements, given each of the four types of constraint.

4.1 The Constant Velocity Constraint

Much previous work has addressed the case of pure translation of objects in the image plane. The early gradient schemes used in computer vision [34,66] assumed that velocity would be constant over a large area of the image. Most correlation and subtraction schemes also embodied this assumption. Marr and Ullman [12,67] proposed a scheme in which each local measurement restricts the true velocity of a patch to lie within a 180^0 range of directions to one side of a segment of the local zero-crossing contour. A set of measurements taken at different orientations along a contour then further restrict the allowable directions, until a single velocity direction is obtained. The scheme that we present in the next section, for analyzing rotation and translation in the image plane, may also be used for the restricted case of pure translation. While these schemes cannot account for the full range of human motion perception, they may be useful for the initial detection and rough measurement of motion in the periphery, or analysis of motion during smooth pursuit eye movements, in which stationary objects translate rigidly with respect to the eye. In computer vision, there are restricted applications for these techniques, such as the tracking of objects along a conveyor, or computation of camera motion [68].

4.2 Rigid Motion in the Image Plane

In this and remaining sections, we will focus on the motion of contours. The results apply, however, to continuous patches in the image as well. First, suppose we have a rigid curve undergoing general motion in space. Its instantaneous motion may be described as the combination of: (1) a rotation with angular velocity ω about a single axis in space, which we will denote by the vector $\mathbf{n} = [n_1, n_2, n_3]^T$ (T denotes the transpose of a vector), and (2) a translation, which we will denote by the vector $\mathbf{d} = [d_1, d_2, d_3]^T$. Let the curve be given parametrically by $C = (x(s), y(s), z(s))$. The location of a point on the curve may be given by the position vector $\mathbf{r} = [x(s), y(s), z(s)]^T$. If we let the optical axis lie along the z-axis, and let the projection of the curve onto the image plane (the (x, y) plane) be orthographic, then the two-dimensional velocity field $\mathbf{V}(s)$ along the contour is given by:

$$\mathbf{V}(s) = M(\mathbf{r} \times \omega\mathbf{n} + \mathbf{d}) = \omega z(s)\begin{bmatrix} n_2 \\ -n_1 \end{bmatrix} + \omega n_3 \begin{bmatrix} -y(s) \\ x(s) \end{bmatrix} + \begin{bmatrix} d_1 \\ d_2 \end{bmatrix} \tag{2}$$

M denotes the matrix which performs the orthographic projection. The first term in the resulting expression describes the component of the velocity field due to rotation in depth about an axis parallel to the image plane (the axis $\mathbf{n} = [n_1, n_2, 0]^T$); the second term is the component due to motion in the image plane (rotation about the axis $\mathbf{n} = [0, 0, n_3]^T$), and the third term is the translation component.

Consider the restricted case of rigid motion in the image plane; the velocity field now corresponds to the combination of a translation, and rotation about the axis $\mathbf{n} = [0, 0, 1]^T$. Thus, $\mathbf{V}(s)$ is given by:

$$\mathbf{V}(s) = \omega \begin{bmatrix} -y(s) \\ x(s) \end{bmatrix} + \begin{bmatrix} d_1 \\ d_2 \end{bmatrix} \tag{3}$$

$\mathbf{V}(s)$ is simply a translation, rotation and scaling of the image curve $(x(s), y(s))$ (Fig.8). In Fig.8a, the curve C_1 undergoes a small rotation and translation in the image plane to yield the curve C_2. The arrows indicate local velocity vectors along the curve. In Fig.8b, these velocity vectors have been translated to a common origin in *velocity space*, where the x and y axes represent the x and y components of velocity. The curve in velocity space has the same shape as the image curve C_1; its size is proportional to angular velocity ω, and it is rotated 90° with respect to C_1 (this relationship is also used in kinematics [69]). The additional translation of the curve in the image yields the same translation of the curve in velocity space. In the case of pure translation, the image of the velocity field in velocity space degenerates to a single vector. In general, the explicit use of the velocity space aids in the visualization of properties of the motion of curves, and provides a tool for establishing theoretical properties of the velocity field.

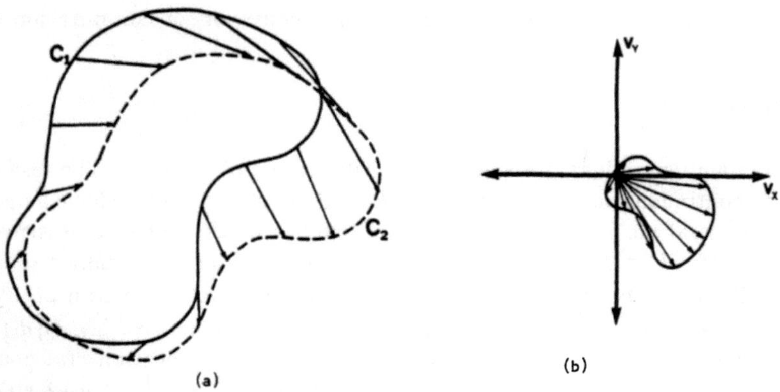

Fig.8 Rigid motion in the image plane. (a) The velocity field in the image (b) The velocity vectors in velocity space

For the simple case of rigid motion in the image plane, this relationship between the shape of the curve and the velocity field is not restricted to continuous motion of the curve. For discrete motion of a curve, we will use the term *displacement field.* If we let σ be a discrete angular rotation of the curve in the image plane, then the displacement field $\mathbf{V}_d(s)$ is given by:

$$\mathbf{V}_d(s) = \begin{bmatrix} \cos\sigma - 1 & \sin\sigma \\ -\sin\sigma & \cos\sigma - 1 \end{bmatrix} \begin{bmatrix} x(s) \\ y(s) \end{bmatrix} + \begin{bmatrix} d_1 \\ d_2 \end{bmatrix} \tag{4}$$

$\mathbf{V}_d(s)$ is also given by a scaling and rotation of the projected image curve $(x(s), y(s))$. In this case, the scale factor k is given by:

$$k = \sqrt{(\cos\sigma - 1)^2 + \sin^2\sigma} = \sqrt{2(1 - \cos\sigma)} \tag{5}$$

The angle of rotation α between the image curve $(x(s), y(s))$ and the corresponding curve in velocity space, is given by:

$$\tan\alpha = \frac{\sin\sigma}{\cos\sigma - 1} = -\cot(\frac{\sigma}{2}) \tag{6}$$

For small σ, $k \approx \sigma$ and $\alpha \approx \pm 90^o$. As before, an additional translation component simply translates the curve in velocity space.

For rigid motion in the image plane, a simple scheme can be used to construct the velocity field. If we know the true direction of velocity for two points on the contour, we can compute the direction of velocity everywhere as follows: (1) construct the lines perpendicular to the direction of velocity at the two known points, (2) compute the intersection of these two lines, (3) from every point p_i along the contour, construct the

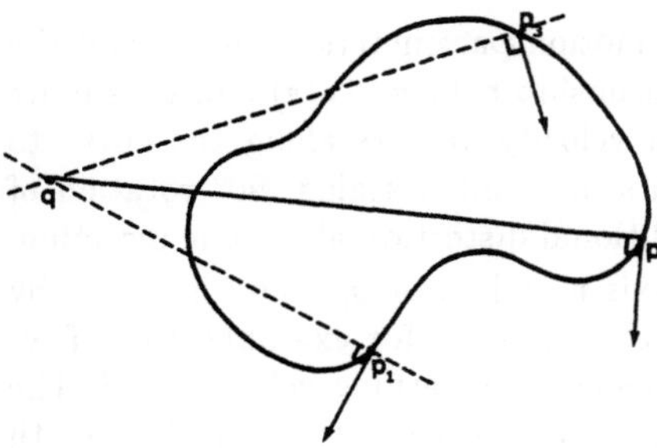

<u>Fig.9</u> Construction of the velocity field for rigid motion in the image plane

line to the intersection point; the true direction of velocity is perpendicular to this line. In Fig.9, we derive the direction at p_2, given known directions at p_1 and p_3.

This construction is simply locating the point about which the motion can be described as pure rotation. For pure translation, the two lines, from points of known direction of velocity, will be parallel, so the direction of motion everywhere will be equal to the direction of motion of the known points. Certainly, if we knew both the direction and magnitude of velocity at two points along the contour, we could compute the global motion parameters, and hence direction and magnitude of velocity everywhere. However, from the direction of velocity alone at two points on the curve, we can compute the direction of velocity everywhere. If we then know the magnitude of the perpendicular components of velocity along the curve, we can compute both direction and magnitude of velocity along the curve.

There are at least two sources for points of known velocity direction in the image. First, identifiable features, such as terminations, may be tracked in two dimensions. Second, for points at which the perpendicular component of velocity is zero, velocity is constrained to lie along the tangent to the curve. For the case of a smooth, rigid, closed curve moving in the image plane, there must exist at least two points on the curve for which the perpendicular component of velocity is zero. Suppose we focused our velocity field computation at the locations of zero-crossings derived from the image. Since zero-crossing contours are generally closed (except at image boundaries), there will usually be sufficient constraint from the image to solve for the velocity field, in the simple case of rigid motion in the image plane.

<u>4.3 The Smoothness Constraint</u>

In this section we will derive a different type of constraint on the velocity field, which will allow us to analyze the projected motion of three-dimensional objects allowed to move freely in space, and deform over time. The specific analysis will assume that we have measured the perpendicular components of velocity along contours in the image. However, the general constraint that we present may be utilized in other motion measurement schemes as well.

The expression (2) related $\mathbf{V}(s)$ to the global motion parameters ω and $\mathbf{n}$, and the shape of the curve $C = (x(s), y(s), z(s))$. The relationship between $\mathbf{V}(s)$ and C is quite simple. If we map the projected two-dimensional velocity vectors along the curve to a common origin in velocity space, their endpoints map out a scaled, 90° rotation of the projected image curve $(x(s), y(s))$, with an additional distortion along one direction. This distortion is directed perpendicular to the axis $\mathbf{n} = [n_1, n_2, 0]^T$, and is scaled by the z component of the curve, $z(s)$. This relationship implies, for example, that if we have a smooth curve in motion, it must generate a smoothly varying velocity field. The real world consists predominantly of solid objects, whose surfaces are generally smooth compared with their distance from the viewer. Thus, intuitively, we seek a velocity field which is consistent with the constraints we derive from the changing image, and which varies smoothly. A single solution might be obtained by finding the velocity field which varies as little as possible [35].

To achieve this, we need some means of measuring the variation in velocity along a contour. For example, we could measure the change in direction of velocity as we trace along the contour. Total variation of the velocity field could then be defined as the total change in direction over the entire contour. A second definition involves measuring the change in magnitude of velocity along the contour. Finally, we could measure the change in the full velocity vector, $\frac{\partial \mathbf{V}(s)}{\partial s}$, incorporating both the direction and magnitude of velocity.

In order to define the variation of the velocity field more formally, first recall the decomposition of velocity into components tangent and perpendicular to the curve (Fig.6 and eqn.1). $\mathbf{e}_1(s)$, $\mathbf{e}_2(s)$ and $v^{\perp}(s)$ can be measured directly from the changing image. $v^{\top}(s)$ is unknown, and must be recovered in order to compute the velocity field $\mathbf{V}(s)$. Aside from knowing $v^{\perp}(s)$ everywhere along the curve, there may also be points at which the direction and magnitude of velocity, and hence both $v^{\perp}(s)$ and $v^{\top}(s)$, are known. In addition, the direction of velocity alone, and hence the ratio $\frac{v^{\perp}(s)}{v^{\top}(s)}$, may be known at points on the curve, for example, where $v^{\perp}(s_i) = 0$ (Section 4.2).

We can now consider a more formal means for measuring the variation in the velocity field. Mathematically, this can be accomplished by defining a functional Θ, which maps the space of all possible vector fields (along the contour), $\mathbf{V}$, into the real numbers: $\Theta{:}\mathbf{V} \mapsto \Re$. This functional should be such that the smaller the variation in the velocity field, the smaller the real number assigned to it. Two candidate velocity fields may then be compared, by comparing their corresponding real numbers. This raises the question of what functional should be used to measure the variation of a velocity field. In the remainder of this section, we will evaluate a set of possible functionals, based on the three measures of variation that we previously presented informally: (1) variation in $\mathbf{V}(s)$, (2) variation in the direction of velocity, and (3) variation in the magnitude of velocity, all with respect to the curve.

(1) Variation in $\mathbf{V}(s)$

A scalar measure of the local variation of $\mathbf{V}(s)$ with respect to the curve is given by $|\frac{\partial \mathbf{V}(s)}{\partial s}|$, shown in Fig.10a. Two nearby velocity vectors along the image curve are

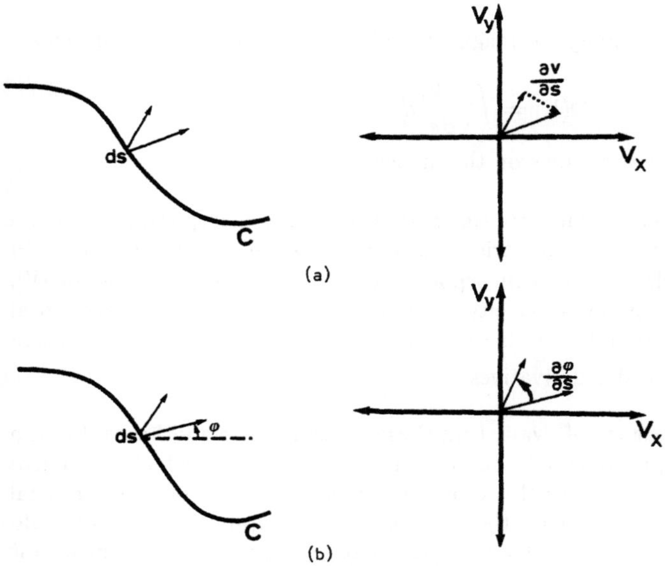

Fig.10 Measuring variation in the velocity field (a) Change in the full velocity vector (b) Change in direction of velocity

translated to a common origin in velocity space, where the vector $\frac{\partial \mathbf{V}(s)}{\partial s}$ is shown with a dotted arrow. For convenience of notation, we wiil omit the argument to $\mathbf{V}(s)$, writing $|\frac{\partial \mathbf{V}}{\partial s}|$. A measure of the total variation of the velocity field along the curve may then be given by the functional:

$$\Theta(\mathrm{V}) = \int |\frac{\partial \mathbf{V}}{\partial s}| ds$$

We may also consider variations on this functional, involving higher order derivatives, or higher powers, such as:

$$\Theta(\mathrm{V}) = \int |\frac{\partial^2 \mathbf{V}}{\partial s^2}| ds \qquad \text{or} \qquad \Theta(\mathrm{V}) = \int |\frac{\partial \mathbf{V}}{\partial s}|^2 ds$$

(2) Variation in Direction
Let the direction of velocity be given by the angle φ, shown in Fig.10b. Total variation of direction along the curve could be given by functionals such as the following:

$$\Theta(\mathrm{V}) = \int |\frac{\partial \varphi}{\partial s}| ds$$

or variations involving higher order derivatives, or higher powers.

(3) Variation in Magnitude

Finally, we could measure the change in magnitude of velocity alone, using functionals such as:

$$\Theta(\mathbf{V}) = \int \frac{\partial|\mathbf{V}|}{\partial s} ds$$

Again, we could also consider variations on this measure.

The functional that we use to measure smoothness may also incorporate a measure of the velocity field itself, rather than strictly utilizing changes in the velocity field along the curve. For example, we could incorporate a term which is a function of $|\mathbf{V}|$. This might be useful if we sought a velocity field which also exhibits the least total motion. In addition, the functional could become arbitrarily complex in its combination of $|\frac{\partial \mathbf{V}}{\partial s}|, |\frac{\partial \varphi}{\partial s}|, \frac{\partial|\mathbf{V}|}{\partial s}$, or higher order derivatives.

We have at least three means of evaluating these measures of smoothness. From a mathematical point of view, there should exist a unique velocity field which minimizes our particular measure of smoothness; this requirement imposes a set of mathematical constraints on our functional. Second, the velocity field computation should yield physically plausible solutions. Finally, if we suggest that such a smoothness constraint underlies the motion computation in the human visual system, this minimization should yield a velocity field consistent with human motion perception. An examination of these smoothness measures from a physical and mathematical point of view suggests that a measure involving the full velocity vector, such as $\Theta(\mathbf{V}) = \int |\frac{\partial \mathbf{V}}{\partial s}|^2 ds$, is most appropriate for the velocity field computation [37]. Of particular importance are the mathematical properties of this functional. It can be shown that, given a simple condition on the constraints that we derive from the image, there exists a unique velocity field which satisfies our constraints, and minimizes $\int |\frac{\partial \mathbf{V}}{\partial s}|^2 ds$. This condition is almost always satisfied by our initial motion measurements. To obtain this result, we take advantage of the analysis used by Grimson [70] for evaluating possible functionals for performing surface interpolation from stereo data. The basic mathematical question is, what conditions on the form of the functional, and the structure of the space of velocity fields, are needed to guarantee the existence of a unique solution? These conditions are captured by the following theorem (see also [71]):

Theorem: *Suppose there exists a complete semi-norm Θ on a space of functions H, and that Θ satisfies the parallelogram law. Then, every nonempty closed convex set $E \subset H$ contains a unique element v of minimal norm, up to an element of the null space. Thus, the family of minimal functions is*

$$\{v + s \mid s \in S\}$$

where

$$S = \{v - w \mid w \in E\} \cap \mathcal{N}$$

and $\mathcal{N}$ is the null space of the functional

$$\mathcal{N} = \{u \mid \Theta(u) = 0\}.$$

It can be shown that the functional $\{\int|\frac{\partial \mathbf{V}}{\partial s}|^2 ds\}^{\frac{1}{2}}$ is a complete semi-norm, which satisfies the parallelogram law. Second, the space of all possible velocity fields, which satisfy the constraints derived from the image, is convex. It then follows from the above theorem that this space contains a unique element of minimal norm, up to an element of the null space. Since our smoothness measure is non-negative, minimizing $\{\int|\frac{\partial \mathbf{V}}{\partial s}|^2 ds\}^{\frac{1}{2}}$ is equivalent to minimizing $\int|\frac{\partial \mathbf{V}}{\partial s}|^2 ds$. The null space in this case is the set of constant velocity fields, since $\int|\frac{\partial \mathbf{V}}{\partial s}|^2 ds = 0$ implies $|\frac{\partial \mathbf{V}}{\partial s}| = 0$ everywhere, which implies $\mathbf{V}(s)$ constant. As a consequence, the following can be shown [37]: If $v^{\perp}(s)$ is known at two points, for which the orientation of the curve is different, then there exists a unique velocity field which satisfies the known velocity constraints and minimizes $\int|\frac{\partial \mathbf{V}}{\partial s}|^2 ds$. An extended straight line will not yield measurements for two different orientations, but in all other cases, there will be sufficient information along a contour to guarantee a unique solution to the velocity field.

We can apply the constraint of least variation and compute a projected two-dimensional velocity field for any three-dimensional surface, whether rigid or non-rigid, undergoing general motion in space. If we measure the variation in the full velocity vector along a contour in the image, using a functional such as $\int|\frac{\partial \mathbf{V}}{\partial s}|^2 ds$, we are guaranteed that there exists a unique solution to the velocity field computation that minimizes this variation. While it is not yet clear that the general smoothness constraint, or the particular measure $\int|\frac{\partial \mathbf{V}}{\partial s}|^2 ds$, is the most appropriate for the motion computation, it is important that this measure satisfies certain essential mathematical requirements, that the other measures do not. For example, the use of a functional incorporating only a measure of velocity direction, which will attempt to make the local velocity vectors as parallel as possible, does not yield functionals which are semi-norms, and consequently, does not lead to a unique velocity field solution. For a scheme to underly the motion computation in the human visual system, it is essential that it be mathematically well-founded.

We should note that an advantage to applying the smoothness constraint along contours is that the minimization of variation in the velocity field is performed along one-dimension, rather than over two dimensions, as in the case of Horn and Schunck's computation of the optical flow [35]. Secondly, to apply the smoothness constraint over an area of the image, it is necessary to specify a neighborhood size, within which constraints will be combined, and smoothness imposed on the velocity field. Unless we can define surface boundaries prior to the velocity field computation, specifying an appropriate area of the image can be difficult. In general, the extent of contours is more highly correlated with single surfaces. The smoothness constraint can be applied to single contours, reducing the problem of integrating motion measurements across object boundaries. Finally, there exist several standard algorithms for the solution of optimization problems such as this (see, e.g. [37]).

4.4 Deriving Additional Constraints from the Image

Additional constraints can be derived from the image if we do not restrict ourselves to the use of instantaneous measurements; for example, we may utilize a second curve, at some time later. If the time interval is small, then the displacement of points along the curve will also be small. We can then require that each point on the first curve project to a point on the second, with a velocity consistent with the instantaneous perpendicular component of velocity $v^{\perp}(s)$; this assumes that velocity is constant over the time interval separating the two curves. In addition, we could still compute the velocity field which exhibits the least variation.

This approach may yield a simpler, more robust algorithm for the velocity field computation, because it utilizes more constraint from the image. However, it has the disadvantage that we may not be able to obtain the theoretical uniqueness results that were possible when we considered the perpendicular components of velocity as a sole source of constraint. A simple example, in which the velocity field solution is not unique is shown in Fig.11. Suppose we are given the initial constraints shown in Fig.11a. The arrows indicate the perpendicular components of velocity along the first curve, and the dotted line indicates the second curve. There are two velocity field solutions consistent with these constraints, shown in Fig.11b,c, corresponding to the two directions of rotation of the circle. Both velocity fields exhibit the same total variation. The use of instantaneous motion measurements alone, together with the additional smoothness constraint, as discussed in Section 4.3, would yield the velocity field given by the vectors in Fig.11a, corresponding to pure expansion of the circle. The additional constraint of the second curve leads to a different solution.

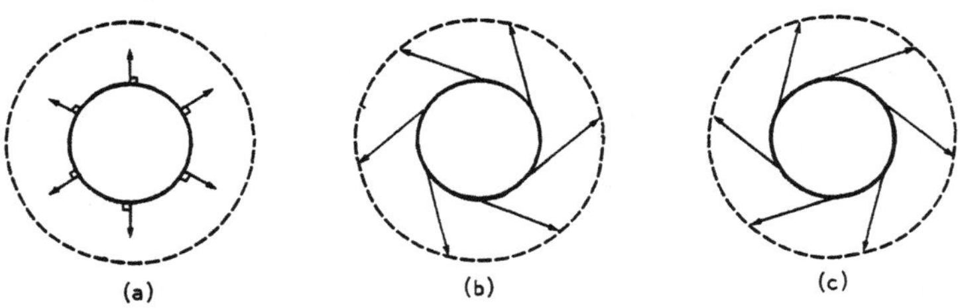

Fig.11 Ambiguity of the velocity field. (a) The initial constraints (b) Rotation of the circle to the left (c) Rotation to the right

The availability of the second curve may simplify the velocity field computation in the following way. The perpendicular component of velocity, measured at a point p on the first curve, constrains the velocity vector at p to project to a point along the line l in the second frame (Fig.12). If in addition, p must project to a point q on the second curve, possible candidates for q may be given by the intersection of l with the second curve. In practice, there will be error in the measurement of $v^{\perp}(s)$, and $v^{\perp}(s)$ may not be constant over small time intervals. As a consequence, we should consider a band in the second frame, to which p must project. Candidates for q are then given by the intersection of this band with the second curve (Fig.12). If the curve has fairly high local curvature, or undergoes rotation, then this intersection alone provides considerable constraint on the velocity field. However, in the worst case of an extended line undergoing pure translation, the second curve offers limited additional constraint. The computation of a precise velocity field requires further analysis of constraints derived from the image, together with additional assumptions. We are presently exploring algorithms which utilize the smoothness constraint for this subsequent computation.

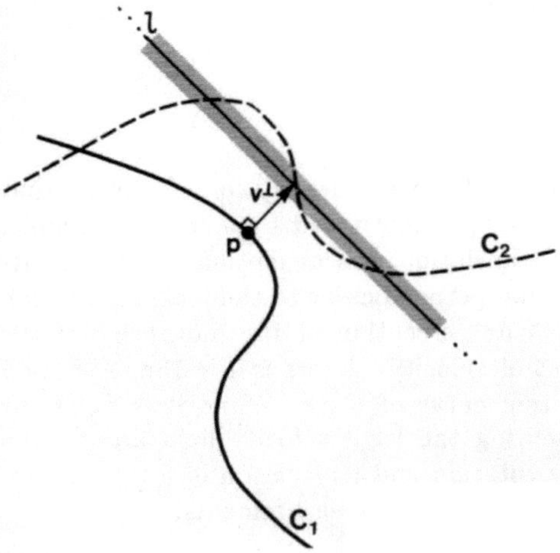

Fig.12 Use of the constraint provided by a second curve

4.5 Summary

We have considered various additional constraints which may be used in the computation of the velocity field from initial motion measurements derived from the changing image. These constraints range from the restricted assumption of pure translation to the general constraint of smoothness of the velocity field, which allows for the arbitrary movement of rigid or non-rigid objects in space. The use of different constraints results in considerable variation in the classes of motion which may be analyzed, the type of algorithm, and the extent of theoretical analysis required to formulate a well-defined computational problem. In analyzing these constraints, we have so far restricted ourselves to addressing purely computational issues. In the next section, we discuss implications for the biological computation of motion. If the human visual system does in fact compute a detailed velocity field, it is likely to use as much constraint as possible from the changing image, together with the least restrictive additional constraints as necessary, to compute a unique velocity field.

5. Some Implications Concerning the Biological Computation of Motion

(i) An underlying assumption of this work is that the local velocity field is explicitly computed and represented. For the human visual system, the idea that there exists an

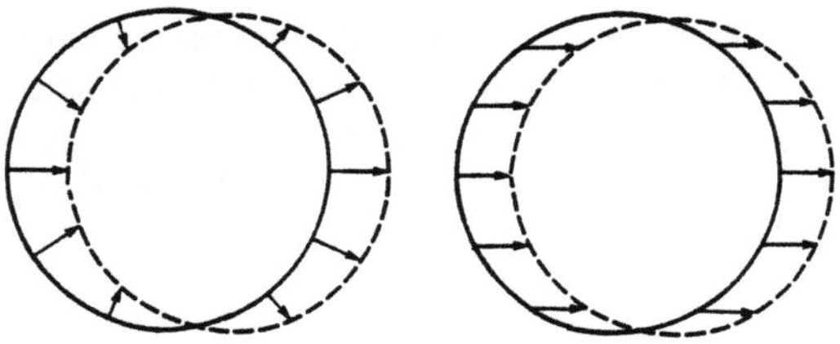

Fig.13 Computing the local velocity field. (a) Initial motion measurements (b) The velocity field corresponding to translation

explicit computation of motion, which is different from the description of motion that could be provided by initial motion detectors, can be motivated by a simple example. In Fig.13, we show a circle undergoing pure translation. Initial motion measurements provide the component of motion in the direction perpendicular to the local orientation of intensity changes in the image (Fig.13a). Our perception of the movement of the circle is pure translation, indicated by the set of velocity vectors in Fig.13b. A second example is that of the rotating and translating curve of Fig.8. While it is not clear whether we are capable of explicitly representing the local velocity field around the contour, we do perceive the movement as the rotation and translation of a rigid curve. Such an interpretation is not explicit in the initial motion measurements.

(ii) The analysis of motion has been separated into two distinct stages; first, the measurement of motion, and second, the use of motion for tasks such as object segmentation and structure from motion. This raises the question of whether the interpretation of three-dimensional structure can influence the computation of the two-dimensional velocity field. For example, does the assumption of rigidity, examined in Ullman's work [13], enter into the velocity field computation? Psychological experiments [13] suggest that the long range motion correspondence is not influenced by the intrepreted three-dimensional structure of a single view. The short range process may be similar.

(iii) We support the idea that there exists two processes for analyzing motion, corresponding to Braddick's long range and short range processes. We suggest that the long range process is based on a token-matching scheme, while the short range process is intensity-based. If this view is valid, it raises the following questions. How do the long range and short range processes interact? Do subsequent computational tasks, such as object segmentation or structure from motion utilize the results of one or the other process? The work of Petersik [72] suggests that the long range process may be crucial to the recovery of structure from motion.

(iv) We suggest that the initial stage of motion analysis consists of the measurement of the perpendicular component of velocity along zero-crossing contours. This can be examined through neurophysiological and psychological studies. In regard to neurophysiology, are the class of directionally-selective simple cells detecting the motion of zero-crossings in their input from the LGN? This is now under investigation [Richter, personal communication]; initial results tend to support this claim. Psychophysical experiments can test whether perceived motion is consistent with the motion of zero-crossings.

(v) We propose that the local motion measurements are then integrated along zero-crossing contours. Again, this may be explored through both neurophysiology and psychology. If the motion integration problem is fundamental to motion analysis, one may expect to find neural mechanisms within the visual system that are involved in this task. Most of the motion sensitive units studied so far do not seem suitable for the integration stage. Motion selective cells in the primary visual cortex of the cat and the monkey respond primarily to edges and bars. To activate such a unit the stimulus must have the preferred orientation, and move in the preferred direction. In contrast, promising candidates for the integration phase would dissociate the effects of orientation and direction of movement, ideally exhibiting specificity for direction of motion but not for orientation. Furthermore, the direction specificity of such a unit is expected to depend on the range of orientations spanned by the stimulus. There are indications for the possible existence of such units in the posterior bank of the superior temporal sulcus of the rhesus monkey [73]. For psychophysical experimentation, there are at least two questions; first, is the motion that we perceive forced to be consistent at least with the sign of the local motion measurements along zero-crossing contours, or can it be overridden, for example, by the long range process, or by the history of the motion? Second, if the integration does take place, are measurements combined over neighborhoods in the image, or along contours? Wallach's [74] demonstrations suggest that the integration may take place along contours.

(vi) Additional assumptions are required for the motion integration problem. Regarding the human system, we may first ask what constraints are derived from the changing image. Does the human visual system strictly utilize instantaneous measurements of velocity, or is a second curve, at some small time interval later, also used to constrain the velocity field. Do we utilize an additional constraint on smoothness of the velocity field, as described here? A constraint such as smoothness may be the least restrictive constraint which allows objects to move freely in space, and deform, but which still allows for the computation of a unique velocity field. Psychophysical experimentation is necessary to determine whether the velocity field that we perceive is the smoothest one possible. Both the short and long range processes face the fact that in general, the motion of elements is not specified uniquely by information in the changing image; do the additional assumptions governing the computation of velocity or correspondence differ in the two processes, or do they differ only in the constraints that are utilized from the changing image?

(vii) Finally, the motion measurement problem has some implications for the interpretation of structure from motion. It has been shown [17] that three-dimensional shape can be recovered locally, from the instantaneous velocity field. The interpretation is sensitive, however, to small errors in the measured velocity. In light of the inherent difficulties in measuring the velocity field precisely, recovery methods that rely solely on the instantaneous velocity field appear unlikely. For the reliable recovery of three-dimensional structure from motion, processes that integrate motion over time are probably required.

Acknowledgements

This article describes research done in the Artificial Intelligence Laboratory of the Massachusetts Institute of Technology. Support for the laboratory's artificial intelligence research is provided in part by the Advance Research Projects Agency of the Department of Defense under the Office of Naval Research Contract N00014-75-C-0643 and in part by National Science Foundation Grant MCS77-07569.

References

1. J. Y. Lettvin, H. R. Maturana, W. S. McCulloch, W. H. Pitts: Proc. IRE. **47**, 1940-1951 (1959)
2. W. Reichardt, T. Poggio: Biol. Cyb. **35**, 81-100 (1979)
3. H. R. Maturana, S. Frenk: Science **142**, 977-979 (1963)
4. H. B. Barlow, R. N. Hill: Science **139**, 412-414 (1963)
5. O. J. Grusser, U. Grusser-Cornehls: "Neuronal Mechanisms of Visual Movement Perception and Some Psychophysical and Behavioral Correlation," In *Handbook of Sensory Physiology*, Vol. VII/3A, ed. by R. Jung, (Springer-Verlag, Berlin, 1973)
6. D. H. Hubel, T. Wiesel: J. Physiol. (Lond.) **160**, 106-154 (1962)
7. D. H. Hubel, T. Wiesel: J. Physiol. (Lond.) **195**, 215-243 (1968)
8. D. N. Lee, P. E. Reddish: Nature **293**, 293-294 (1981)
9. H. Wallach, D. N. O'Connell: J. Exp. Psych. **45**, 205-217 (1953)
10. G. Johansson: Perception and Psych. **14**, 201-211 (1973)
11. G. Johansson: Sci. Am. **232**, 76-88 (1975)
12. D. Marr, S. Ullman: Proc. Roy. Soc. Lond. B **211**, 151-180 (1981)
13. S. Ullman: *The Interpretation of Visual Motion* (MIT Press, Cambridge and London, 1979)
14. S. Ullman: Proc. Roy. Soc. Lond. B **203**, 405-426 (1980)
15. K. Prazdny: Biol. Cyb. **36**, 87-102 (1980)
16. W. F. Clocksin: Perception **9**, 253-269 (1980)
17. H. C. Longuet-Higgins, K. Prazdny: Proc. Roy. Soc. Lond. B **208**, 385-397 (1981)
18. E. A. Smith, D. R. Phillips: IEEE Trans. Comp. **c-21**, 715-729 (1972)
19. J. A. Leese, C. S. Novak, V. R. Taylor: Patt. Recog. **2**, 279-292 (1970)
20. R. L. Lillestrand: IEEE Trans. Comp. **c-21**, 654-659 (1972)
21. K. Wolferts: Proc. Int. Joint Conf. Patt. Recog. **2** 1-2 (1974)
22. H. H. Bell, J. S. Lappin: Perception and Psych. **11**, 263-268 (1973)
23. A. J. Pantle, L. Picciano: Science **193**, 500-502 (1976)
24. J. T. Petersik, K. I. Hicks, A. J. Pantle: Perception **7**, 371-383 (1978)
25. R. Jain, W. Martin, J. K. Aggarwal: Comp. Graph. Image Proc. **11**, 13-34 (1979)
26. R. Jain, D. Militzer, H. H. Nagel: Proc. IJCAI **5** 612-618 (1977)
27. R. Jain, H. H. Nagel: IEEE Trans. PAMI **PAMI-1**, 206-214 (1979)

28. H. H. Nagel: Comp. Graph. Image Proc. **7**, 149-194 (1978)
29. B. Hassenstein, W. Reichardt, W.: Chlorophanus. Z. Naturf. **IIb**, 513-524 (1956)
30. H. B. Barlow, R. W. Levick: J. Physiol. (Lond.) **173**, 477-504 (1965)
31. R. C. Emerson, G. L. Gerstein: J. Neurophysiol. **40**, 136-155 (1977)
32. T. Poggio, W. Reichardt: Quart. Rev. Biophys. **9**, 377-438 (1976)
33. V. Torre, T. Poggio: Proc. Roy. Soc. Lond. B **202**, 409-416 (1978)
34. C. I. Fennema, W. B. Thompson: Comp. Graph. Image Proc. **9**, 301-315 (1979)
35. B. K. P. Horn, B. G. Schunck: Artif. Intel. **17** 185-203 (1981)
36. L. Davis, Z. Wu, and H. Sun: In *Proc. ARPA Image Understanding Workshop*, ed. by L. S. Baumann (Science Applications Inc., Arlington, VA, 1979)
37. E. Hildreth: In *Proc. Workshop on Rep. and Control.* (IEEE Computer Society Press, Los Angeles, 1982)
38. J. P. Frisby: Vis. Res. **12**, 1145-1166 (1972)
39. P. A. Kolers: *Aspects of Motion Perception* (Pergamon Press, New York, 1972)
40. A. J. Pantle: J. Opt. Soc. Am. **63**, 1280 (1973)
41. J. Ternus: Psych. Forsch. **7**, 81-136 (1926) Translated in: *A Source Book of Gestalt Psychology*, ed. by W. D. Ellis, (Humanities Press, New York, 1967)
42. D. Navon: J. Exp. Psych. (Hum. Perc. Perf.) **2**, 130-138 (1976)
43. S. Ullman: Perception **7**, 683-693 (1978)
44. S. Ullman: Perception **9**, 617-626 (1981)
45. J. L. Potter: IEEE Trans. Sys., Man, Cyb. **SMC-5**, 390-394 (1975)
46. J. L. Potter: Comp. Graph. Image Proc. **6**, 558-581 (1977)
47. J. K. Aggarwal, R. O. Duda: IEEE Trans. Comp. **c-24**, 966-976 (1976)
48. W. K. Chow, J. K. Aggarwal: IEEE Trans. Comp. **c-26**, 179-185 (1977)
49. J. W. Roach, J. K. Aggarwal: IEEE Trans. PAMI **PAMI-1**, 127-135 (1979)
50. D. Lawton: Comp. Graph. Image Proc. *in press* (1982)
51. D. Marr: Phil. Trans. Roy. Soc. Lond. B **275**, 483-524 (1976)
52. D. Marr: *VISION* (W. H. Freeman Co., San Francisco, 1981)
53. M. Wertheimer: Zeitchrift fur Psych. **61**, 161-265 (1912)
54. W. Neuhaus: Archiv fur die Gesamte Psych. **75**, 315-348 (1930)
55. S. M. Anstis, B. J. Rogers: Vis. Res. **15**, 957-961 (1975)
56. S. M. Anstis: Phil. Trans. Roy. Soc. Lond. B. **290**, 153-168 (1980)
57. O. J. Braddick: Vis. Res. **13**, 355-369 (1973)
58. O. J. Braddick: Vis. Res. **14**, 519-527 (1974)
59. O. J. Braddick: Phil. Trans. Roy. Soc. Lond. B. **290**, 137-151 (1980)
60. J. T. Petersik, K. I. Hicks, K. I., A. J. Pantle: Perception **7**, 371-383 (1978)
61. M. Riley: *The Representation of Image Texture* (MIT AI TR-649, 1981)
62. E. C. Hildreth: Comp. Graph. Image Proc. *in press* (1982)
63. D. Marr, E. C. Hildreth: Proc. Roy. Soc. Lond. B. **207**, 187-217 (1980)
64. D. Marr, T. Poggio: Proc. Roy. Soc. Lond. B. **204**, 301-328 (1979)
65. J. Richter, S. Ullman: Biol. Cyb. **43**, 127-145 (1982)
66. J. O. Limb, J. A. Murphy: Comp. Graph. Image Proc. **4**, 311-327 (1975)
67. J. Batali, S. Ullman: In *Proc. ARPA Image Understanding Workshop*, ed. by L. S. Baumann (Science Applications Inc., Arlington, VA, 1979)
68. A. Bruss, B. K. P. Horn: *Passive Navigation* (MIT AI Memo 661, 1981)

69. R. Hartenberg, J. Denauit: *Kinematic Synthesis of Linkages* (McGraw-Hill Co., New York, 1964)
70. W. E. L. Grimson: *From Images to Surfaces. A Computational Study of the Human Early Visual System* (MIT Press, Cambridge and London, 1981)
71. W. Rudin: *Functional Analysis*, (McGraw-Hill Co., New York, 1973)
72. J. T. Petersik: Perception **9**, 271-283 (1980)
73. S. M. Zeki: J. Physiol. **236**, 549-573 (1974)
74. H. Wallach: "On Perceived Identity: The Direction of Motion of Straight Lines," In *On Perception*, ed. by H. Wallach, (Quadrangle, N.Y., 1976)

Visual Coding of Position and Motion

S. Anstis

Department of Psychology, York University,
Downsview, Ontario M3J 1P3, Canada

1. Visual Structure and Visual Coding

1.1 The Eye and the Optic Nerve

Instead of writing a broad review about all the different features which are coded by the visual system, I shall concentrate on only two stimulus dimensions: retinal position, and motion across the retina. These are a good pair to pick, because they embody coding principles which are much used by the visual system. I shall start with hints which we glean about coding schemes from studying the constraints which are imposed by the structure of the visual system.

A physicist might evaluate the performance of the visual system by comparing it with a perfect device and seeing how far actual visual performance departs from perfection. I shall take the humbler approach of a development engineer who estimates merely how much better the actual system is than the prototype device that he might have thought of first - how far the visual system has evolved from mediocrity instead of how far it still would have to go to reach perfection.

It turns out that the visual system extracts excellent performance from limited resources by means of a division of labour in which the parts are specialised for different tasks, as in a well-developed economic system.

In most vertebrates the retina is highly non-uniform in its spatial layout, consisting of a foveal region of very high acuity surrounded by a periphery whose acuity falls off progressively with eccentricity. In fact there is a linear increase in the size of the minimum separable target measured psychophysically (ANSTIS 1974), in the size of receptive fields, and in the separation between receptive fields (FISCHER 1973). As we shall see below, the receptive field will be taken as the unit of visual resolution. So the graded acuity across the retina can be accounted for by the graded coarseness of the retinal mosaic. The structure of the retina may be compared to an imaginary sunflower head with small seeds in the middle and progressively larger seeds further out from the centre.

This layout of the retina is an ingenious compromise between the conflicting demands of field of view and resolution. DRASDO (1977) has estimated that there are about 33,000 receptive fields per degree of visual angle in the human fovea. To endow the whole retina with the acuity of the fovea centralis would require several hundred million fibres in the optic nerve. On the other hand, if the 100M receptors and 1M ganglion cells were distributed uniformly across the existing retina, the eye's acuity would fall by a factor of several hundred. A retina with a fovea surrounded by a lower-acuity periphery can be compared to a low-magnification finder telescope with a large field of view, which would find any interesting target and then steer on to it a high-powered main telescope with a very small field which could examine the target in more detail.

Other sensory surfaces show the same non-uniform division of labour as the retina. The finger tips can resolve two tactile points which are only 1-2 mm apart. This increases progressively up the arm to 10 mm on the upper arm (WEINSTEIN 1968).

1.2 Overlapping Receptive Fields can Improve Visual Acuity

If the receptive fields of ganglion cells abutted each other like tiles in a pavement, then the black and white elements of the finest resolvable pattern would have to fall on to neighbouring tiles. However, overlapping receptive fields can resolve a pattern which is finer than the retinal tiles. Three abutting tiles A, B, C give four discriminable signals (if we include the null signal), but three tiles which overlap like a Venn logic diagram add four new discriminable areas, namely the crescent shaped Boolean intersections A.B, B.C, C.A, and the triangular shaped intersection A.B.C. (ANSTIS 1975). In the limiting case, n overlapping fields could define up to 2^n separate intersecting areas, if each field were carefully arranged to bisect every other field. In practice the gain would be far less than this, but it can still be substantial.

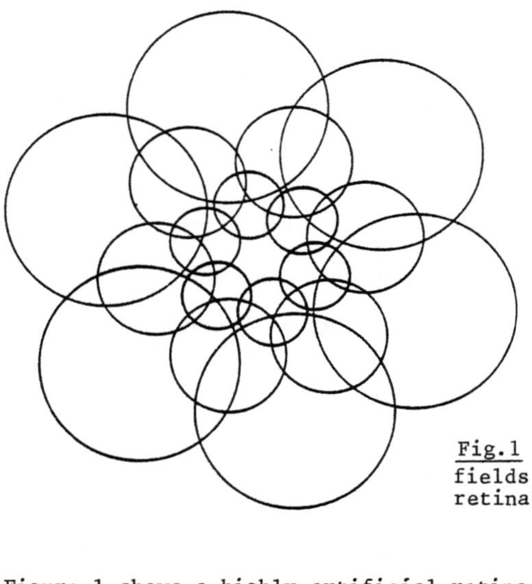

Fig.1 Retina with overlapping receptive fields which increase in size with retinal eccentricity

Figure 1 shows a highly artificial retina of intersecting circles in a hexagonal array, which increase linearly in size with distance from the centre as real ganglion receptive fields do. There are about six times as many intersecting areas as there are circles. Analysing such an array would require substantial circuits to decode the Boolean signal such as 'A and B and not C' into the desired information about the stimulus position. And indeed the visual cortex does contain about 510 times as many neurons as the optic nerve.

In fact acuity is far better than this because neurons have an adjustable firing rate which carries far more information than the one-bit Boolean choice of either 'on' or 'off'. A neuron with a positional tuning curve will fire maximally for a stimulus in the optimal position, and gradually less briskly as the stimulus moves away in any direction from the optimal position. Position is now coded by the ratio of firing rates in two neighbouring neurons, which can give 'infinite' positional acuity, subject only to the noise level which limits the discrimination of firing rates.

2. Visual Coding of Retinal Position

2.1 Some Problems

I should like to conjecture how the visual system codes the retinal position of stimuli. At first it might seem obvious that the position of a stimulus on the retina is coded simply by the local sign of the neurons which it stimulates, so that we know where a stimulus lies on the retina because we know where the neurons which it stimulates are (HERING 1899). This is probably true so far as it goes, but it does not go nearly far enough. I shall take the story far enough to throw some light on various puzzling findings about the sensing of position:

1. Vernier acuity and stereo acuity, or two of the hyperacuities as WESTHEIMER (1976) has called them, are five or ten times finer than grating acuity and can resolve displacements which are considerably smaller than the diameter of a foveal cone (Fig. 2).

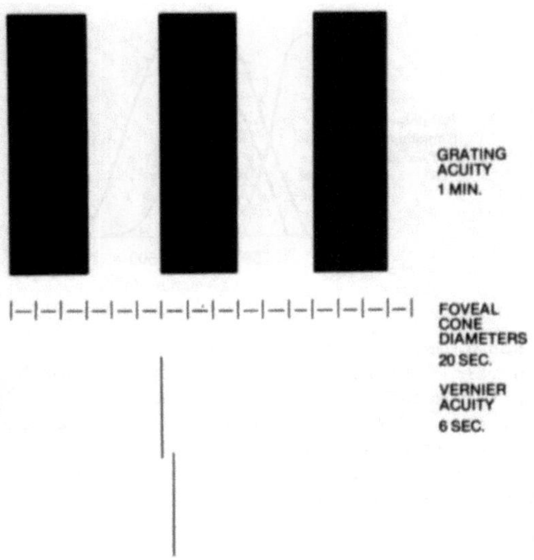

Fig. 2. Diagram, to scale, of: a, just resolvable grating: each black bar is one arc min wide. b, diameters of foveal cones (20 arc sec). c, just noticeable vernier offset of 6 arc sec

2. Two adjacent points, which lie too close together to be resolved when presented simultaneously, can easily be resolved when presented successively, and they will give a sensation of apparent motion. In other words the displacement threshold is lower than the spatial resolution threshold, both in the fovea (WESTHEIMER & MCKEE 1975, 1977b: LEGGE & CAMPBELL 1980: HADANI, GUR, MEIRI & FENDER 1980) and in the retinal periphery (THORSON, LANGE & BIEDERMAN-THORSON 1969).

3. The position sense can be adapted by a figural aftereffect of 'contour repulsion' (KOEHLER AND WALLACH 1944), which cannot be explained in terms of the more recently discovered, but more intensively studied, aftereffect of apparent shift in spatial frequency (BLAKEMORE AND SUTTON 1969).

Findings 1 and 2 suggest that favourable configurations in space or in time can greatly improve the resolution of position, and finding 3 suggests that perceived positions can be distorted by prior adaptation. None of these results are readily explained by HERING'S (1899) local-sign model for the coding of visual position, which is still followed by most textbooks. They can be better understood if the neural coding for position has some analogy to coding for colour, in which the visible spectrum is represented in the responses of three mechanisms with broad, overlapping tuning curves to wavelength (Fig. 3).

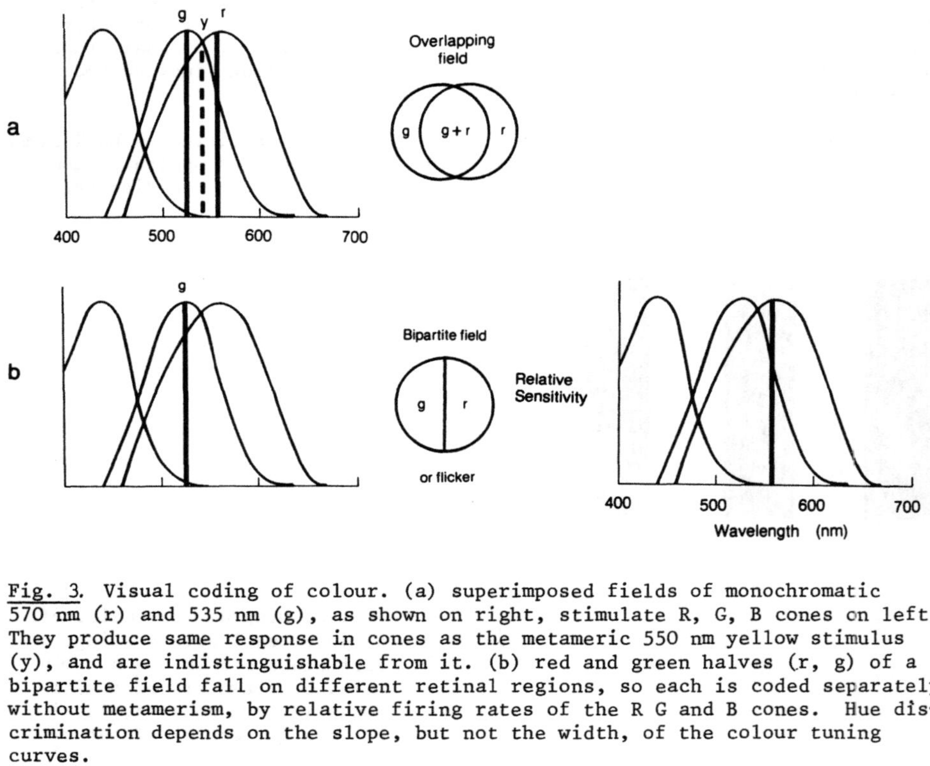

Fig. 3. Visual coding of colour. (a) superimposed fields of monochromatic 570 nm (r) and 535 nm (g), as shown on right, stimulate R, G, B cones on left. They produce same response in cones as the metameric 550 nm yellow stimulus (y), and are indistinguishable from it. (b) red and green halves (r, g) of a bipartite field fall on different retinal regions, so each is coded separately, without metamerism, by relative firing rates of the R G and B cones. Hue discrimination depends on the slope, but not the width, of the colour tuning curves.

2.2 Position Coding by Spatial Sensitivity Profiles

Retinal receptors converge on to later neurons, the retinal ganglion cells. I suggest that the position of a point stimulus is coded not directly by the retinal receptor which it excites but by the ratio of firing which it provokes in the numerous ganglion cells whose receptive fields it stimulates. A small displacement of a single line will alter this firing ratio. The displacement threshold corresponds to the hue discrimination threshold in colour vision, and is limited not by the width of the tuning curves but only by the visual system's ability to discriminate changes in firing rate. However, the minimum separable is limited by the width of the positional tuning curves: stimuli which are separated by less than one tuning-curve width cannot be spatially resolved, because they stimulate the same set of tuning curves and will be neurally blurred together by 'positional mixing'. They will be perceived as a single point or blob located at the mean position of the two stimulus points.

The receptive fields of retinal ganglion cells, which have been plotted by physiological means in many species, can be regarded as positional tuning curves. So can the line-spread functions, measured psychophysically. Two parallel, subthreshold lines summate when they are within 5-10 min arc but inhibit each other when they are somewhat further apart (RENTSCHLER AND FIORENTINI 1974: HINES 1976: WILSON, PHILLIPS, RENTSCHLER AND HILZ 1979). When they are much further apart they do not interact. A similar 'mexican hat' neural line-spread function may also be derived by applying a Fourier transform to the human contrast sensitivity function (Fig. 4). (WILSON AND ROBSON discuss elsewhere in this volume whether the CSF is mediated by a single broadly tuned channel or is the envelope of many more narrowly tuned channels).

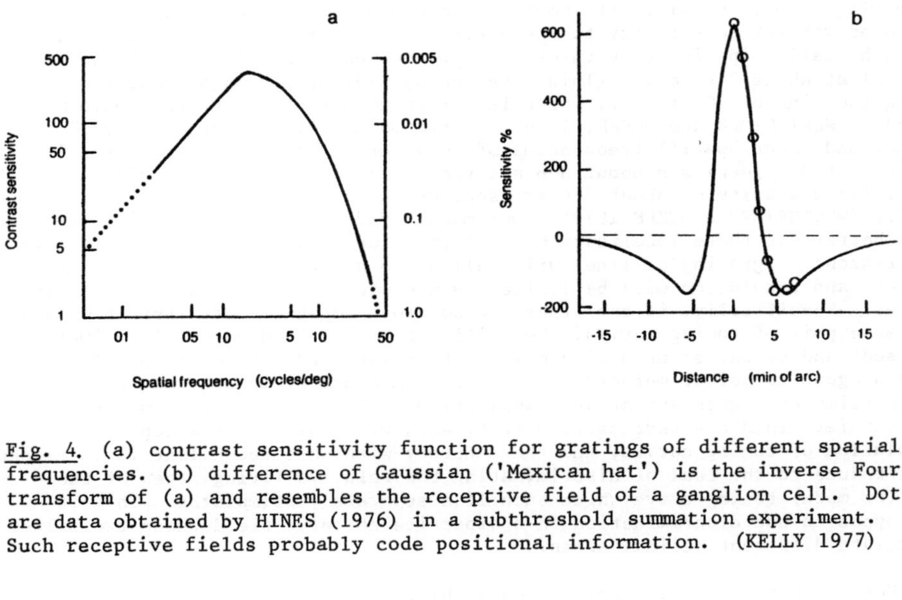

Fig. 4. (a) contrast sensitivity function for gratings of different spatial frequencies. (b) difference of Gaussian ('Mexican hat') is the inverse Fourier transform of (a) and resembles the receptive field of a ganglion cell. Dots are data obtained by HINES (1976) in a subthreshold summation experiment. Such receptive fields probably code positional information. (KELLY 1977)

These neural line-spread functions should not be confused with the purely optical line-spread function measured by KRAUSKOPF (1962), CAMPBELL and GUBISCH (1966) and GORRAND (1979), which has a central core only about 1 min wide if the pupil size is optimised to 2.3 mm, and no inhibitory flanks. This optical line-spread function is generally taken as an unavoidable limitation of the eye, but it might be used to the advantage of acuity. Receptors in the fovea centralis each have a private line up the optic nerve, with no neural convergence in the ganglion cell layer. But the optical line-spread function will cause each receptor to respond maximally to a stimulus centred on it, and progressively less as the stimulus moves about 1 arc min away from it. So the firing ratio between two adjacent receptors will vary systematically as a stimulus moves between them, allowing vernier resolution which is much finer than the inter-cone spacing. It makes no difference to the brain whether the positional tuning curves of foveal cones are produced by neural convergence or by optical diffraction. In this view, vernier resolution (but not grating resolution) can be finer than either the retinal mosaic or the retinal optics would suggest.

2.3 Hyperacuity in space: Vernier and stereo acuity

A vernier offset of only 5-6 arc sec between two vertical lines is readily detected (WESTHEIMER & MCKEE 1977). Performance is best when the tips of the lines are separated by 2-5 arc sec, and is not significantly changed when the targets are three dots whose colinearity has to be judged (LUDVIGH 1953: BECK & SCHWARTZ 1979). Remarkably, vernier acuity is unaffected by movement of targets across the retina. For example, vernier acuity for a 200 msec exposure, which is too brief to permit eye movements, is substantially the same whether the target is stationary or moving at a velocity of 4 deg/sec or even faster (WESTHEIMER & MCKEE 1975, 1976: FAHLE AND POGGIO 1981). This means that a subject can detect the direction of offset when a pair of lines is offset by less about one-fifth of a receptor diameter, while the whole pattern is moving across the width of about 90 receptors in 200 msec. TYLER (1973) measured vernier sensitivity to the curvature of a sinusoidal wiggly line, which he called 'periodic' vernier acuity. He found that sensitivity was maximal at about 3 cy/deg. (This gives an optimal peak-trough separation along the line of 10 arc sec, which is larger than the 2-5 arc sec optimum found by WESTHEIMER and MCKEE, 1976). Performance fell off steeply for both higher and lower spatial frequencies of curvature. The displacement sensitivity (at 3 cy/deg) was about 3-6 arc sec. Stereo acuity can resolve a binocular disparity of about 3-6 arc sec, which is comparable to vernier acuity (WESTHEIMER & MCKEE 1979). Our model predicts that acuity will be much better for these localisation tasks than for resolution, for the following reason: A grating of lines will fall on to a set of positional tuning curves, and resolution will be limited by the width and spacing of the tuning curves. Specifically, if two lines are so close together that they stimulate the same pair of tuning curves, they will not be resolved but will be 'metamerised' and perceived as a single line at an intermediate position. This is analogous to colour metamerism in which a mixture of monochromatic red and green stimulus lights appear as a subjective yellow in Fig. 3. However, if two vernier lines are vertically offset, each will fall on to a separate population of tuning curves, and the accuracy with which each line is localised will reduce to the task of discriminating the relative firing rates of the two (or more) tuning curve which each line separately stimulates. This is analogous to hue discrimination in a bipartite coloured field whose two halves stimulate different retinal areas.

Notice that the model implies neural blurring or interpolation along the direction of the vernier offset. It does not imply any averaging along the length of the vernier lines, as suggested by HERING (1899). The fact that vernier acuity is as good for three dots as for a conventional vernier target (LUDVIGH 1953: BECK & SCHWARTZ 1979) is evidence against HERING'S theory, but not against mine.

2.4 Hyperacuity in Time: The Fine Grain Motion Illusion

THORSON et al. (1969), BIEDERMAN-THORSON et al. (1971) and FOSTER et al. (1981) flashed two spots in sequence in the retinal periphery. Each spot was illuminated for 15 msec, with a 50 msec dark interval between flashes. The spots were too close together to be resolved when stationary, but when flashed in sequence they gave easily resolvable apparent movement with the illusory impression that a single dot moved over a path of considerable extent. Thus, a pair of dots which were actually separated by only 12 arc min appeared to traverse a path 2 deg long when viewed at an eccentricity of 10 deg, and 6 deg long at an eccentricity of 24 deg. When mapped on to the cortex by the human cortical magnification factor (COWEY AND ROLLS 1974), the illusory path length spanned a patch of cortex about 3 mm dia, regardless of stimulus eccentricity: such a region in primate visual cortex corresponds approximately to the locus of cortical cells that 'see' a given retinal point.

Once again, the two stationary spots fell on the same array of positional tuning curves. They could not be resolved but were confounded together by positional metamerism. However, each of the two successive spots in the display was presented to the same array of tuning curves, but at different times, so it could be successfully localised without interference between spots. Hence resolution of the stationary spots was limited by the spatial width and separation of the tuning curves, but localisation of the jumping spots reduced as for a vernier target, to the task of discriminating firing rates in different tuning curves, which is limited only by noise. This is analogous to hue discrimination in a flickering field in which two hues such as red and green are presented at different times, i.e. in alternation. Assuming that the flicker is slow enough to be temporally resolved, the red and green are not subject to metamerism and do not fuse into subjective yellow.

2.5 Reversed Apparent Motion

ANSTIS (1970) discovered an illusion of apparent motion when two black and white pictures were projected on a screen, overlapping and out of exact register by only a few min arc. If one picture faded down as the other picture faded up, strong apparent motion (AM) was seen. However, if one picture was the photographic negative of the other, with the blacks and whites reversed in luminance, then strong AM was seen in a direction opposite to the physical displacement (ANSTIS & ROGERS 1975 and in press). Similar effects were found in stereoscopy (ROGERS & ANSTIS 1975): If the left eye saw the left half of a black and white stereo pair, whilst the right eye saw the right half plus a dim superimposed negative of the left half, then reversed stereo was seen, and stimuli with crossed disparity, which usually look nearer, actually looked further away. GREGORY & HEARD (1982) have recently made an extensive parametric study of some related phenomena in motion, stereo depth, and vernier offset.

These effects can be explained by the positional tuning curve model, which predicts that the contours consisting of a positive edge with a superimposed and slightly displaced negative edge will be shifted by the neural blurring inherent in the tuning-curve representation of visual space.

2.6 Figural After-effects

Visual after-effects can give clues to visual coding (ANSTIS 1975). Following adaptation to lines which are curved or tilted away from the vertical, a vertical test line appears to be curved or tilted in the opposite direction (GIBSON 1937). Following adaptation to a grating of medium spatial frequency, a coarse test grating seems even coarser, and a fine test grating seems even finer than it really is (BLAKEMORE & SUTTON 1969). These after-effects show a 'distance paradox'. The maximum tilt after-effect occurs not when the orientations are almost the same, but when they differ by 10-15 degrees, and the maximum apparent shift in spatial frequency occurs when the adapting and test frequencies differ by about half an octave. These after-effects are probably caused by adaptation of visual channels tuned to orientation or spatial frequency. I wish to suggest that the figural aftereffects (FAE) of contour repulsion (KOEHLER & WALLACH, 1944: ROBINSON, 1972) can also be explained by adaptation of tuning curves which are tuned for retinal position.

If a line is fixated for a time and then replaced by another line, close to the position of the adapting line, then the second line will appear displaced away from it (Fig. 5). KOEHLER and WALLACH reported a distance paradox for the positions of the test and adapting contours.

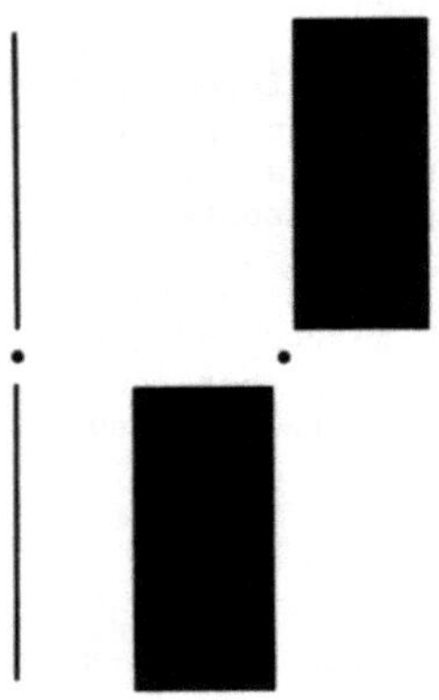

Fig. 5

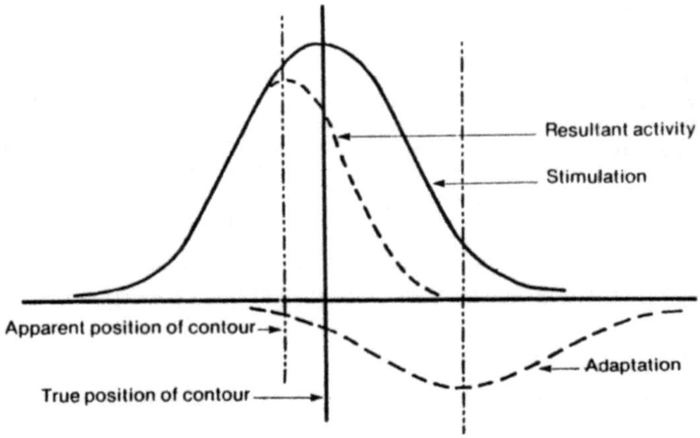

Fig. 6

Fig. 5. Figural after-effect of contour repulsion (after KOEHLER & WALLACH 1944). Fixate the spot between the two black blocks for about one minute. Transfer your gaze suddenly to the right-hand spot, and the upper test line will look apparently displaced to the right, the lower test line to the left. This is predictable from positional tuning curves

Fig. 6. Model of figural after-effects. Test line normally sets up Gaussian pattern of activity in a population of positional tuning curves (upper curve). Note that this is not a single tuning curve, but is a population response. Adaptation to a line displaced to the right (lower dashed line) skews the response of the adapted population when exposed to the test line. This skewing, to the left, accounts for the figural after-effect. (After ROBINSON 1972)

Figure 6 suggests how an adapting contour would stimulate a population of receptive fields, or positional tuning curves, and excite in them a pattern of activity which is the mirror image of the shape of an individual tuning curve. A slightly shifted test contour exposed to this adapted population will produce a response whose peak is skewed away from the position of the adapting contour. Models of this type have been applied to after-effects of tilt and spatial frequency, as reviewed by OVER (1971), COLTHEART (1971) and ANSTIS (1975).

OSGOOD and HEYER (1952) based their theory of FAEs on the idea of MARSHALL and TALBOT (1942), that nystagmic eye movements led to a line or contour on the retina being represented in the visual cortex as a normal distribution of neural activity. This theory is no longer much believed in its original form, since it is now known that nystagmic eye movements do not improve acuity (TULUNAY-KEESEY 1960) and are not required for the FAE (HOCHBERG AND HAY 1956: KRAUSKOPF 1960). They wrote before the discovery of receptive fields but my suggestion that position is coded by Gaussian, or difference-of-Gaussian, sensitivity profiles, relies on the same geometry as theirs.

3. Visual Coding of Motion

3.1 Algorithms to Code Short-Range Apparent Motion

To a physicist, motion is simply change of position over time. To a psychophysiologist matters are not so simple; the visual system probably uses more than one kind of algorithm to extract motion from a scene (ANSTIS 1978, 1980). In what follows I shall discuss the coding of visual motion, especially the simplified case of intermittent or apparent motion (AM), in which a spot or a pattern of spots jumps from one position to another.

BRADDICK (1974) has suggested that AM is perceived by two different visual processes; a short-range process, probably mediated by hard-wired neural motion detectors, which responds only to small jumps of about 15 arc min or less: and a long-range process, probably more central and with perhaps a cognitive component, which responds to much larger jumps, up to tens of degrees. Short-range AM is probably coded by hard-wired neural motion detectors. Such neurons have been found in the rabbit by BARLOW and LEVICK (1965), and in many other vertebrate species. BARLOW and LEVICK noted that the direction-selective neurons which they found would also respond to AM, i.e. to a pair of dots flashed in sequence within the neural receptive field. They conjectured that such a neuron might code motion by receiving inputs from two adjacent retinal areas A and B. Region A would excite the neuron, and B would inhibit it through a built-in neural delay. A spot moving in the neuron's null direction would provoke a signal from A which would be countermanded by B, preventing any neural output. However, a spot moving in the cell's preferred direction would excite A, and the (delayed) inhibitory signal from B would arrive too late to countermand A's signal, so the neuron would fire. GRUESSER and GRUESSER-CORNEHLS (1973) have reviewed other models of this type, and REICHARDT (1961) proposed a broadly similar model for the insect eye, in which the signals from areas A and B were multiplied or correlated instead of subtracted one from the other. POGGIO discusses mechanisms of this kind elsewhere in this volume.

These delay-and-correlate or delay-and-subtract algorithms in effect compare an image at one instant with that at a slightly later instant and display the differences between the two, as in Fig. 7. MACKAY (1959) made a positive and negative print of the same 16 mm movie, shifted the two prints a few frames relative to each other and projected both prints together, to enhance the changes over time which occurred in x-ray movies of speech and bodily motion. CANDY, FRANKE, HASKELL & MOUNTS (1971) noted that most of a TV picture is highly redundant, since the stationary parts of a picture are the same in each successive frame, and they found they could transmit TV pictures with eight times the usual efficiency by transmitting only those picture elements which changed between successive frames - about 4% of picture elements for people conversing on a picturephone. This achieved substantial bandwidth

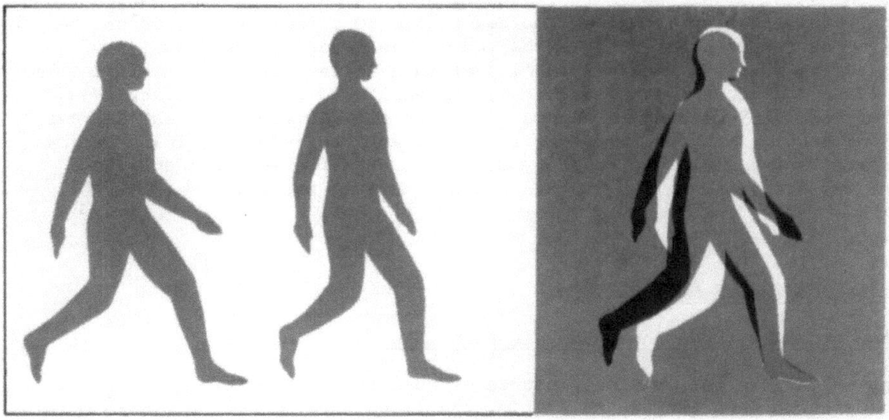

Fig. 7 a, b, pictures at times T1 and T2. c, difference picture made by subtracting a from b; stationary objects disappear but moving or changing objects leave a record

compression. Unfortunately, a cut to a new scene could be transmitted only very slowly as the buffer took a long time to fill up.

The subtraction method detects the presence of change (a temporal event). To discriminate the direction and velocity of motion (a spatiotemporal event), one image needs to be shifted around spatially over the other to obtain the best match. Piecewise local motions will require piecewise shifts. BILLINGSLEY (1973) has done this with a computer. Traffic moving along a highway was filmed from overhead by a helicopter. When one moved frame was digitally subtracted from its predecessor, stationary cars cancelled out and disappeared, whereas cars which had moved between frames showed double-images where the positive was displaced from the negative. If one frame was shifted prior to subtraction by an amount corresponding to 30 mph, then cars travelling at the speed disappeared from the difference picture. A series of shifts and subtractions would reveal the velocities of all the cars. Clearly the amount of computing needed is large, and increases dramatically as one expands the range of picture shifts which are searched.

This point-by-point, shift-and-subtract algorithm extracts motion from successive images. Its biological analogue is to expose the image to a parallel array of hard-wired Barlow (or Reichardt) motion detectors. Different spatial shifts of the two pictures are represented by Barlow units with different separations between their retinal inputs. As for a computer, the neural costs increase sharply if one searches across too great a spatial range, because more Barlow units will be needed. BRADDICK (1974) has found evidence that neural motion detectors are sensitive to jumps of only 15 min arc or less, whereas larger jumps are handled by a back-up process, perhaps more cognitive and more central, which he terms the 'long-range process'.

It is harder to simulate the long-range process than the short-range process because the algorithm would involve pattern recognition rather than image processing. To indicate some of the problems that such an algorithm needs to solve, I shall describe some studies I have made on coding and decision processes in long-range apparent motion. In terms of coding I have found that the inputs to the long-range motion process include luminance and contrast, but not colour. In terms of decision processes V.S. RAMACHANDRAN and I have examined the choices made by the visual system among equally possible motion paths in an ambiguous display consisting of four jumping dots.

3.2 Inputs to Long-Range AM: Luminance and Contrast

If an isolated white bar flashes first in one position then in another, the correspondence problem is easy to solve: the two bars are matched up perceptually because they have the same luminance. However, the two bars need not have the same luminance contrast. A white bar followed by a displaced black bar gives perfectly good AM, and if a white bar is followed by a pair of bars - a white bar above it together with a black bar below it - the first bar appears to split into two parts, one moving up as the other moves down. In other words when AM between bars of the same contrast is put in competition with AM between bars of opposite contrast, the former does not suppress the latter: instead, both co-exist in the form of split motion (Fig. 8a). But now consider an upper dark bar lying parallel to a lower light bar. If the bars suddenly change places, will an observer perceive a dark bar jumping downward, or a light bar jumping upward, or both at once? Each percept is equally compatible with the signals of changing luminance. The previous demonstration might lead one to expect two opposed motions, with the light and dark bars apparently jumping in opposite directions. But this is seen only rarely. Instead, the bar which has the higher contrast against the

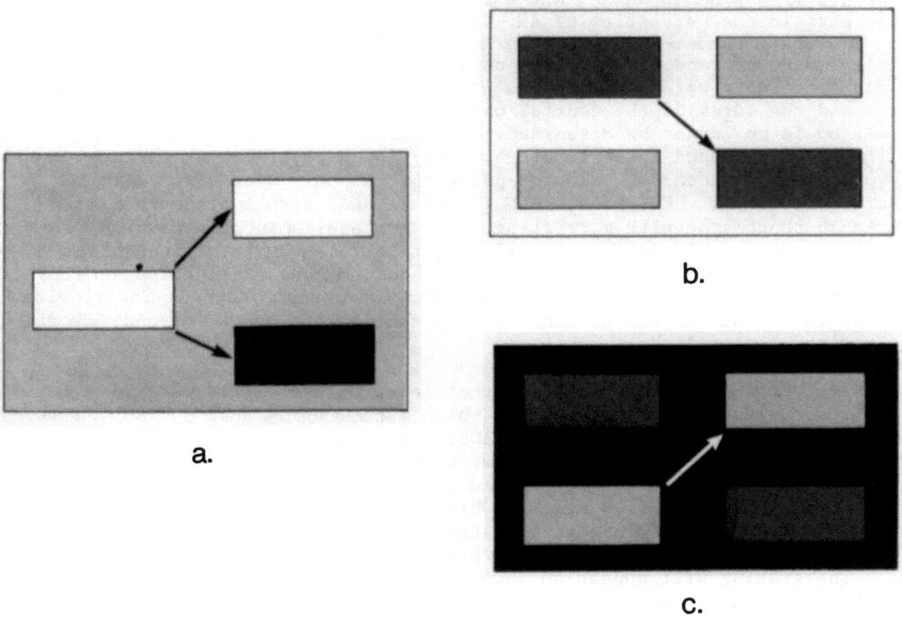

Fig. 8. Luminance and contrast are inputs to the long-range apparent motion process. (a) A white bar at time T1 was followed at time T2 by a white bar above and a black bar below. Note that the bars were actually aligned vertically; the bars at T2 have been offset to the right in the Figure for clarity only. (b,c) If a dark upper bar and a light lower bar exchange places, the direction of apparent motion depends upon the background luminance

background is perceived as moving. On a white ground it is the dark bar, and on a black ground it is the light bar, that appears to move. The bar which is seen as moving looks like a 'figure', and the two apparently empty, stationary slots between which it appears to jump look like a 'ground'.

Very occasionally one sees a light bar and a dark bar jumping in opposite directions simultaneously in a reversible figure/ground picture. Occasionally, also, the motion fluctuated, with sometimes the light bar and sometimes the dark bar moving. SIGMAN and ROCK (1974) found that when an opaque occluder moved to and fro over a pair of stationary bright spots, subjects correctly saw the motion of the occluder and did not report AM on the spots, even though the time intervals chosen would give AM in the dark when the occluder could not be seen. SIGMAN and ROCK regarded this AM as an intelligent perceptual solution to the problem posed by the stimulus. However, our AM appeared to be determined by the luminance contrast of the stimulus, not by higher-level cognitive factors. TYLER (1973b) viewed a bright spot in a dark room which jumped back and forth between two positions to give AM. He found that sometimes when the spot actually jumped from left to right, subjects reported seeing a dark shadow or fog jumping from right to left. TYLER called this 'omega movement', and found that its spatio-temporal properties were different from AM of the spot itself. An analogy might be the motion of holes in a transistor, which move opposite to the flow of electrons.

3.3 Inputs to Long-Range Apparent Motion: Little Input from Colour

ANSTIS (1970) and RAMACHANDRAN and GREGORY (1978) found that colour plays little part in the perception of AM. They presented in alternation a pair of black and white random-dot kinematograms (JULESZ 1971), in which an oscillating central square of randomly arranged dots could be seen. When equiluminous red and green dots were used instead of black and white, the AM was much less marked and the edges of the central oscillating square, which were defined by the AM, could no longer be discerned. CAVANAGH, TYLER and FAVREAU (1982) have confirmed and extended these findings. They found that weak AM could be seen in equiluminous random-dot coloured patterns, but only over a very limited range of conditions. CAVANAGH, BOEGLIN and FAVREAU (1982) discovered a remarkable phenomenon with a drifting red-green equiluminous sinusoidal grating, such as one could make by interlacing a red/black with a green/black sinusoidal grating. The motion of such an equiluminous grating appeared to be solwed down by 60% to 100%, sometimes looking almost stopped. The slowing down was most dramatic for gratings of low spatial frequency (<1 cpd) which were moving slowly (<1 Hz).

We have obtained further evidence that colour plays little part in motion perception, by developing a technique which uses minimum apparent motion as a criterion to assess equiluminance (ANSTIS AND CAVANAGH 1983). Consider two coloured square-wave gratings, one made of red and green bars, the other of dark and light yellow bars. If the red/green grating is suddenly replaced by the dark yellow/light yellow grating which is displaced by one quarter cycle (half a bar width) to the right, then the grating will appear to jump sideways, in a direction which depends upon the brightness relations between the gratings. The grating will appear to jump to the left if the red bars are darker than the green bars, but to the right if the green bars are darker than the red. If the red and green bars are made equiluminous, no consistent AM will be seen. Subjects were asked to adjust the relative luminance of red and green to the point of minimum motion. This procedure was easy to carry out and gave equiluminance settings which compared favourably with the existing techniques of minimum flicker and minimum border (WAGNER AND BOYNTON 1972). Thus, information about motion is mediated almost entirely by visual channels which respond to luminance, not to chrominance.

3.4 A Vector from Split Motion of Random Dots

If a single dot is followed by two flanking dots placed symmetrically on either side of it, it will appear to split into two, with one fragment moving to the left, the other to the right (ULLMAN 1979). The three dots can be in an L instead of in a straight line; split motion is still seen, and it does not matter whether the two arms of the L are equal or unequal. However, RAMACHANDRAN and I filled a TV screen with 500 sparsely scattered random dots, and replaced them with 500 L-shaped pairs. In these conditions, split motion was never seen, but instead the whole field of dots seemed to move obliquely as a unit. Moreover, unlike the isolated trio of dots, the apparent motion of the field was now strongly influenced by the relative dot separation. If the two arms of the L were the same length, then the field appeared to move at 45 deg, i.e. along an average or vector of the horizontal and vertical component. If the two arms of the L were made unequal so that the vertical jump was twice the horizontal jump, then again the entire field was seen moving as a whole without splitting. The direction of the apparent motion was still a vector, but this time it was not fixed but variable: it now depended on the size of the display, as controlled by the viewing distance. Seen from close up (large retinal stimulus) the apparent motion was horizontal, but as the viewing distance was increased (smaller stimulus) the apparent direction of motion became first oblique and then nearly vertical (Fig. 9).

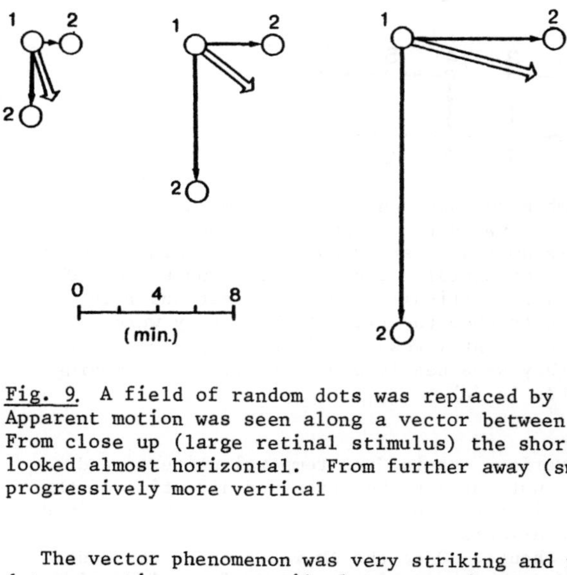

Fig. 9. A field of random dots was replaced by 'L-shaped' pairs of dots. Apparent motion was seen along a vector between the horizontal and vertical. From close up (large retinal stimulus) the short AM dominated and the AM looked almost horizontal. From further away (smaller stimulus) the AM looked progressively more vertical

The vector phenomenon was very striking and robust, and persisted when the dot separations were easily large enough to resolve. Possibly apparent motion is mediated by a system with rather low spatial acuity which averaged or metamerised the motion components in the stimulus, even when the dots which carried the motion could be readily resolved by the position-sensing system.

3.5 Post-Coding Decision Processes: Ambiguous AM of four Dots

If two similar pictures, such as two successive frames of a movie, are flashed in sequence, apparent motion can be seen. How similar must the pictures be to produce AM? This is the "correspondence problem" (ULLMANN 1979). The visual system must decide which point in one picture corresponds to a given point in the other picture. Failure to solve this problem will lead to false motions between non-corresponding points, or else to the complete breakdown of apparent motion.

RAMACHANDRAN and I have studied the correspondence problem, using an ambiguous motion display consisting of four dots which are located at the corners of an imaginary square (Fig. 10a,b). The two dots on the positive diagonal are flashed at time T1, alternating with the two dots on the negative diagonal which are flashed at time T2. The cycle repeats indefinitely. The visual system has to decide which dots should be treated as corresponding, and can pair them off either as an upper and lower pair, or else as a left and right pair. In either case an observer sees two dots moving back and forth, either horizontally along the top and bottom edges of the square, or else vertically along the left and right sides of the square. The percept reverses spontaneously from time to time.

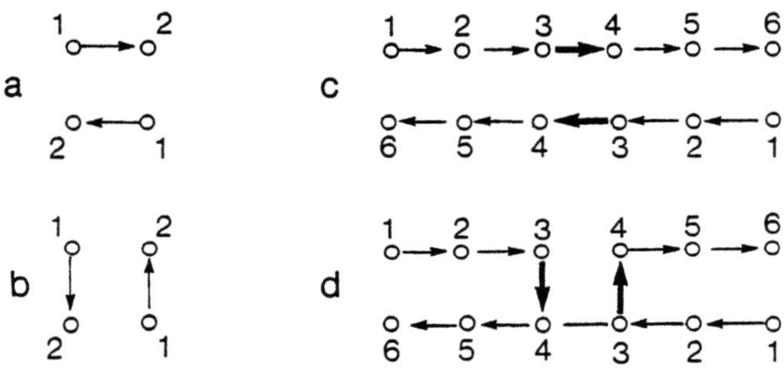

Fig. 10. Decision processes in ambiguous apparent motion. Two dots were flashed at time T1, alternating with two dots at opposite corners of an imaginary square at time T2. This ambiguous stimulus could be seen as moving in two ways: (a) two dots moving horizontally along the top and bottom edges of the square, or (b) two dots moving vertically along the left and right edges of the square. Numbers by dots show the time at which they were presented. When embedded in two horizontal rows of six dots, presented in the order shown by the numbers, they were nearly always perceived as moving horizontally (c), not vertically as in (d)

We embedded the four dots in various spatial and temporal contexts and found that some did, and some did not, affect the perceived direction of the AM. Notice that the stimulus configuration of the four dots always remained the same. We found the following effects:

1. Hysteresis, or reluctance to change. Once the AM was seen in one mode (vertical or horizontal), it tended to persist in that mode for a long time and be resistent to change.
2. Covering two dots has little effect. Suppose an observer reports vertical AM. Covering up the lower pair of dots will constrain him to see horizontal motion between the upper pair. However, when the four dots are then uncovered there is little tendency to keep seeing the horizontal motion: he is likely to flip back to seeing vertical motion.
3. A conditioning motion had little effect. Another way to produce obligatory horizontal motion was to interpose a pair of flashed dots midway along the top and bottom sides of the square. Five cycles of this conditioning

motion were alternated with five test cycles of the ambiguous four-dot display. However, there was little carry over from the unambiguous conditioning AM to the ambiguous test AM: following exposure to the horizontal AM, an observer was equally likely to see the ambiguous AM as horizontal or vertical.

4. A spatial array of many dot-quartets all tended to oscillate in step. An array of 8 x 8 dot-quartets were made to flash in synchrony. It was found that all of them tended to move in step: if one of them moved horizontally (or vertically), they all did, and it was extremely rare to see some of the dot-quartets moving vertically and others moving horizontally. Some columns of the array were mirror images of the others, so that they moved to the right (or down) when the other columns moved to the left (or up). These columns remained locked in step with the others, showing that the important variable was the axis of the apparent motion (vertical vs. horizontal), not its direction (left vs. right, or up vs. down).

5. Visual momentum. In Fig. 10, the four dots on their own could be seen with equal ease as moving horizontally (a) or vertically (b). But if the four dots were embedded in a horizontal sequence of flashing dots which mimicked two spots moving in opposite directions along parallel horizontal paths, there was a strong tendency to see the horizontal AM at the expense of the vertical. Fig. 10 shows the two possible percepts: (c) If the AM of the four embedded dots was horizontal, two dots were perceived as moving past each other along linear paths with eight stations along each path. This was called 'streaming'. (d) If the AM of the embedded dots was vertical, then two dots were perceived as approaching each other and bouncing back along U-shaped paths. This was called 'bouncing'. It was found that subjects had a strong tendency to see horizontal AM or streaming. This could be measured by pitting it against the cue of proximity, i.e. by moving the two rows of dots closer together, which favoured 'bouncing' by proximity.

Thus there is a stong tendency to perceive motion as persisting along a straight line. Perhaps the visual system eases its coding task by incorporating decision rules based upon physical constraints and redundancies which exist in the physical world, such as Newton's First Law of motion.

Acknowledgements

This work was supported by Grant A 0260 from the Natural Science and Engineering Council of Canada (NSERC). The ideas on positional metamerism were developed during conversations with I.P. HOWARD. I collaborated with P. CAVANAGH on the colour and motion experiment, and with V.S. RAMACHANDRAN on the experiments with four dots. I thank I.P. HOWARD, B.P. MOULDEN, H. ONO and C.W. TYLER for commenting on various drafts of this manuscript.

REFERENCES

Anstis, S.M. (1970) Phi movement as a subtraction process. Vision Res. 10, 1411-1430.

Anstis, S.M. (1974) A chart demonstrating variations in acuity with retinal position. Vision Res. 14, 589-592.

Anstis, S.M. (1975) What does visual perception tell us about visual coding? In Handbook of Psychobiology (ed. M. Gazzaniga & C. Blakemore). New York: Academic Press.

Anstis, S.M. (1978) Apparent movement. In: Handbook of Sensory Physiology, vol. VIII: Perception (ed. R. Held, H. Leibowitz & H.-L. Teuber). New York: Springer Verlag.

Anstis, S.M. (1980) The perception of apparent motion. Phil. Trans. R. Soc. Lond. B. 290, 153-168.

Anstis, S.M. & Cavanagh, P.C. (1983) A minimum motion technique for judging equiluminance. In: J. Mollon and E. Sharpe (Eds): Proceedings of the NATO Conference on Colour Vision, Cambridge. London: Academic Press.

Anstis, S.M. & Rogers, B.J. (1975) Illusory reversal of depth and movement during changes of contrast. Vision Res. 15, 957-961.

Anstis, S.M. & Rogers, B.J. (in press) Oscillating positive-negative patterns give illusion of continuous apparent movement. Vision Res.

Barlow, H.B. & Levick, W.R. (1965) The mechanism of directionally selective units in the rabbit's retina. J. Physiol. Lond. 178, 477-504.

Beck, J. & Schwartz, T. (1979) Vernier acuity with dot test objects. Vision Res. 19, 313-320.

Biederman-Thorson, M., Thorson, J. & Lange, G.D. (1971) Apparent movement due to closely spaced sequentially flashed dots in the human peripheral field of vision. Vision Res. 11, 889-904.

Billingsley, F.C. (1973) Digital image processing for information extraction. J. Man-Machine Studies 5, 203-214.

Blakemore, C. & Sutton, P. (1969) Size adaptation: A new aftereffect. Science, 166, 245-247.

Braddick, O. (1974) A short-range process in apparent motion. Vision Res. 14, 519-528.

Campbell, F.W. & Gubisch, R.W. (1966) Optical quality of the human eye. J. Physiol. 186, 558-578.

Candy, J.C., Franke, M.A., Haskell, B.J. & Mounts, F.W. (1971) Transmitting television as clusters of frame-to-frame differences. Bell System Tech. J. 50, 1889-1917.

Cavanagh, F., Boeglin, J. & Favreau, O. (1982) Perception of apparent motion in the chrominance channel. (Ms. in preparation).

Cavanagh, P., Tyler, C.W. & Favreau, O. (1982) Perception of velocity of sinewave gratings in the chrominance channel. (Ms. in preparation).

Coltheart, M. (1971) Visual feature-analysers and aftereffects of tilt and curvature. Psych. Rev. 78, 114-121.

Cowey, A. & Rolls, E.T. (1974) Human cortical magnification and its relation to visual acuity. Expl. Brain Res. 21, 447-454.

Drasdo, N. (1977) The neural representation of visual space. Nature 266, 554-556.

Fahle, M & Poggio, T. (1981) Visual hyperacuity: spatiotemporal interpolation in human vision. Proc. R. Soc. Lond. B 213, 451-477.

Fischer, B. (1973) Overlap of receptive field centres and representation of the visual field in the cat's optic tract. Vision Res. 13, 2113-2120

Foster, D.H., Thorson, J., McIlwain, J.T. & Biederman-Thorson, M. (1981) The fine-grain movement illusion: A perceptual probe of neuronal connectivity in the human visual system. Vision Res. 21, 1123-1128.

Gibson, J.J. (1937) Adaptation with negative aftereffect. Psych. Rev. 44, 222-244.

Gorrand, J.M. (1979) Diffusion of the human retina and quality of the optics of the eye on the fovea and the peripheral retina. Vision Res. 19, 907-912.

Gregory, R.L. & Heard, P. (in press) Visual dissociation of movement, position, and stereo depth: some phenomenal phenomena. Quart. J. Exp. Psychol.

Gruesser, O.J. & Gruesser-Cornhels, U. (1983) Neuronal mechanisms of visual movement perception and some psychophysical and behavioral correlations. In: Handbook of Sensory Physiology, Vol. VII/3a: Central Processing of Information. Jung, R. (ed). New York: Springer Verlag.

Hadani, I., Gur, M., Meiri, Z. & Fender, D.H. (1980) Hyperacuity in the detection of absolute and differential displacements of random dot patterns. Vision Res. 20, 947-952.

Hering, E. (1899) Ueber die Grenzen der Sehschaerfe. Ber. math.-phys. Cl. d. Koenigl. Saechs. Gesell. Wiss. Leipzig. Naturwiss. Teil, pp. 16-24.

Hines, M. (1976) Line spread function variation near the fovea. Vision Res. 16, 567-572.

Hochberg, J.E. & Hay, J. (1956) Figural aftereffect, after-image, and physiological nystagmus. Am. J. Psychol. 69, 480-482.

Howard, I.P. (1982) Human visual orientation. Chichester: John Wiley & Sons.

Julesz, B. (1971) Foundation of cyclopean perception. University of Chicago Press.

Kelly, D.H. (1977) Visual contrast sensitivity. Optica Acta 24, 107-129.

Koehler, W. & Wallach, H. (1944) Figural aftereffects: An investigation of visual processes. Proc. Am. Phil. Soc. 88, 269-357.

Krauskopf, J. (1960) Figural aftereffects with a stabilized image. Am. J. Psychol. 73, 294-297.

Krauskopf, J. (1962) Light distribution in human retinal images. J. opt. Soc. Am 52, 1046-1050.

Legge, G.E. & Campbell, F.W. (1981) Displacement detection in human vision. Vision Res. 21, 205-214.

Ludvigh, E. (1953) Direction sense of the eye. Am. J. Ophthalm. 36, 139-142.

MacKay, R.S. (1959) Display of moving parts of a scene. Science 130, 223-224.

Marshall, W.M. & Talbot, S.A. (1942) Recent evidence for neural mechanisms in vision leading to a general theory of sensory acuity. In Visual Mechanisms, Biol. Symp. (ed. Kluver, H.), pp. 117-164.

Osgood, C.E. & Heyer, A.W. (1952) A new interpretation of figural aftereffects. Psychol. Rev. 59, 98-118.

Over, R. (1971) Comparison of normalization theory and neural enhancement explanation of negative aftereffects. Psychol. Bull. 75, 225-243.

Ramachandran, V.S. & Gregory, R.L. (1978) Does colour provide an input to human motion perception? Nature 275, 55-56.

Reichardt, W. (1961) Autocorrelation, a principle for the evaluation of sensory information by the central nervous system. In: Rosenblith, W.A. (Ed.): Sensory Communication. Cambridge, Mass.: MIT Press.

Rentschler, I. & Fiorentini, A. (1974) Meridional anisotropy of psychophysical spatial interactions. Vision Res. 14, 1467-1473.

Robinson, J.O. (1972) The psychology of visual illusion. London: Hutchinson University Library.

Rogers, B.J. & Anstis, S.M. (1975) Reversed depth from positive and negative stereograms. Perception 4, 193-201.

Sigman, E. & Rock, I. (1974) Stroboscopic movement based on perceptual intelligence. Perception 3, 9-28.

Thorson, J., Lange, G.D. & Biederman-Thorson, M. (1969) Objective measure of the dynamics of a visual movement illusion. Science 164, 1087-1088.

Tulunay-Keesey, U. (1960) Effects of involuntary eye movements on visual acuity. J. opt. Soc. Am. 50, 769-774.

Tyler, C.W. (1973a) Periodic vernier acuity. J. Physiol. Lond. 228, 637-647.

Tyler, C.W. (1973b) Temporal characteristics in apparent movement: Omega movement vs. Phi movement. Quart. J. Exp. Psychol. 25, 182-192.

Ullmann, S. (1979) The interpretation of visual motion. Cambridge, Mass.: MIT Press.

Wagner, B. & Boynton, R.M. (1972) Comparison of four methods of heterochromatic photometry. J. opt. soc. Am. 62, 1508-1515.

Weinstein, S. (1968) Intensive and extensive aspects of tactile sensitivity as a function of body part, sex, and laterality. In: Kenshalo, D.R. (Ed): The Skin Senses. Springfield, Ill.: Charles C. Thomas

Westheimer, G (1976) Diffraction theory and visual hyperacuity. Am. J. Optom. physiol. Optics, 53, 262-364.

Westheimer, G. & McKee, S.P.(1975) Visual acuity in the presence of retinal image motion. J. opt. Soc. Am. 65, 847-850.

Westheimer, G. & McKee, S.P. (1967) Integration regions for visual hyperacuity. Vision Res. 16, 1-5.

Westheimer, G. & McKee, S.P. (1977) Spatial configurations for visual hyperacuity. Vision Res. 17, 941-947.

Wilson, H.R., Phillips, G., Rentschler, I. & Hilz, R. (1979) Spatial probability summation and disinhibition in psychophysically measured line-spread functions. Vision Res. 19, 593-598.

Designing and Implementing an Algorithm to Extract Motion Information from Images

C.J. Taylor, J.N.H. Brunt, and R.N. Dixon
Department Medical Biophysics, University of Manchester,
Manchester, United Kingdom

P.J. Gregory
Joyce Loebl Ltd., Gateshead, United Kingdom

Abstract

Extracting information from an image frequently involves combining a priori structural information with data from the image. In cine sequences of time varying events it is often also necessary to combine information from different images in order to build up a self-consistent description of what is happening. This paper describes an analysis method which uses a continuously modified model, allowing temporal, spatial and structural information to be combined in a simple way with high computational efficiency. The example of its application to cineangiocardiograms is discussed in detail. The techniques used in this and other applications suggest data structures and language constructs which are important in allowing problem solutions to be written down simply in high level language. Some of these ideas are described and the influence they have had on our design of image analysis hardware is discussed.

Introduction

This paper addresses the problem of extracting motion information from a time sequence of grey-level images. Since we are interested in practical applications of automated image analysis we start with the assumption that any particular task will involve analysing the motion of specific image structures. We also assume that there will be some prior knowledge constraining the form and motion of the structures of interest.

The simple approach to this problem is to treat each image independently, extracting a description of the structures of interest, possibly with the help of some prior knowledge [1]. When all images have been thus analysed it is necessary to establish the equivalence of points from frame to frame so that motion information may be collected. The obvious disadvantage of this method is that although it can make use of the high degree of spatial redundancy normally present in an image it cannot make use of the temporal redundancy which is also present in movie sequences. In addition the task of identifying equivalent points is not trivial nor is it easy to make effective use of prior knowledge.

Continuously Updated Model

We have developed a technique which attempts to avoid the problems which have been discussed and involves constructing a model to describe the form and motion of the structures of interest. The model is used to predict the features of interest in each image based on the result for the previous image. The grey-level data in each image is used to modify the prediction of this continuously updated model so that only changes from what is expected need be dealt with. This method has the advantage that it allows spatial and temporal information to be combined to achieve a solution which is consistent with all the data. It is robust because the model can be designed to predict a reasonable solution for each image even for frames where there is little useful grey level information. Prior knowledge can be used both in designing the form of the model and in testing grey level derived modifications to the model prediction. Finally the method is computationally efficient since only changes are significant. The model can often be used to identify limited areas of the image in which the grey levels need be inspected. In the next section we describe the application of these ideas to a specific task.

Application to Cineangiocardiograms

A cineangiocardiogram is a sequence of X-ray images of the left ventricle, the main pumping chamber of the heart [2]. The procedure involves introducing into an artery in the leg or arm a long flexible tube, or catheter, which is manipulated until it enters the left ventricle via the aorta. Once in position X-ray opaque dye is injected via the catheter. A sequence of X-ray images is taken at about 50 frames per second to record, on film, the projection of the chamber during the next few heart beats before the dye is exhausted. Fig.1 shows a sequence of frames from such a procedure.

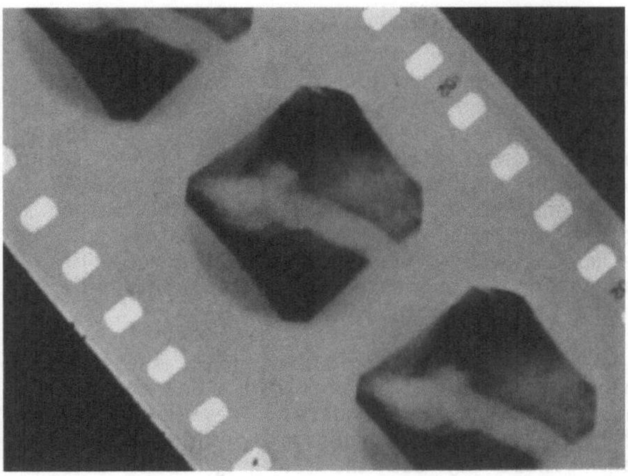

Fig.1. Frames from an angiocardiogram

The purpose of taking the pictures is to view, in detail, the pumping action of the heart when other evidence suggests that it is not operating correctly. Ultimately it is desirable to characterise the motion of small segments of the heart wall to identify those regions which are intact and contracting normally and those which have been affected by partial or complete interruption of the blood supply.

It is difficult to form a very clear impression of what is happening in an abnormal heart simply by showing the X-ray sequence as a movie since the motion can be quite complex. It is our aim to use automatic image

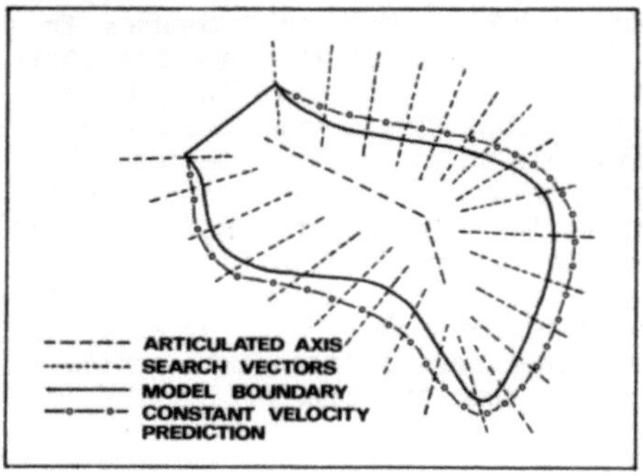

Fig.2. Model of the left ventricle

analysis to follow the motion of the heart wall from frame to frame and to provide the cardiologist with an analysis of the motion.

The main image analysis problems are illustrated in Fig. 1. First the dye is not thoroughly mixed with the blood, which leads to apparent inconsistencies between frames. Next other organs cast shadows which tend to obscure parts of the edge of the ventricle. Finally the X-ray imaging system introduces nonstructured noise which is significant compared to the contrast of the "signal". These difficulties suggest that an updating model method should be advantageous because it allows all of the spatial and temporal information to be used.

The algorithm which we have developed for this application uses a simple model which describes the boundary of the ventricle. Fig. 2 shows how points on the model boundary are constrained to lie on search vectors emanating radially from an articulated axis. In the absence of other information each point on the boundary is assumed to move at its own constant velocity based on its immediate history. The fixed search directions are used to avoid the instability which could be caused by allowing the directions to depend on the current model. The method of generating the search vectors attempts to match the anatomy of the ventricle so that they are approximately normal to the heart wall. In practice the temporal extrapolation is augmented by spatial extrapolation which we have described in detail elsewhere [3].

Having obtained, for a new frame of the angiocardiogram, the boundary predicted by the model, a grey level profile is obtained along each search direction in the neighbourhood of the model boundary. Grey-level criteria can be applied to each profile to detect potential modifications which should be made to the model. These may use the raw grey-level data or, for instance, a cubic fit to it. Suggested boundary points are points of maximum gradient, maximum second derivative, or indeed local thresholding may be used and the point at a fixed fraction of the local grey range selected. An average of the points suggested by the various criteria, weighted to represent their relative importance, is found, together with a quality parameter. The quality parameter indicates the reliability of the grey-level decision. If the confidence is sufficiently high the model is modified so long as it meets anatomical constraints which limit the acceptable curvature of the boundary and its acceleration. If the anatomical constraints are not met the model point is not modified. Fig. 3 shows the fitted boundary at one point in the development of the model and Fig. 4 shows the set of boundaries during one complete cycle of the heart.

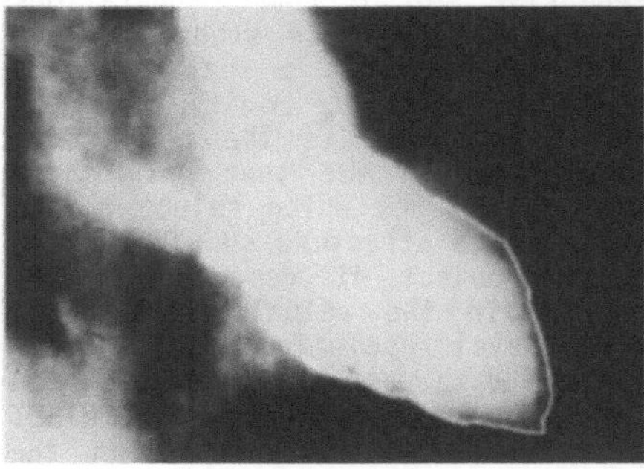

Fig.3. A model boundary

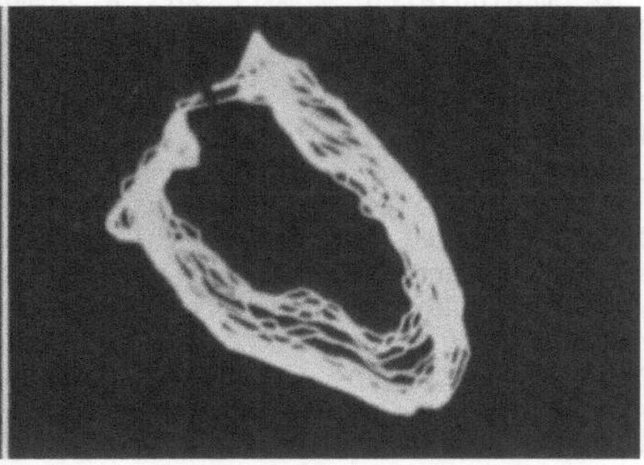

Fig.4. Set of ventricle boundaries

Finally it should be noted that the analysis described so far becomes somewhat simpler if some attempt is made to remove static structures from the image. Fig 5a shows a typical frame including shadows from other organs which do not change very much during one heart beat. If an image is obtained before the introduction of contrast medium this may be subtracted from each opacified frame, leaving grey level data relating mainly to the ventricle itself as shown in Fig 5b. An alternative which also has advantages is to subtract from each frame the previous frame as shown in Fig. 5c.

Language

Now we have outlined the continuously updated model technique and described a specific example of its application we turn to the problem of expressing the solution in machine intelligible form. The main factors limiting any practical application of automated image analysis are the ease with which the solution may be expressed and the efficiency with which it may be executed. For most applications it would be impractical to develop special hardware for the particular task so we limit our attention to general purpose hardware. The programming language for a general purpose image analysis system must thus allow algorithms to be expressed as simply as possible whilst allowing efficient execution.

The main elements of a language which allow image analysis algorithms to be described efficiently are the ability to refer to the main "objects"

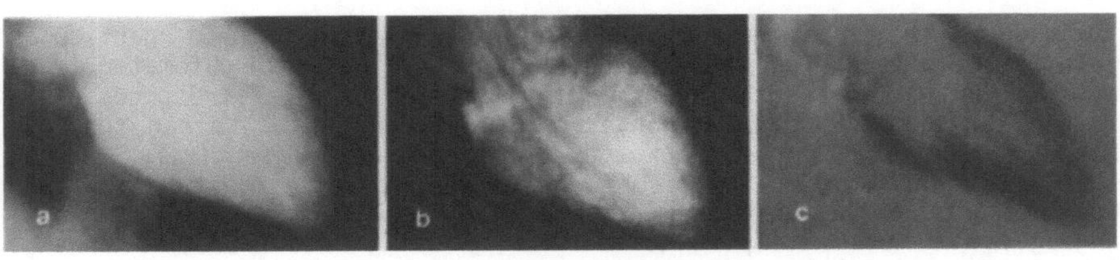

Fig.5. (a) Original angiocardiogram. (b) Original with unopacified frame subtracted. (c) Original with preceding frame subtracted

to be manipulated simply and a comprehensive set of common operations which involve them. If we consider our example the main steps involved are: to subtract one image from another, to construct a set of search vectors, to sample the image along each search vector, to modify the model using the grey-level data and to predict and generate the next model boundary. If we examine these operations we can see that only three important types of data are used: images, structures which select points within an image, and grey-level profiles. A data structure should contain all the information which relates to the object it describes. The structure IMAGE should, for instance, contain the actual image data, parameters to describe the spatial and grey-level resolution used and some method of describing the useful area of image data once the image has been manipulated. We group together under the generalised structure POINTSET all objects which select points within an image. Some examples of the form a POINTSET might take are illustrated in Fig. 6. These include a single picture POINT, an open curve or ARC and a closed curve or BOUNDARY. Although the internal representations of these structures may be quite different, they all describe a set of points in an image and are treated as variants of the same type. In the application we have considered we have used ARCs to describe the search vectors, POINTs have been found using grey-level criteria on the grey-level profiles and these have been used to modify the model stored in the form of a BOUNDARY. The grey profile is an example of a tabulated function of one variable, a type which is common to many applications and can usefully be generalised to allow manipulation using standard procedures. The general form is called PROFILE .

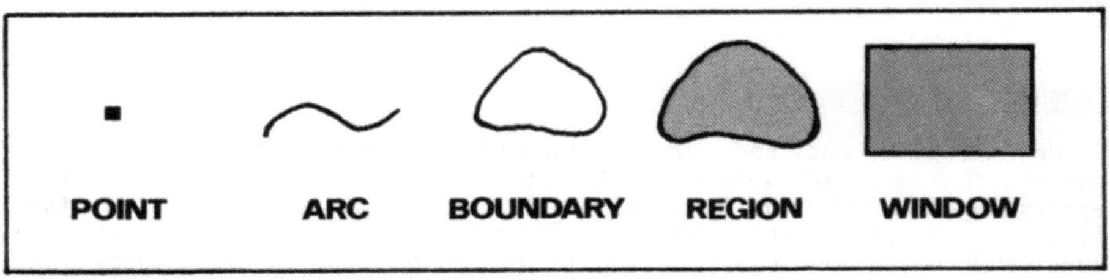

Fig.6. Variant forms of the data structure POINTSET

Another important point which should be noted from the application which we have considered is that various interactions are required between the data structures. Subtraction of an unopacified frame is an IMAGE to IMAGE transformation. Taking the model and generating search vectors is a POINTSET TO POINTSET transformation. We also have a transformation involving a POINTSET (search vectors) and an IMAGE to give a grey PROFILE, and so on.

Although it cannot be demonstrated by reference to a particular application it turns out that most ideas currently used in automated image analysis can be represented by a fairly small set of data structures and a limited set of basic types of interaction between them. We can conclude that in order to apply automated image analysis effectively to new problems it is important to be able to express these data structures and the interactions between them very clearly and simply in high level language. We have attempted to do this by using Pascal [4], a language with good support for structured data, to predefine all the commonly used data types. Standard procedures handle all the common interactions between the data types.

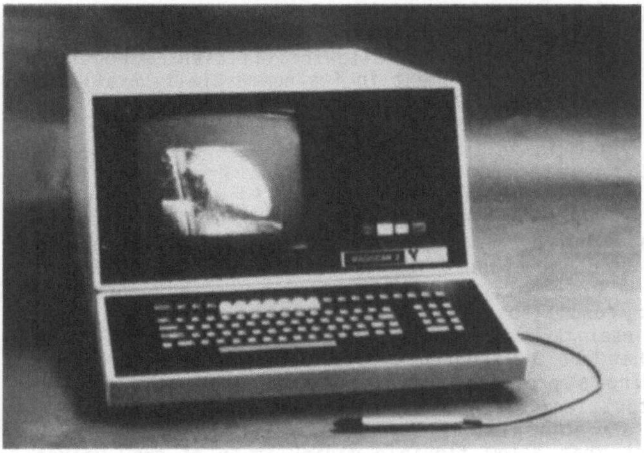

Fig. 7. General view of the Magiscan 2 Image Analysis System

Hardware

While this language framework facilitates program development, it is obviously also important in any practical application that the programs written can be executed efficiently. This should be true for all parts of a program and for all types of data manipulation. It is not sufficient, for instance, for IMAGE to IMAGE transformations to be performed efficiently at the expense of other equally important types of operation (5). This view has influenced our design of hardware and has led to a system which has a fairly conventional microprogrammed serial

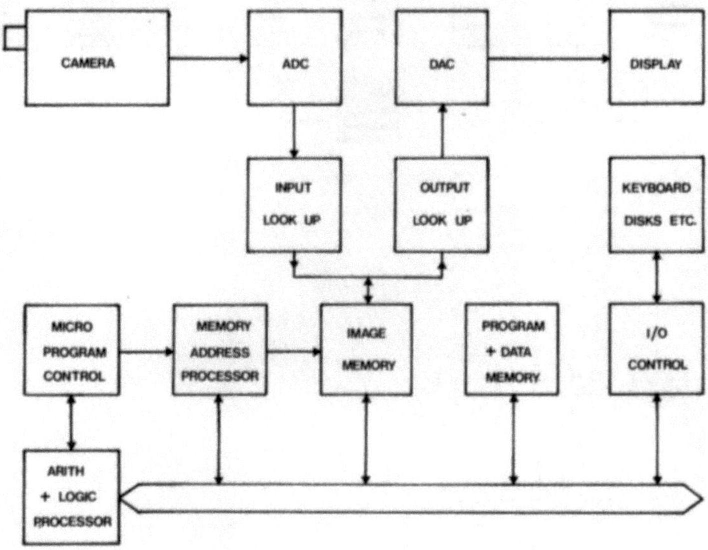

Fig.8. Block diagram of the Magiscan 2 Image Analysis System

architecture, optimised for the types of operation we have been considering. Interactions between data structures are handled by special purpose hardware and microprogram firmware, allowing efficient execution. Fig.7 shows a general view of the instrument in its commercially available form (6). Fig 8 is a block diagram of the hardware.

At the heart of the system is an Image Memory. Because of its special importance this is kept separate from a general purpose Program and Data Memory. There is an Arithmetic and Logic Processor (ALP) which is capable of manipulating data in either memory. Because of the importance of memory addressing when handling image data, a special processor is provided to compute image memory addresses (MAP). The ALP and MAP are controlled in synchronism by a Micropgrogram Controller. The controller contains Read Only Memory and also contains Writeable Control Store which can be used to perform user defined operations at microcode level. The interconnections show that the ALP controls data interchange between the different units, including a provision for data driven operation of the MAP. General input output facilities are provided by an I/O Controller. Image input and display are in real time direct into and out of the Image Memory, via lookup tables that allow flexible manipulation of grey values during loading and display. The normal input device is a TV camera although the I/O channel can be used to communicate with other types of imaging system.

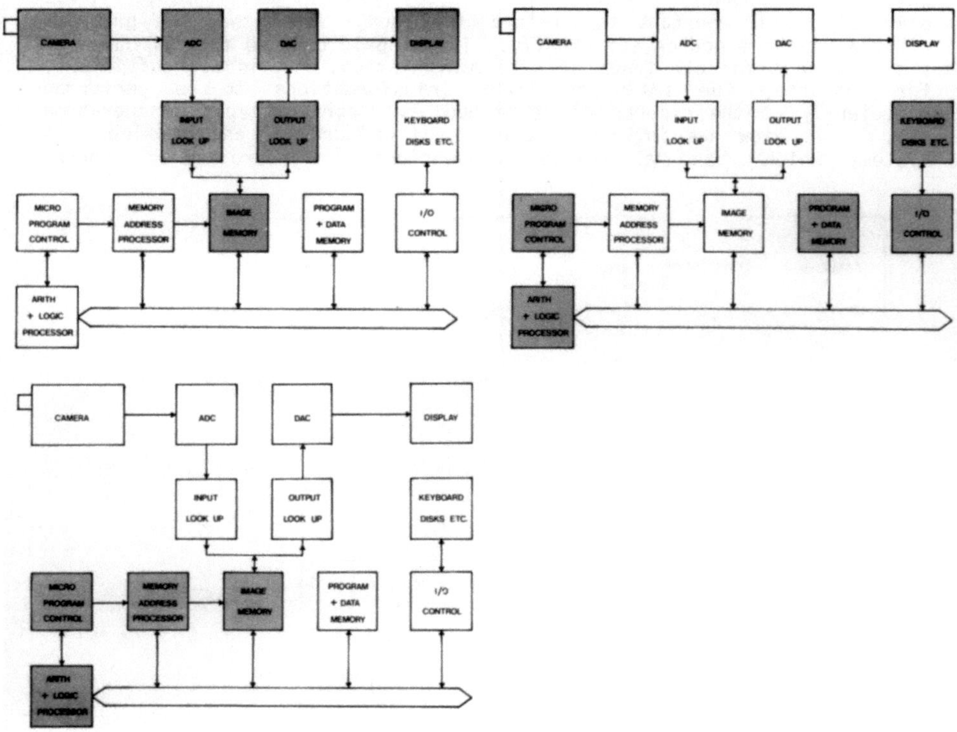

Fig.9. (a) Raster graphic group. (b) Minicomputer group. (c) Image Processor group

To understand better how the system is used it can be usefully considered as having several personalities as illustrated in Fig 9. The grouping of Microcontroller, ALP, Program + Data memory and I/O controller is basically a modern minicomputer capable of executing high level language programs. The Video I/O and Image Memory together form a raster graphics display system. The group of Microcontroller, ALP, MAP and Image Memory is an image processor capable of high speed image to image transformations.

Summary

We have described the outline of a general technique for extracting motion information from movie sequences. The method assumes prior knowledge of the general form of the image structures of interest. We have described how the method has been applied successfully to the particular problem of analysing cineangiocardiograms. We have discussed some features of the general purpose image analysis system used to implement the solution and indicated how the hardware design was influenced by high level language considerations. Clinical aspects of the work have been discussed elsewhere [3, 7]

References

1. R.W. Smalling, M.H. Skolnick. D. Myers, R. Shabetai, J.C. Cole, D. Johnston, "Digital boundary detection, volumetric and wall motion analysis of left ventricular cine angiograms", Comput. Biol. Med., 6, pp. 73-85. 1976.

2. D. Verel, R.G. Grainger, Cardiac catheterization and angio-cardiography. An introductory manual, Churchill Livingstone, 1977.

3. J.N.H. Brunt, C.J. Taylor, D.J. Rowlands, "A software based system for geometrical analysis of left ventricular cineangiograms", Computers in Cardiology (Ed. K.L.Ripley, H.G. Ostrow) IEEE Computer Society, pp. 437-40, 1979.

4. K. Jensen, N. Wirth, Pascal: User Manual and Report, Springer-Verlag 1975.

5. Multicomputers and Image Processing. (Eds. K. Preston and L. Uhr) Academic Press, 1982.

6. "Magiscan 2 Users Manual" Joyce Loebl (a Vickers company). Marquisway, Team Valley, Gateshead, Tyne & Wear, UK.

7. H.G. Love, J.N.H. Brunt, D.J. Rowlands, "The acquisition of left ventricular cineangiograms of adequate quality for interpretation including quantitative analysis: problems and solutions", Computers in Cardiology 1981 (Ed. K.L. Ripley) IEEE Computer Society, 1981, pp. 313-6.

Stereopsis

J. Mayhew

Department of Psychology, University of Sheffield,
Sheffield S10 2TN, United Kingdom

The two eyes receive slightly different views of the world, and stereopsis is the process that recovers a description of the 3D structure of the world from these different images. In general, an object in the world projects to different positions on the retinae of the left and right eyes. The differences in the horizontal and vertical coordinates of the two eyes' projections are called respectively the horizontal and vertical disparities. Considered as a computational problem, stereopsis has characteristically been regarded as consisting of two sub-problems: (i) the correspondence problem which is the problem of measuring the disparity of each point in the two eyes' projections, and (ii) the interpretation problem which is the use of disparity information to recover the orientation and distance of the surfaces in the scene.

For obvious reasons there is considerable commercial and military interest in the development of efficient stereo-algorithms. It is also a "classical" low level vision problem that has much exercised some visual psychophysicists and physiologists, partly for historical reasons but also because it is essentially a knowledge-free low level visual problem seemingly independent of semantics and the difficult and little understood issues of object perception. That is, despite years of research that proved the contrary, stereopsis was thought to be a tractable problem.

The Correspondence Problem

The articulation and formulation of the correspondence problem owes much to the seminal work of JULESZ [1], who realised the implications that the breaking of camouflage by stereopsis had, both for the level of visual processing that provided the input to the stereo-process and for the nature of the problem of stereopsis itself. Firstly, a camouflaged object in front of a similarly textured background, though monocularly invisible, could be immediately revealed in depth when viewed binocularly. This natural observation was examined in the laboratory using computer-generated random dot stereograms portraying simple and complex surfaces. The results supported the original observation. Because the perceived shape existed only as local point-for-point correlations between two halves of the stereogram, these demonstrations argue that stereopsis is an early visual process occurring before the level at which shape and texture regions are made explicit (see also [2]). Secondly, since in a random dot stereogram each dot in one eye's view could potentially match with any dot in the other eye's view, the random dot stereogram clearly raised an important question. How did the brain solve this massive ambiguity problem (the solution to which was called "global" stereopsis by JULESZ to distinguish it from "local" stereopsis, which he called the process of matching the individual dots)? Of course in the case of sparse and simple line stereograms the ambiguity problem was reduced to trivial proportions if it existed at all. While the problem had been recognised before, JULESZ's random dot stereograms provided an important methodological paradigm for the psychophysical investigation of the correspondence problem.

The solution of the correspondence problem is a representation making explicit the local disparity information for every point in the scene, and requires answers to the following questions:
a) The choice of matching primitive. How should we define the 'points' that are to be matched from one eye's view to the other?
b) The matching and disambiguating rules. How are we to decide which point in one eye's view matches with which point in the other eye's view, and how are we to resolve ambiguous or false matches?

Obviously the answers to these questions are not necessarily independent, and thus many solutions to the correspondence problem can be characterised as locating the peak in some form of 'piecewise cross-correlation' defined over particular matching primitives. Possibly the simplest form of the solution uses some metric of grey-level cross-correlation over a small window, to find for each pixel in one view the coordinates of the corresponding pixel in the other image. Such a method (area correlation) works well in strongly textured images with few surface discontinuities. An alternative general strategy uses not grey level values but linear edge segments as the matching primitive. Descriptions of edges (e.g. position, orientation, contrast,etc.) are first obtained from the left and right eyes' views in an image preprocessing stage to provide a less locally ambiguous input to the correlational-matching stage of the algorithm.

Much ingenuity has been directed to engineer algorithms to minimise the cost (e.g. coarse to fine processing cones) and the effectiveness (e.g. the exploitation of non-linear iterative relaxation processes to optimally resolve problems of local ambiguity) of both area and edge based correlations. Thus two important ways in which algorithms differ are the extent to which information from previously solved or unique matches is iteratively utilised to resolve ambiguities and eliminate false matches, and also the way in which processes on different areal scales and resolutions are integrated in both the matching and disambiguating stages of the process. But a characteristic of most algorithms has been the following assumptions (not always made explicit):
i) Depth and hence disparity varies smoothly almost everywhere. For area correlation, the assumption is that the surface from which the disparity measurements arise does not differ markedly in each eye's view, i.e. there is photometric correspondence between the two eyes' views. For edge correlation, the expression of the assumption is that the edges of surfaces and surface markings vary continuously in depth along their length and this is preserved as the geometric invariance between the two images.
ii) The camera geometry is known, or the projection is orthographic and/or the lines of sight or viewing directions of the two eyes are effectively parallel. The result of this assumption is that many theories and algorithms have been concerned only with measurements of horizontal disparities and limit the search for matches on corresponding horizontal rasters in the two eyes' views. A possible use of vertical disparity measurements will be discussed later.

The remainder of the paper presents a theory of stereopsis which exploits the insights both computational and psychophysical, gleaned by FRISBY and myself over the last five years, some of it from our own research. We owe an immense intellectual debt to the late D. MARR and his colleague T. POGGIO, though we think their theory [3] almost completely wrong; and the one we think most in accord with ours was developed totally independently by BAKER and BINFORD [4].

Yet Another Partially Correct Solution to the Correspondence Problem

The starting point for our solution to the correspondence problem was what we called the BRPS conjecture [5,6]. This, in its original formulation, proposed that the processes of human binocular combination integrally relate the extraction of disparity information with the construction of the raw primal sketch assertions. Which is to say that the low level description of the scene computed from the 2D grey level image(s) makes explicit local information about the nature and position of luminance discontinuities in a 3D coordinate system. As we see it the strength of this conjecture was that it suggested a starting point in the search for answers to the questions: what are the matching primitives? and what are the matching rules? without making any assumptions about the contents or characteristics of the world that were peculiar to stereopsis. If you can see it, you can, within the resolution of the system, see it in depth! The matching primitives will be the very same as those used in the first stage of computing the primal sketch; and the matching rules will be derived from the grouping rules used to produce the raw primal sketch descriptions. Part of the motivation was a challenge to an established orthodoxy which exploited nearest neighbour disparity matches ("similar disparity facilitation") along the same raster as a fundamental part of the solution to the ambiguity problem. The weakness of the BRPS conjecture was its failure to recognise and hence exploit a heuristic which the human visual system almost certainly uses. This is the use of nearest neighbour disparity matches along the same raster to prevent the fusion of objects with a disparity gradient greater than 1 (a powerful disambiguating restriction which can be derived from, but is not entailed by the geometry of stereo projection). Even if you can see it, you sometimes cannot see it in depth!

The Matching Primitive

The requirement [7] that the matching primitive stand in reasonably direct relationship to a surface marking or other physical entity is clearly met by exploiting the primitives used in the computation of the object's description. MARR points out that this requirement is not met by the grey-level values in the image (as used in most area correlation algorithms) which only rarely stand in a unique relationship to a physical object. In contrast, this requirement is met in algorithms using a form of edge correlation. A matching primitive used by both MARR and POGGIO [3] and ourselves is the zero-crossing in the convolutions of the image with a difference-of-Gaussians (DOG) operator at different scales. This operator corresponds to the spatial frequency tuned channels for which there is considerable psychophysical evidence for their importance in visual information processing and stereopsis (see [8] for a review), and an elegant theoretical justification of the use of this operator for the early measurement processes of the primal sketch computation is given by MARR and HILDRETH [9]. Apart from their obvious suitability to locate edges and changes in contrast (which is why we chose to use them), there are also good theoretical reasons for using zero-crossings as the matching primitive [10,3]. Furthermore recently WATT and MORGAN [11], in a visual acuity task, found psychophysical evidence strongly suggesting that zero crossings are used in human visual processing. Nevertheless for surfaces where changes in luminance are correlated with (possibly caused by) changes in depth (e.g. a rippled, untextured matt surface viewed from directly above but illuminated from one side), zero-crossings are in principle unsuitable stereo matching primitives. This follows from the fact that in the convolutions of the images of such a surface, the zero-crossings in each eye's image will be localised at the positions corresponding to the places where the luminance value was equal to the space average luminance value. The disparity carried by corresponding zero-crossing matches will be the average of the disparity variation over each luminance change; hence a flat rather than rippled surface will be seen. Stereograms presenting such surfaces are readily generated and the depth percepts of human observers are not as would be predicted if zero-crossings were the only matching primitive extracted from the spatial frequency channel convolutions of the image [6]. An alternative primitive, for which there is little other evidence, is the peaks in the channel convolutions.

Thus to summarise one answer to the first part of the correspondence problem: How should we define the points that are to be matched from one eye's view to the other? We proposed that stereopsis utilises the same primitives as are used in the computation of the raw primal sketch, namely zero-crossings (and possibly peaks) extracted from the convolutions of the images with DOG operators at different scales, or different spatial frequency channels. We turn now to the second problem, what are the matching rules and how are false and ambiguous matches resolved?

The Matching Rules

The primary assumption is that the stereo-process attempts to minimise the number of unmatched points in each eye's view subject to a number of constraints derived from the geometry of the projection of opaque objects and the computation of their description. The fundamental assumption is that stereo-perspective projection almost always preserves the order, the continuity, direction, and in large measure, the scale, of luminance gradients in each eye's image. Matching rules can be derived to exploit this. There are two further constraints: the epipolar constraint which is a restriction on where one should look for matches, and the opacity constraint, which at its most general is a restriction on membership of the solutions to the problem of global stereopsis. Dealing with them in reverse order:-

1) The opacity constraint, as originally formulated in [5,6] was that visual objects masked from both eyes' views cannot be seen (!), and furthermore only rarely will points project only to one eye's image. Most natural visual objects, surfaces and surface markings are opaque (exceptions may be still waters and some insect wings). From this we derived the rule that matches of raw primal sketch descriptions (carrying as they do implications of surface properties at their different scales) cannot be assigned depth locations from which they would be invisible (thus only one matching entity is possible on a line of sight). It should be noted that this rule does not preclude the fused percept of several transparent planes (containing opaque markings) lying at different depths [1,12] since it is not derived directly nor deals explicitly with the properties of surfaces beyond the spatial extent of the raw primal sketch description of the particular visual entity (e.g. blob, edge or bar) under consideration. The opacity constraint is more a reason why the ambiguity problem should be solved rather than a solution to it. It requires only that geometrically consistent configurations of otherwise acceptable matches be allowed (and amongst these, those that minimise the number of unmatched points are to be preferred). Thus it has much the same role in our theory as the uniqueness constraint has in the MARR and POGGIO [3]

theory but is derived from somewhat different conceptions of the problem. As we shall see it allows too much for the human visual system. Some configurations of matched points, though geometrically correct and within normal disparity range, cannot be fused [13].

2) The epipolar constraint. This derives directly from the geometry of stereoscopic projection from which it follows that a plane passing through the optical centres of the eyes projects to a straight line in each eye's image. This constraint has been exploited, even if not explicitly and often incorrectly (MAYHEW and FRISBY [14] and many others; see [15] for an explanation and criticism) in stereo algorithms that restrict the search for matches to a single raster in the two eyes' views. Often they have used corresponding horizontal rasters. This is only really justified if the optical axes of the eyes are parallel or effectively so, as when fixating distant objects. In the case when the eyes are converged to a near point and/or with asymmetrical gaze the epipolar lines in the two images may diverge considerably from the horizontal. In practice the deviation from horizontal of the short length of raster over which matches are sought may not be so significant as the vertical disparity between the appropriate rasters in each eye's image. That is, the local region of raster may be for all practical purposes regarded as effectively horizontal, but one will be at a different height in one eye than the other. If the viewing system geometry, such as convergence angle of the eyes or the direction and distance to the point of fixation, is known then it is possible to restrict the search for matches to the appropriate rasters. Ambiguous matches may be rejected if they are incompatible with viewing system geometry, but otherwise the search must be over a 2D region. In the section on the interpretation of disparity it will be shown that there is a very simple method for computing the distance and direction of the fixation point from disparity information, which some evidence suggests may be exploited by the human visual system [16,17]. It is thus possible to use iteratively the disparity of matched points to compute the viewing system parameters and then exploit this global information to direct the search for the remaining matches to the appropriate regions of the image. However, in the absence of information about the viewing geometry, the rule derived from the epipolar constraint is that matches for neighbouring edge descriptions along a raster should seek their matches along a raster in the other eye and though the rasters may be at different heights, locally the vertical order of rasters is preserved.

3) Edges of surfaces and surface markings are spatially continuous. They will tend therefore to project to continuous luminance changes in the two eyes' images. Disparity will tend to vary smoothly along descriptions of such changes but may be discontinuous across them. Matches between descriptions (zero-crossing segments) that maintain figural continuity are to be preferred. The maximum orientation differences between corresponding segments of edge description in the two eyes' images will occur when the edge lies in the same plane as the nose (i.e. the median sagittal plane), and only very rarely will the differences between the eyes be greater than 60 degrees. This will be discussed below.

4) A given location in space can hold only one object at any one time, and because stereoscopic projection almost always preserves the direction and scale of luminance gradients, this object will project to luminance changes with similar descriptions along the corresponding lines of sight in the two eyes' views. The exception to this occurs when a nearer object masks a contrast change produced by an object behind it from one eye's view, in which case depending on context, there will be an extra, unmatchable and/or rivalrous primitive. However the problems produced by such "accidentals" are

confined to their immediate neighbourhood by the rule that matches of similar descriptions of luminance change are preferred over rivalrous matches, and by the ordering rule below. Our implementation of this rule exploits the principles that MARR [7] and MARR and HILDRETH [9] proposed for combining the zero-crossings from different spatial frequency channels to compute the descriptions of edges, blobs, and other raw primal sketch elements, and in particular the principle of spatial coincidence. This states that if the zero-crossings from different spatial frequency channels are spatially coincident, then this is strong evidence that they are produced by the same luminance change. We exploit this argument to justify the use of cross-channel correspondences among zero-crossings as a matching constraint. This rule essentially embodies what may be regarded as a version of the coarse-to-fine processing strategy used in many AI domains to narrow the search space. However, implicit in the above rationale is that where applicable one matches a "spectrum" or a spatially localised pattern of zero-crossings as an entity. That is, the matching across different scales is done in parallel, and not as is commonly the case, as a process of serial refinement. This is an important distinction because it is possible to design stereograms for which a serial search strategy (cf. [3,18] possibly also [4]) stabilises at depth assignments different from those obtained with a parallel implementation and, also from those reported by human observers [5,6]. It should be emphasised that such a stereogram would be unlikely to occur outside the laboratory and thus this result may be of interest only to those concerned to understand human stereo vision.

5) The ordering constraint. Stereo projection almost always preserves the order and the non-metrical adjacency relationships of luminance gradients along the appropriate rasters in the two eyes' images (the recovery of the metrical differences in adjacency relationships may be regarded as the purpose of the whole endeavour, namely to measure disparity). Consider a small blob or thin bar. Since stereoscopic projection almost always preserves the direction of luminance gradients projected into each eye's image, its projection will almost everywhere along its length give rise to a pair of adjacent opposite signed edge descriptions along the appropriate rasters. Similarly a series of reflectance changes will give rise to a series of same signed edge descriptions along the corresponding rasters in the two eyes' images. Thus this ordering of the signs of adjacent zero-crossings or edge descriptions can be used as the basis of a matching rule whose effect will be somewhat similar to the rules derived directly from considerations of surface continuity. The matching rule is that, where possible, matches are preferred that maintain the ordering and non-metrical adjacency of successive pairs of edge descriptions or objects in each eye's image. (Notice that the ordering constraint is not confined to the horizontal; matches along one epipolar line immediately partition the search space for matches on other epipolar lines.)

The ordering constraint is inherent in almost all stereo algorithms using variations of edge based and even areal correlation that exploit nearest neighbour disparity facilitation to resolve ambiguities. For example, in JULESZ's spring-coupled-magnetic-dipole metaphor/analogue model of stereopsis [1], the spring coupling between adjacent dipoles in each eye's representation "physically" prevented its violation. Of course it is explicitly used as a constraint by BAKER and BINFORD [4] and it is more general than MARR and POGGIO's continuity constraint [3], which they derived from their assumption of surface smoothness to justify, not disambiguation by figural continuity, but the use of areal matching statistics. Though the stereo-projection of non-malicious textured surfaces will preserve the order of luminance changes produced by surface markings, the other contents of the world may not entirely justify the importance MARR and POGGIO give (in both their theory and the algorithm implementing it) to assumptions of surface smoothness over a spatial extent much greater than warranted by the scale of the matching primitives. It may be that such assumptions are more appropriately utilised at a stage of processing exemplified by GRIMSON's surface interpolation algorithm [19] than in stereopsis.

There is definitely one situation where the enforcing of the ordering rule will prevent a veridical depth assignment. This occurs when two objects lie directly or almost directly behind one another and each eye can "see" the most distant object round the nearer object. Then, though the 'within object' ordering of the edges is maintained, the 'between object' ordering is reversed in each eye's projection. This is a very difficult situation for the human visual system to fuse and generally one of the objects is seen diplopically. The exception seems to occur when the two objects are sufficiently similar that a consistent matching of non-corresponding edge descriptions is possible and then the ordering constraint results in a fused percept in which the objects are seen, not lying one behind the other (the cyclopean opacity constraint!) but lying side by side. This illusion produced by the visual system electing to accept the ghosts or false matches rather than the veridical points is known as the nail illusion [20] and is suggestive evidence for the importance of the ordering constraint in human stereo-vision.

There is evidence that both the ordering and continuity constraints described above may be implemented indirectly in the human visual system as a limit on the maximum disparity gradient that can be fused. BURT and JULESZ [13] showed psychophysically that the limiting disparity gradient that can be fused is about 1 and pairs of points presenting such a gradient will be seen diplopically even though the actual disparity is well within Panum's fusional area and neither the ordering or continuity constraints are violated. They measure disparity gradients for a pair of points as follows: Imagine two points (1 and 2) projecting to locations L1, L2 and R1 and R2 on the corresponding rasters in the two eyes' images (it could be that 1 and 2 are vertically separated as well as horizontally but there is no vertical disparity between corresponding points) then it seems that the disparity gradient is given by

((L1-R1)-(L2-R2))/SQRT((((L2+R2)/2)-((L1+R1)/2))**2)+V**2) ,

where V is the vertical separation of the two dot pairs. With this formulation the limiting disparity gradient that a vertical surface can produce is 2 (i.e. it projects to a line in one eye's view if viewed "end on"). The limitation demonstrated by BURT and JULESZ is not a geometrical limitation but an isotropic restriction in the implementation which can also explain the nail illusion as an example of the general case. The nail illusion configuration presents disparity gradients greater than 2; these cannot be fused. It also presents non-veridical matches with disparity gradients that can be fused, the human visual system does the best it can and the illusion occurs. (Of course, order reversals in the two eyes necessarily present disparity gradients equal or greater than 2). Furthermore, as BURT and JULESZ point out, the allowable orientation differences that limit the matching rules derived from the figural continuity constraint can also be seen as a disparity gradient limitation. As implemented by POLLARD and MAYHEW [21] (and only for points on the same raster) this disparity gradient limit requires that the matches of a pair of points in the left eye with separation S must be a pair of points in the right eye with minimal separation S/3 and maximum separation Sx3. This is essentially a data driven Panum's fusional area.

The Interpretation Problem

It has commonly been thought that only horizontal disparities contain useful qualitative or relative depth information, and that in order to recover the quantitative 3D structure of the scene the disparities must be scaled by some absolute depth information. This has generally been assumed to be extra-retinal in origin and the favourite candidate has been information derived from the convergence angle of the eyes (see [22] for a review, also [19,23]), though HELMHOLTZ thought otherwise. Little is known about how this extra-retinal information is obtained, whether it has sufficient resolution or how it is accessed by the processes using it; however, it is in principle possible to recover the 3D structure of the scene without recourse to any information other than that provided by retinal disparity [24,25,26,27,16,17]. Of these methods the simplest, at least as far as I understand it, is by MAYHEW [17] and this is the method that will be briefly described.

It has long been recognised that only the horizontal disparities carried useful information about the local depth variations in the scene. Vertical disparities on the other hand, because they are so insensitive to depth, can be used to extract the viewing system parameters of angle of gaze and viewing distance without requiring strong assumptions about the nature of the surface from which they originate. From an approximation to the geometry of stereoscopic perspective projection and assuming that the eyes are free to rotate only around their horizontal and vertical axes (an assumption totally unjustified by the psychophysical data! (PRAZDNY personal communication)), it is possible to derive the following expression relating the vertical disparity to the direction and distance of the point of fixation:

$$V = Ir(c+g)/d \quad ,$$

where r and c are the retinal row and column coordinates (assumed known) of the match in one eye and V is the measured vertical disparity of the match in the other eye. I is the interocular distance (assumed known) and g and d are the unknown viewing system parameters of gaze angle and distance. The corresponding expression for the horizontal disparities H, derived from a locally planar surface is:

$$H = I(c2 + gc + Pc + Qr + K/d)/d \quad ,$$

where P and Q are the gradients of the surface in the horizontal and vertical directions and K is the intersection of the plane with the z axis, i.e. the line joining the point of fixation to midpoint of the line between the ocular centres of the eyes. The full derivation of these expressions is given in [17] (see the appendix by LONGUET-HIGGINS for some mathematical respectability, also [16]). Obviously once the viewing system parameters of gaze and distance (g and d) have been obtained the interpretation of the horizontal disparities to recover the 3D structure of the scene is relatively straightforward, at least in so far as the disparity representation can now be transformed into a piecewise local depth map of the scene. However as can be seen in the expression for the horizontal disparities the coefficient for the unknowns g and P is the same. Thus there is no unique and complete solution though it is possible to solve for the viewing distance (note that if the terms for g and P are treated as a single unknown, an approximate solution gives the horizontal gradient of the surface with respect to the direction the nose points, rather than the direction of gaze). However the expression for horizontal disparities relies on the assumption that the surface is locally planar, and a solution would only be possible if this assumption is satisfied for all the points required to solve the system of simultaneous equations. On the other hand, the expression for the vertical disparities makes no surface assumptions but only requires that the locations from which they derive are not greatly different in depth. Furthermore, the expression involves only two unknowns so it is in principle possible to solve for the distance and gaze from vertical disparity measurements from only two correctly matched points. Computer simulation has shown that the above approximate expression can be used to derive satisfactory estimates of viewing system parameters over a wide range of viewing conditions and surfaces [17, 28].

There is however a difficulty; the range and magnitude of vertical disparities is very much smaller than for the horizontal disparities, so the question arises: could the visual system use them, and if so, does it? There is a well known visual phenomenon known as the induced effect [29] which is the illusory tilt of a fronto-parallel surface produced by expanding one eye's view in the vertical direction. The horizontal disparities are unchanged by this optical distortion but nevertheless the surface appears tilted around the vertical axis, in the opposite direction but with equivalent magnitude to the tilt produced by expanding the same eye's image in the horizontal direction. In the latter case, called by OGLE the geometric effect, the optical distortion produces horizontal disparities and the tilt is as would be expected on purely geometric grounds. OGLE extensively investigated these phenomena and used them to obtain estimates of stereo-acuity over a wide range of retinal eccentricity. His estimates of stereo-acuity of about 0.2% magnification per degree eccentricity in both the vertical and horizontal dimensions are comparable with estimates of acuity found using different psychophysical methods. Using the expression for the vertical disparity given above it is algebraically trivial to show that the effect of vertically magnifying one eye's view has no effect on the estimate of distance, but changes the gaze term by an amount scaled by the distance. When values of these terms are substituted in the expression for horizontal disparity, it can be readily appreciated that, because the coefficient for the gaze and horizontal gradient of the surface P is the same, the error in the estimate of gaze will be absorbed in the algebra to give an equal and opposite error in the estimate of P (the tilt of the surface around the vertical axis). At a viewing distance of 40 cm, OGLE found that the increase in perceived surface tilt per % increase in magnification was between 3 and 3.5 degrees. Assuming an interocular distance of 6.5 cm, and a similar viewing distance, the expression above predicts a value of 3.53 degrees (for a more detailed mathematical treatment see [16,17]). Thus the data from OGLE's psychophysical investigation of the Induced Effect can be used to argue that the human visual system is sufficiently sensitive to vertical disparities to be able to use them to estimate the direction of gaze and possibly also, the viewing distance. If so, the fact that the expressions for the vertical and horizontal disparities above are in such excellent quantitative agreement with the psychophysical data suggests that humans may implement the same approximations to recover the 3D structure of the scene. Whether vertical disparity information is used by human observers to recover the viewing distance is currently being investigated at this laboratory, but whatever the results, the method may be of some application in the design of visually competent automata. It may be of interest in this regard that the question of resolution is not as important as one might at first think. Our answer to the problem of poor resolution leading to difficulties in numerical methods for solving simultaneous equations is to use the peak in a 2D Hough-like accumulator where the two dimensions are distance (in units of interocular distance) and gaze (units of 0.5 degree). Even with resolution limited to the nearest pixel (1 pixel subtending 1.2 min/arc) with simulated data, estimates of distance and gaze for near viewing distances (less than 60 cms) have been accurate to within a few cms [28]. Whether the human visual system utilises retinal information, extra retinal information or a mixture of both to recover the viewing system parameters is a matter for investigation, but it is well known that the viewing distance has a marked and systematic effect on the perceived surface orientation from a given disparity gradient. We turn now to this next stage of disparity processing.

In the above theory the output of the stereo-process is a relatively sparse low level 3D description of the visual scene. Only minimal local grouping processes have been used at the level of simple visual descriptive entities such as edge-segments, bars, blobs, etc. It is assumed that this structure is then the input to a subsequent stage of processing that makes explicit the higher order 3D structure of the scene in terms of descriptions of surface orientation, discontinuities, extremal boundary regions, etc. Such a representation has been called by MARR and NISHIHARA [30] the 2 1/2D sketch and is assumed to be a dynamic memory with considerable computing power (!) that serves to integrate surface information derived from numerous sources (e.g. stereopsis, motion parallax, shading, texture, surface contours, etc.) in a viewer-centred coordinate system. There seems as yet little real computational advance in understanding this problem. The emphasis seemingly over the last few years has been on the discovery of methods for recovering surface information from images. Now that there are several at least approximate methods for recovery of this information [31], research may turn to the problem of utilising it in shape recognition schemes. There is however, some psychophysical evidence concerning the next stage of processing of disparity information.

Evidence for processes operating on the disparity "image", or output representation, has been obtained using different psychophysical paradigms (see [8] for a review). These processes have been called hypercyclopean channels to distinguish them from spatial frequency channels. Spatial frequency channels measure luminance gradients over different spatial scales, hypercyclopean channels measure disparity gradients over different spatial scales. Thus a possible role for the hypercyclopean channels may be in the computation of a representation making explicit the surface discontinuities, analogous to the role of the spatial frequency channels in the computation of the primal sketch. In support of this idea that the human visual system may exploit similar strategies in both luminance and disparity domains, ANSTIS, HOWARD and ROGERS [32] found a stereo analogue of the Craik-Cornsweet brightness illusion. In the latter the effect of a cusp-like luminance discontinuity changes the apparent brightness of the surfaces on each side as though the visual system has extrapolated/interpolated from the magnitude of the discontinuity or differential over the actual luminance values. Similarly in the stereo analogue the small (below threshold?) disparity gradients to the side of the discontinuity are not perceived but rather the surface is seen as planar, with the depth on each side being determined by the magnitude of the disparity discontinuity rather than the actual disparity values. Whether this is an important, or merely a superficial and possibly misleading, analogy must await further computational and psychophysical evaluation, but ROGERS, GRAHAM and ANSTIS [33,34,35] have evidence that there may be processes with very similar characteristics operating on the output of the optical flow/motion parallax computations, and recently presented evidence that they may in fact be one and the same. This similarity between surface perception from stereopsis and motion parallax is further supported by the fact that the expressions above for the horizontal and vertical disparities can be derived from the rather more complete analysis by LONGUET-HIGGINS and PRAZDNY [36] of the problem of interpreting moving retinal images (see appendix to [17] by LONGUET-HIGGINS).

In Conclusion

This paper has made no reference to any physiological data. This is not entirely due to restrictions on space or to a psychologist's ill-informed prejudice. The implementation of stereopsis in the human (and other animal visual systems) is incredibly complicated, ill-understood and, as the clinical literature shows, goes wrong in an enormous number of ways (for access to this area see [37,38]). However interesting in its own right, I think much of this literature reports answers to questions not yet correctly formulated and possibly technically before their time. If it is an acceptable criterion for understanding that one can make a machine to do it, I think it is fair to say that physiological insights have informed almost none of the recent developments in computer vision. If the first step to understanding stereopsis is to make explicit the nature of the task and devise methods for solving it, then the repeated discovery of disparity sensitive cells in the visual cortex (or equivalent structure) in different species of animal [39] has, as a matter of fact, been of less importance than the recognition and articulation of the correspondence problem and the discovery of methods for its solution (let alone the question of the interpretation problem). At its crudest the argument is as follows: finding a cell that recognises one's grandmother does not tell you very much more than you started with; after all, you know you can recognise your grandmother. What is needed is an answer to how you, or a cell, or anything at all, does it. The discovery of the cell tells one what does it, but not how it can be done.

This paper has presented a theory of stereopsis. Several algorithms, by several people including ourselves, implementing various aspects of it seem to demonstrably "solve" the stereo computation. I assert then, with some confidence, but otherwise totally without proof, that the implementation of the first four constraints (opacity, epipolar, figural continuity, spectral continuity and coincidence) will recover all (and only) the geometrically and physically possible solutions. The opacity and epipolar constraints enforce the requirements that the global solutions be physically possible, the other two local matching constraints enforce the requirement that the matching image entities be derived from the same physical object. The theory makes only minimal assumptions about the nature of the world. On the other hand, the ordering constraint is of a different status to the others. It effects a restriction on the set of correct solutions (of course, it also very effectively eliminates a lot of incorrect solutions). I (re)discovered it (it has certainly been exploited in some form or other before [1,4], and for an extensive treatment in other problem domains see [40]) in an attempt to solve the question of why a horizontal row of dots in the fronto-parallel plane should appear like that, and not like any of the other possible (and sensible) configurations, without making any assumptions about the world. This configuration is a "classic" demonstration of the correspondence problem, i.e. which image of which dot matches with which? and why do we only see this particular configuration when the others are equally possible? The answers usually justify some form of cooperative "mechanism" of nearest-neighbour-similar-disparity-facilitation. There is a point about the demonstration that often has not been fully appreciated, namely that it is not just a demonstration of local ambiguity, it is also a demonstration of global ambiguity. There are deep, in-principle ambiguities that cannot be resolved; i.e. there are stereograms to which there cannot be a unique solution. If there are in-principle ambiguities, a general information processing strategy that the brain may adopt is to implement the least restrictive assumption that will resolve the ambiguities. As a research strategy; if something is impossible and yet the brain solves it, what assumptions is it making? Recently GRIMSON [19], following MARR, used the above demonstration as evidence for power of the assumption that surfaces of objects are smooth and hence disparity varies smoothly. The same demonstration can be used to support the ordering constraint which I prefer over the MARR-POGGIO-GRIMSON smoothness constraint because it is more general and makes no assumptions about the contents of the world. However against it, the psychophysical results of BURT and JULESZ [13] show that the ordering constraint may only be indirectly implemented in the human. The disparity gradient limit prevents its violation at the expense of occasionally excluding other "legitimate" matches. Though this may not, in practice, be important it does raise difficult questions about the relation of theories of information processing tasks and their status as theories of human function. Is the gradient limit the brain's way of conservatively implementing the ordering constraint, or is it an implementation restriction whose origins derive from another source? Does the human visual system know a better theory than we do, or is it satisfying the same theory but exploiting a saving in doing it this way? More research is needed, send money! Finally, in this paper psychophysical data has been exploited where it seemed useful or apposite; phenomena which I do not understand, or for which there does not, at this time, seem to be any good reason, I have omitted, as well as a lot more. There does seem to be an awful lot of it and for a review of most of it see [8,38].

Acknowledgements

This research was supported by SERC Grant No. GR/B43265. Thanks are due to my colleagues J.P. Frisby, S. Pollard and S. Peek.

References

1 B. Julesz, *Foundations of Cyclopean Perception*, Univ. of Chicago Press, Chicago (1971)

2 J.P. Frisby and J.E.W. Mayhew, *Perception 8*, 153-156 (1979)

3 D. Marr and T. Poggio, *Proc.Roy.Soc.Lond., B 204*, 301-328 (1979)

4 H. Baker and T.O. Binford, *Int. 1981 Jt.Conf.AI., 6*, 631-636 Van. BC. (1981)

5 J.E.W. Mayhew and J.P. Frisby, *Proc. AISB-80 Conf.* (1980)

6 J.E.W. Mayhew and J.P. Frisby, *Artif.Intell., 17*, 349-387 (1981)

7 D. Marr, *Phil.Trans.Roy.Soc.Lond., B 275*, 483-524 (1976)

8 B. Julesz and R.A. Schumer, *Ann.Rev.Psychol., 32*, 575-627 (1981)

9 D. Marr and E. Hildreth, *Proc.Roy.Soc.Lond., B 207*, 187-217 (1980)

10 D. Marr, S. Ullman and T. Poggio, *J.Opt.Soc.Am., 69*, 914-916 (1979)

11 R.J. Watt and M.J. Morgan (1982 in press)

12 J.P. Frisby and J.E.W. Mayhew, *Perception 7*, 431-436 (1978)

13 P. Burt and B. Julesz, *Perception 9*, 671-682 (1980)

14 J.E.W. Mayhew and J.P. Frisby, *Perception 9*, 69-86 (1980)

15 J. Ninio, *Perception 6*, 627-643 (1977)

16 J.E.W. Mayhew and H.C. Longuet-Higgins, *Nature Lond., 297*, 376-379 (1982)

17 J.E.W. Mayhew, *Perception* (1981 and still in press)

18 W.E.L. Grimson, *Phil.Trans.Roy.Soc.Lond., B 292*, 217-253 (1981)

19 W.E.L. Grimson, *From Images to Surfaces*, MIT Press, Camb. Mass., (1981)

20 J.D. Kroll and ven de Grind, *Perception 9*, 651-669 (1980)

21 S. Pollard and J.E.W. Mayhew (in preparation)

22 J.M. Foley, *Psych.Rev., 87*, 411-434 (1980)

23 D. Marr, *Vision: A Computational Investigation into the Human Representation and Processing of Visual Information*, Freeman, San Francisco (1982)

24 E.H. Thompson, *Photogrammetric Rec., 3* (14), 152-159 (1959)

25 H.C. Longuet-Higgins, *Nature Lond., 293*, 133-134 (1981)

26 H.C. Longuet-Higgins, *Perception* (1982 in press)

27 K. Prazdny, *Fairchild AI Lab.Memo.* (personal communication 1982)

28 S. Peek and J.E.W. Mayhew (in preparation)

29 K.N. Ogle, *Binocular Vision*, W.B. Saunders Co., Philadelphia and London (1950)

30 D. Marr and H.K. Nishihara, *Proc.Roy.Soc.Lond., B 200*, 269-294 (1978)

31 M. Brady, *Computing Surveys, 14* (1), 2-71 (1982)

32 S.M. Anstis, I. Howard and B.J. Rogers, *Vis.Res., 18*, 213-217 (1978)

33 B.J. Rogers and M. Graham, *Vis.Res.* (1981)

34 B.J. Rogers, M. Graham and S.M. Anstis, *Invest.Opthal.Vis.Sci.(Suppl.), 20*, 223 (1981)

35 B.J. Rogers and M. Graham, Paper read to the E.P.S. London meeting (1982)

36 H.C. Longuet-Higgins and K. Prazdny, *Proc.Roy.Soc.Lond., B 208*, 385-397 (1980)

37 J.I. Nelson, *Documenta Opthalmologica, 51*, 3-100 (1981)

38 J.I. Nelson, *J.Theo.Biol., 49*, 1-88 (1977)

39 D. Whitteridge and P.G.H. Clarke (Eds), *Proc.Roy.Soc.Lond., B 203*, 379-515 (1979)

40 J.R. Ullmann, *Conference Series 44, Inst.Phys.Lond.*, 50-66 (1979)

Viewer-Centered Intensity Computations

R.J. Woodham

Forestry/Computer Science, University of British Columbia, Vancouver, Vancouver, B.C. V6T 1W5, Canada

1. Introduction

Computational vision is the study of computational systems that interpret images. That is, it is the study of systems that produce symbolic descriptions of a world from images of that world. A scene domain consists of objects whose visible surfaces are defined in three spatial dimensions. An imaging device projects rays of light onto a plane. The image domain consists of a spatially varying brightness function (image irradiance) defined over a bounded planar region. The problem is to reconstruct a three-dimensional representation of the scene from its two-dimensional projection onto the image plane.

Substantial progress has been made both to define the levels of representation required and to describe specific modules working at each level. Four levels of representation emerge. Using the terminology of MARR and his colleagues [1], they are:

1. the primal sketch
2. the raw 2 1/2-D sketch
3. the full 2 1/2-D sketch
4. 3-D representation

The primal sketch represents locations in an image where intensity is changing abruptly. Explicit surface properties are assigned to the locations described in the primal sketch to form the raw 2 1/2-D sketch. Explicit surface properties include: reflectance changes, illumination changes, discontinuities in depth and discontinuities in surface orientation. The surface properties determined at specific image locations are interpolated so that the surface properties are defined everywhere in the image. This becomes the full 2 1/2-D sketch. This level of representation has also been called intrinsic images [2]. When the surface property in question is surface orientation, it has been called a needle map [3]. These first three levels of representation are all defined in a viewer-centered coordinate system. That is, they all define spatially varying functions over the plane forced by the imaging geometry. The fourth level produces a full 3-D description of the objects in view, independent of viewer position.

This paper discusses a photometric approach to early visual processing. An image irradiance equation is developed to model how the physical world determines what we see. The goal is to go directly from the image domain to a representation at the level of the full 2 1/2-D sketch. Unfortunately, the image domain does not uniquely determine the surface properties of interest. Image irradiance results from the interaction of several factors, some of which are properties of the objects in view and some of which are not. The effects of those which are, shape and surface material, must be separated from each other and from the artifacts of those which are not, illumination, shadows and viewing direction.

2. The Tools

Image formation is modeled by an image irradiance equation defined in a viewer-centered coordinate system. This development was first given by HORN [4] and is now a standard part of the literature in computational vision. The required details are summarized here. There are two aspects to consider: one deals with the geometry of image projection, the other with the radiometry of image formation.

2.1 Image Projection

Optical systems, in general, perform a perspective projection. If the size of the objects in view is small compared to the the viewing distance, then the perspective projection can be approximated by an orthographic projection as illustrated in Fig. 1. It is convenient to choose a left-handed Cartesian

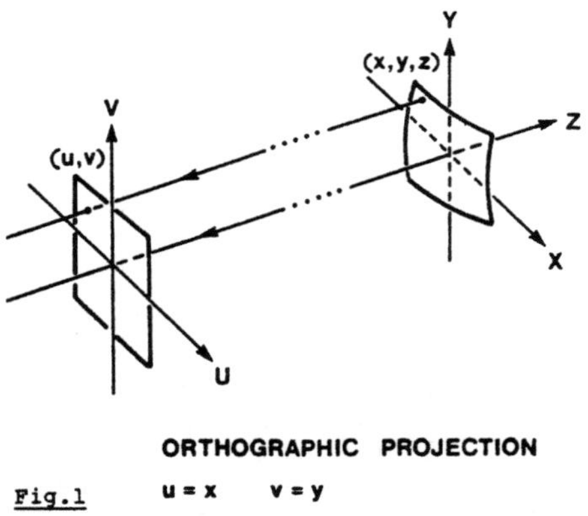

Fig.1

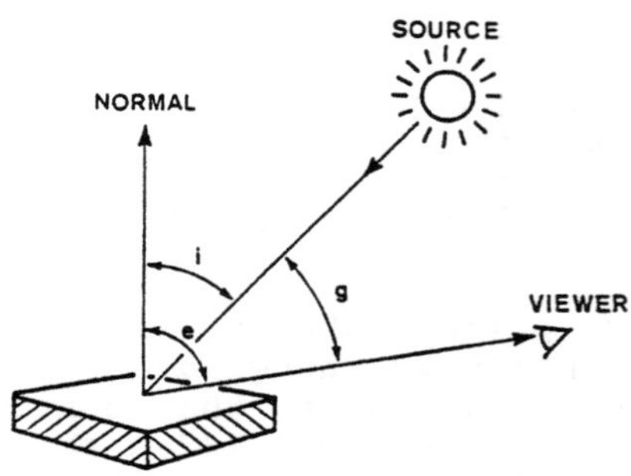

Fig.2

Fig.1. An orthographic projection. In an orthographic projection all rays from object surface to image are parallel

Fig.2. Defining the angles i, e and g. The incident angle i is the angle between the incident ray and the surface normal. The view angle e is the angle between the emergent ray and the surface normal. The phase angle g is the angle between the incident and emergent rays coordinate system such that the viewing direction is aligned with the negative z-axis. Without loss of generality, the image plane is scaled so that surface point (x,y,z) projects to image point (x,y).

2.2 Surface Reflectance

The amount of light reflected by a surface element depends on its optical properties, on its microstructure and on the distribution of incident illumination. For most surfaces, the fraction of incident illumination reflected in a particular direction depends only on the surface orientation. The reflectance characteristics of such a surface can be represented as a function $\phi(i,e,g)$ of the three angles i, e and g defined in Fig. 2. The reflectance function $\phi(i,e,g)$ determines how bright a surface element appears when viewed from a particular direction. As such, it corresponds to a viewer-centered definition of surface reflectance. Surface orientation, as given by the surface normal vector, determines the required angles i and e.

2.3 Surface Orientation

If the equation of a surface is given explicitly as

$$z = f(x,y) \tag{1}$$

then a surface normal is given by the vector

$$\left[\frac{\partial f(x,y)}{\partial x}, \frac{\partial f(x,y)}{\partial y}, -1\right] \tag{2}$$

If parameters p and q are defined by

$$p = \frac{\partial f(x,y)}{\partial x} \text{ and } q = \frac{\partial f(x,y)}{\partial y} \tag{3}$$

then the surface normal can be written as [p,q,-1]. The quantity (p,q) is called the gradient of f(x,y) and gradient space is the two-dimensional space of all such points (p,q). Gradient space is a convenient viewer-centered representation of surface orientation.

Another way to represent surface orientation is to consider points on a unit sphere, called the Gaussian sphere. Each point corresponds to the unit vector formed between the center of the sphere and that point. If the viewer is located far above the north pole looking southward and if a plane is constructed tangent to the sphere at the north pole, then gradient space is the projection of the Gaussian sphere formed by extending each unit vector outwards until in intersects the plane. Thus, gradient space represents the northern hemisphere of the Gaussian sphere.

2.4 Surface Curvature

Three parameters are required to specify the curvature at a point on a surface. One representation is based on the second partial derivatives, with respect to x and y, of the surface $z = f(x,y)$. Let H be the 2x2 matrix given by

$$H = \begin{bmatrix} \frac{\partial^2 f(x,y)}{\partial x^2} & \frac{\partial^2 f(x,y)}{\partial x \partial y} \\ \frac{\partial^2 f(x,y)}{\partial y \partial x} & \frac{\partial^2 f(x,y)}{\partial y^2} \end{bmatrix} \qquad (4)$$

H is called the Hessian matrix of the function $z = f(x,y)$. Hessian matrices are used extensively in nonlinear programming. H is indeed a three parameter function of x and y since, for smooth surfaces, the order of differentiation can be interchanged (i.e., H is symmetric).

2.5 The Reflectance Map

A reflectance map $R(p,q)$ determines image irradiance as a function of gradient coordinates p and q. One way to interpret the reflectance map is as a transformation of a surface reflectance function $\phi(i,e,g)$ into a function of p and q. Expressions for $\cos(i)$, $\cos(e)$, and $\cos(g)$ can be derived using normalized dot products of the surface normal vector $[p,q,-1]$, the vector $[ps,qs,-1]$ which points in the direction of the light source and the vector $[0,0,-1]$ which points in the direction of the viewer. To illustrate, one idealized surface reflectance function is given by

$$\phi(i,e,g) = \rho \cos(i) \qquad (5)$$

This reflectance function corresponds to the phenomenological model of a perfectly diffuse (lambertian) surface which appears equally bright from all viewing directions. Here, ρ is a reflectance factor and the cosine of the incident angle accounts for the foreshortening of the surface as seen by the source. The corresponding reflectance map is given by

$$R(p,q) = \frac{\rho(1 + p\,ps + q\,qs)}{\sqrt{1 + p^2 + q^2}\,\sqrt{1 + ps^2 + qs^2}} \qquad (6)$$

It is convenient to represent $R(p,q)$ as a series of contours of constant brightness in gradient space. Fig. 3 illustrates the reflectance map defined above for the case $\rho = 1.0$, $ps = 0.7$ and $qs = 0.3$.

Reflectance maps can be modeled analytically, as above, or measured empirically. A calibration object of known shape can be used to determine the reflectance map for a particular imaging situation. The reflectance map can then be used to analyse other objects made of the same material and viewed under the same conditions of illumination. Determining the reflectance map in this way has the added benefit of automatically correcting for the transfer characteristics of the imaging device.

2.6 The Image Irradiance Equation

The image irradiance equation becomes

$$I(x,y) = R(p,q) \tag{7}$$

where $I(x,y)$ is image irradiance and $R(p,q)$ is the corresponding reflectance map. This is a nonlinear first-order partial differential equation. The photometric approach "solves" this equation by finding surfaces $z = f(x,y)$ which satisfy eq. (7). The development of this approach is discussed in section 3.

Alternatively, one can simply view eq. (7) as one equation in two unknowns p and q. Image irradiance determines a relation that p and q must satisfy. But, in order to uniquely determine p and q, additional constraint must be provided. Properties of surface curvature are considered in section 4. Multiple images in section 5.

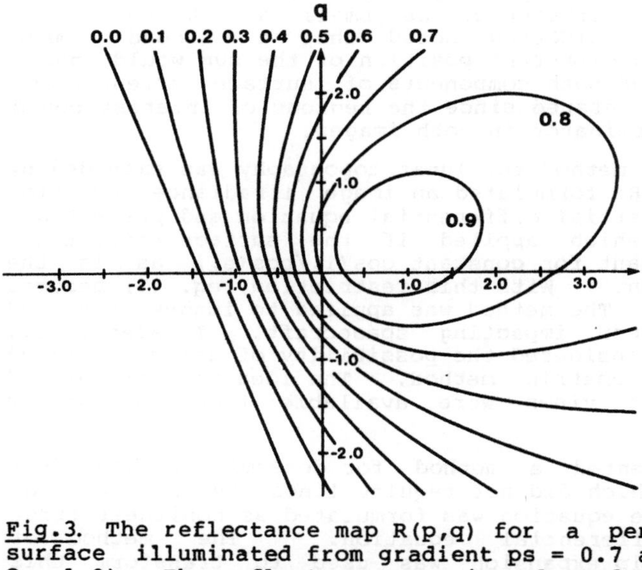

Fig.3. The reflectance map $R(p,q)$ for a perfectly diffusing surface illuminated from gradient $ps = 0.7$ and $qs = 0.3$ (with $\rho = 1.0$). The reflectance map is plotted as a series of contours spaced 0.1 units apart

3. Development of the Photometric Approach

The study of the photometric properties of a surface is a standard method of investigation in astronomy, especially as applied to the lunar surface. The moon has remarkable photometric properties arising from the special structure of its surface. At full moon when the source is directly behind the viewer (i=e), the distribution of brightness over the disk is approximately the same everywhere. Galileo considered this particular property of the moon three centuries ago in his "Dialogues on Two Systems of the World".

Since that time, many investigators have carefully measured the surface reflectance function $\phi(i,e,g)$ of the moon. Other investigators have experimented with hundreds of terrestrial materials and surface microstructures in an attempt to replicate the photometric properties of the moon. These experiments led to detailed predictions of the composition, particle size and porous nature of the lunar soil. Two reviews of this work give details [5,6].

The idea that photometric properties can be used to derive quantitative information about surface shape seems to have originated with VAN DIGGELEN [7]. This work did not make use of the surface reflectance function $\phi(i,e,g)$ but exploited properties of surface points near the terminator (the boundary between the light and dark hemispheres). Only one component of surface orientation could be determined. The reconstructed one-dimensional elevation profiles for hills in the maria delineated height differences as small as 10 to 20 m. Interestingly, VAN DIGGELEN noted that a second image corresponding to a different position of the sun would enable the determination of both components of surface orientation. This idea was not pursued since the regions of interest could not be near the terminator in both images.

The photometric method for lunar topography was extended by RINDFLEISCH [8]. He formulated an image irradiance equation as a first-order partial differential equation and presented a solution method which applied if the surface reflectance function was constant for constant cos(i)/cos(e), as is the case for the moon. With this restriction, eq. (7) becomes linear in p and q. The method was applied to images returned by the three Ranger impacting spacecraft. Interestingly, RINDFLEISCH also considered the possibility of using a second image in his photometric method. The idea was not pursued since only single views were available from the Ranger spacecraft.

HORN [9] presented a method for determining shape from image irradiance which did not require linearity in p and q. An image irradiance equation was formulated as nonlinear first order partial differential equation. The method of characteristic strip expansion was used to transform this equation into a set of five ordinary differential equations which were solved numerically. The method was successfully applied to a number of test images including one of a human face. Subsequently, using the gradient space, HORN re-expressed his image irradiance equation in the form given above by eq. (7).

HORN's method now had a simple interpretation. The Hessian matrix H determines the change in surface orientation [dp,dq] corresponding to a small movement [dx,dy] in the image. That is

$$[dp,dq]' = H[dx,dy]' \tag{8}$$

(' denotes vector transpose). If two linearly independent [dx,dy]'s and the corresponding [dp,dq]'s are known then H is completely determined. By taking partial derivatives of the image irradiance equation with respect to x and y, two equations are obtained which can be written as the single matrix equation

$$[Ix,Iy]' = H[Rp,Rq] \tag{9}$$

(Ix and Iy denote partial derivatives of I(x,y) with respect to x and y. Rp and Rq denote partial derivatives of R(p,q) with respect to p and q.) Thus, the image determines one correspondence. Suppose a point (x,y) is known to have gradient (p,q). Choose a small movement [dx,dy] in the direction [Rp,Rq]. Then, the change in surface orientation [dp,dq] is in the direction [Ix,Iy]. More precisely

$$\text{if } [dx,dy] = [Rp,Rq]ds \text{ then } [dp,dq] = [Ix,Iy]ds \tag{10}$$

where ds is a differential element of path length. Note that [Rp,Rq] defines a normal to the contour of constant reflectance at the current (p,q) and [Ix,Iy] defines a normal to the contour of constant intensity at the current (x,y). Iterating this procedure using the updated point [x+dx,y+dy] with gradient [p+dp,q+dq] generates a path through the image for which the gradient is determined. This path is a characteristic strip.

Several questions remained. How are initial conditions established? What conditions require termination of characteristic strip expansion? How is the surface z = f(x,y) interpolated between characteristic strips? What conditions must R(p,q) and I(x,y) satisfy to guarantee the existence of a solution surface z = f(x,y)? Under what conditions is this solution unique? BRUSS has recently provided some formal answers to these questions [10]. These results are highly technical in nature and are not reproduced here. Rather, the next section presents several results which integrate the photometric approach with specific surface properties known a priori or determined at the level of the raw 2 1/2-D sketch.

4. Exploiting Additional Surface Properties

Properties of surface curvature can be combined with information at the level of the raw 2 1/2-D sketch to reconstruct the full 2 1/2-D sketch representation of object shape. Several cases have been considered.

4.1 Planar Surfaces with Trihedral Corners

Objects whose surfaces are planar form the familiar polyhedral domain. HORN has demonstrated that when three visible planes meet at a point the orientation of each plane can be determined locally [4]. The technique combines the quantitative approach to interpreting line drawings of polyhedra, developed by MACKWORTH [11], with the quantitative constraint provided by image irradiance.

To illustrate, suppose the three planes are A, B and C. Let GA, GB and GC be the corresponding gradients. MACKWORTH has shown that the line in gradient space joining the gradients of two planes is perpendicular to the line in the image formed by the intersection of the two lines. This provides three constraints, but it is not enough to determine the position and scale of the triangle in gradient space formed by the three gradients GA, GB and GC. Three additional constraints are provided by measurements of image intensity for the three planes. The six constraints, taken together, are generally sufficient to determine the orientation of the three planes A, B and C.

4.2 Developable Surfaces

Many surfaces have the property that through every point on the surface there passes at least one straight line lying entirely on the surface. If the surface has the additional property that all points on that line have the same tangent plane then the surface is called developable. Now, a surface $z = f(x,y)$ is developable if and only if the determinant of the corresponding Hessian matrix H is everywhere zero. WOODHAM has shown that this constraint on the Hessian matrix H is sufficient to determine surface orientation locally from image intensity [12].

4.3 Occluding Boundaries

Another approach makes use of an object's silhouette. Parts of a silhouette may correspond to sharp edges on the surface, others to places where the surface curves around smoothly. The latter parts are referred to as occluding boundaries. At occluding boundaries the viewing angle e is 90 degrees. IKEUCHI and HORN have used occluding boundaries as initial conditions for a numerical solution to the image irradiance equation [13].

Occluding boundary information cannot be expressed in the gradient space since (p,q) becomes unbounded as e approaches 90 degrees. One solution is to use an alternate projection of the Gaussian sphere to represent surface orientation. If the center of projection is the south pole, not the center of the sphere, the stereographic projection is obtained. In the stereographic projection, points on an occluding boundary lie on the circle of radius 2 centered at the origin of stereographic space.

Determining a surface $z = f(x,y)$ which satisfies the image irradiance equation becomes a constrained optimization problem where occluding boundaries provide initial conditions and the image irradiance equation and smoothness constraints determine an objective function to be minimized. The method has been shown to work reasonably when the reflectance map is only a crude approximation.

5. Using Multiple Images

Another approach is to use multiple images to provide additional constraint. These images are taken from the same viewing direction, but under different conditions of incident illumination. This technique is called photometric stereo [14] and it allows one to determine surface orientation locally without smoothness assumptions.

Suppose two images I1(x,y) and I2(x,y) are obtained by varying the direction of incident illumination. Since there is no change in the imaging geometry, each picture element (x,y) in the two images corresponds to the same object point and hence to the same gradient (p,q). This means that one does not have the problem of first identifying corresponding points in multiple views, as is the case with binocular stereo. The effect of varying the direction of incident illumination is to change the reflectance map R(p,q) that characterizes the imaging situation.

Let the reflectance maps for the two imaging situations be R1(p,q) and R2(p,q) respectively. Consider a particular point (x0,y0) with measured intensities I1(x0,y0) = a and I2(x0,y0) = b. One obtains two (nonlinear) equations for the gradient (p,q)

$$R1(p,q) = a \quad \text{and} \quad R2(p,q) = b \qquad (11)$$

The intersection of the corresponding iso-brightness contours in gradient space determines the surface orientation at (x0,y0). There may be more than one solution, in which case additional information (such as a third image) may be needed to determine the unique answer.

In some cases, three (or more) images also allow the determination of the reflectance factor ρ at each (x,y). For reflectance given by eq. (6) above, three views are sufficient to uniquely determine both the surface orientation and the reflectance factor ρ at each image point, provided the three directions of incident illumination are not coplanar.

Photometric stereo is easily implemented. The stereo computation, after an initial calibration step, is purely local and may be implemented by table lookup, allowing real-time performance. Photometric stereo is a practical scheme for environments, such as industrial inspection, in which the nature and position of the incident illumination can be controlled.

The multiple images required for photometric stereo can be obtained by explicitly moving a single light source, by using multiple sources calibrated individually or by rotating the object and imaging device together to simulate the effect of moving a single light source. An equivalent to photometric stereo can also be achieved in a single view by using multiple sources which can be separated by colour.

Several practical implementations of photometric stereo have been attempted. High accuracy has been achieved under conditions of precise calibration [15]. Using extended light sources to alter the shape of the reflectance map is useful when dealing with specular surfaces [16]. Qualitatively correct reflectance maps seem to give qualitatively correct results [17]. So far, however, experimentation has been confined to single convex objects to avoid inaccuracies due to the effects of secondary reflections.

6. 3-D Representation

It is possible to define a function for three-dimensional objects which assigns to each point on the Gaussian sphere a weight proportional to the frequency of occurrence of that orientation on the object's surface. This function is called the extended Gaussian image and for convex objects it is known to be unique, up to translation.

Modeling objects in terms of their extended Gaussian images may have some advantages [3]. The viewer-centered methods described in this paper determine at most one hemisphere of an object's extended Gaussian image. Matching the visible hemisphere against the extended Gaussian image representation of known objects can be used to determine object identity and orientation in the scene [18].

7. Discussion

The photometric approach can be considered along three dimensions: Is it correct? Is it practical? Does it relate to human vision? To the extent that it adequately represents the laws of physical optics, the image irradiance equation must, of necessity, be correct. At this level, it represents a theory of the problem, but not a theory of solutions to that problem. The theory confirms that an image underconstrains the scene. Examples have been constructed to demonstrate that a given object made of a given material and illuminated in a given way generates an identical image to that of a different object made of a different material illuminated in a different way [12]. There are trade-offs between shape, surface reflectance and direction of illumination that cannot be resolved in a single view.

There are two paths to follow. One path involves restricting the scene domain. Generic knowledge of surface properties and illumination conditions can lead to precise scene reconstructions in a variety of circumstances. This is the path taken by the photometric approach, the results of which have been summarized in this paper. The other path involves discovering general perceptual principles which apply regardless of scene domain. Constraints imposed by the perceiver may imply a unique reconstruction even though the physics of the problem does not. It seems to me that this must, of necessity, apply in human vision.

The photometric approach provides a basis for understanding effects that occur in real images. HORN has shown that an analysis of intensity profiles across edges in the polyhedral domain can often lead to their direct interpretation as convex, concave or obscuring [4]. GRIMSON uses the image irradiance equation in his work on binocular stereo to constrain surface interpolation in areas which did not produce assertions at the level of the raw 2 1/2-D sketch [19].

Computer-based techniques in remote sensing often assume that image irradiance can be related directly to ground cover. This has been successful for wheat and other crop inventories in the prairies of North America. But it has been largely unsuccessful for forest inventory in British Columbia and other areas of rugged terrain. It is possible to determine conditions of illumination and terrain for which variations in image irradiance due to topography dominate those due to ground cover [20]. The ability to predict image features for known terrain and illumination conditions has been used to automatically register Landsat images to digital elevation models (DEM's) [21]. Attempts are being made to use DEM's to remove variations due to topography [22].

Any algorithm for scene reconstruction which determines a unique solution must either embody assumptions about the scene domain or about the perceiver. The tools summarized in this paper can help to make those assumptions explicit. They help to make computational vision a theoretical science as well as an experimental one.

8. References

1. Marr, D., Vision, W. H. Freeman, 1982.

2. Barrow, H. G. & J. M. Tenenbaum, Recovering intrinsic scene characteristics from images, in Computer Vision Systems, A. R. Hanson & E. M. Riseman (eds.), pp. 3-26, Academic Press, 1978.

3. Horn, B. K. P., Sequins and quills - representations for surface topography, in Representation of 3-Dimensional Objects, R. Bajcsy (ed.), Springer-Verlag, (in press).

4. Horn, B. K. P., Understanding Image Intensities, Artificial Intelligence (8)201-231, (1977).

5. Minnaert, M., Photometry of the moon, in The Solar System Vol. III: Planets and Satellites, G. P. Kuiper & B. M. Middlehurst (eds.), pp. 213-248, Univ. Chicago Press, 1961.

6. Hapke, B., Optical properties of the lunar surface, in Physics and Astronomy of the Moon (2nd edition), Z. Kopal (ed.) pp. 155-211, Academic Press, 1971.

7. Van Diggelen, J., A photometric investigation of the slopes and the heights of the ranges of hills in the maria of the moon, Bulletin of the Astronomical Institute of the Netherlands Vol. XI, No. 423, (1951).

8. Rindfleisch, T., Photometric method for lunar topography, Photogrammetric Engineering (32)262-276, (1966).

9. Horn, B. K. P., Obtaining shape from shading information, in The Psychology of Computer Vision, P.H. Winston (ed.) pp. 115-155, McGraw-Hill, 1975.

10. Bruss, A. R., The image irradiance equation: its solution and application, AI-TR-623, MIT AI Laboratory, Cambridge, Mass. (1981).

11. Mackworth, A. K., Interpreting pictures of polyhedral scenes, Artificial Intelligence (4)121-137, (1973).

12. Woodham, R. J., Analysing images of curved surfaces, Artificial Intelligence (17)117-140, (1981).

13. Ikeuchi, K. & B. K. P. Horn, Numerical shape from shading and occluding boundaries, Artificial Intelligence (17)141-184 (1981).

14. Woodham, R. J., Photometric method for determining surface orientation from multiple images, Optical Engineering (19)139-144, (1980).

15. Silver, W. M., Determining shape and reflectance using multiple images, SM Thesis, Dept. Electrical Engineering and Computer Science, MIT, Cambridge, Mass., (1980).

16. Ikeuchi, K., Determining surface orientation of specular surfaces by using the photometric stereo method, IEEE-PAMI (3)661-669, (1981).

17. Coleman, Jr., E. N. & R. Jain, Obtaining 3-dimensional shape of textured and specular surfaces using four-source photometry, CGIP (18)309-328, (1982).

18. Ikeuchi, K., Reconstruction of 3-d objects using the extended Gaussian image, Proc. IJCAI-81, pp. 595-600, Vancouver, B.C., (1981).

19. Grimson, W. E. L., From Images to Surfaces, MIT Press, 1981.

20. Horn, B. K. P. & B. L. Bachman, Using synthetic images to register real images with surface models, CACM (21)914-924, (1978).

21. Little, J. J., Automatic registration of Landsat MSS images to digital elevation models, Proc. Workshop on Computer Vision: Representation and Control, pp. 178-184, Rindge, New Hampshire, (1982).

22. Woodham, R. J., Using digital terrain data to model image formation in remote sensing, Proc. SPIE Vol. 238, pp. 361-369, (1980).

Part IV

Pattern Recognition

Statistical Image Processing

J. Kittler

Technology Division, SERC Rutherford Appleton Laboratory,
Chilton, Didcot, OX11 0QX, England

1. Introduction

The title of the paper suggests coverage of a very broad spectrum of problems in image processing which could be approached from the statistical point of view. Indeed, the spectrum could include topics ranging from image preprocessing, through image enhancement, to image understanding. However, the theme of this meeting - the analysis and interpretation of images - considerably narrows the subject of statistical image processing. It automatically excludes the problems of filtering, image reconstruction, enhancement and coding from consideration, although the solution of these may be a necessary prerequisite to successful machine perception of imagery data.

The subject of this paper is even more selective, as it only focusses on statistical decision making in general and on statistical pattern recognition in particular, and on its role in analysing pictorial data.

It is generally accepted that statistical pattern recognition is a useful tool in image analysis and interpretation. Certainly in some areas such as character recognition and remote sensing it is traditionally considered as an essential tool for solving the classification tasks involved. In other image processing application areas, however, the importance of this tool seems less apparent, at least judging from the image processing literature. The emphasis appears to be on ad hoc approaches and the common tactics seems to be to devise algorithms capable of solving a problem at hand without worrying about the general applicability of the solution.

The purpose of this paper is to argue that statistical pattern recognition could and should play a much wider role in image interpretation irrespective of the field of application. It will be shown in Section 2 that many stages in image analysis could be formulated as a statistical decision making problem. The statistical pattern recognition methodology will be briefly reviewed in Section 3. Section 4 is devoted to a more detailed treatment of one specific problem to illustrate the usefulness of the advocated approach. In particular, edge detection will be posed as a statistical pattern recognition problem. The threshold on the edge signal magnitude of the template edge detector [1], which minimises the edge detection error, will then be derived.

2. Pattern Classification and Image Description

The purpose of image analysis is to provide a concise and meaningful description of the scene captured in an image. But what constitutes a concise and meaningful description? We would certainly agree without hesitation that a list of pixel data does not convey any useful information on which another machine (or biological system without vision capabilities) could act. The situation would not improve much, even if we coded the data in the list using, for instance, run length coding. It is true that the coding would lead to a more efficient description but this still is far from what we would understand as concise and meaningful. Unfortunately it is not so easy to be positive. We can only attempt to define these two terms very loosely and hope that even a vague elaboration will help at least a little to pinpoint some of the essential ingredients of concise and meaningful image description.

Let us commence by observing that the semantics of an image will depend on the context. The context is defined by the activity in connection with which observations are made. The activity perhaps involves some decision making which is based on information about the interacting environment or subject and it therefore defines the scope of observation. Depending on the activity, the information sought from observations may be more specific or general as illustrated on the following examples. For instance, in automatic mail sorting we may be interested in reading only the town name to which a letter is addressed so that it can be crosschecked with the postal code, i.e. a very specific information. More general information is required in imaging the human lungs for cancer where we aim at detecting malignant lesions. In scene analysis for a free roving robot initially we may seek a very general scene description to decide on a route. But once this decision has been made, the information sought from the image is likely to be much more specific. It will be mainly related to the selected route and its vicinity for the purpose of obstacle negotiation and progress monitoring.

Thus image description is meaningful, if it is relevant to the information sought. It is pertinent to ask now, how concise a meaningful description can be. Obviously the achievable brevity of description will be a function of the amount of information sought. However, this still leaves a lot of room for manoeuvre. We can, for instance, describe a malignant lesion in the lungs as a convex segment with pixel intensity values or texture distinct from that of the surrounding tissue. Alternatively, we can just give it a name - tumour. Thus labelling or naming objects in the image is the most concise way of describing it.

Needless to say that we can go even further and label the whole image if that information is sufficient. For instance in the case of lung cancer diagnosis we may be interested simply in labelling the image either normal or abnormal. For mass screening purposes this description may be adequate and it is questionable whether any other description could be more concise and clearer than that. It amounts to reducing the multimegabit pixel information present in the original image into a meaningful single bit description.

Labelling or naming objects or patterns is nothing but object or pattern classification. It is the process of assigning a pattern into a category or class of patterns with similar characteristics or attributes. Thus pattern recognition is an essential tool for describing images in a very concise manner.

In other situations, a binary classification will not provide an adequate description. For instance, we may be interested in the extent of the disease as characterised by the number of tumours and their size. For the purpose of the patient treatment, we may want to know the exact tomography of the tumours. In other words we are interested in image description giving increasing levels of detail. The more detailed descriptions are conventionally referred to as

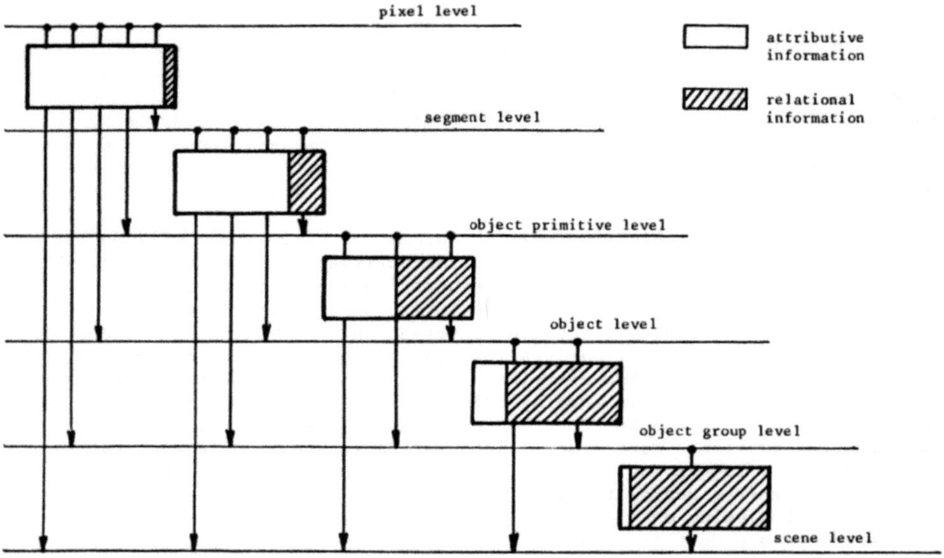

Fig. 1. Image description levels and image description generation routes

low level descriptions while categorical (interpretative) descriptions are called high level descriptions.

Let us now consider a possible hierarchy of descriptions as illustrated in Fig.1. A number of points is worth noting. In principle, any description level can be reached from any lower level provided the mapping process is compatible with the information required - image semantics. We can proceed from the pixel level directly to the scene level, if, for instance, the required labelling is potentially binary. By a direct mapping from low level to high level descriptions we have no access to intermediate levels of descriptions, should supplementary (more detailed) information be required.

At lower levels, the most important information for generating higher level descriptions is attributive or also referred to as semantic. The higher the description level from which the mapping commences, the greater the relational information component. The description generating process is basically a decision making process. At lower levels decisions are based primarily on attributes, at higher levels the shift of emphasis in decision making is towards rules governing the relations between image entities. It should be noted that the lower levels can convey meaningful description only when the high level interpretation has been completed. Thus for detailed image descriptions, low levels are necessary but not sufficient.

As a typical example illustrating the various levels of description let us consider the problem of machine reading the string of characters STRATFORD ON AVON. At the lowest level we have the grey-level image containing these three words. After thresholding and region growing we get the segmented characters of the text. These can be decomposed into strokes, loops, etc. to obtain the object primitive representation. The actual objects (characters) are identified by studying the detected primitives and their relations.

The grouping of characters leads to meaningful object groups - in our particular case, words. Finally by relating these groups we can interpret the original image.

We could, of course, attempt to decide right at the pixel level what the meaning of the image was, for instance, by matching the image with possible templates. Alternatively we could try to recognise individual words or less ambitiously, individual characters or even the structural primitives of characters and proceed to higher levels of description as outlined earlier.

Having established that pattern recognition is the most appropriate tool for generating image descriptions regardless of their level, we shall now consider it in more detail. It has been argued that at higher levels we are concerned more with recognition of relations between picture entities rather than their intrinsic properties. The mathematical apparatus suitable for analysing pattern relations or structures is syntactic pattern recognition. However, as this particular topic is beyond the scope of the present paper, we shall refer the reader to the major references in the literature [2-6].

In the next section we shall concentrate on pattern recognition based on attributes which is particularly relevant for generating descriptions derived from the data at lower levels. However, the approach is not relevant exclusively to pattern attribute processing. It also plays a role in pattern relation processing where, for instance, stochastic grammars need to be used to handle ambiguities in the production rules.

3. Attribute Pattern Recognition

The processes involved in machine recognition of patterns can be represented by a three stage model given in Fig.2.

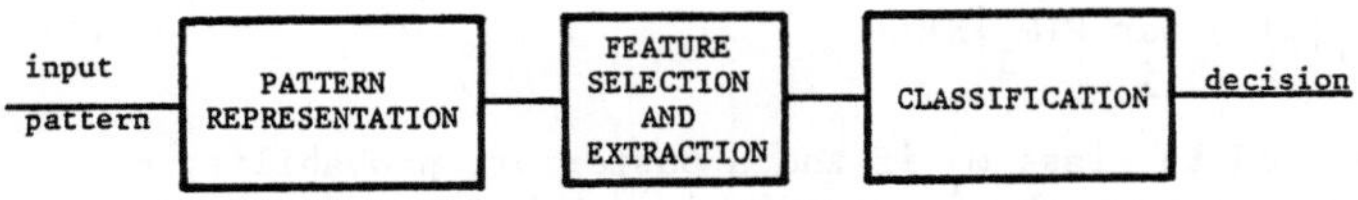

Fig.2. The pattern recognition system model

The first stage embodies the processes involved in pattern representation. Apart from any necessary preprocessing, these include the extraction of measurements and possibly various transforms. For instance, in edge detection for image segmentation, the representation pattern is a set of the outputs of edge masks which measure the edge signal in various edge directions [7-9].

If the representation pattern contains redundant or irrelevant information, it will be necessary to derive from it only features which are essential for pattern classification. The feature pattern then constitutes the input to the final, classification stage. This process is referred to as feature selection and extraction. It will be particularly important in generating high level descriptions directly from low level ones.

Finally the classification stage assigns the feature pattern into an appropriate category. In other words it gives the input pattern a name or a label.

In the image processing context we may be interested in at least two types of pattern recognition. The most important and of wide applicability is recognition of the generic identity of patterns - pattern classification. A more specialist problem is recognition of specific patterns, e.g. human faces. The latter can be handled only by comparing an unknown pattern with stored patterns and select the one which best matches the processed pattern. The matching of two patterns can be carried out by correlation.

There are more options available when solving the conventional pattern classification problem. Numerous solutions suggested in the literature have been derived to cope with specific types of pattern variability. In view of this, ULLMANN [13] suggested generalised correlation as a unified approach to pattern recognition, capable of handling all admissible distortions of patterns. However, as generalised correlation cannot be implemented in hardware using the currently available technology, from the practical point of view it does not change the state of the art in the pattern recognition system design.

The most systematic pattern recognition methodology has been developed for pattern variability that can be handled by statistical models. In other words, in situations where pattern X can be considered as a continuous multivariate random variable having a unique conditional probability distribution for each class ω_i, $i=1,2,\ldots m$. It will be convenient to characterise the probability model for X in terms of the set of class conditional probability density functions $p(X|\omega_i)$ and the a priori class probabilities $P(\omega_i)$, $\forall i$.

3.1 Statistical Pattern Classification

The above probabilities constitute all the essential ingredients for making decisions about class membership of patterns X, provided it is appropriate to weigh all classification errors equally. The corresponding optimal decision rule, known as the Bayes minimum error rule, states:

$$\text{assign } X \text{ to } \omega_i \text{ if } P(\omega_i|X) = \max_j P(\omega_j|X) \tag{1}$$

Thus pattern X is assigned to class ω_i if the a posteriori probability of class ω_i given X is greater than the a posteriori probability of any other class. The functions, $P(\omega_j|X)$ can be computed using the Bayes formula relating the conditional probabilities as

$$P(\omega_j|X) = \frac{P(\omega_j)\ p(X|\omega_j)}{p(X)} \tag{2}$$

where p(X) is the mixture density given as

$$p(X) = \sum_{j=1}^{m} P(\omega_j)\ p(X|\omega_j) \tag{3}$$

The Bayes decision rule (1) has been in existence in one form or another for more than two centuries. The recent and current research in statistical pattern recognition relating to this topic is concerned with the problems of its implementation. The problems lie in the fact that we do not know the probability functions $p(X|\omega_j)$. The only information assumed available for the recognition system design is a set of training patterns the class membership of which is known.

The absence of knowledge of these functions has opened a wide spectrum of problems for the statistical pattern recognition community. The use of practically implementable decision rules which are inferred from small sample sets raises the questions of their

- asymptotic performance
- small sample performance
- computational complexity
- storage requirements
- dependence on the distance measure employed

and many related problems.

There are basically two approaches to implementing the rule in (1). Either we can estimate the probability functions involved or we can determine the class discriminating surfaces or functions.

The discriminant function method of the classifier design has not witnessed many new developments over the last decade. The classical textbooks [e.g. 10,11] and the more up-to-date text [12] provide the state of the art coverage of this design approach. Among the recent contributions worth noting is the work of DO TU and INSTALLE on algorithms minimizing the error probability [14].

The approach based on estimating the class conditional distributions poses particular difficulties when these distributions are nonparametric [12]. The k-nearest neighbour (k-NN) probability density function estimation method and the related k-NN decision rules have been intensively studied in this context since the early nineteen seventies. As these developments are very significant they will be briefly reviewed in the following.

The asymptotic performance of NN rules was first studied by COVER and HART [15]. Later works [16,17] showed that the error bounds on the asymptotic performance of these rules hold even without any regularity conditions on the distributions of pattern classes. Attempts to improve the error bounds derived in [15] have also been made [19,20] and they are reviewed in [18].

An interesting family of NN rules is that which allows a reject option. The error-reject relationship was first studied by CHOW [32] in the context of the Bayesian decision theory. Its application to the NN approach is due to DEVIJVER [22]. The trade-off between error acceptance and reject probabilities is discussed in detail in [23] and [24].

The implementation aspects of NN rules have been reviewed extensively by DASARATHY and SHEELA [21]. To reduce the computational requirements, a number of fast algorithms for searching nearest neighbours has been devised. However, the computational requirements of even the most ingenious schemes are still excessive because of the large data bases required to characterise the classes. Moreover, these algorithms do not alleviate the considerable storage requirement. These considerations motivated the work on training set editing and condensing which would automatically reduce the computational burden of the nearest neighbour search. The idea is to characterise each class by a small representative subset of training patterns guaranteeing a performance for certain rules which is at least as good as that on the complete set.

The idea of condensing the training data set was first suggested by HART [28]. It inspired a number of algorithms which are described in GOWDA and

KRISHNA [29]. Unfortunately, these algorithms do not guarantee that the decision boundaries of the NN rule on the reduced and the complete training sets coincide. However, the recent work on computational geometry [30,31] has led to solutions which allow for exact reconstitution of the decision boundaries.

While the main purpose of editing the training data set is to improve the performance of the NN rule [26,27], it also results in a data set reduction. Recently the editing technique was refined by DEVIJVER and KITTLER [25]. It not only overcomes the problems arising from the particular handling of data as suggested in [26] but also permits a repeated application of the editing procedure to further improve the classification performance. When no further performance improvement is possible, one of the condensing techniques mentioned earlier could be used to achieve a maximum data reduction.

An important issue relating to the application of NN rules is how to measure the distance between two patterns. SHORT and FUKUNAGA [33] have recently suggested that in the two class case the distance should be measured along the direction linking the local means of the classes. The means corresponding to a point X are inferred from the observations of its k-nearest neighbours. This approach minimizes locally the mean squared error between the infinite sample and small sample probabilities of misclassification of the NN rule.

3.2 Feature Selection and Extraction

The main purpose of feature selection and extraction is to reduce the dimensionality of pattern representation without incurring any loss of discriminatory information. In feature selection dimensionality reduction is achieved by selecting among the measurements providing pattern representation those which contain most of the discriminatory information. The feature extraction approach on the other hand is to use all available information about patterns and map this information into a lower dimensional feature space.

A dimensionality reduction may be necessary for a number of reasons. First of all, the classifier complexity is usually a nonlinear function of the data dimensionality. Second, recognition performance has been observed to improve when the sample size to dimensionality ratio increases. This behaviour is known as the peaking phenomenon and it has been discussed in detail by DUIN [34]. As the sample size is usually fixed, the control over this ratio is vested with the dimensionality of pattern representation.

Another benefit of feature evaluation is that it provides the means for assessing the achievable performance of the recognition system. Feature extraction may in addition be the most effective way to compress information for transmission purposes while feature selection offers a unique opportunity to minimize the cost of the measurement extraction.

The problem of feature evaluation involves the determination of the

- form of the mapping
- feature evaluation criterion
- feature space dimensionality
- optimisation procedure

A detailed and most up-to-date treatment of these aspects can be found in DEVIJVER and KITTLER [12].

4. The Application of Statistical Pattern Recognition to Edge Detection

As an example of the use of statistical pattern recognition in image analysis we shall consider the problem of edge detection. This particular problem has been chosen for two reasons. First of all, it is relatively simple and, second, it is a typical example of decision making tasks which are not normally formulated as a statistical pattern recognition problem.

Discontinuities in pixel intensity values in the image of a scene are usually indicative of the presence of edges of objects in the scene, the luminance value of which differ from the background. There is a number of approaches to edge detection that can be adopted, ranging from global methods, through line following, to sophisticated relaxation algorithms [35-45]. The most popular and very effective are local edge detection methods which involve operation on the immediate neighbours of a pixel being investigated.

A local edge detector is defined in terms of n x n masks where n can be any integer greater than or equal to two, but considerably less than the number of image sampling points in each of the orthogonal axis of the image frame, in order to retain its local character. For instance, in BROOKS' detector [43] the kernel size is 2 x 2 while n in MARR's masks [45] is of the order of 30.

For every pixel of the analysed image, the masks of an edge detector centred at the pixel are correlated with the image. Let the output of the j - th mask be denoted x_j. The edge signal magnitude X is then a function of these mask outputs x_j, i.e.

$$X = G(x_1, x_2, \ldots x_q) \tag{4}$$

where q is the number of masks defining the edge detector. Each mask is usually sensitive to edges in one specific direction.

If the processed image was noise free, it would be relatively easy to detect the presence of edges in the image by deterministic methods. In practice, however, image data is almost invariably affected by noise and consequently the edge detection problem becomes a classical statistical decision making problem.

There are many ways in which the edge detection problem can be approached depending on the complexity of the function G and on our knowledge of the noise statistics. In the absence of any model for the statistical process characterising the noise which corrupts the image data, it is most appropriate to determine the probability distributions for the random variable X conditioned on the two possibilities - edge present, and edge absent - using nonparametric methods.

The pixel noise can however be often modelled quite adequately as being additive and uncorrelated, with a Gaussian distribution having a zero mean and variance σ^2. Under these assumptions the class conditional distributions of X can be sometimes derived analytically, depending on the forms of G and of the edge detector masks. ABDOU and PRATT [44] obtained the class conditional distributions for individual mask outputs x_i of several linear operators. These results however apply only to the case of a constant edge direction.

Here we shall derive the density functions $p(X|\omega_i)$, i=1,2, where ω denotes - no edge- and ω - edge present -, for the template edge detector.

It is defined in terms of four masks shown in Fig.3. The edge magnitude of the detector is given as

$$X = \max_j | x_j | \qquad j = 1,2,..4. \tag{5}$$

The robustness of this detector has been demonstrated in [44] and its accuracy studied in [1], where it has been shown that in the noise-free case the detected edge magnitude signal is within 1.5% of the actual edge signal irrespective of the edge direction.

-1	0	1	0	1	2	1	2	1	2	1	0
-2	0	2	-1	0	1	0	0	0	1	0	-1
-1	0	1	-2	-1	0	-1	-2	-1	0	-1	-2

Fig.3. The template edge detector masks

For any edge angle θ the magnitude of the edge signal X will be identical to the output of one particular mask. Let us denote by $\Delta\theta_j$ the interval of edge angles for which the j-th mask yields a maximum output. As the output of each mask is defined by its correlation with a corresponding image window, X is a linear function of the pixel values. Thus we can apply the well-known results in probability theory relating to the calculation of the probability distribution of a linear combination of random variables.

Let us denote the j-th mask coefficient values by a_{jkl}, k = 1, 2, 3, l = 1, 2, 3. Then the output of the mask, and hence the signal X in the interval $\Delta\theta_j$ is given as

$$X = \left| \sum_{k=1}^{3} \sum_{l=1}^{3} a_{jkl} F_{kl} \right| \tag{6}$$

where F_{kl} are the pixel values in the image window correlated with the mask. To simplify the analysis we shall assume that the pixel values F_{kl} are such that, in the noise-free case, they will give rise only to non-negative values of the double summation. Then, in the absence of noise, X will take the following values:

$$X = \begin{cases} 0 & \text{no edge} \\ E & \text{edge} \end{cases} \tag{7}$$

where E is almost constant everywhere in $\Delta\theta_j$ as already pointed out. It should be noted, however, that this property holds only for the edge passing through the centre of the image window [1].

In the presence of noise, X in (6) with the absolute value operation ignored, will be a normally distributed random variable with the class conditional means

$$m_1 = 0 \qquad \text{for } X \varepsilon \omega_1 \tag{8}$$

and

$$m_2 = E \qquad \text{for } X \varepsilon \omega_2 \tag{9}$$

The variance, Σ^2, of each conditional distribution is given by

$$\Sigma^2 = \sigma^2 \sum_{k=1}^{3} \sum_{l=1}^{3} a^2_{jkl} \tag{10}$$

Taking the absolute value operation in (6) into account will have the effect of folding the probability density function for the no-edge case, but practically no effect on $p(X|\omega_2)$. Thus we have

$$p(X|\omega_1) = 2\,[2\pi\Sigma^2]^{-\frac{1}{2}} \exp\left\{-X^2/2\Sigma^2\right\} \tag{11}$$

and

$$p(X|\omega_2) = [2\pi\Sigma^2]^{-\frac{1}{2}} \exp\left\{-(X-E)^2/2\Sigma^2\right\} \tag{12}$$

Due to symmetry, the lifting of the restrictions on F_{kl}, which will result in the double summation in (6) assuming any real value, will not affect the density functions in (11) and (12). Moreover, it is easy to verify that the conditional distributions of X are identical in all intervals $\Delta\theta_j$. Thus we have derived the required distributions and the decision rule (1) can now be applied.

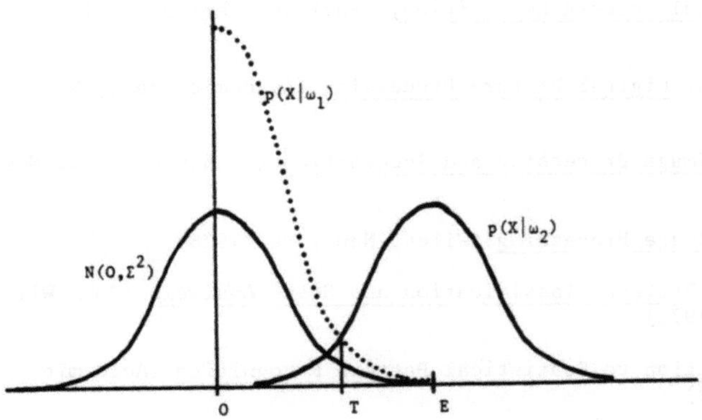

Fig.4. Showing the class conditional probability density functions

The effect of folding is a very interesting and somewhat unexpected conclusion. It implies that the optimal threshold which divides the edge detector magnitudes X into the classes of edge and no-edge signals is closer to E than to 0, i.e. it will always be greater than E/2. This is apparent from Fig.4. The optimal value of the threshold could be obtained by substituting (11) and (12) into (2) and applying the decision rule (1). $P(\omega_i)$ should be set to 0.5 if we assume that, a priori, the edge presence or absence is equally likely. Taking the logarithm of the decision rule then yields

$$\text{assign } X \text{ to } \omega_1 \text{ if } X < \frac{E}{2} + \frac{\Sigma^2}{E}\log 2 \tag{13}$$

Thus the optimal threshold T is defined as

$$T = \frac{E}{2} + \frac{\Sigma^2}{E}\log 2 \tag{14}$$

References

1. J Kittler: To be published in Issues in Acoustic Signal/Image Processing and Recognition, Ed. by C H Chen (Springer Verlag, Berlin 1983)

2. K S Fu: Syntactic Methods in Pattern Recognition (Academic Press, New York 1974)

3. K S Fu, ed.: Syntactic Pattern Recognition, Applications (Springer Verlag, New York 1977)

4. R C Gonzalez and M G Thomason: Syntactic Pattern Recognition, An Introduction (Addison Wesley, Reading MA 1978)

5. K S Fu: Syntactic Pattern Recognition and Applications (Prentice Hall, Englewood Cliffs, NJ 1982)

6. T Pavlidis: Structural Pattern Recognition, (Springer Verlag, Berlin, 1977)

7. A Rosenfeld, A C Kak: Digital Picture Processing (Academic Press, New York, 1976)

8. E L Hall: Computer Image Processing and Recognition (Academic Press, New York, 1979)

9. W K Pratt: Digital Image Processing,(Wiley, New York, 1978)

10. R O Duda, P E Hart: Pattern Classification and Scene Analysis (John Wiley and Sons, New York, 1973)

11. K Fukunaga: Introduction to Statistical Pattern Recognition (Academic Press, New York, 1972)

12. P A Devijver, J Kittler: Pattern Recognition a Statistical Approach (Prentice Hall, Englewood Cliffs, 1982)

13. J R Ullmann: in this volume

14. H Do Tu and M Installe: IEEE Trans. C-27, 648 (1978)

15. T M Cover and P E Hart: IEEE Trans. IT-13, 21 (1967)

16. C J Stone: Ann. Statist., 5, (1977)

17. L Devroye: IEEE Trans. PAMI-3, 75 (1981)

18. P A Devijver: Pattern Recognition in Practice, E S Gelsema, and L N Kanal Eds. (North Holland, Amsterdam 1980)

19. L Györfi and Z Györfi: IEEE Trans. IT-24, 512 (1978)

20. P A Devijver: IEEE Trans. IT-25, 749 (1979)

21. B V Dasarathy and B V Sheela: Proc. 1977 IEEE Int. Conf. Cybernetics and Society, Washington DC, pp. 630-634 Sept. 1977

22. P A Devijver: in Pattern Recognition Theory and Application, K S Fu and A B Whinston Eds. (Noordhoff, Leyden 1977)

23. P A Devijver: MBLE Res. Rept. R-346, Brussels 1977

24. P A Devijver: in Aspects of Signal Processing, Part 2, G Tacconi Ed. (Reidel, Dordrecht 1977)

25. P A Devijver and J Kittler: Proc. Fifth Int. Conf. Pattern Recognition, Miami Beach, Fl., pp.72-80, Dec. 1980

26. D L Wilson: IEEE Trans. SMC-2, 408 (1972)

27. C S Penrod and T J Wagner: IEEE Trans. SMC-7, 92 (1977)

28. P E Hart: IEEE Trans. IT-14, 515 (1968)

29. K C Gowda and G Krishna: IEEE Trans. IT-25, 480 (1979)

30. G T Toussaint: in Pattern Recognition Theory and Applications, J Kittler, K S Fu and L F Pau Eds. (D Reidel, Dordrecht 1982)

31. P A Devijver and M M Dekesel: Philips Research Laboratory Memo No 13, Brussels (1981)

32. C K Chow: IEEE Trans. IT-16, 41 (1970)

33. R D Short and K Fukunaga: Proc. Fifth Int. Conf. Pattern Recognition, Miami Beach, Fl., pp. 81-86, Dec. 1980

34. R P W Duin: On the Accuracy of Statistical Pattern Recognisers (Dutch Efficiency Bureau, Pijnaker 1978)

35. L S Davis: Computer Graphics and Image Processing, 4, 248 (1974)

36. S Levialdi: in Digital Image Processing, J C Simon and R M Harralick Eds. (D Reidel, Dordrecht 1981)

37. R A Kirsch, L Cahn, C Ray and G H Urban: Proc. Eastern Joint Computer Conference, pp. 221-229, 1957

38. M H Hueckel: Journal A C M, 18, 113 (1971)

39. J W Modestino and R W Fries: Computer Graphics and Image Processing, 6, 409 (1977)

40. P Lenkin: Computer Graphics and Image Processing, 9, 150 (1979)

41. B J Schachter, A Lev, S W Zucker and A Rosenfeld: IEEE Trans. SMC-7, 813 (1977)

42. J M Tennenbaum, A C Kay, T Binford, G Falk, J Feldman, G Grape, R Paul, K Pingle and I Sobel: Proc. IJCAI, pp. 521-526 (1969)

43. M J Brooks: Computer Graphics and Image Processing, 8, 277 (1978)

44. I E Abdou and W K Pratt: Proc. IEEE, 67, 753 (1979)

45. D Marr: Phil. Trans. R. Soc. London, B275, 483 (1976)

Memory Networks for Practical Vision Systems

I. Aleksander

Department of Electrical Enginerring and Electronics,
Brunel University, Uxbridge, United Kingdom

1. INTRODUCTION

For most, neural modelling as an enterprise in the design of image recognition systems has a historical beginning in 1943 with the work of McCULLOCH and PITTS [1] and a quasi-death in 1969 with the condemnation by MINSKY and PAPERT [2] of ROSENBLATT's development of such a model into a "perceptron" [3]. The argument in [2] centers on the blindness of such systems to geometrical and topological properties of images for which the sort of calculational powers offered by classical computer structures is better suited. It was also seen that the inferential power of some Artificial Intelligence programs offered more to image recognition (or "scene analysis") than the neural modelling approach.

This paper is a report on work which might, in some sense, have appeared to be "heretical" in the late 1960s [4], as it suggested that the neural model could be used to design systems which would benefit from novel solid state architectures. Indeed, the practical recognition system called WISARD* [5] has been based on the use of 32,000 random access memory (RAM) silicon chip devices as "neurons" in a single-layer "neural net". This system will be addressed indirectly in parts of this paper, while the central aim is to provide a new assessment of such nets in image recognition tasks that are important in robotics. This is a major application area for WISARD, as it can take decisions on 512 x 512 pixel images at the rate of 5-10 per second. Robotic vision tasks require decisions at this order of speed on rotated and translated versions of objects and sometimes demand the recognition of combinations of objects even if occlusion takes place. The behaviour of such networks in this area will be assessed using a theoretical methodology somewhat different from empirical and problem-oriented methods which are common in this field. The methodology adopted here is to lay a general theoretical foundation for the operation of the system, show its sensitivity to key parameters and then employ an analytically tractable group of simplified shape-analysis problems to illustrate the way in which problem definition leads to the choice of parameters.

The paper ends with a description of current work on dynamic versions of such nets, and expectations for the future.

2. THE RAM/NEURON ANALOGY

Although this has been published before, it is included here in a different format for completeness. The central function of most neural models is the discriminant function:

* Wilkie, Stonham's, and Aleksander's Recognition Device

$$\sum_i w_i x_i = \theta \qquad (2 - 1)$$

where, x_i is the i-th, binary synaptic input of the neuron
and w_i is its associated synaptic "weight", $-1 < w_i < 1$
while, θ is a threshold such that the neuronal axon fires

iff $\sum_i w_i x_i \geqslant \theta$

and does not otherwise.

If the input consists of N synapses, i.e. $1 \leqslant i \leqslant N$, $i = 1,2,3...$ (2 - 1) (2 - 1) can be written in the more general form of a mapping:

$$X \xrightarrow{\bar{W}} \{0, 1\} \qquad (2 - 2)$$

where X is the set $\{\bar{x}_1, \bar{x}_2, \bar{x}_3 ..... \bar{x}_{2N}\}$ of binary patterns and $\bar{W}$ is the weight vector $(w_1, w_2, ...w_j...w_N)^T$ while {0, 1} indicate firing, 1 or not, 0 T indicating a vector transposition.

A Random-Access Memory, on the other hand performs the mapping

$$X \xrightarrow{\bar{M}} \{0, 1\} \qquad (2 - 3)$$

where $\bar{M}$ is a binary vector of the form $(m_0, m_1 ... m_{2N-1})$ which is, in fact, the content of the memory, where the "inputs" of the memory are N "address" terminals, and the output is a single binary terminal. Thus m_j is a single-bit "word" addressed by the j-th pattern from X.

One notes the following differences between the model represented by (2 - 2) and that represented by (2 - 3).

a) In the former, the logical function is stored as a vector $\bar{W}$ of N continuous variables while in the latter it is stored in a binary vector $\bar{M}$ of $2^N - 1$ variables.

b) The former can achieve only "linearly separable" functions, which is a number much inferior (but hard to calculate [6]) to all the functions (2^N of them) that the latter can achieve, each being characterised by a distinct $\bar{M}$ vector in the latter case.

None of this so far describes the way that the $\bar{W}$ or $\bar{M}$ values are built up, that is, the way the model "learns". In most cases the procedure from the point of view of a user is much the same. The system is "given" the desired output for specific $\bar{x}_j$ patterns present at the input (from a "training" set). Some machinery needs to be put in motion within the system to bring about the desired mapping. In neural models using $\bar{W}$ this can be quite a complex procedure needing many iterations [6]. In the latter models the bits m_j of $\bar{M}$ are set directly, with the "desired" response being fed to the RAM through its "data input" terminal.

When "perceptrons" were central to image recognition work it was thought that the differences mentioned in (a) gave the system some of its "intelligence" since fewer examples of the $X \xrightarrow{\bar{W}} \{0,1\}$ mapping need to be shown to the system in order for it to 'fix' its function. Now one suspects that this 'fixing' is an arbitrary process which is as likely to worsen performance as to improve it. As will be seen, it is the structure of the network that gives one a measure of control over the discriminatory powers of the system, and not so much the generalisation effects within the elements of the net. Thus the total lack of generalisation within RAMs is of little consequence, what is important is the way they are deployed in networks and the way such networks are trained.

On a technological point one notes that $\bar{W}$ calls for storage of analogue or highly multi-valued data. This has led some experimenters to build machines in which $\bar{W}$ was implemented with roomfuls of potentiometers or special registers. The RAM system, on the other hand, leads to highly cost-effective systems which can either be implemented directly on conventional machines (leading to one or two decision/second on a 256 Kilobit vector), or be implemented by totally parallel hardware (which could give up to 10^6 decisions per second) or with semi-parallel WISARD-like systems [7], [8] which give tens of decisions per second at a relatively low cost.

3. SINGLE LAYER, RANDOMLY-CONNECTED NETS : PARAMETERS AND DEFINITIONS

R is the number of bits in the image to be recognised.
N is the number of address (synaptic) inputs to each RAM, known as N-TUPLE
A discriminator is a group of K RAMs where K usually is R/N as a discriminator is made to "cover" the image just once, in a one-to-one- manner. The connection is made at random.
C is the number of classes to be distinguished in a particular problem. This implies the existence of C discriminators, one per class.

Training is the process of presenting the system (at the overall input) with examples of patterns in the i-th class and "writing" into the i-th discriminator, logical 1's at all of its K "data in" terminals. It is assumed that all the memories are reset to 0 at the start of a training run.

A response r_j of the j-th discriminator to an arbitrary pattern $\bar{x}_i$ is obtain- as follows. First one notes that each discriminator generates an integer c_j

$$0 \leqslant c_j \leqslant K \quad , \quad \text{for all j from 1 to C}$$

The response is then quoted as a percentage of K,

$$\text{i.e. } r_j = c_j/K \ \%$$

In this connection one also defines a response vector to $\bar{x}_i$

$$Z(\bar{x}_i) = (r_1 \ , \ r_2 \ .... \ r_c)^T$$

A decision is made on the basis of $Z(\bar{x}_i)$, the decision being a singular mapping:

$$Z(\bar{x}_i) \xrightarrow{\delta} \{d \ / \ d \in \{1,2... \ K\} \cup \emptyset\}$$

where d is the class assigned to the unknown pattern and $\emptyset$ is a symbol for a "don't know" decision.

Several decision strategies, may be employed for example:

a) The maximum response strategy, where d is the value of j for the maximum r_j in $Z(\bar{x}_i)$. This is written $d = JMAx[Z(\bar{x}_i)]$

while the actual response of the discriminator in question is written:

$$f = MAX[Z(\bar{x}_i)] \quad (3 - 1)$$

b) The confidence conditioned strategy, where the highest response $Z(\bar{x}_i)$ with d removed exceeds a pre-specified threshold T . This is written

$$d = JMAX[Z(\bar{x}_i)]$$

$$\text{if} \quad MAX[Z(\bar{x}_i)] - MAX[Z(\bar{x}_i) - d] > T_C \quad (3 - 2)$$

or else $d = \emptyset$.

c) The average conditioned strategy, where the average response must exceed a pre-specified threshold Ta before the decision is noted. This is written

$$d = JMAX[Z(\bar{x}_i)]$$

$$\text{if} = AVE[Z(\bar{x}_i)] > T$$

else $d = \emptyset$

There are many others used in specific applications.

One notes that the system treats the image as an R-bit vector. Should there be grey levels involved, and should the image consist of p pixels of b bits each

$$R = p \times b$$

The above description should be read as being "notional". This means that its implementation could either be in the form given, in which case a fast, parallel system would result (as already mentioned in 2) or, both semi-parallel and serial re-organizations of the "notion" are possible [7], [8].

4. ANALYSIS OF SYSTEM TRAINING AND RESPONSES

The theory and early understanding of such systems was provided not by those interested in neural modelling but by workers in pattern recognition. This originated with work on N-tuple techniques by BLEDSOE and BROWNING to which much was added in the context of character recognition by ULLMAN [10]. Some general analysis may be found in work by the author [11], [12], again largely in the area of character recognition.

The objective here is to provide a simplified theoretical background relevant in situations such as robot vision, where the data is originally derived from shapes, and variations and provided by rotation translations or distortions of such shapes.

Three effects will be pursued:

a) The probable response of a single discriminator as a function of N and training

b) The probable response of several discriminators as a function of N and training
c) The probable saturation of the system as a function of N and training.

The equation which is common to all three of the above effects in the probability of siting a randomly selected N-tuple is a specific field of Q points in a total field of R points. We call this P(N,Q,R) where,

$$P(N,Q,R) = \left(\frac{Q!}{N!(Q-N)!}\right) \div \left(\frac{R!}{N!(R-N)!}\right)$$

$$= \frac{Q!(R-N)!}{R!(Q-N)!} \qquad (4 - 1)$$

This holds as long as $1 \leqslant N \leqslant Q \quad N = 1,2,3...$

while $P(N,Q,R) = 0$ when $N > Q$

4.1. Single Discriminator Analysis

For clarity we simplify the analysis to a system that has been trained on only two patterns $\bar{x}_1$ and $\bar{x}_2$ and is tested with a third $\bar{x}_3$. We define the similarity between two patterns (say $\bar{x}_1$ and $\bar{x}_2$)

$$\bar{s}_{12} = \bar{x}_1 \,\theta\, \bar{x}_2 \quad ;$$

θ is the equivalence operator such that

s_i of $\bar{s}_{12} = 1$ iff x_i of $\bar{x}_1$ is equal to x_i in $\bar{x}_2$

We further define the area of similarity S_{12} related to $\bar{s}_{12}$ as the number of 1's in $\bar{s}_{12}$.

One can now compute the probability $P(\bar{x}_3)$ of generating an output of 1 from a specific RAM in response to $\bar{x}_3$. This is achieved when its N-tuple is sited either in the area of similarity between $\bar{x}_3$ and $\bar{x}_1$ or the area of similarity between $\bar{x}_3$ and $\bar{x}_2$ but not both (to avoid double counting). This may be written using (4 - 1) as:

$$P(\bar{x}_3) = P(N,S_{13},R) + P(N,S_{23},R) - P(N,\{\bar{s}_{13} \cap \bar{s}_{23}\}, R) \quad ; \qquad (4 - 2)$$

note that x_i in $\{\bar{s}_{13} \cap \bar{s}_{23}\}$ is 1 iff x_i is 1 in $\bar{s}_{13}$ and x_i is 1 in $\bar{s}_{23}$

It is clear that Boolean algebra applies to the input vectors and it may be shown that the area of $\bar{s}_{13} \cap \bar{s}_{23}$ is the total overlap between $\bar{s}_1$, $\bar{s}_2$ and $\bar{s}_3$. Let this be S_{123}.

Hence:

$$P(\bar{x}_3) = \frac{(R-N)!}{R}\left[\frac{S_{13}!}{(S_{13}-N)!} + \frac{S_{23}!}{(S_{23}-N)!} - \frac{S_{123}!}{(S_{123}-N)!}\right] \qquad (4 - 3)$$

Note that this leads directly to the most likely response of the discriminator (say the jth one)

$r_j = P(\bar{x}_3)$, or, for a better way of relating r_j to $\bar{x}_3$ this may be written: $r_j(\bar{x}_3) = P(\bar{x}_3)$.

To begin to appreciate the effect of N, first let N = 1. It may be shown that (4 - 3) collapses to

$$r_j(\bar{x}_3) = \frac{\text{Area Of}[\bar{x}_3 \cap (\bar{x}_1 \cup \bar{x}_2)]}{R}$$

The significance of this is that $r_j(\bar{x}_3)$ reaches a maximum (i.e. 100%) when $\bar{x}_3 = \bar{x}_1 \cup \bar{x}_2$. That is, the discrimination is so poor that the system totally confuses $\bar{x}_1 \cup \bar{x}_2$ with the individual $\bar{x}_1$ and $\bar{x}_2$ (the latter two also giving a 100% response).

Now assume that $\bar{x}_3 = \bar{x}_1 \cup \bar{x}_2$ and that $1 < N \leqslant R$.

In computing $r(\bar{x}_1)$ one notes that $S_{11} = R$, $S_{12} = S_{123}$,

hence (4 - 3) gives

$$r_j(\bar{x}_1) = \frac{(R - N)!}{R!} \left[\frac{R!}{(R - N)!} + \frac{S_{12}!}{(S_{12} - N)!} - \frac{S_{12}!}{(S_{12} - N)!} \right]$$

$$= 100\%.$$

By the same reasoning

$$r\ (\bar{x}_2) = 100\%\ (\text{irrespective of N}) .$$

It may be shown using (4 - 3) that $r_j(\bar{x}_3)$ rapidly and monotonically decreases to zero as N increases, certainly having reached it by the time N reaches S_{13} or S_{23}, whichever is the greater.

These principles of increasing discrimination against combinations of training patterns clearly apply to greater numbers of training patterns, while, as can be imagined, the equivalents of equation (4 - 3) become a rather messy series of irreducible terms.

The main fact remains that a single discriminator has a discriminatory power which increases sharply and monotonically with N.

4.2. Multi-Discriminator Analysis

Again only a simple case with two discriminators is considered in order to demonstrate trends which occur in more complex situations.

Assume that there are two discriminators giving responses r_1 and r_2 and that each is trained on one pattern, say $\bar{x}_1$ and $\bar{x}_2$ respectively. The system is tested with a third pattern $\bar{x}_3$. Then the responses to the third pattern are in full notation:

$$r_1(\bar{x}_3)$$

and $r_2(\bar{x}_3)$.

According to (3 - 1) the decision goes to

$$d = \mathrm{JMAX}[r_1(\bar{x}_3),\ r_2(\bar{x}_3)]$$

$$f = \mathrm{MAX}[r_1(\bar{x}_3),\ r_2(\bar{x}_3)].$$

Without loss of generality let $d = 1$, $f = r_1(\bar{x}_3)$.
It is convenient to redefine the confidence on d in (3 - 2) in the following relative way

$$\mathrm{CON}(d) = [r_1(\bar{x}_3) - r_2(\bar{x}_3)] \div [r_1(\bar{x}_3)]. \qquad (4 - 5)$$

Noting that, from (4 - 3) the responses are

$$r_1(\bar{x}_3) = \frac{(R-N)!}{R!} \cdot \frac{S_{13}!}{(S_{13} - N)!},\quad r_2(\bar{x}_3) = \frac{(R-N)!}{R!} \cdot \frac{S_{23}!}{(S_{23} - N)!} \qquad (4 - 6)$$

then we have

$$\mathrm{CON}(d) = 1 - \frac{S_{23}(S_{23}-1)(S_{23}-2)\dots(S_{23}-N+1)}{S_{13}(S_{13}-1)(S_{13}-2)\dots(S_{13}-N+1)} \qquad (4 - 7)$$

To clarify the role of N, one notes that generally $N \ll S_{13}$ or S_{23} and S_{13} or $S_{23} \gg 1$. Then, to observe trends one can say very roughly that

$$\mathrm{CON}(d) \cong 1 - \left(\frac{S_{23}}{S_{13}}\right)^N \qquad (4 - 7a)$$

Thus, not only does increasing N improve the discrimination properties of a single discriminator, but also provides, for improved discrimination within a multidiscriminator system in an exponential way.

However, one recalls that the storage costs of the system increase exponentially with N too, and it is also possible to degrade the performance of the system through too high a discriminatory power, hence the choice of N is usually a compromise between these factors [10], [12]. Optimal values of N for the recognition of hand-printed characters ase central to [10].

4.3. Saturation

Starting with all the RAMs at zero, the total number of logical 1s in the memory of a discriminator is a function of the number of training patterns seen by that discriminator. Let the average relative no. of 1s in a single RAM be

$$M(t) = [\text{no. of 1s}]/2^N$$

then one can expect that the most likely response to an arbitrary (or noisy) input pattern $\bar{x}_a$ is

$$r_j(\bar{x}_a) = M(t). \qquad (4 - 8)$$

This is called the degree of saturation of the discriminator.

One can then define an absolute confidence factor CON(a), as the confidence in relation to the response to $\bar{x}_a$.

Hence:

$$\mathrm{CON}(a) = 1 - \frac{M(t)}{\mathrm{MAX}[Z(\bar{x}_i)]} \qquad (4 - 9)$$

As training increases $M(t)$ approaches 2^N asymptotically while $MAX[Z(\bar{x}_i)]$ approaches 1. Thus CON(a) tends to 0 with training at a rate determined by $1/2^N$. (Note that in tending for 0, CON(a) can temporarily assume negative values, depending on MAX $]Z(\bar{x}_i)]$). This again shows how the confidence in the system depends heavily on N. However, and this may be surprising, in practical cases with WISARD, for R = 512 x 512, it was never necessary to have N greater than 8.

Finally, it becomes of interest to estimate the rate of growth of M(t) with t by means of a highly approximate analysis.

Assume that a discriminator trained on a set of patterns $T = \{\bar{x}_1, \bar{x}_2, \bar{x}_3 \ldots \bar{x}_{t-1}\}$ is now tested with the t-th pattern $\bar{x}_t$ giving a response $r(\bar{x}_t)$. This response is a measure of the N-tuples already set, hence if the discriminator is now trained on $\bar{x}_t$, the total additional number of bits set in the entire discriminator memory is $R(1 - r(\bar{x}_t))$. Therefore the increment in M(t)(i.e. $\Delta M(t)$) is given by

$$\Delta M(t) = \frac{1}{2^N} \cdot \frac{1}{K} \cdot R(1 - r(\bar{x}_t))$$

$$= \frac{1}{2^N} \cdot N(1 - r(\bar{x}_t)) \quad (\text{Since } K = \frac{R}{N}) \qquad (4 - 10)$$

From this one can calculate

$$M(t) = \frac{N}{2^N} \sum_{i=1}^{t} (1 - r(\bar{x}_i)) \qquad (4 - 11)$$

To get an even better approximate feel for this relationship, using the same argument as in (4 - 6), (4 - 7) and (4 - 7a) one notes that

$$\Delta M(t) \cong \frac{N}{2^N} \left(1 - \left(\frac{S_{st}}{R}\right)^N\right) \qquad (4 - 12)$$

where S_{st} is the similarity between the t-th pattern and $\bar{x}_s$ where $\bar{x}_s$ is the pattern in T most similar to $\bar{x}_t$.

The table below indicates some rates of saturation $\Delta M(t)$.

S_{st}/R	N=2	4	8	
.9	.095	.086	.018	$\Delta M(t)$
.5	.375	.234	.03	

Thus one notes again the beneficial effect of increasing N as for a given degree of similarity S_{st} the rate of saturation is reduced with increasing N.

5. ANALYSIS OF RESPONSES TO SHAPES

This section refers to the discriminatory powers of a RAM single-layer system with respect to "shapes". These can be described as sets of patterns

$A = \{\bar{a}_1, \bar{a}_2, \bar{a}_3 \ldots\}$ for shape A

$B = \{\bar{b}_1, \bar{b}_2, \bar{b}_3 \ldots\}$ for shape B where $\bar{a}_j \in A$ and $\bar{a}_k \in A$ are patterns of the same "shape" translated and/or rotated to a new position. We wish to address the following problems

a) Position-free recognition
b) Coping with several shapes and occlusion
c) Finding the position and orientation of simple shapes.

To simplify matters, a particular example will be followed throughout this section and that is that there are two specific shapes related as shown in Fig. 1.

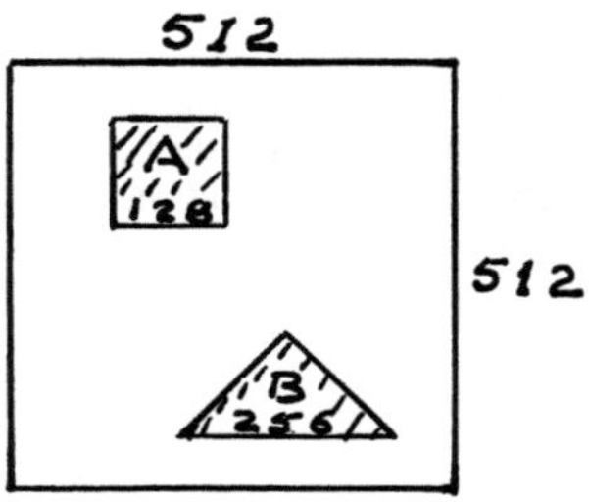

Fig. 1. Example of Shapes

Thus set A is the set of all 128 x 128 squares that could appear in a 512 x 512 frame and B a set of triangles dimensioned as shown.

One notes that the two shapes have the same area so that one removes the possibility of mere area measurement. One notes further that the black-white shading and geometrically definable nature of the objects is arbitrary. The system is equally capable of dealing with textured objects which would defeat most geometircally oriented recognition techniques such as [13].

5.1. Position-free recognition

For simplicity, one assumes that the above two shapes can only appear in their "stable" positions with respect to the lower edge of the frame. This gives one orientation for the square A and three for the triangle B. The first step in selecting N is deciding on the size of the training set. This is guided by the fact that an arbitrarily placed object must provide an area of overlap with an element of its appropriate training set which is greater than the maximum overlap between the two objects. A simple but tedious calculation shows that for this condition to hold, the discriminators can be trained in a scan-like fashion with a total of 81 patterns for discriminator A (the square) while B requires 114 triangles with the long side horizontal and 242 with the short side horizontal.

N is selected so as to keep the system out of saturation. Using equation (4 - 11) and an approximation for $r_j(\bar{x}_j)$ as used in (4 - 12) but this time applied to (4 - 3), one can derive an equation for the rate of saturation $\Delta M(t)$, based on the overlap of the next training pattern and two others. These are, the one just above and the one just to the left, for the scan going left to right and top to below.

Thus

$$\Delta M(t) \simeq \frac{N}{2N}\left[1 - \left(\frac{S_1}{R}\right)^N - \left(\frac{S_2}{R}\right)^N + \left(\frac{S_{12}}{R}\right)^N\right] \qquad (5 - 1)$$

where S_1 is the overlap area of the new training pattern with the one above
S_2 is the overlap area of the new training pattern with the one to the left
and S_{12} is the common area between S_1 and S_2 as before.

This expression can be summed for the entire scan from a knowledge of the areas of overlap which need to be calculated for each stable position. Taking the triangle as the more stringent case it is interesting to tabulate M(t), the % saturation as a function of N in a significant range:

N =	8	9	10	11
M(t) triangle =	>100%	>100%	80%	52%

This is a highly pessimistic calculation as the approximations themselves are pessimistic. However, the result indicates that N = 10 is the least value that can be chosen. The confidence with which the systems will distinguish between triangles and squares for N = 10 can be checked by using the worst overlap with its opposing class and the "worst fit" with its own class in (4 - 7a). It turns out that the system responds with a minimum confidence level of

82% in response to a square
and 81% in response to the most "difficult" triangle.

In practice one would not necessarily adopt a scan-training technique (presented here because it is analytically tractable). Indeed, one would start by selecting a feasible value for N and adjusting the training so as to keep the system out of saturation, increasing N only if the eventual confidence levels fall near or even below zero (giving the odd erroneous classification). The training patterns would be determined by operator presentation of objects, which is particularly attractive in robotic situations where the operator, in general, communicates with the robot by means of examples.

5.2. Effect of several shapes and occlusions

Much can be done by studying the responses of the system given constraints on combinations of patterns that can appear in the field of view. Here the previous example is continued but the class of patterns that can be viewed is extended to the presence of any two objects in the field of view. Again, this is done for analytic simplicity, and initially it is assumed that no overlaps occur. Also, the response equation is simplified for taking into account the two nearest training patterns as in section 5.1. The equation for the confidence factor is that of (4 - 7a). N is taken as 10. Using this it may be noted that the response of the "square" discriminator to a single square in the image, $r_s(x_s)$, is

$$82.4\% \leqslant r_s(\bar{x}_s) \leqslant 100\% \text{ (with a minimum of 82\% confidence).}$$

Similarly for the "triangle" discriminator and a single triangle, $r_t(\bar{x}_t)$, is

$$81.8\% \leqslant r_t(\bar{x}_t) \leqslant 100\% \text{ (with a minimum of 81\% confidence).}$$

Now, if two objects are in the field of view simultaneously (a square and a triangle, non-overlapping) we have for this $\bar{x}_{st}$, say:

$$45.5\% \leqslant r_s(\bar{x}_{\ t}) \leqslant 54.2\% \text{ (with a maximum confidence of 18.9\%)}$$

and similarly for the triangle

$$42.3\% \lesssim r_t(\bar{x}_{st}) \lesssim 54.2\% \text{ (with a maximum confidence of 19.27\%).}$$

The range of responses arises from the degree of overlap between the test pattern and the nearest training pattern. If the two objects were of the same kind, say both squares (i.e. $\bar{x}_{ss}$), these responses are modified as follows

$$55.1\% \lesssim r_s(\bar{x}_{ss}) \lesssim 66.4\% \text{ (with a confidence range 21.8\% to 50\%)}$$

$$33.1\% \lesssim r_t(\bar{x}_{ss}) \lesssim 43.1\% \text{ (with the above confidence range).}$$

Very similar results are obtained if both objects are triangles ($\bar{x}_{tt}$), with $r_t(\bar{x}_{tt})$ being the highest.

Clearly, these results can be distinguished in two steps: first on the basis of the confidence level which identifies the situation group (one object, two similar objects, two dissimilar objects) and second the response itself which identifies the object(s) in the first two groups.

Considering now the question of occlusions, take the worst possible overlap between the triangle and the square, which is the case where the right angle of the triangle is coincident with one of the corners of the square. Predicted responses for the two discriminators turn out to be identical as the similarity areas with nearest training elements are the same. The response itself is (to $\bar{x}'_t$), say)

$$r_s(\bar{x}'_{st}) = r_t(\bar{x}'_{st}) = 78.5\% .$$

One notes that the predicted confidence level is zero and this fact alone despite the much higher response, in comparison with $r\ (\bar{x}_{st})$, allows one to detect the occlusion or quasi-occlusion as being a specific case of the presence of two objects.

Although in practice one might increase the number of discriminators which would then be trained separately for multiple object classes and occlusions, the above results are indicative of the discrimination which is available without taking this step.

5.3. Position and Orientation detection

In this section we merely point to the fact that the response of several discriminators, each trained on different positions or orientations of an object, may indicate with a reasonable degree of accuracy the actual position of the object. To take an almost trivial example, imagine a 16-discriminator 512 x 512-bit system trained on 128 x 128 squares, each discriminator having seen just the one square in one of the 16 possible disjoint positions. In use, with a square in an arbitrary position, the response of the discriminators will depend on the overlap area between an input square and those in the training set, allowing one to compute the position of the square with some accuracy. This extends even to N = 1 when one is merely doing a parallel mask-matching exercise. However, if one wishes to carry out the position-finding exercise at the same time as retaining some discrimination between several objects, a higher valued N can be used.

With higher N, one can also reduce the number of discriminators in the system in the same way that each can be trained on a judiciously selected half of significant positions, allowing the system to provide a binary code for the actual position.

Clearly, all that has been said about position applies to orientation. On the whole, the system designer has merely to select a relatively low number of discriminators to pin-point key orientations and the value of N according to the principles discussed above and use the system as an interpolatory device to provide precision.

6. ALTERNATIVE STRUCTURES FOR FUTURE DEVELOPMENT

So far in this system the structure that has been analysed is of a "feedforward" kind. That is, it consists of two sub-systems; the net itself and the decision machinery (which calculates MAX $[Z(\bar{x}_i)]$ where the former having received a "stimulus" image feeds data forward to the latter, which outputs a "response", that is, the decision itself. Once the adaptation period is over this "stimulus-response" action is combinational, that is, the action of a trivial finite state machine with only one state. Much current work is aimed at the incorporation of the above system into a dynamic structure where feedback between the "response" and the "stimulus" is introduced. Part of the "response-domain" thus becomes the "state-space" of a finite state machine, and, if one views the image spaces as some sort of Turing machine tape, it becomes clear that the system extends its computational powers much beyond that of a single-layer net (where single-layer networks are static) and that the extent of this needs researching. Thus these systems are "dynamic".

To be specific, work is progessing on using the decision domain to control the size and position of the viewing window, as well as for recognition. This follows the work of REEVES [14] and DAWSON [15], the latter having shown that letting the system control the zoom enables one to label a scene of two-dimensional objects in terms of size, position and orientation. However, in these cases position control was achieved either by a preprogrammed scan [15] or by small increments in horizontal and vertical positions [14]. In current work, one seeks to control the position of the window by saccadic jumps, which enables a system to take a series of "closer looks" before providing a final decision. Being a learning system, the saccadic jumps become context-dependent providing a powerful tool not only in object labelling in multi-object scenes but also in the analysis of similarly shaped objects wich differ in local detail.

A second line of enquiry into dynamic systems follows some earlier work on short-routed feedback [16], the design of adaptive language acceptors [17] and more recently, dichotomizers with increased sensitivity through feedback [18] and artificial perception in array-structured systems [19]. The common theme running through the above investigations is that a tendency towards stability in adaptive dynamic systems results in interesting processing properties. In current work this property is being harnessed through giving the decision domain an "alphabet" and translating the stable interaction with an input into linguistically structured statements. For example, the system may provide decision statements such as, "The triangle is above the square" or, "There are two triangles". This provides a direct, fast solution to some old scene analysis problems, and may provide the basis for the design of useful inductive learning engines for "expert systems" of the future. In passing, one notes that such feedback systems overcome "perceptron limitations" [2] and may have something to contribute to an explanation for the success of living brains which, after all, have to get along on the basis of perceptron-like devices without the benefit of "fifth generation" computer languages and structures.

7. REFERENCES

[1] McCULLOCH, W.S. and PITTS, W.H. : "A Logical Calculus of the ideas Imminent in Nervous Activity ", Bull. Math. Biophys. 5, pp. 15 - 33, 1943.

[2] MINSKY, M.L. and PAPERT, S : "Perceptrons: An Introduction to Computational Geometry", MIT Press, Cambridge, Mass. 1969.

[3] ROSENBLATT, F. "Principles of Neurodynamics", Spartan, Washington 1962.

[4] ALEKSANDER, I. and ALBROW, R.C. "Pattern Recognition with Adaptive Logic Circuits", Proc. IEE - NPL Conf. on Pattern Rec. IEE Publication 42, pp. 1 - 10, 1968.

[5] ALEKSANDER, I, STONHAM, T.J. and WILKIE, B.A. "Computer Vision Systems for Industry", Digital Systems for Industrial Automation, 1, No. 4, 1982.

[6] WIDROW, B. : "Generalisation and Information Storage in Networks of Adaptive Neurons", in Self-Organising Systems, Yovits et al. Eds., Spartan, New York, 1962

[7] WILKIE, B.A. : "Design of A High-Resolution Adaptive Pattern Recogniser", Ph.D. Thesis, Brunel University - In Preparation.

[8] British Patent Application No. 8135939, November 1981.

[9] BLEDSOE, W.W. and BROWNING, I.: "Pattern Recognition and Reading by Machine", Proc. East. J.C.C., pp. 225 - 232, 1959.

[10] ULLMAN, J.R. : "Experiments with the N-tuple method of Pattern Recognition. IEEE Trans. Comp., p.1135, 1969.

[11] ALEKSANDER, I. and ALBROW, R.C., "Microcircuit Learning Nets : Hamming Distance Behaviour", Electronics Letters 6 pp.134 - 5, 1970

[12] ALEKSANDER, I, and STONHAM T.J. : "A Guide to Pattern Recognition Using Random Access Memories", IEE Proc. on Computers and Digital Techniques, 2, pp.29 - 36, 1979.

[13] DESSIMOZ, J.D., and KAMMENNOS, P.: "Software for Robot Vision", Digital Systems for Industrial Automation 1, pp. 143 - 160, 1982.

[14] REEVES, A.P.: "Tracking Experiments with an Adaptive Logic System", Ph.D. thesis, University of Kent, Electronics Laboratories, 1973.

[15] DAWSON, C, : "Simple Scene Analysis Using Digital Learning Nets", Ph.D. thesis, University of Kent, Electronics Laboratories, 1975.

[16] FAIRHURST, M.C. : "Some Aspects of Learning in Digital Adaptive Networks", Ph.D. thesis, University of Kent, Electronics Laboratories, 1973.

[17] TOLLEYFIELD, A.J.: "Some Properties of Sequential Learning Networks", Ph.D. thesis, University of Kent, Electronics Laboratories, 1975.

[18] ALEKSANDER, I, : "Action-Oriented Learning Networks", Kybernetes, 4 pp. 39 - 44, 1975.

[19] WILSON, M.J.D., "Artificial Perception in Adaptive Arrays", IEEE Trans. on Systems, Man and Cybernetics, 10 pp.25 - 32, 1980.

Part V

Spatially Analogue Processes

Matching Elastic Templates

D.J. Burr

Bell Laboratories, Holmdel, New Jersey 07733, USA

Abstract

This paper describes a new efficient algorithm for elastic template matching. The algorithm is based on a coarse to fine matching strategy which iteratively improves the correspondence between images. The high cost of elastic matching is greatly reduced by incorporating efficient techniques for image representation and search. Recognition accuracy of 99% was achieved when using the elastic technique for handprinted character recognition.

1. Introduction

Coarse-fine processing strategies have recently been popular in the field of computer vision as means for reducing the *cost* of search [1]. Not yet well understood is the potential for coarse-fine strategies to also improve *accuracy* of search. This is particularly relevant to disparity analysis, for which models have been independently proposed by BURR [2] and by MARR and POGGIO [3]. Both models use variable size Gaussian operators for blurring either the image or the disparity field, with coarse disparities guiding the selection of finer estimates.

BURR [2] models an image as an elastic membrane which can be stretched into correspondence with another image. The algorithm compares two images by first segmenting them into parts and then finds matching parts in each image. The matches provide disparity information which is interpreted as a *force* field. The forces stretch the image a number of times to reduce disparity.

The stretching model was successfully applied to the recognition of handprinted numerals [4]. However, there were some difficulties: (1) It was slow. (2) Quantization of patterns to reduce search sometimes resulted in unacceptable matching error. (3) Size and rotation adjustment took too many iterations. This paper describes a modified elastic stretching algorithm which eliminates all of these problems. The new algorithm is applied to the more difficult problem of recognizing handprinted capital letters.

The computational efficiency has been dramatically improved by using novel search techniques which examine only highly likely disparity candidates. Preadjustment of size and slant allow fewer elastic iterations to accomplish the same goal [5]. Efficiencies are also incorporated into the template search.

The quantization problem is reduced by using coarse quantization for searching, and cubic interpolation for actual disparity measurement. Quantization error is improved also by permitting both patterns to stretch onto one another, thus minimizing segmentation differences.

The paper first reviews the elastic matching algorithm in Sections 2-4. Sections 5 and 6 discuss the modified elastic algorithm, and Sections 7 and 8 discuss experiments in character recognition using the modified algorithm.

2. Elastic Stretching Algorithm

Given two edge images I_1 and I_2, centroids are first computed for each image and then the images are aligned concentric to one another. Local disparities are then found between I_1 and I_2 and vice versa, forming two fields of two-dimensional vectors. Both disparity fields act in concert to stretch one of the images, usually selected beforehand as the *elastic* image.

The stretched image undergoes the same process once again, serving now as the input elastic image. Two disparity fields are again computed and the elastic image is stretched once more. After several iterations of this process, the final image shapes are compared by template matching. Alternatively, the trajectories followed by each edge endpoint during stretching could be saved and used as improved estimates of disparity.

Figure 1 shows a block diagram of the algorithm. The modified elastic stretching algorithm has the same overall form, except that I_2 becomes elastic for the final step and it is stretched onto I_1^N. Also, stretches $k = 0$ and $k = 1$ are replaced by size and slant normalization.

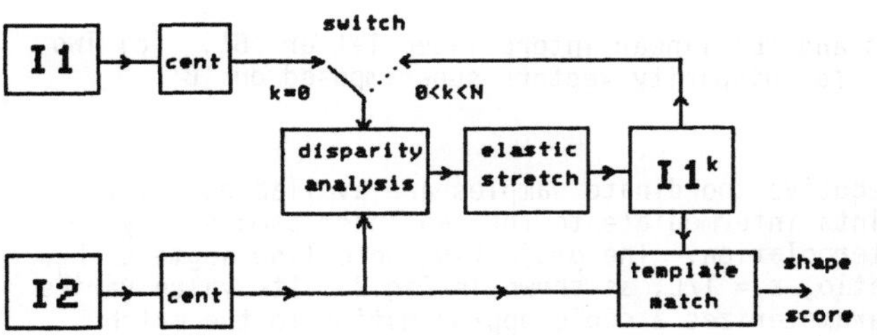

Fig. 1. Block diagram of the elastic stretching algorithm

3. Estimation of Disparity

Each image consists of a set of sampled coordinate pairs, $\{(x_i,y_i,f_i): 1 \le i \le N\}$, with a corresponding flag f_i to indicate whether the pen is up or down. If the pen is down for two consecutive points, $(i,i+1)$, then the points are connected by a line segment L_i.

Disparity is defined between coordinate point i and line segment L_j. The length of the perpendicular projection from the point to the line defines the positional disparity, $P(i,j)$. Whenever this projection falls outside the line segment, then $P(i,j)$ becomes the distance to the *nearest* endpoint of L_j (see Fig.2).

The measure could be enhanced by making the selection sensitive to angle difference between lines L_i and L_j [4]. If we let $A(i,j)$ represent the absolute difference between the angles of the two lines, then the total distance $D(i,j)$ is the weighted sum of the squares:

$$D(i,j) = A^2(i,j) + wP^2(i,j) \tag{1}$$

The minimum with respect to j determines the corresponding line segment, indicated by j^*. This is represented symbolically by:

$$j^* = \underset{1 \le j \le N-1}{\text{argmin}} \; [D(i,j)] \; , \tag{2}$$

where *argmin* stands for the argument j which minimizes $D(i,j)$ for $1 \le j \le N-1$. The disparity vector is determined by projection onto line L_{j*} as illustrated in Fig.2. An example of the disparity field between two images of the letter 'B' is shown in Fig.3.

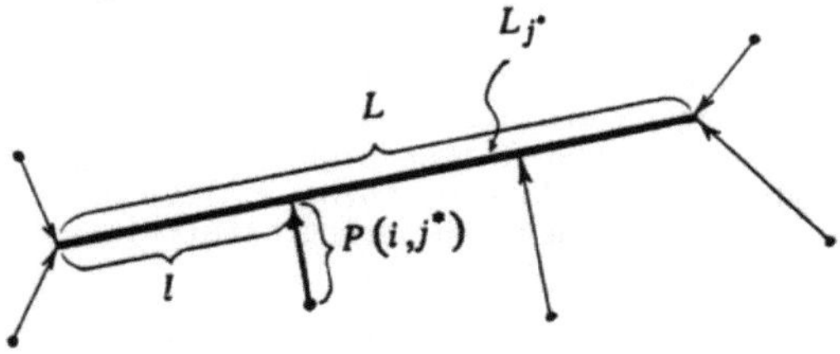

Fig.2

Fig. 2. Some examples of disparity vectors relative to line L_{j*}, and projection ratio, $s = l/L$

Fig.3

Fig.3. (a) Sampled points and (b) linear interpolated letter 'B'. (c) Two superimposed characters. (d) Disparity vectors superimposed on 'B'

If at least four consecutive coordinate samples are available, (x_1,y_1) through (x_4,y_4), then points intermediate to the two inner samples may be approximated by cubic interpolation. The projection onto line segment L_{j*} defined the projection ratio, $s = l/L$, as shown in Fig.2. Its value varies from 0 to 1 and it parameterizes a cubic approximation to the match point (x,y) :

$$x = d + as + bs^2 + cs^3 \; . \tag{3}$$

The constants a,b,c,d are found by solving the system of four linear equations for $s = -1,0,1,2$ and $x = x1,x2,x3,x4$. The expression for the y coordinates takes a similar form. Examples of both linear and cubic interpolated characters are shown in Fig.4.

4. Elastic Image Model

Disparity fields can be interpreted as *force* fields which are used to stretch the elastic image. Stretching is done by performing a local weighted average of the disparity vectors within the region about the stretched point.

The basis for the elastic image approach is that errors in the disparity field can be reduced by cooperation. This cooperative mechanism is a direct result of the elastic stiffness of the image, which does not allow a large

Fig.4. (a) Sampled points, (b) linear, and (c) cubic interpolation of some characters

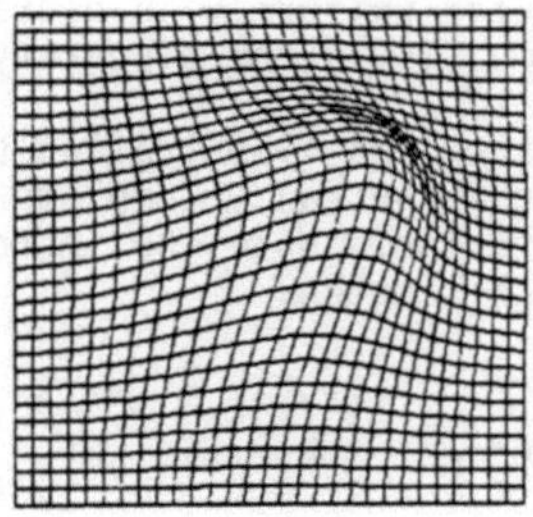

Fig.5. Elastic stretching of a grid by a single force

change from one point to the next. Because it is difficult to find one elastic value which produces the correct disparities, stretching is iterated. The desired result is achieved by using a sequence of stretches each with decreasing stiffness.

Given a single disparity d the elastic deformation of the sheet E due to that force has a radial Gaussian profile:

$$E = dG(r) , \tag{4}$$

where r is the radial vector connecting the evaluation point and the force location. $G(r)$ is defined by:

$$G(r) = \exp\left(\frac{-r \cdot r}{\sigma^2}\right) , \tag{5}$$

where σ is the range of the Gaussian, corresponding to elastic stiffness. The composite stretch E_c due to a set of disparities $\{d_i: 1 \le i \le N\}$ is just the Gaussian-weighted sum of the individual deformations:

$$E_c = \frac{\sum_{i=1}^{N} d_i G(r_i)}{\sum_{i=1}^{N} G(r_i)} \tag{6}$$

This is simply Green's theorem which says that the elastic stretch can be computed by a linear superposition of unit stretch functions. The Gaussian represents the unit response function, or the deformation resulting from a single force located at a single point.

Figure 5 shows the results of applying (4) to a cartesian grid. The force, d , is toward the upper right corner, and σ is 25% of the grid edge length.

The measure used for template matching throughout the paper is:

$$T = \sum_{i=1}^{N} \frac{d_i \cdot d_i}{L_s} , \tag{7}$$

where L_S is the sum of lengths of all strokes in image I_2. Disparities d_i are computed uniformly in distance along the strokes rather than uniformly in time, as in the disparity analysis.

Figure 6 shows an illustration of a sequence of stretches (b-f) performed on an elastic image (a). Template match measures and size of σ are shown below each stretch. Note the improvement in the template match measure as a result of each stretch.

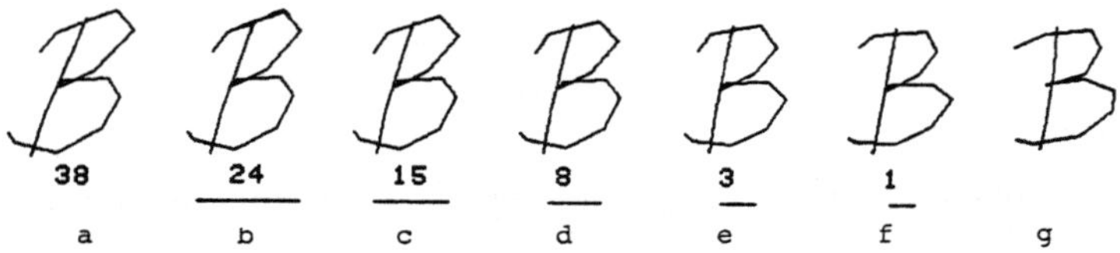

Fig.6. Example of the original elastic stretching algorithm

5. Search Reduction Techniques

5.1 Disparities

Both the disparity search and elastic stretching have mn cost, where m and n are the number of edges in the two images. Whether mn is large (stereo problem) or the number of image comparisons is large (character recognition), the cost is great. Therefore efficient techniques are essential.

Because of image continuity the matched points of two neighbours are likely to be neighbours in the other image. This is true in character recognition for objects in the same class. Continuity also holds in stereo matching except at locations of rapid depth change [6].

If continuity is assumed, then considerable reduction can be achieved in the search for correspondences. Rather than searching all line segments, only those neighbours of the previous match are searched. Global search is permitted only on two occasions: (1) the first disparity computed in a particular image pair, and (2) whenever the *length* $|d_i|$ of the current disparity exceeds a fixed threshold, T_d, then a re-search is performed globally. However, immediately after global search, then local search is again resumed. For images which have limited deformation between the pair this is a powerful heuristic. The resulting search is mostly local, resulting in near linear cost in n. The local search requires that (2) be modified to read:

$$j^* = \underset{(j^{**}-\nu)\le j\le(j^{**}+\nu)}{\operatorname{argmin}} [D(i,j)] \tag{8}$$

where j^{**} represents the index of the previous match and ν is the neighbourhood interval.

Figure 7 shows a comparison between disparities computed using several different thresholds, and the original unrestricted search $(T_d = 0)$. It can be seen, as expected, that the optimum threshold (about 20% of image size) is that which just exceeds the maximum expected disparity length. This produces minimum distortion and maximum search reduction. Some error is encountered in the disparity field and the size of the error is proportional to the threshold. Note the large errors when a threshold of 40 is used.

a b c d e

Fig.7. Disparity vectors computed using reduction thresholds, T_d = (b) 0 , (c) 20, and (d) 40 . The image height is approximately 100 units

5.2 Elastic Stretching

A comparable cost reduction can also be achieved in elastic stretching. All disparity forces normally contribute cooperatively to the elastic stretch computation (6), but forces close to the evaluation point contribute most. Therefore, we can approximate this by excluding disparities outside a proximity window. Because the handwriting images are stored as strokes, it is easy to limit the computation to neighbouring disparities on the same stroke. Thus the limits in (6) are changed from $1 \le i \le N$ to $(i'-u) \le i \le (i'+u)$, where i' represents the sample closest to the evaluation point, and u is the window size.

Figure 8 shows an example of elastic stretching using different values of u . These are compared also to the nonreduced elastic stretch (7a). The limiting case for small windows is that the elastic stretch degrades to the disparity estimate. This case produces the smallest difference, but it is misleading. In general, some smoothing should be used as this reflects the low-frequency distortions which occur between examples of the same character.

For experiments such as handwriting recognition, highly accurate disparity may not be necessary. Only enough correct matches are needed to discriminate between characters of the recognition set. In stereo vision, however, disparity is used directly to evaluate depth. If the viewing angle is small, then even a slight disparity error becomes a large depth error. Therefore, stereo requirements put more severe limitation on the search reduction methods that can be used.

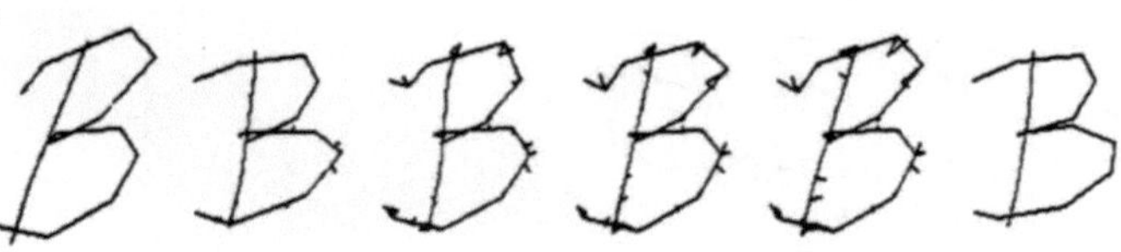

1 7 14 18

a b c d e f

Fig.8. Elastic stretching cost reduction for various window sizes: u = (b) 0 (no smoothing), (c) 2, (d) 4, and (e) all points. The images (b-e) represent stretched versions of (a), with image (f) as the goal

6. The Modified Elastic Stretching Algorithm

The initial disparity field contains information useful to size and slant adjust the image. To accomplish size and slant adjustment the following averages are computed over the disparity vector field:

$$a_1 = \sum_{i=1}^{N} (|r_i+d_i|-|r_i|) \tag{9}$$

$$a_2 = \sum_{i=1}^{N} |r_i| \tag{10}$$

$$a_3 = \sum_{i=1}^{N} d_{ix}\ \mathrm{sign}(r_{iy}) \tag{11}$$

$$a_4 = \sum_{i=1}^{N} |r_{iy}| \tag{12}$$

where d_i is the disparity vector of the ith point, and r_i is its location relative to the image centroid. Size is adjusted by multiplying all points by the size factor $f = 1+(a_2+a_1/2)/(a_2-a_1/2)$. Slant is adjusted by replacing x-coordinates with $x+2y(a_3/a_4)$.

Two elastic stretches followed this adjustment using (6). Both images were stretched onto each other using the same elastic constant. This dual stretch approach makes the images less sensitive to segmentation differences.

Size and slant adjustment methods based on disparities are superior to absolute methods which adjust each image independently. In practice, angle differences of up to 80 degrees have been removed using the adjustment methods described above.

Examples of the modified elastic stretching algorithm are shown in Fig.9. Parts (a) and (f) show the input patterns. The result of size and slant normalization are shown respectively in 9(b) and 9(c). 9(d) shows the result

Fig.9. Examples of the 4-stage modified elastic stretching algorithm

of the first elastic stretch and 9(e), the second elastic stretch. The second row of figures for the 'R-R' combination and the 'R-B' combination show the superimposed disparity vectors. The superimposed cubic interpolations are shown in (d) and (e) for the corresponding elastic stretches above. The numbers indicate the template matches at each step, relative to image (f). The number in column (e), however, is the template match relative to (d).

The search reduction techniques described in the previous section were implemented in the modified algorithm. In addition, an efficient template search technique was implemented [7] which proposed candidates to the elastic

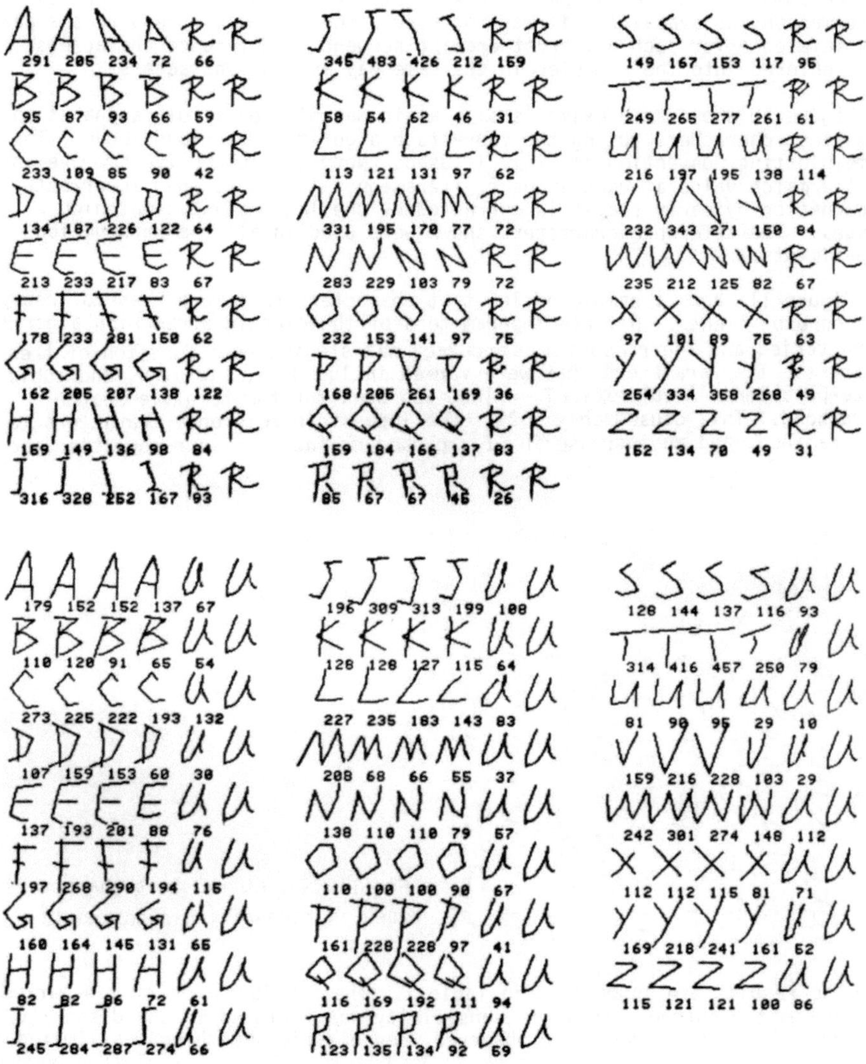

Fig.10. Character recognition examples

matcher based on the output of the normalizer alone. By monitoring the template match value in progress, it was possible to reject a given reference character whenever that value exceeded a fixed threshold T_t. In addition, only those references were tested which had the same number of strokes as the test character.

7. Experiments in Handprinted Character Recognition

Characters were input using a conventional pen digitizer. A signal was sent to the computer when each character was completed. Though time sequence information was available, it was not used since that would require a separate reference for each different drawing sequence. Reference characters were entered into the computer in the same way as test characters.

Figure 10 shows some examples of elastic matching of a full alphabet to some test characters, using the four-stage algorithm described above. The same plotting convention as Fig.9 is used. Note the successive improvement in the match value at each stage of iteration. Also note that incorrect recognition by early stages is corrected by the later elastic matching stages. The reference characters shown were used in all the recognition experiments.

Figure 11 shows a sample of 100 test characters that were used for recognition experiments. Different parameters of the elastic stretching algorithm were varied and the recognition accuracy was studied as a function of these changes. The parameters that were varied included σ, T_d, u, w, and cubic interpolation. A value of $T_t = 100$ was used for the template search threshold. This caused only 12% of all possible reference characters to be examined, and an increase in T_t caused no increase in recognition accuracy.

Fig.11. Set of 100 test characters used in recognition experiments

The results are summarized in Table 1. The conclusion is that a marked improvement occurred when angle sensitivity was included in the disparity search. Improvement was also observed when elastic smoothing was limited to neighbour points on the same stroke. Absence of cubic interpolation degraded the recognition but only slightly, from 99% to 97%. There was no sig-

Table 1. Elastic Matching Parameter Variations

Cubic Interpolation	Elastic Stiffness (sigma)	Disparity Threshold (Td)	Stretch Neighbourhood (u)	Angular Dependence (w)	Recognition Rate
Yes	55	0	4	0.2	97
Yes	100	0	4	0.2	98
Yes	55	0	4	0.4	95
Yes	55	0	4	0.2	99
Yes	55	0	4	0.0	98
Yes	55	0	4	0.1	97
Yes	55	0	all	0.1	93
Yes	55	20	4	0.2	98
No	55	0	4	0.2	97

nificant loss in accuracy when disparity search was limited to a local 20% window. The best combination of parameters included cubic interpolation, 20% angular sensitivity, elastic smoothing including only 8 nearest neighbours, and $\sigma = 50$ (about 1/2 image size). This combination produced 99 correct recognitions out of the 100 characters tested.

8. Discussion

The potential high accuracy of elastic matching has been demonstrated as well as some methods for reducing its cost. The recognition speed averages a few seconds per character on a VAX-11/780 running C. This could be improved by optimizing the code and by overlapping recognition with character input. Even greater speed increases are possible by enhanced preprocessing, perhaps using fast parsing techniques [8].

Another simple savings which was not tried is to rank order the candidates prior to elastic matching. By testing the most likely first, the remaining candidates could be quickly eliminated based on a threshold set by the early match.

The elastic matching algorithm is potentially attractive for stereo disparity analysis. By limiting correlation to edges rather than the full gray scale picture, the modified elastic algorithm could achieve significant cost savings [6]. Further work must be done, though, to see whether sufficient accuracy exists for the more demanding requirements of stereo matching.

9. Conclusions

The modified elastic matching algorithm can achieve high (99%) recognition accuracy and is computationally efficient. This is achieved by combining efficient techniques for image representation and matching with coarse-fine strategies for searching. Coarse-fine techniques have the potential for both efficient and accurate recognition and thus deserve more attention in the future.

References

1 S. Tanimoto and A. Klinger (eds.), *Structured Computer Vision*, Academic Press, New York (1980)

2 D.J. Burr, 'A Dynamic Model for Image Registration', *Computer Graphics and Image Processing, 15*, 102-112 (1981), also *Proceedings Pattern Recognition and Image Processing Conference*, Chicago, Ill. (1979)

3 D. Marr and T. Poggio, 'A Computational Theory of Human Stereo Vision', *Proceedings of the Royal Society of London B, 204*, 301-328 (1979)

4 D.J. Burr, 'Elastic Matching of Line Drawings', *IEEE Transactions on Pattern Analysis and Machine Intelligence, PAMI-3, 6*, 708-713 (November 1981)

5 D.J. Burr, 'A Normalizing Transform for Cursive Script Recognition', *Proc. Sixth International Conference on Pattern Recognition*, Munich, Germany (October 19-22, 1982)

6 D.J. Burr, 'On Computer Stereo Vision with Wire Frame Models', Ph.D. Thesis, Department of Electrical Engineering, University of Illinois, Urbana, Illinois (January 1978)

7 D.I. Barnea and H.F. Silverman, 'A Class of Algorithms for Fast Digital Image Registration', *IEEE Transactions on Computers, C-21, 2,* 179-186 (February 1972)

8 T. Pavlidis and F. Ali, 'A Hierarchical Syntactic Shape Analyzer', *IEEE Transactions on Pattern Analysis and Machine Intelligence, PAMI-1, 1,* (2-9 January 1979)

Analogue Representations of Spatial Objects and Transformations

L.A. Cooper

Learning Research and Development Center, University of Pittsburgh,
Pittsburgh, PA 15260, USA

Questions concerning the nature of the internal representation or coded form of visual/spatial information have been prominent in the growing field of cognitive science. In particular, considerable discussion and debate have centered around the issues of the extent and the nature of structural and/or functional correspondences between internal representations and their external visual counterparts. In this paper, evidence both for and against analogue representations of spatial objects and transformations will be presented and evaluated. Following SHEPARD, COOPER, and their co-workers [e.g., 1-4], a distinction between "analogue" representations and processes and "nonanalogue" representations and processes will be drawn in the following way: An analogue representation or process is one in which the relational structure of external events is preserved in the corresponding internal representations. A non-analogue representation or process is one for which this is not the case. Note that this distinction incorporates a relatively weak criterion for the demonstration of the existence of an analogue process or representation. Strict continuity of an internal process is not required, nor is a "first-order" isomorphism between the structure of an internal representation and the structure of an external object or event. Furthermore, no stand has been taken on the question of the neural mechanisms that might underlie analogue representations and processes.

It is argued below -- in the context of particular experimental findings -- that to qualify as an analogue process or representation, it must be demonstrable that the internal process or representation goes through an ordered series of intermediate states that have a one-to-one relationship to the intermediate stages in the corresponding external process or object. And, this one-to-one correspondence need entail only that the internal process or representation allow the individual carrying out such a mental transformation to be particularly prepared for the presentation of the appropriate external object at the appropriate location in space, if such an external object were to be presented.

Considerable experimental evidence purporting to demonstrate the existence of analogue internal representations and processes, in the sense sketched above, has accumulated during the past decade. These experimental efforts have in common the use of a powerful psychophysical technique -- mental chronometry -- permitting inferences about internal, mental events and processes from purely observable behavioral data. The general idea underlying the use of chronometric measures is that the time required to solve a visual/spatial problem -- when considered in relation to specified, measurable properties of test stimuli -- can provide information that places significant constraints on possible internal representations and processes used to solve the problem. In much of the relevant research, the amount of time needed to determine whether two visual objects are identical in shape or different (e.g., mirror images) has been found to depend systematically on the extent of spatial difference between the orientations of the two objects. On the basis of a pattern such as this, it can be argued that the underlying internal representations and processes must be of a class that would produce the appropriate orderly relationship between solution time and extent of spatial displacement.

In a seminal experiment, SHEPARD and METZLER [5] asked subjects to determine as rapidly as possible whether two simultaneously displayed perspective drawings of three-dimensional objects were the same in shape or mirror images of each other. In addition to a possible difference in shape, the objects could also differ in portrayed orientation, either in the picture plane or about an arbitrary axis in depth. The results of this experiment were clear and striking. Both for the group data and at the level of individual observers the time required to determine that the two portrayed objects were the same in shape increased in a remarkably linear fashion with increases in portrayed angular difference between the orientations of the objects. Furthermore, both the slope and the intercept of the linear time functions were roughly equal for both picture-plane and depth angular differences. This central and elegant finding led the investigators to suggest that the task was performed by "mentally rotating" an internal representation of one object into congruence with the orientation of the other object and then comparing the two representations for a match or mismatch in shape. This process of mental rotation -- which SHEPARD and METZLER conceived of as an analogue of an external rotation -- appeared to have an average limiting rate of about 60° per second for these subjects and these particular visual materials.

This orderly linear relationship between response time and extent of angular displacement has been well documented for a variety of visual stimuli and task modifications. For example, linear response-time functions have been obtained when a single rotated stimulus must be compared with a canonically oriented pattern in memory [6, 7]. Similar functions have been found when alphanumeric characters [7] or random two-dimensional polygons [6] are used as stimulus materials. This same orderly relationship has been obtained when rotations must be carried out before the presentation of a test shape, in order to prepare for its appearance in a specified orientation [6, 7]. It is also found when rotations are combined with additional spatial transformations [8] and when the "same-different" discrimination is altered to include subtly different distractors, as well as mirror images, as test stimuli [9]. (Reviews of these and other studies of mental rotation can be found in [1] and [2].)

The host of experiments demonstrating a linear relationship between time and portrayed difference in the orientations of visual objects has consistently been interpreted as providing evidence for analogue representation of spatial objects and spatial transformations. This is because the linearity indirectly fulfills the chief criterion for an analogue process -- viz., a correspondence between internal and external relational structures and transformational distance and time suggests that the internal process used to solve rotation problems passes through a trajectory of intermediate states which have a one-to-one correspondence to the intermediate stages in the rotation of an external object. That is, the linear reaction-time function indicates that the time required to compare two visual stimuli presented at two sufficiently different orientations A and C is an additive combination of the times needed to compare those stimuli at orientations A and B and at orientations B and C. SHEPARD, COOPER, and their collaborators have argued that such a finding constitutes indirect evidence for the claim that the internal process underlying comparison of the visual stimuli presented in orientations A and C passes through an intermediate state corresponding to orientation B [1-7, 9].

More direct experimental demonstrations of the analogue nature of mental rotation have been provided by COOPER and SHEPARD [7, Experiment II] and by COOPER [4]. These experiments are powerful in that they show the required one-to-one correspondence between intermediate stages in an external rotation and intermediate states of an internal process in the specific sense of an individual engaged in the mental rotation being especially ready to respond accurately and rapidly to the appearance of an external test stimulus that is presented at just the appropriate orientation in the transformational trajectory.

In one of these studies, COOPER [6] initially trained subjects to discriminate arbitrarily-defined "standard" from "reflected" versions of random two-dimensional polygons at one particular orientation in the picture plane. Following the discrimination training, individual polygons were presented at one of six orientations, and subjects were required to determine as rapidly as possible whether each was a standard or a reflected version. Reaction time increased linearly with angular departure of the shapes from the (canonical) orientation at which training had occurred. As in the earlier SHEPARD and METZLER [5] study, the linear time function was taken as indirect evidence that subjects were mentally rotating the test polygon into the trained orientation in order to achieve the standard-reflected discrimination. Rotation rates were computed separately for individual subjects and individual stimuli.

In a subsequent experiment, COOPER [4] tested the same subjects using the same visual stimuli, but modified procedure. Subjects were simply asked to imagine a particular shape from the previous experiment rotating from the trained orientation about a circle at a "natural" rate. On the basis of the data from the earlier experiment, it was possible to compute at what orientation in the internal transformational trajectory a given subject's (rotating) representation of a given shape would be at a particular point in time. On half of the trials, a test shape was presented at one such "expected" orientation at an unpredictable moment in time, and the subject was required to determine as rapidly as possible whether the probe was a standard or a reflected version of the shape being imagined. On the other half of the trials, a test polygon was presented in an orientation inferred <u>not</u> to be congruent with the orientation of the subject's (rotating) internal representation, and the same discrimination was required.

The results of this experiment were clear and elegant. Reaction times were shortest and uniformly rapid when the test shape was presented in the orientation that corresponded to the assumed spatial position of the rotating internal representation. This was true (a) regardless of the angular departure of this expected orientation from the initially-trained position, and (b) even for intermediate orientations at which test stimuli had never been previously presented in the initial experiment. Furthermore, when test shapes were presented in unexpected or incongruent orientations, reaction times increased linearly with the angular departure of the test probe from the inferred expected orientation.

These results are consistent with the notion that mental rotation is an analogue of a physical rotation in the specific sense of the internal and external processes having the required relation of readiness for responding to the appropriate external object at the appropriate stage in the tranformational trajectory. Contrast this, for example, with an internal transformation such as matrix multiplication by which a computer might compute the new coordinates for a rotated system of points [cf., 1]. This process would not qualify as analogue because the intermediate stages in the calculation would not have a one-to-one correspondence with intermediate orientations, in the sense that the computer would not be in a state of heightened readiness for responding to the presentation of intermediate orientations at intermediate times.

In addition to these studies of mental rotation, many other chronometric studies have frequently been interpreted as providing evidence for analogue internal representations and processes. Most extensive, perhaps, is the body of experimental work produced by KOSSLYN and his co-workers. (For a comprehensive review of this research effort and a theoretical perspective, see KOSSLYN [10].) Unlike the mental rotation experiments, in KOSSLYN'S experiments subjects are generally instructed to generate internal representations that they would call "mental images" and to use these representations as a basis for performing a variety of visual/spatial tasks. A typical task is one used by KOSSLYN [11] and KOSSLYN, BALL, and REISER [12]. In these experiments, subjects were instructed to form visual images of previously viewed objects or spatial layouts (in the case of [11], objects with a horizontal or a vertical orientation, and in the case of [12], a map of a fictional island containing a number of identifiable landmarks). They were then asked to "mentally focus" on one end of a particular imagined object or on one landmark on the imagined map. The central experimental task required subjects to verify as rapidly as possible whether a certain property was present on the image of the object or whether a certain landmark was contained on the image of the map. The chief manipulation was the distance between the attended end of the object and the probed property or the distance between the attended and the tested locations on the map.

In both sets of studies, KOSSLYN [11,12] found that verification times increased linearly with the distance between the focussed end of the object or location on the map and the property or landmark that was tested. This function relating verification time to extent of spatial distance is open to an interpretation similar to that made on the basis of the data from the mental rotation experiments discussed above. That is, just as the linear relationship between reaction time and angular differences between two visual objects provides indirect evidence for a correspondence between intermediate internal processing states and intermediate external orientations, so too the linear relationship between verification time and distance may suggest that an internal representation of space -- when subjected to the transformation of scanning -- has a structural correspondence to the external depiction of distance between objects or properties which is required of an analogue representation.

Chronometric data of the sort reviewed above have been interpreted by many as providing evidence for analogue representation and processing of visual/spatial information; however, this interpretation has been challenged by various investigators on both theoretical and empirical grounds. One form of the challenge has consisted of presenting experimental results that appear to infirm the notion of analogue representation and processing of visual information. Another type of challenge has centered around the development of alternative, nonanalogue accounts of the internal representations and processes that could underlie the relevant chronometric findings. These two sorts of objections are discussed below in turn.

One set of experimental findings that has refined the account of the nature of the representations and processes involved in mental rotation and which, to some extent, has questioned the analogue interpretation favored by SHEPARD, COOPER, and their co-workers [1-7, 9] has come from the work of CARPENTER and JUST [13, 14]. These investigators have measured not only the time required to compare the shapes of two disoriented visual objects, but also the pattern and duration of eye fixations that subjects produce while making the comparison. Their analysis of eye fixations suggests that three component operations are carried out in achieving comparisons between visual figures like those used by SHEPARD and METZLER [5]. The first operation is one of search, in which parts of the two figures that potentially correspond to each other are located. The second operation is one of transformation and comparison, in which corresponding segments of the figures are mentally rotated and sequentially compared. The third operation is one of confirmation, in which other parts of the figures are compared in order to determine whether they are congruent as a result of the mental rotation.

The theoretical account that these investigators offer on the basis of both their chronometric and their eye fixation data is slightly different from the analogue account offered by SHEPARD and COOPER [1-7, 9] in two respects. First CARPENTER and JUST [13, 14] suggest that rotations are carried out sequentially, on individual segments of visual objects, rather than on an integrated, holistic internal representation which we might expect of an analogue process. Second, their eye fixation analysis has led them to suggest that the rotation operation operates on these portions of visual figures in discrete steps of about 50°. This view, while emphasizing the possibility that rotations may be performed on discrete parts of visual representations in discrete steps, is still consistent with the criterion for an analogue process outlined in [1-7, and 9]. For, during a mental rotation of, e.g., 150°, the internal process passes through intermediate stages corresponding to intermediate external orientations of 50° and 100°.

A second set of experimental findings that challenges the analogue account of mental rotation has been presented by PYLYSHYN [15]. PYLYSHYN has argued elsewhere [16] that to qualify as analogue, an internal process must not be capable of being shown to be "cognitively penetrable." What this means, in essence, is that if the operation of a cognitive process can be shown to be influenced by goals, beliefs, knowledge, or other external factors, then that process is cognitively penetrable and hence cannot be regarded as a primitive, knowledge-independent process. PYLYSHYN [15, 16] further argues that to qualify as analogue a representation or a process must exhibit a lack of penetrability. For, if an internal process can be influenced by knowledge or other symbolic factors, then that process and the representations on which it operates must themselves be symbolic in nature.

PYLYSHYN [15, 16] has attempted to show that so-called analogue processes such as mental rotation [1-7] and mental scanning [10-12] are cognitively penetrable in two ways. First, he has presented experimental evidence putatively demonstrating that factors such as stimulus structure and practice can affect the rate of these mental processes. Second, he has suggested that the response-time patterns that have commonly been taken as support for analogue processes and representations may better be accounted for by appealing to subjects' "tacit knowledge" of the demands of the experimental situation. By "tacit knowledge" PYLYSHYN [16] refers to the notion that subjects have knowledge of the nature of physical transformations and, further, interpret the experimental situation as inviting the simulation of what they know the appropriate physical transformation to be.

The experiments in which PYLYSHYN [15] has attempted to show cognitive penetrability are ones that required subjects to determine as rapidly as possible whether or not a test probe, presented simultaneously with a visual figure and in various different orientations on different trials, was a part of the figure. As in other mental rotation experiments, PYLYSHYN found that reaction time for the part-verification task increased as the angular difference in the orientations of the figure and the part increased. However, the "goodness" or coherence of the test probes as parts of the figures -- which was the chief experimental manipulation -- produced variations in the slopes of these reaction-time functions, indicating that more structurally coherent parts could be mentally rotated more rapidly than less coherent parts. From these data, PYLYSHYN concluded that since mental rotation rate can be influenced by aspects of stimulus structure, it is a cognitively penetrable, and hence nonanalogue, process.

Whether or not one accepts PYLYSHYN's criterion of nonpenetrability as a condition for an analogue process, his conclusions should nonetheless be viewed with caution [cf., 1, Chapter 8]. The method of stimulus presentation and the judgment required of the subject were sufficiently different from those used in other studies of mental rotation [1-7, 9] to question the comparability of the experimental findings. Furthermore, the processing time measured by PYLYSHYN included components other than rotation -- such as encoding and comparison of the figures and the parts -- which themselves could have produced the reaction-time differences associated with part "goodness." Nonetheless, while PYLYSHYN's experiment is not conclusive, exploration of the influence of stimulus and judgmental factors on the rate of operations like rotation and scannning seems a fruitful avenue to pursue. The results of such experiments may yet prove, and, in the case of one study discussed below [9] already have proved to be theoretically incisive.

PYLYSHYN's [16] second attempt to invoke cognitive penetrability has involved arguing that subjects use tacit knowledge of the nature of physical transformations in making the timed judgments required in mental rotation and mental scanning experiments. Hence, the linear reaction-time data do not necessarily reflect the operation of an internal analogue process; rather, these data may reflect nothing more than subjects' beliefs about the time course of physical operations coupled with a desire to conform to the implicit demands of the experimental situation. This is a difficult argument to evaluate since introspective evidence might not reveal the use of such "tacit" knowledge, in the absence of a conscious strategy on the part of the subject. KOSSLYN [17] has, however, provided a twofold rejoinder to PYLYSHYN's argument based on the use of tacit knowledge. KOSSLYN's criticisms of this position can be summarized as follows: First, results consistent with an analogue account can be obtained in the absence of explicit instructions to generate visual images or to engage in processes analogous to physical transformations. Second, subjects will sometimes exhibit performance that indicates that their internal representations of spatial objects and transformations have properties similar to those found in perceptual experiments and which, further, are counterintuitive. These results are most likely not attributable to tacit beliefs about physical objects and events.

A second line of objection to the usefulness of postulating internal analogue representations and processes has proceeded in a somewhat different direction than that of the empirical work discussed above. In essence, it has involved the attempt by various investigators -- most notably, ANDERSON [18, 19] -- to produce discrete, nonanalogue systems that mimic the behavior of analogue systems, and thus to account for certain of the critical chronometric findings. ANDERSON [18] argues further that nonanalogue accounts are to be preferred on grounds of parsimony. That is, only one symbolic representational system need be postulated to explain the coding and processing of both linguistic and spatial information. (For a somewhat modified view, see ANDERSON [19].)

This is a difficult argument to evaluate, for as ANDERSON [18] shows formally, it is possible with the proper set of representational and processing assumptions to design a nonanalogue system that will indeed produce performance frequently attributed to an analogue system. However, in response to this position, two interrelated points should be noted. The first -- attributable to KOSSLYN and POMERANTZ [20] -- is that certain discrete, propositional theories may be too powerful, in that any empirical result may be explained after the fact. Such theories, though, may have insufficient constraints to predict a particular experimental outcome in advance. In contrast, versions of an analogue account can make rigorous experimental predictions that place serious constraints on the nature of the representations and processes underlying certain empirical phenomena.

Consider, for example, the study by COOPER [4] discussed above, in which points in a hypothesized underlying transformational trajectory were probed with test forms at appropriate temporal intervals. An analogue account of the representations and processes tapped in this experiment makes a clear prediction concerning the relative speed of "probe-expected" and "probe-unexpected" trials. The prediction, which was confirmed in the experiment, is that reaction times should be rapid and virtually constant for all orientations on "probe-expected" trials, but should increase with increasing angular difference between the probed and the expected orientations on "probe-unexpected" trials. While some form of a discrete, propositional model may be able to accomodate such results, there seems little or no priori reason for such a system to behave in the manner described above. Still another experimental finding that, minimally, constrains the class of nonanalogue models that can account for so-called analogue phenomena has been reported by COOPER and PODGORNY [9]. Using a restricted set of visual stimuli, these investigators were able to show that the complexity of the stimuli had no influence on the rate of mental rotation. While this finding may prove to be highly task, stimulus, and practice specific, it nonetheless is consistent with an analogue interpretation of the underlying representations and processes and is handled with difficulty only by certain versions of nonanalogue accounts [cf., 18].

In summary, then, analogue models for visual/spatial representations and transformations have recently been challenged on both empirical and theoretical grounds. The empirical challenges can be shown to be inconclusive and further experimental work is needed to resolve the inconsistencies. The theoretical challenges have taken the form of proposing nonanalogue models that may account for at least some of the findings generally regarded as supporting the need for positing analogue representations and transformations. While there appears at present to be no satisfactory way to test alternative models, consideration of an entire body of converging experimental findings may help to constrain the class of models capable of providing an adequate account of visual representation and processing.

The theoretical account provided thus far of analogue representations and transformations underscores the similarity between perceptual and imaginal operations and codes, e.g., between carrying out a mental rotation and perceiving an external physical rotation. (See SHEPARD and PODGORNY [21] for a more complete description of relevant experimental evidence.) After SHEPARD and COOPER [1] and NEISSER [22], the internal representation undergoing an analogue mental transformation might be viewed as an anticipation for perceiving the appropriate external object in the appropriate external spatial location, were it actually to appear. Given this theoretical framework, the question naturally arises as to whether correspondences can be found between the time course and nature of mental transformations and those of more strictly perceptual processes.

One experimental paradigm that has recently been introduced by COOPER and VALLONE [23] explored the question of the continuity between perceiving transformations on depictions of objects in space and internally extrapolating such transformations. Their procedure involves displaying a line drawing of an object undergoing a specified transformation (e.g., a rigid rotation about an axis in the picture or plane or in depth) at some specified rate. At some unpredictable moment during the transformation, the object disappears from view, and at some later and again unpredictable time the same object reappears at a new position in the transformational trajectory. The observer's task is to determine whether or not the object has reappeared at the appropriate spatial location, were it to have continued transforming during the blank period, at the rate and in the fashion originally viewed. Experiments using this procedure have been initiated only recently and results are as yet quite preliminary [23]. Nonetheless, the paradigm appears to have promise for investigating questions concerning the limiting conditions for performing mental simulations of perceived spatial transformations in terms of the type and rate of the transformation, the correspondence between the rate of transformation and the rate of spatial transformations generally encountered in the environment, the distance along the transformational trajectory along which the mental extrapolation occurs, and the nature of the object being transformed.

Another experimental situation, which has been used by SHEPARD and his co-workers [1, 24, 25] to examine the relationship between mental transformations and certain perceptual phenomena, presents observers with visual displays shown in rapid alternation. In an initial experiment SHEPARD and JUDD [24] found that when perspective views of the same three-dimensional object were shown alternately in two different orientations, observers reported an illusion of rigid rotation of a single object under certain conditions. As in the case of the earlier mental rotation experiments, using the same visual displays, the minimum exposure time required to produce coherent apparent rotational motion of a single three-dimensional object increased linearly with the angular difference between the portrayed orientations of the two separate views. Again in accordance with findings from mental rotation experiments, the slope of the minimum cycle durations, as a function of angular difference, was equivalent for rotations in the picture plane and in depth. (Similar results have been obtained using random polygons as stimuli when those polygons are asymmetric [25].)

One notable difference between this apparent rotational motion and the mental rotation task is found in the time necessary to produce a perceptual illusion of rigid rotation and the time required to undertake a purely imaginal rotation of an internal representation of a visual object. In the case of the former, perceptually-driven type of transformation, critical times are on the order of only hundreds of milliseconds whereas in the case of the latter, effortful type of transformation -- in which an internal representation of the spatial structure of an object continually preserved during the transformational process -- times are on the order of several seconds. Nonetheless, the striking similarities in the pattern of results obtained in the two situations have led SHEPARD [1, 24, 25] to argue that both the perceptual and the imaginal transformations reflect the operation of underlying analogue processes and representations.

In summary, this paper has presented a brief and incomplete overview of experimental evidence that argues for the necessity of positing an analogue system for the representation and processing of visual/spatial information in certain situations. The evidence has come from psychophysical experiments in which the time course of performing mental transformations on representations of visual objects has been shown to correspond to the time course of performing analogous physical transformations on external objects. The level of description at which internal processes have been characterized as analogue is sufficiently abstract to avoid any speculation concerning underlying neural mechanisms. Rather, an internal process can qualify as analogue if it can be shown that the intermediate states in the processes have a one-to-one relation to the intermediate stages of the corresponding external process. And, this one-to-one relationship is of a behavioral disposition or of a readiness to perceive and respond to a particular external object in a particular external spatial position, if such an object were actually to be presented. Conceived in this fashion, the nature of analogue representations and transformations can be assessed in behavioral experiments that have the potential of elucidating the way in which mental processes simulate or model external operations and events.

References

1. R.N. Shepard, L.A. Cooper: Mental Images and Their Transformation (MIT Press, Bradford Books, Cambridge, Massachusetts 1982)

2. L.A. Cooper, R.N. Shepard: In Handbook of Perception, Volume VIII, Perceptual Coding, ed. by E.C. Carterette (Academic Press, New York 1978)

3. R.N. Shepard: In Perceptual Organization, ed. by M. Kubovy and J.R. Pomerantz (Lawrence Erlbaum, Hillsdale, New Jersey 1981)

4. L.A. Cooper: Percept. Psychophys. 19, 296 (1976)

5. R.N. Shepard, J. Metzler: Sci. 171, 201 (1971)

6. L.A. Cooper: Cog. Psychol. 7, 20 (1975)

7. L.A. Cooper, R.N. Shepard: In Visual Information Processing, ed. by W.G. Chase (Academic Press, New York 1973)

8. M. Kubovy, P. Podgorny: Percept. Psychophys. 30, 24 (1981)

9. L.A. Cooper, P. Podgorny: J. Exp. Psychol.: Hum. Percept. Perform. 2, 503 (1976)

10. S.M. Kosslyn: Image and Mind, (Harvard University Press, Cambridge, Massachusetts 1980)

11. S.M. Kosslyn: Percept. Psychophys. 14, 90 (1973)

12. S.M. Kosslyn, T.M. Ball, B.J. Reiser: J. Exp. Psychol.: Hum. Percept. Perform. 4, 47 (1978)

13. M.A. Just, P.A. Carpenter: Cog. Psychol. 8, 441 (1976)

14. P.A. Carpenter, M.A. Just: In Eye Movements and the Higher Psychological Functions, ed. by J.W. Senders, D.F. Fisher, and R.A. Monty (Lawrence Erlbaum, Hillsdale, New Jersey 1978)

15. Z.W. Pylyshyn: Mem. & Cog. 7, 19 (1979)

16. Z.W. Pylyshyn: Psychol. Rev. 87, 16 (1981)

17. S.M. Kosslyn: Psychol. Rev. 88, 46 (1981)

18. J.R. Anderson: Psychol. Rev. 85, 249 (1978)

19. J.R. Anderson: Technical Report #ONR-82-1, (1982)

20. S.M. Kosslyn, J.R. Pomerantz: Cog. Psychol. 9, 52 (1977)

21. R.N. Shepard, P. Podgorny: In Handbook of Learning and Cognitive Processes ed. by W.K. Estes (Lawrence Erlbaum, Hillsdale, New Jersey 1978)

22. U. Neisser: Cognition and Reality (W.H. Freeman, San Francisco 1976)

23. L.A. Cooper, R. Vallone: Unpublished data (1979)

24. R.N. Shepard, S.A. Judd: Science 191, 952 (1976)

25. J.E. Farrell, R.N. Shepard: J. Exp. Psychol.: Hum. Percept. Perform. 7, 477 (1981)

Part VI

Higher Level Representations in Image Processing

Contrast of Structured and Homogenous Representations

G.H. Granlund and H. Knutsson

Picture Processing Laboratory, Linkoeping University,
S-581 83 Linkoeping, Sweden

1. Introduction

A large part of the information that we want to process is of structural or spatial nature. This means that the information is provided partly by data points in a space and partly by the structural relationships between these data points. As examples of such information we can mention speech, image information, complex decision problems and in general what is included in the field of Artificial Intelligence [1].

The situation that information is represented partly as data and partly as structure leads to a vicious circle. Data points cannot be referenced and processed before the structure has been identified. On the other hand, the structure is defined by the data points, and cannot be identified until data points have been detected and processed. One example is the analysis of ECG signals. See Fig. 1.

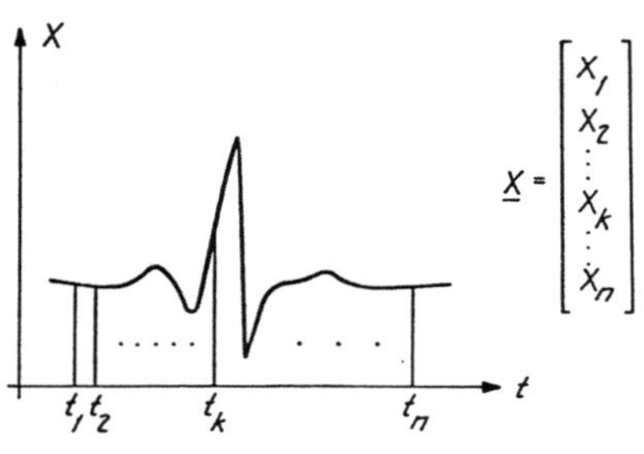

Fig. 1. Example of sampling of an ECG signal

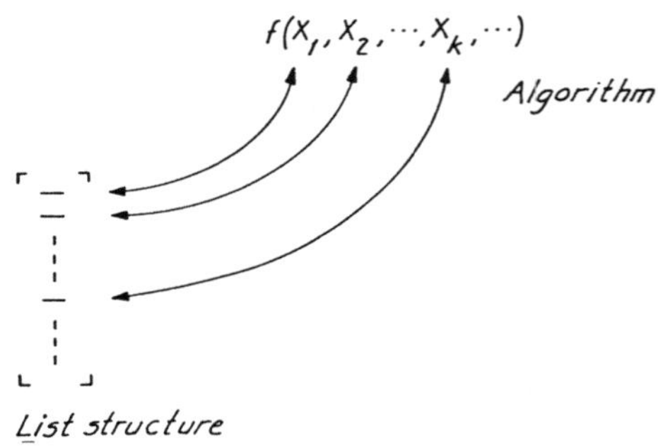

Fig. 2. Correspondence between algorithm and data structure

We may start the processing by time-sampling the signal at equidistant points, according to Fig. 1, obtaining in a given situation the samples indicated in the figure. The crucial problem is that the sample that has been labeled x_k in one observation, may not be comparable to the sample labeled x_k in another observation.

This is due to a fundamental limitation in conventional computing procedures, which assume a situation like the one illustrated in Fig. 2.

An algorithm is somehow defined, in which references are made to particular data in a data structure. The references are generally assumed to be simple, which means that we know where a particular variable x_k is stored, although we do not know its value.

The concept of separated algorithm and data set provides the basis for the von Neumann structure of computer. This is in conventional computers represented by a separation of programs and data. The program provides the contextual frame and the operations to work on the data. The programs internally have a hierarchical structure by which a complex set of operations are interrelated. The data is also conventionally arranged into some form of structure which may be hierarchical with appropriate levels of reference pointers.

Given these properties, what makes the conventional von Neumann machine serial is the fact that the communication between programs and data is performed through a single channel. This implies that the program hierarchy has to establish a single level at which it can communicate with the data set. The access to the data set also has to be established at a particular level, and after entry at that level a certain unit of data can be accessed often using a sequence of reference pointers. See Fig. 3 for an intuitive illustration of this situation. We can visualize this aspect of computation as one of unfolding a hierarchical program structure and a hierarchical data structure to reach some common suitable level of communication, at which exchange of information can take place.

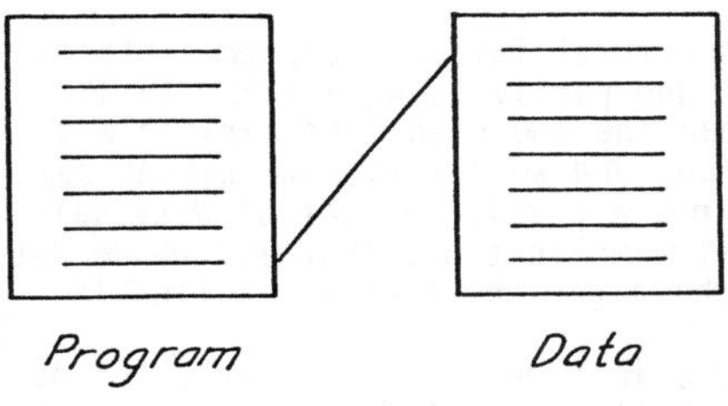

Fig. 3. Illustration of communication between program and data hierarchies

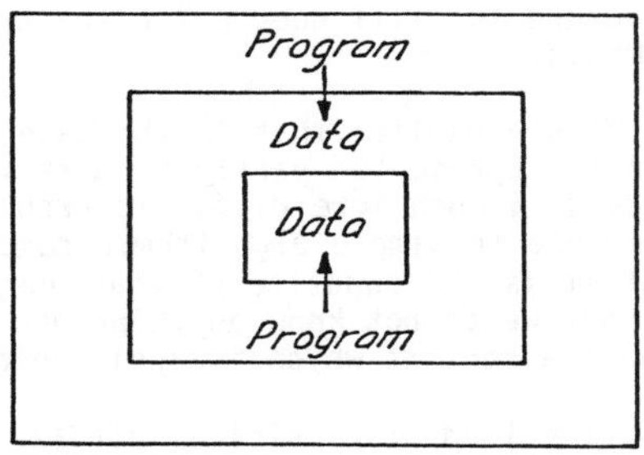

Fig. 4. Illustration of the concept of multiple contextual frames

The slanted line of communication in Fig. 3 illustrates the fact that the communication is effected by the conceptually lowest levels of the program hierarchy, while the access to the data structure generally is done through the top of its hierarchy. This has some unwanted effects which will be discussed later.

There is indeed mixture of program and data aspects in conventional computing, where a program considers itself data and makes modifications in itself. Although such operations are qualitatively very important, they are very limited in terms of occurrence.

Another view of the situation is given in Fig. 4. The data set is enclosed in a program set, which operates predominantly inwards upon the data set. The program provides the contextual frame and the operations to work on the data set. However, a computer program may very well appear as data, which it does to a compiler. This means that we can establish a second contextual frame, where the compiler acts upon the program as data. See Fig. 4.

If we want to generalize, we can define a succession of partially or totally enclosing contextual frames such as the operating system, the computer, the programmer, the computer manufacturer, the economical system of the user country, etc. For whatever size system we want to consider, we can always conceive of a frame surrounding that system providing influences that may be interpreted as context.

This observation, which is by no means new, leads to the conclusion that there are no clear cut distinctions between programs or operations and data. A forced choice would tend to favor the view that the information seen outwards in the contextual frame is program, while the information seen inwards can be interpreted as data.

This problem of how information may appear as operand as well as an operation simultaneously seems to be the key question in computation and information processing. However, there is so far no general theory available as guidance.

Various attempts have been made to devise distributed processing systems with different types of control structures allowing a high degree of parallelism. One of the best known instances is the so-called data flow computer [2]. Although operations are data controlled, an elaborate control structure is still needed for the transfer of data among the processing elements.

We saw earlier that in the case of structural information, the information is represented partly by data values and partly as structure. This leads to a much more difficult problem than the one discussed earlier with reference to simple algorithmic computation, and we may to some extent say that it is the opposite of that one; we know a particular set of data values but we do not know anything about their implication, that is, we do not know the context which is applicable or the algorithm that is applicable.

This leads to a vicious circle, because in order to interpret a set of data we have to know the structure in which it is located. However, there is no way that the structure can be identified unless we start evaluating data. This vicious circle of interrelationship is given an intuitive illustration in Fig. 5.

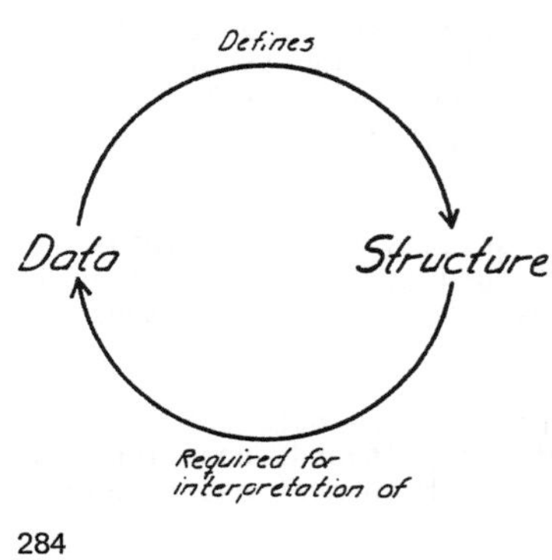

Fig. 5. The vicious circle of structural information

This illustrates one of the key problems in Artificial Intelligence or processing of structural information: How do we find the context (appropriate set of algorithms) for a particular data set in a given situation, before we know what the data set represents? Can this vicious circle be avoided or somehow broken, and the problem be solved? The discussion of this question will be deferred to section 3, where we will find that there is no easy solution to the problem.

2. Clues from the visual system

From the preceding discussion it is apparent that a conventional computer cannot work without an intelligent being structuring the processing situation in an appropriate manner. Consequently there exist processing structures which can handle this type of information, and one possible approach to the problem is to look at biological systems for useful clues. A number of the features of the hierarchical computing structure to be described have been suggested by the existing knowledge concerning the visual system. It would carry too far in this context to go through the entire background for these considerations. Only the more important features will be discussed briefly.

It should be emphasized that the computing structure is not intended to imply a model of the visual system, although it is believed that certain properties would be relevant in a model of the visual system.

- Spatially ordered projections

  It has been found to be a universal feature of the anatomical organisation of the visual system [3], that it is organised in terms of spatially ordered projections. This indicates that spatial relationship is important, and it speaks against a use of global transforms, at least at lower levels.

- Modularity

  Several examples have been found of modularity in the nervous system such as the hypercolumn structure [4, 5]. It can be presumed that even Nature has had to economize and standardize the structures used.

- A standardized information representation

  HUBEL and WIESEL found that cortical cells which are sensitive to the orientation of a stimulus are grouped in columns or slabs of common orientation [6]. This provides what Hubel and Wiesel term "sequence regularity". It is here postulated that such a sequence regularity or closure may be a universal feature of the information representation used in the nervous system, as it provides a number of computational advantages as will be discussed later. This is also the representation being used for color in the visual system. An important property of the information representation is that it allows a continuous grading of the certainty of a particular class membership, thereby indicating the degree of commitment to a particular interpretation of an event [7].

- Standardized symmetry operations

  At low levels of the visual system there have been found cells which respond to lines and edges of different orientations [4]. Complementarity in terms of aftereffects has also been discovered for circularly symmetrical patterns [8]. It is postulated that classes of symmetry operations are employed in the processing of the visual system. Symmetry operations have attractive semantic as well as syntactical properties.

Hierarchical computing structure

A number of physiological findings in the visual system, as well as in the nervous system in general, indicate the existence of different levels of computation [9, 10]. It is postulated that these levels may form a more or less strict hierarchy.

3. Computing structures

As indicated in section 1 data and structure are intertwined in a vicious circle. See Fig. 5. This vicious circle can not be broken into in any clear-cut and unambiguous way. Still the problem has to be attacked in some way. It appears that the only way to approach the problem is to break into the circle at the position of structure. This means that a hypothetical structure or model is assumed and this model is applied upon the data. The data is evaluated on the basis of this model hypothesis, and the hypothesis is supported or rejected to a greater or lesser extent. See Fig. 6.

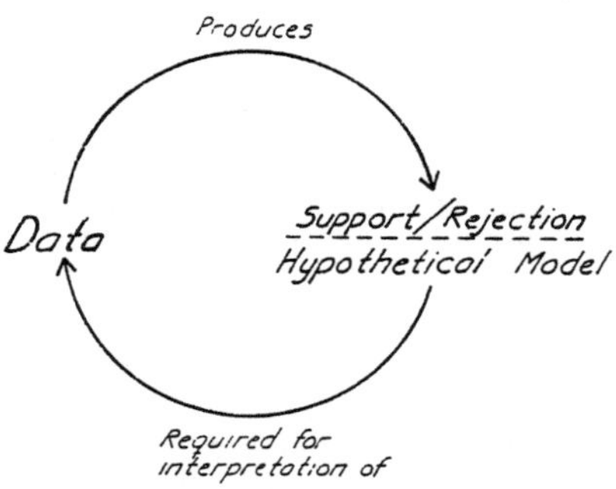

Fig. 6. Breaking up the vicious circle of structural information

It is apparent that the number of possible hypotheses increases with the size or the complexity of the data set. It is obvious that there is an infinite number of hypotheses concerning the content of a data set representing an image and that a search among alternatives would never end. Thus the number of alternatives has to be limited. This implies that the scope of the hypotheses can not be too wide, from which follows that a particular, limited set of hypotheses can only be used to describe a small data set.

In order to determine the structure of the data well enough to allow a computation in conventional ways, heuristic methods traditionally have been used to invoke the a priori knowledge about the actual problem. Heuristic methods can generally be viewed as producing particular projections of the whole data set. One effect of this projection is a reduction of the data to a smaller amount, which is hopefully representative of the relevant aspects of the problem. Another effect of the projection is that structural features of the data set are emphasized, to serve as references for other data points to produce the desired, unambiguous list structure.

There are three fundamental limitations of the heuristic approach:

1 Heuristic methods can only be used when there are strong, unambiguous structural features of the data, such as dominant peaks, separation of object and background, etc.

2 Heuristic methods are generally specific for each problem, and cannot be utilized for another one.

3 In most problems it is not possible to formulate a simple heuristic strategy for resolving structural markers, because the structural features may be embedded in the data in a very subtle way to a depth of several levels. One important consequence of this fact is the hypothesis that it is in general not possible to obtain one single global projection or transformation which resolves structural features but that relevant projections are only valid within limited size regions, and are furthermore dependent upon context.

In more general terms the process of interpretation has to somehow follow the assumed process of data generation. This gives a form of the vicious circle as indicated in Fig. 7. A set of available hypotheses or models is applied upon a given data set. This gives an estimation of properties of the data set in terms of model parameters. The outcome of the estimation procedure reflects three different issues:

1 The parameters of the data set in terms of the models used

2 How well the data set can be fitted to the model

3 The relevance of the set of models used for the description.

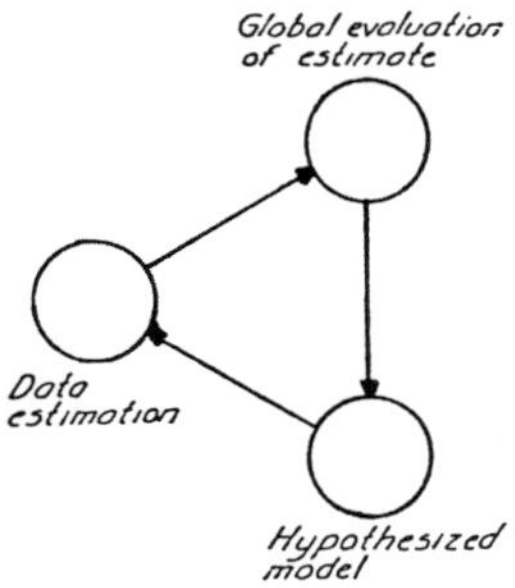

Fig. 7. The circular model of interpretation

These three issues represent a substantial part of the difficulties encountered in information processing. A more general discussion of these matters is outside the scope of this presentation. We will only observe that the main problem is how hypotheses or models can be modified using the outcomes of estimation from previous models. This might be considered the fundamental problem in Artificial Intelligence. No general solutions can be provided to this problem. The GOP hierarchical computing structure to be described later is one attempt to formalize a system for regularized evaluation and modification of hypotheses in a hierarchical structure.

4. Hierarchical structures

It is apparent that processing of structural information has to be made in a way that allows a simultaneous analysis of data values and an identification of the inherent structure of the data. There are several reasons to turn to a hierarchical process in this case. One reason is that a description can only be made for a limited size subspace or neighborhood. This limitation is partly due to computational aspects. With increasing size of the neighborhood, the number of derivable states or features within the neighborhood increases in an explosive fashion. Another more fundamental aspect is that there appears to be no such thing as one complete interpretation of the state of a subspace or neighborhood. The effect is that an aspect of an interpretation becomes relevant only in a particular contextual situation. Another formulation is to say that there is a large set of true statements that can be made about the state within a neighborhood, but not all of these are of interest in a given situation.

As a particular set of hypothesis can only deal with a limited size data set, it will be necessary to combine results from several such data subsets. For a processing system to be feasible, the setting up and evaluation of hypothesis has to be organized in some well structured way. The combination of results of local hypotheses into more global hypotheses can be viewed as an operation at a higher level. An appealing structure is a hierarchical one, where results from hypothesis tests at a lower level are subjected to hypothesis tests at a higher level.

Hierarchical structures have been proposed for the processing of structural information [11,12]. Similar models for visual perception have also been proposed, which perform transformations in homogenous layers [13]. So far it has proved difficult to devise effective hierarchical structures for use in information processing. A major problem has been to combine the requirements of meaningful descriptions with communication between levels in the hierarchy. If information from a higher level is to communicate with

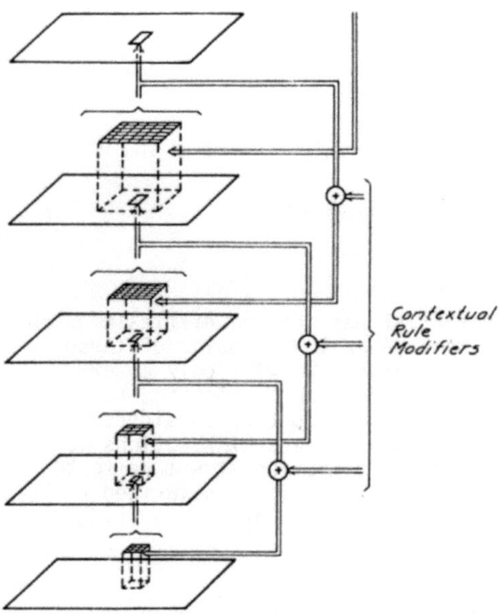

Fig. 8. Hierarchical structure relating information representations and operations

and make sense at a lower level in the function of context, there are very strong restrictions upon the representation of information as well as upon the operations that separate different levels. The operations will among other things have to act as interfaces between different levels of the hierarchy in bottom-up as well as in top-down direction.

5. The GOP hierarchical computing structure

The GOP (general operator processor) implies an algorithmic structure which allows operations within hierarchical structures [14, 15]. Due to limitations in space it will only be possible to give a qualitative discussion of the more fundamental properties. Certain aspects of the presentation will be in terms of 2-dimensional signals, but the principles are applicable to signals of any dimensionality.

In the processing of image information, the hierarchical structure may appear as in Fig. 8. The original image appears at the lowest level. An operation with a particular size neighborhood produces a transformed image at the next level. Another operation uses this transformed image to produce a higher level transform, and this gives a sequence of bottom-up operations. There is also a top-down flow of information which enables information from a higher level to control operations at a lower level.

On each neighborhood a set of operation kernels are applied, each one measuring a particular feature of the event within that neighborhood. As an example at low levels in an image hierachy, the operation can measure the edge content in four different directions. As edges of all orientations cannot be expected within a particular, small neighborhood, we can infer the mutual exclusion of edges of perpendicular orientation. This leads to a 2-component vector representation, where perpendicular edges will be represented by vectors of opposing directions.

This interpretation at low levels can be extended to a general principle for information representation at all levels. Features derived from a particular neighborhood are mapped into a 2-dimensional space and arranged in a circular fashion, where conflicting or incompatible features are located at opposing positions, while relatively independent features are at right angles. The feature measurements are weighted together to form a vector describing the event within the neighborhood. The direction of the vector indicates the dominant class membership of the event, and the magnitude gives the certainty of this choice. See Fig. 9.

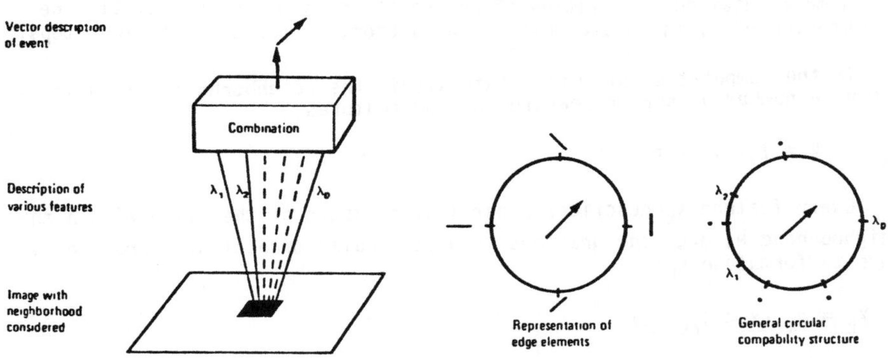

Fig. 9. Illustration of feature generation and information representation

This information representation is consistently used through all levels of the hierarchy. However, the meaning of the information varies with the level of the hierarchy. We may say that information is represented in terms of relative similarity - dissimilarity.

6. Structure of operations

The general structure of an operation is indicated in Fig. 10.

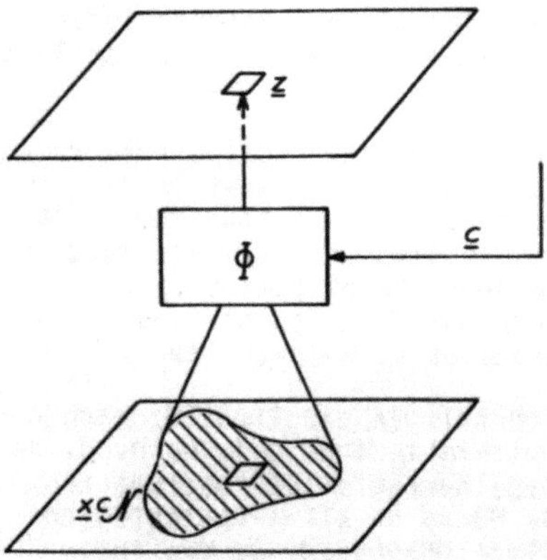

Fig. 10. General structure of operation

An output vector $\underline{z}$ is computed as a function of the data elements $\underline{x}$ within the neighborhood N and the context vector $\underline{c}$

$$\underline{z} = \Phi\ (\underline{x} \subset N,\ \underline{c})$$

The data elements, $\underline{x} \subset N$ within the neighborhood N, the output data $\underline{z}$ and the context descriptor $\underline{c}$ are all assumed to be vectors. In the ensuing discussion we will in particular assume that they are two-dimensional vectors, which does not imply any fundamental limitations of the generality of the approach.

A more detailed formulation of an operation is given in Fig. 11. The computation of $\underline{z}$, the state description vector, is performed in two steps.

In the computation of $\underline{z}$ the state within the neighborhood N is described by a number of scalar measurements or features

$$X_k \qquad k = 1, \ldots, n$$

Every feature X_k describes a particular aspect of the state within the neighborhood N, e.g. the gradient in a particular direction. There are two ways to formulate X_k:

1 $X_k = g_k\ (\underline{x} \subset N_{rs},\ \underline{c})$

2 $X_k = g_k\ (\hat{\underline{x}}_r,\ \hat{\underline{x}}_s,\ \underline{\Delta}_{rs},\ \underline{c})$

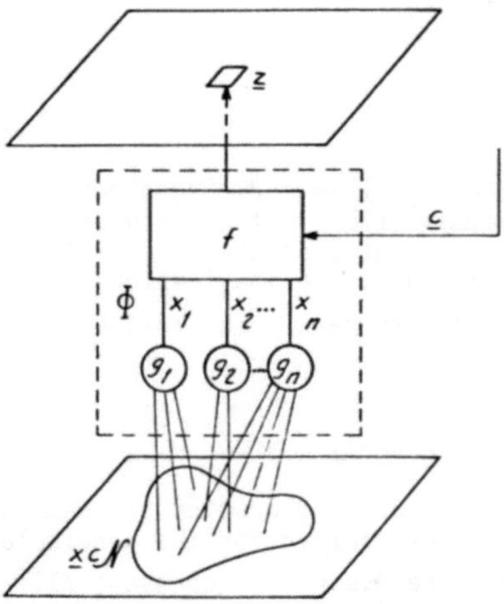

Fig. 11. More detailed structure of operation

In the first formulation X_k is viewed as a function of image elements $\underline{x}$ within a subregion N_{rs}, part of the entire neighborhood N. In the second formulation X_k is a function of two complex valued variables $\hat{\underline{x}}_r$ and $\hat{\underline{x}}_s$, which are each estimates describing events within the neighborhoods N_r and N_s, where $N_r \cup N_s = N_{rs}$. See Fig. 12.

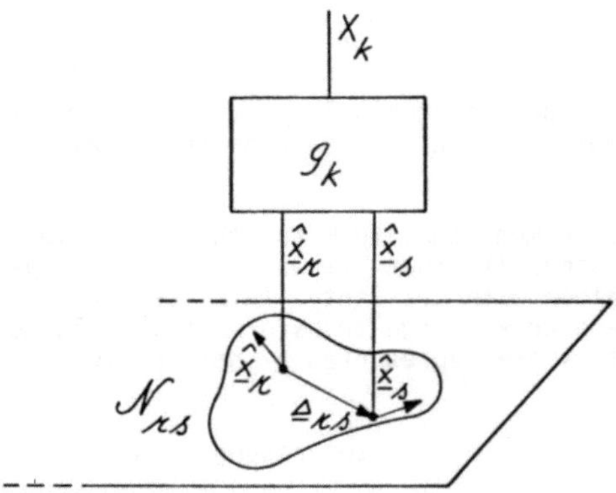

Fig. 12. Features derived as functions of events

The vector $\underline{\Delta}_{rs}$ is the spatial relative vector between events $\hat{\underline{x}}_r$ and $\hat{\underline{x}}_s$; $\underline{\Delta}_{rs} = \Delta_{rs} \exp j\varphi$. In general the second formulation will be used. An event represented by an estimate $\hat{\underline{x}}_r$ is assumed to be a condensed representation of the data within a sub neighborhood or region $N_r \subset N$. See Fig. 13 for an intuitive representation of a neighborhood with subregions.

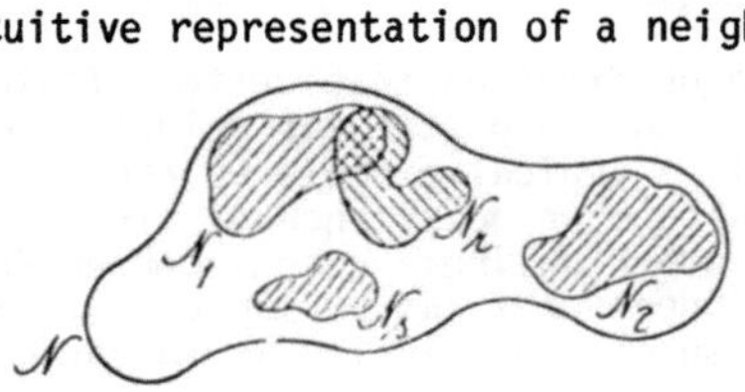

Fig. 13. Representation of a neighborhood with subneighborhoods

It is assumed that the estimate $\hat{\underline{x}}_r$ is a weighted sum of the data $\underline{x}_h$ within the region N_r

$$\hat{\underline{x}}_r = \sum_{N_r} \underline{w}_{rh} \cdot \underline{x}_h \qquad \underline{x}_h \subset N_r$$

The estimates $\hat{\underline{x}}_r$ may well not be explicity available during a particular computation, but only implicit in the computation of the features X_k. The estimates $\underline{x}_r$ have sampling properties dependent upon the size of the neighborhood and of the weighting coefficients $\underline{w}_{rh}$. The implications of these sampling properties have been discussed elsewhere. See KNUTSSON et al. [16].

7. Information representation and context dependency

As seen in the preceding section the first part of an operation produces a set of n features X_k which will generally all be nonzero for the particular neighborhood considered. A priori it would be necessary to represent all the components as a description of the state within the neighborhood.

The general principle with regard to information representation is that the original n features produced, X_k, are condensed into only 2, which means that the output is represented as a complex valued variable $\underline{z}$. We will already now suggest the notion that arg ($\underline{z}$) points out a dominant class membership of the state within the neighborhood considered. The magnitude of $\underline{z}$

$$z = |\underline{z}|$$

can be viewed as a deterministic magnitude, a probability, or some form of confidence or certainty. The representation can be viewed as an expression of fuzzy memberships [17].

The main reason for the choice of dimensionality 2 for the output vector is that this is the lowest dimensionality that allows a continous representation of class membership together with certainty. This does not imply that the description of an image or an image transform is limited to two variables, but that the descriptors are represented by sets of two variables.

This reduction of dimensionality is based upon two assumptions:

1 The assumption of internal compatibility restrictions between features given a particular contextual situation;

2 The assumption that only certain aspects of the information within a particular neighborhood are relevant in a particular contextual situation.

The first assumption is built upon the observation that certain features are not likely to occur simultaneously within a particular neighborhood given a particular contextual situation. As a typical example we can mention the case at low levels where the simultaneous occurrence of lines and edges at perpendicular orientations is a rather unlikely event. Consequently, two features which a priori could be independent can be mapped into a single variable with opposite signs, when such a restriction is valid.

The second assumption is based upon the belief that an interpretation of an image can only involve certain aspects of the image at a given time. The full discussion of the reasons for these fundamental restrictions is outside the scope of this paper. The assumption is consequently made that in all but trivial situations context has to be invoked to restrict the information propagated to the next level.

An important question is: How can contextual information be represented and in what way does it restrict the information propagated? In this presentation we will have to restrict the treatment of this question to a formal one. It is postulated that the effect of context is to produce a transformation matrix $\underline{C}$ as a function of the context vector $\underline{c}$

$$\underline{C} = [f(\underline{c})]$$

where C is an n x 2 matrix .
The output vector $\underline{z}$ is then defined

$$\underline{z} = \underline{C}\ \underline{X} = [f(\underline{c})]\ \underline{X}$$

where $\underline{X}$ is the feature vector or a function thereof.

It is apparent that the expression implies a mapping from n to 2 dimensions of the feature vector $\underline{X}$. This can be visualized as the context vector $\underline{c}$ defining a plane in n-dimensional space upon which the feature vector $\underline{X}$ is projected to produce the 2-dimensional output vector $\underline{z}$.

The general strategy for selection and implementation of the context vector $\underline{c}$ is outside the scope of this paper and still an open question. Strategies that have proved useful have been selection of features that support dominant trends around the neighborhood, and selection among feature sets which at a higher level have the ordinary relationship of compatibility and incompatibility. These strategies have been used for image enhancement and coding as has been described elsewhere. See KNUTSSON et al. [18] and WILSON et al. [19].

We will now take a look at the operations producing the feature components X_k. The output vector $\underline{z}$ from an operation can also be viewed as a vector sum

$$\underline{z} = \sum_k X_k \exp j\phi_k$$

where the magnitude function is

$$X_k = g(\underline{\hat{x}}_r, \underline{\hat{x}}_s, \underline{\Delta}_{rs}, \underline{c})$$

and the argument function is

$$\phi_k = f(\theta_r, \theta_s, \arg \underline{\Delta}_{rs}, \underline{c}) \quad ,$$

where

$$\underline{\hat{x}}_r = \hat{x}_r \exp j\theta_r \qquad \text{and} \qquad \underline{\hat{x}}_s = \hat{x}_s \exp j\theta_s$$

One standard form of the magnitude function is

$$X_k = \hat{x}_r \cdot \hat{x}_s \cdot g_\theta(\theta_r, \theta_s, \underline{\Delta}_{rs}, \underline{c})$$

We can see that the magnitudes of the individual events form a product with the remaining part of the function. One view of the magnitudes x_r and x_s is that they represent a probability of existence of the events. Consequently $x_r \cdot x_s$ would imply the joint probability of the simultaneous events. A feature is a function of two events in a data space of some dimensionality; typically dimensionality two in the case of image information.

As one example of symmetry operation we will give the following:

$$\underline{z} = \Sigma X_k \cdot \exp j\phi_k$$

where disregarding context

$$X_k = g(\underline{\hat{x}}_{kr}, \underline{\hat{x}}_{ks}) = \hat{x}_{kr} \cdot \hat{x}_{ks} \left|\sin \frac{\theta_{kr}-\theta_{ks}}{2}\right| \text{ and}$$

$$\phi_k = 2\varphi_k \quad .$$

We observe that the magnitude of a particular feature X_k is proportional to the product of the event magnitudes x_{kr} and x_{ks} and dependent upon the difference between the class memberships expressed by θ_r and θ_s. Used at low levels in image processing it provides the dominant orientation and magnitude of a line or edge going through the neighborhood.

The symmetry operations define a hierarchy of information primitives. In the case of 2-dimensional signals the image primitives may imply lines and edges, texture, curvature and divergence, convexity and concavity, partial closure, regions, combinations of regions, and so on.

8. Homogenous representations and operations

The computing structure discussed in the preceding sections represents a clear departure from conventional computing structures in that program and data are mixed to quite a different degree. We can view this intuitively as the hierarchical structures of operations and data respectively being interlaced. See Fig. 14.

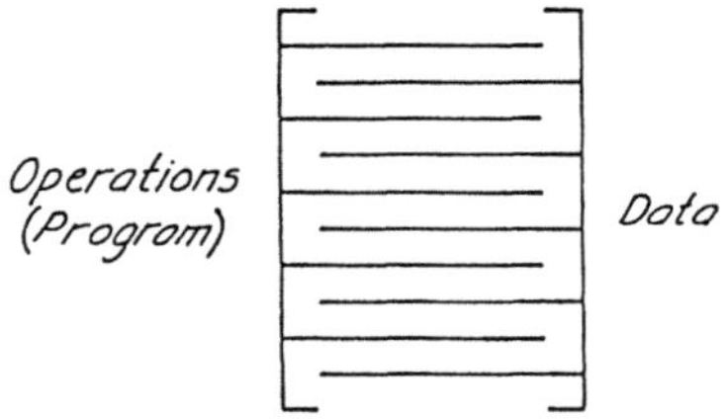

Fig. 14. Intuitive illustration of hierarchical operation and data structure interlaced

This does not mean that they are merely "sandwiched" together in some arbitrary fashion, but that operations and data to some extent imply the same thing. If we want to make an identification with traditional concepts, we could for a first approximation say that data is information flowing from bottom up, while operations constitute information flowing downward. The

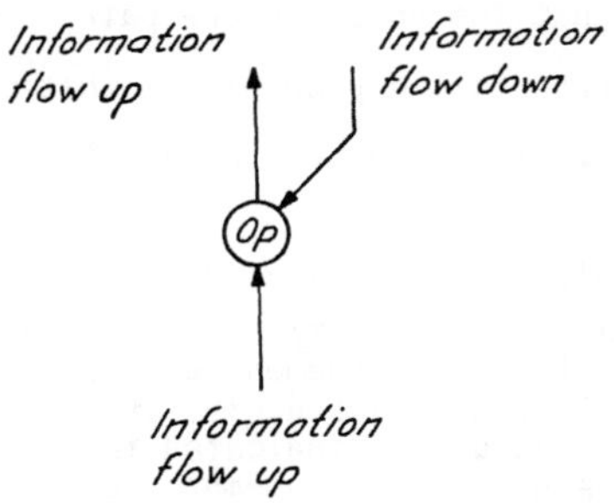

Fig. 15. Illustration of operation and information flow

data flowing downward provides the contextual modification of a standard set of operations. See Fig. 15.

The GOP information representation can be viewed as a hierarchical, distributed data structure. We can view the vectorial representation of information as a continuous representation of a stylized data structure containing four elements only p_1 to p_4.

Every one of the four pointer elements points to a predetermined pair of data structures at the lower level of the hierarchical data structure. See Fig. 16.

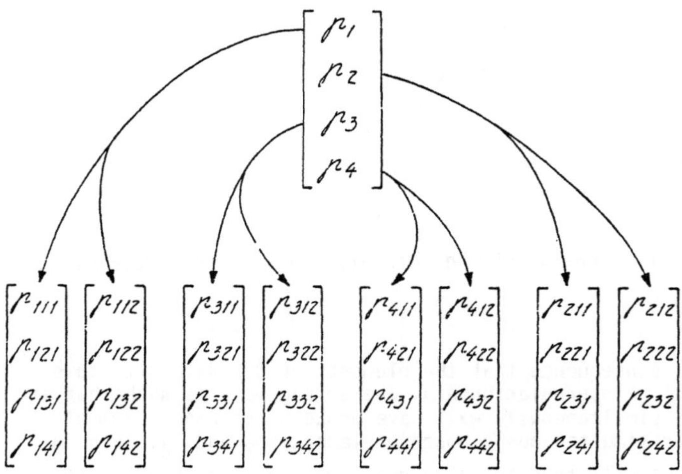

Fig. 16. Pointer arrangement in hierarchical data structure

The effect of the values p_1 to p_4 is to indicate the "degree of pointing" or degree of reference to the predetermined pair of low-level data structures assigned to the particular pointer elements. The pointer element values can be viewed as probabilities where

$$\sum_{i=0}^{4} p_i = 1$$

The element p_i represents the probability that the substructure pair will be referenced. The element $p_0=1-\sum_{i=1}^{4} p_i$ represents the probability that no reference is made to the lower level data structure that could be referenced.

The particular arrangement of the pointer elements p_i can be viewed as a "circularization" of a data structure in terms of similarity and dissimilarity. See Fig. 17. Assuming an originally arbitrary arrangement of elements with respect to their content as in Fig. 17a, we can conceive of a reshuffling of elements to produce a new data structure as indicated in Fig. 17b. In this new data structure we have the desired arrangements of elements p_k in terms of similarity of occurrence or semantic content as described earlier. The ordering structure of similarity is a circular one, implying that p_1 and p_4 are next to each other in terms of similarity or probability of simultaneous occurrence. We can consequently consider the ends of the list being joined to each other, as indicated by the dashed line.

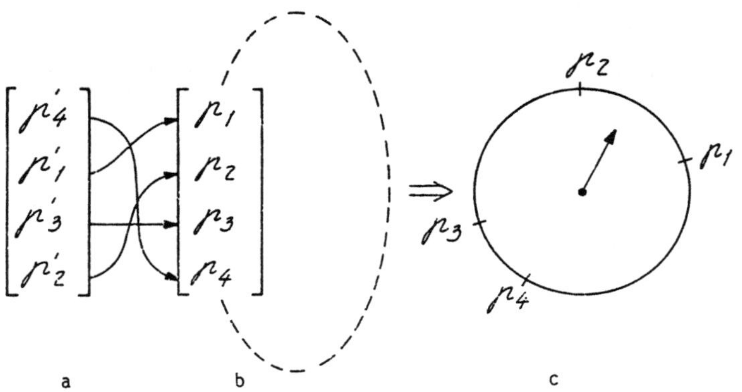

Fig. 17. Intuitive illustration of the "circularization" of a data structure

This gives as a consequence that the elements of the data structure have a very particular interrelationship. The arrangement is such that it is unlikely that we simultaneously will have nonzero reference element values p_1 and p_3, or simultaneously nonzero elements p_2 and p_4. This implies that references will tend to go to one lower level pair only. See Fig. 18. One implication of this sparse appearance of nonzero elements in the data structures is that a high degree of data compression can be obtained.

Referring to the GOP information representation discussed earlier we can now make the generalization that the list with discrete elements implementing pointers to other data structures is in fact represented by a complex valued reference pointer. See Fig. 17c. This has several advantages compared to the discrete version used in the introductory discussion. It allows a continuous representation of the degree of similarity between lower level data structures. The orientation of the vector determines what

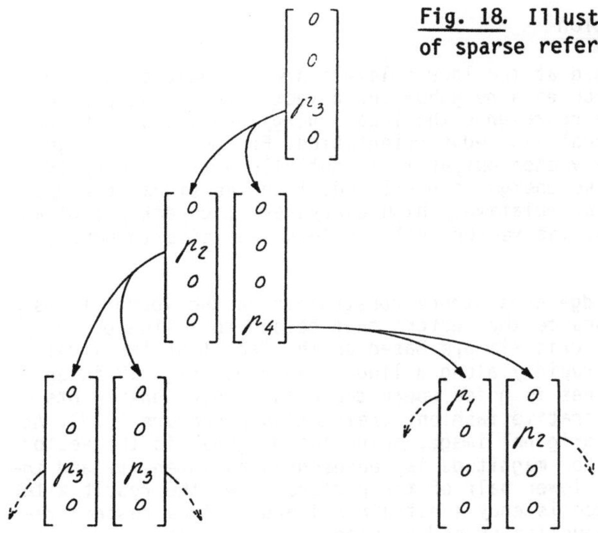

Fig. 18. Illustration of the mechanism of sparse references

particular lower level data structure pair that is pointed out, and the magnitude of the vector indicates the degree of reference or the probability of reference.

This continuous representation of the data structure allows a high degree of information compression. One component in this information compression is due to the arrangement of reference pointers in terms of relative compatibility, which implies that a number of pointers can be mapped into only two dimensions.

In addition to a continuous representation of reference pointers, this structure allows a continuous allocation of space for the data. This feature is very important in that it allows the implementation of selforganizing or learning systems. It is also important for the implementation of highly redundant systems.

An important aspect of this data structure, is that it can be viewed as the original data being broken up into a hierarchical structure consisting of reference pointers only. In each reference pointer are described the symmetry properties of the nearest lower level data structures.

9. Experimental results

In the preceding sections we have discussed a hierarchical computing structure with a distributed data structure together with operations. This is a computing structure that allows fully parallel computation and use of associative functions, where the latter issue will have to be omitted in the discussion. A great deal of research remains to be done in order to fully evaluate the possibilities and solve the problems with this approach. However, the first version of computer using this approach, the GOP Image Processor which has been operative since 1979, has proved to be extremely effective in the processing of image information, and experiments performed have supported the arguments given.

9.1 Line and edge description

The symmetry operations used at the lowest levels are equivalent to line and edge detectors [14]. For each neighborhood a vector output is produced, where the vector magnitude represents the local energy and the vector angle represents the dominant local line/edge orientation. For medium and high signal to noise ratios the vector output is essentially unaffected by the noise. However, if the noise energy is nearly equal to the signal energy, the vector magnitude will be relatively high everywhere (correctly indicating the presence of noise). The vector angle is less sensitive to noise, and unbiased.

By imposing line and edge consistency constraints on the vector transform, is is possible to enhance the vectors that belong to a line or an edge [20]. The consistency criteria are based on the fact that the local orientation does not vary rapidly along a line or an edge. Fig. 19 is an example of detection of lines in noise where the consistency constraints have been imposed in an iterative fashion, (relaxation procedure [21]). At the top in Fig. 19 is the original image, below the original is the vector output image where the vector magnitude is represented as intensity and the vector angle as color. The lower half of the picture shows the result after iteratively imposing the consistency constraints 1 and 4 times respectively. Regrettably, the reproduction procedure used does not allow a presentation in color of this and the following figures.

9.2 Texture description

Using the same type of operations as for line and edge detection, it is possible to obtain an effective texture description. Two features of high

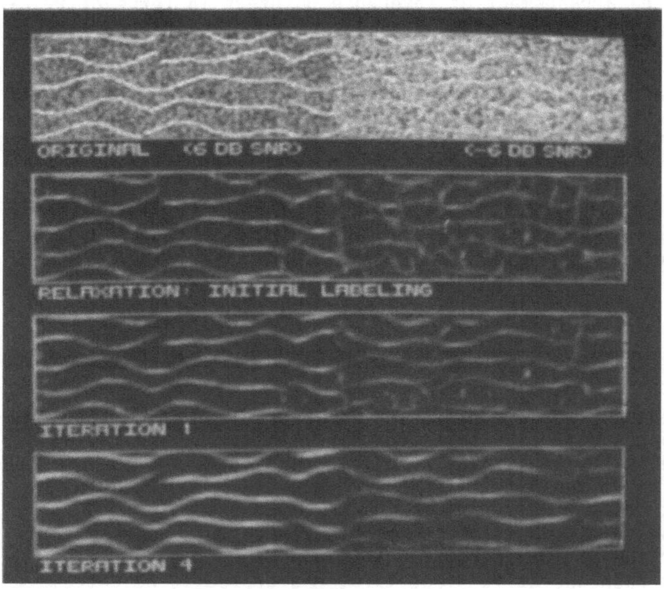

Fig. 19. Example of consistency operation applied upon lines in noise

significance in texture description are orientation dominance angle and frequency content. A local orientation estimate is produced in a way identical to that of the line/edge case. An estimate of the frequency content can be obtained by combining the outputs from line/edge operators used at different scales, i.e., having different frequency response. Both the orientation and the frequency estimates are in most cases much too localised to be used directly as texture descriptors, (this will always be the case if the operators are smaller than one texture element). This problem is easily overcome since both the orientation and the frequency estimates are locally represented as vectors in such a way that a vector averaging operation is well behaved. Consequently averaging of the two estimates can be carried out over an area sufficiently large to produce a reliable output.

Fig. 20a shows a (512 x 512) image where four different textures (from BRODATZ's book on Textures [22]) have been "cut" together, the different textures are seafan, cork, calfskin and water. In Fig. 20b the orientation estimate, averaged over a 64 x 64 window, is shown represented as color. Fig. 20c is the corresponding output for the frequency estimate. Fig. 20d shows how, by using the outputs in Fig. 20b and Fig. 20c, the image can be

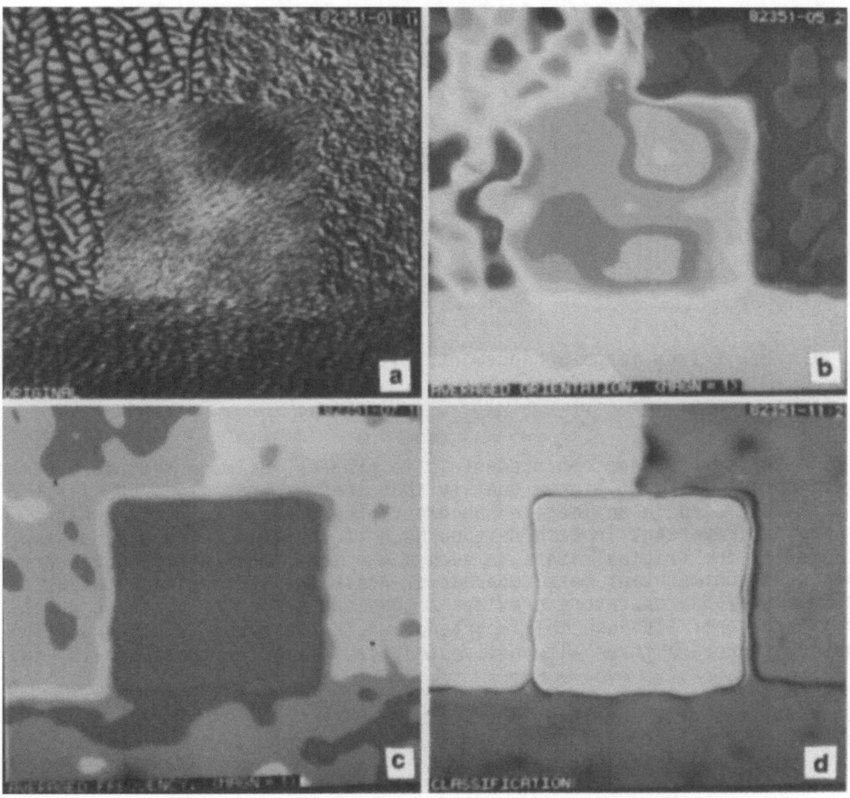

Fig. 20. Example of texture description. a) Original image. b) Orientation estimate. c) Frequency estimate. d) Result of segmentation

classified into four distinct categories which coincide well with the original different textures. Color indicates the class membership.

9.3 Image enhancement

Fig. 21 is an example of image enhancement; the noisy (10 dB SNR) original image being to the left and the enhanced image to the right.

Fig. 21. Result from image enhancement. a) Original image. b) Enhanced image

The purpose of image enhancement is to produce an image that to a human viewer appears to be of higher quality than the original [18]. The first and difficult step in an image enhancement procedure is to decide what features are important in each neighborhood of the image. The second step is to modify the original image in such a way that important features are enhanced and unimportant ones suppressed. Again the vector image output from the line/edge operators provides a good basis for the decision procedure in step one. Although the importance of the features is continuously weighted, there are three alternative decisions made for the image in Fig. 21.

1 Only average gray level is important
2 Edge/line consistency (one dimensionality) is important
3 Each pixel value is important.

1 This alternative is chosen if the vector image is locally of low magnitude and is line/edge inconsistent e.g., shoulder, cheek and background

2 This alternative is chosen if the vector image is locally line/edge consistent, e.g., edge of shoulder, edge of hat

3 This alternative is chosen if the vector image is locally of high magnitude and is line/edge inconsistent, e.g., eyes, eyelashes.

With the type of information representation used the descriptors representing these alternatives imply different level image primitives. The second step in the enhancement procedure is relatively straight forward. A context-dependent filtering scheme is used which performs:

1 Lowpass filtering

2 Low pass filtering along the line/edge direction and all pass filtering perpendicular to the line/edge direction

3 All pass filtering.

An illustration of iteratively applied line/edge consistency is provided by Fig. 22. It shows the type of image that is produced if white Gaussian noise is filtered as if it were an image of type 2. The result is shown after 1, 2, 4 and 8 iterations of filtering and it is apparent that the image is forced to be locally line/edge consistent. The control information for the filter orientation was in this case obtained from a finger print image.

Another experiment that was carried out to examine the effects of forced line/edge consistency is shown in Fig. 23. Here the first step in the enhancement procedure, i.e., the interpretation, was carried out using the original image (upper left in Fig. 23). Then a large amount of noise was added to the original (upper right). Step 2 in the enhancement procedure was then carried out iteratively 1 (lower left) and 4 times (lower right) on the noisy image. The result after 4 iterations is intriguing and is reminiscent of the structure of a painting or a sculpture.

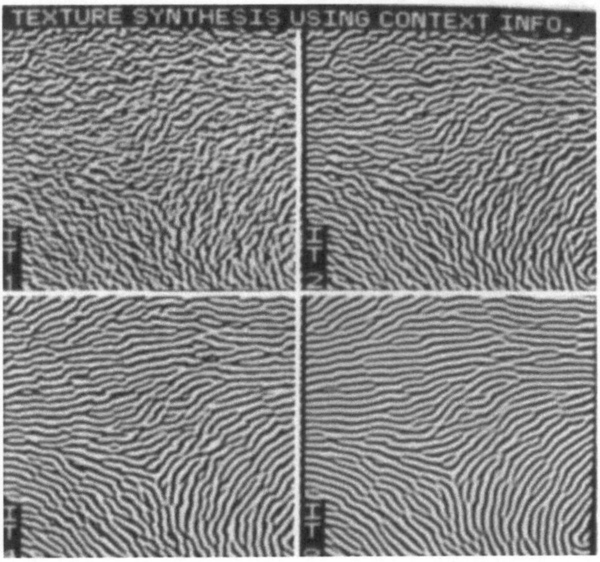

Fig. 22. Context-dependent grouping of noise elements. Result after 1,2,4 and 8 iterations

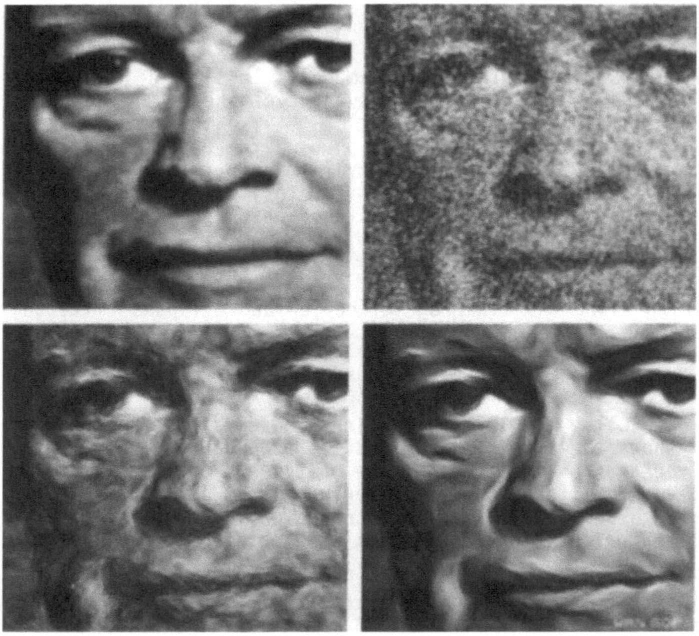

Fig. 23. Result from enhancement of a very noisy image

Fig. 24. Result from image coding

9.4 Image coding

Image coding and image enhancement are closely related topics. The fact that the final receiver in both cases is a human being is maybe the most important link between the two.

A good prodedure for enhancement would then be advantageous to incorporate in an image coding scheme. This is indeed the case [19]. Fig. 24 shows the original image (left) and the image coded at a rate of 0.27 bits/pixel (right). The coder used here is a hybrid transform predictive coder with the predictor based on the ideas put forward in the image enhancement section [18].

10. Acknowledgements

This research was supported by the National Swedish Board for Technical Development. The authors also want to express their appreciation of the enthusiastic work done by the GOP group.

References

1. P. H. Winston: Artificial Intelligence. (Addison-Wesley Publishing Company, 1977).
2. J. B. Dennis, G. A. Boughton, K. C. Leung, Proceedings of 7th Ann. Symp. Computer Architecture, pp. 1-8.
3. J. H. Allman, J. Kaas, Brain Res. 35, 89-106, (1971).
4. D. H. Hubel, T. N. Wiesel, J. Physiol. 160, 106-154, (1962).
5. D. N. Hubel, T. N. Wiesel, M. P. Stryer, J. Comp. Neurol. 177, 361-380, (1978).
6. D. H. Hubel, T. N. Wiesel: J. Comp. Neurol. 158, 295-302, (1974).
7. D. Marr, A. I. Memo No. 340, Massachusetts Institute of Technology, (1975).
8. D. M. Mackay: Perception and Brain Function, Second Neuroscience study Program, (M.I.T. Press, Cambridge, Mass. 1970), pp 303-315.
9. J. I. Allman, J. H. Kaas, Brain Research, 76, 247-265 (1974).
10. S. M Zeki, J. Physiol. 277, 273-290 (1978).
11. S. Tanimoto, T. Pavlidis, Comput. Graph. Image Proc. 4, 104-119, (1975).
12. A. Rosenfeld, Proceedings of 5th Int. Joint Conf. on Pattern Recognition, Miami, USA, (1980), pp. 802-811.
13. H. Marko: Biomedical Pattern Recognition and Image Processing, (Verlag Chemie GmbH, Weinheim 1979) pp. 269-296.
14. G. H. Granlund, Comp. Graph. and Image Proc. 2, 154-173, (1978).
15. G. H. Granlund, H. Knutsson, Proceedings of IEEE Int. Conf. on Acoustics, Speech and Signal Processing, Paris, (1982), pp. 11-16.
16. H. Knutsson, G. H. Granlund, Proceedings of the 5th International Conf. on Pattern Recognition, Miami, Florida, (1980), pp. 45-48.
17. L. A. Zadeh, Information and Control, No. 3, Vol. 8, 338-353, (1965).
18. H. Knutsson, R. Wilson, G. H. Granlund, IEEE Trans. on Communications, In Press (1983).
19. R. Wilson, H. Knutsson, G. H. Granlund, IEEE Trans. on Communications, In press (1983).
20. M. Hedlund, Proc. IEEE CompSoc Conf. on Pattern Recognition and Image Processing, Las Vegas (1982) pp. 93-96.
21. A. Rosenfeld, R. A. Hummer, S. W. Zucker, IEEE Trans. on Systems, Man and Cybernetics, No. 6, Vol. SMC-6, 420-433, (1976).
22. P. Brodatz: Textures (Dover, New York (1966)

Figure/Ground: Segmentation and Aggregation

T.O. Binford

Artificial Intelligence Laboratory, Stanford University,
Stanford, CA 94305, USA

Introduction

Previously, we presented inference rules for interpreting images as surfaces [BINFORD 81; LOWE and BINFORD 81]. We extend that analysis to the figure/ground problem, which we formulate as the discovery of non-random structure in images, whether interpreted as surfaces in three spaces, or as patterns and texture in the plane. We consider the organization of images as images, i.e. the formation of canonical groupings which do not have direct interpretations as surfaces in three space. This analysis is based on the same perceptual principles as the previous work. A uniform mechanism based on lateral inhibition unifies segmentation and aggregation.

We have a practical objective which is to implement more general interpretation in ACRONYM. ACRONYM employs a model-driven mechanism: it predicts appearances of parts of objects as ribbons or ellipses; it matches predicted ribbons or ellipses with observed ribbons or ellipses; it tests whether clusters of image elements satisfy object constraints in three space. To emphasize, ACRONYM predicts and matches in images, and tests in three space. Model-driven interpretation is useful in utilizing special knowledge, especially in searching for members of a small class of objects, or in taking advantage of viewpoint-specific information. Model-driven interpretation is not a good paradigm for very general vision. Its weakness lies in predicting images and matching images. It is difficult to predict useful image observables from general knowledge. ACRONYM does a good job of viewpoint-insensitive prediction: it predicts images of parts of objects using quasi-invariant observables as symbolic constraint expressions involving variables.

ACRONYM has a strong weakness; it tests single ribbons against predicted ribbon constraints, then it tests all pairs of ribbons, aggregating clusters of ribbons which satisfy constraints. This is hopelessly wasteful and an esthetic disaster. Immediately we think of introducing proximity; how do we make a quantitative, general, and reliable definition of proximity? We would like to implement proximity and would like to include stronger conditions for pairing ribbons. We also need better means for forming ribbons. We chose to seek fundamental underpinnings for grouping operations on image elements.

Model-based prediction and image matching is only of part of the ACRONYM design. We mentioned above that a major advantage for some applications and a major limitation of generality of ACRONYM is image prediction and image matching. We emphasize that the major thrust in ACRONYM is not toward model-based prediction and image matching, but toward three space prediction and matching. Inference rules in ACRONYM provide a three space matching and interpretation mechanism that is more general and more powerful [LOWE and

BINFORD 81]. The new analysis is expected to apply to a much broader class of images.

The principle that the image is insensitive to the observer implies that grouping operations should be performed at a dense sample of positions, a dense sample of orientations if directional, and a dense sample of scales. Structures occur at multiple levels; approximations are multi-level. In the ocean, for example, the water surface can be approximated at one scale by ripples and eddies superimposed on waves; at a larger scale, it is approximately circular; at the scale of the universe it is a point.

We choose a principle of limited total computational complexity; we define total complexity as the product of time and parallelism. Total complexity is equal over equal numbers of cells of the retina. Diameter-limited computation would satisfy our complexity limit, but diameter-limited grouping is not effective here; what is appropriate is complexity-limited grouping, limited by the number of neighbours. If density of features is constant over equal numbers of retinal cells, then the condition is equivalent to complexity linear in number of features. Complexity linear in the number of features also implies grouping of each element with a constant number of other elements. We make the wild conjecture that time for computation depends on the complexity of the local feature, the number of levels it passes through, and not on the number of such local features. In this sense, computation time is independent of density and depends on image content.

This complexity principle makes a commitment that perceptual operators are local on a scale depending on the density of background features. In Figure 1, from [LOWE and BINFORD 82], a row of five dots which are clearly grouped when alone are not spontaneously distinguishable in a field of dots. They are highly unlikely as an accidental alignment. If the human perceptual grouping process were global, the row of dots would be obvious.

We discriminate elements which are distinguishable from random assortment. Not all demonstrable differences are distinguishable within complexity limits and information limits, as experiments in texture discrimination demonstrate. We test a dense set of aggregations for each size on a cover of neighbourhoods on an n-dimensional metric space. We segment each potential aggregation by computing a combined likelihood based on computing differences from the local average and comparing them with distributions of such differences for the particular area of the particular picture. Non-parametric rank tests of significance are used. Adaptation phenomena are unfortunately overlooked here. This is a parameter-free approach. Noise is determined by the process. A dense set of structure levels are obtained.

Figure is not merely absence of segmentation, but the discovery of positive evidence for non-random alignment. Positive evidence includes colinearity, density, parallelism, and others, grouping is based on fairly simple properties. The interpretation is that non-random alignments in an image imply alignment in space; common properties in an image imply common properties in space, a common mechanism.

Image Modelling vs Scene Modelling

We discuss two principles to generate perceptual features, image modelling and scene modelling. Scene modelling aims to provide approximation mechanisms for a class of scenes, and for the class of images of that class of scenes. Image modelling aims to provide approximation mechanisms for a class of images. We argue that scene modelling subsumes image modelling for our purposes, that is, image modelling has nothing to say about choosing the class of images to approximate. Scene modelling determines an effective class of images. From these, we consider two paradigms for segmentation, image approximation, and feature detection.

Image modelling means choosing a set of operators, called observables, on the image intensity surface. The optimum choice of operators depends on the structure of the intensity surface. Constraints on the optimum choice come from three sources: computational limits of the perceptual system; information limits of the signal; and image structure determined by the scene.

If we consider only the image itself, there is no basis for choosing the class of images we should deal with. Properties of the image itself are abstractions of sensor behaviour. Properties such as discrete pixels and digitized intensity values put weak constraints on the image structure and on the choice of image operations.

The perceptual process has limited computational resources. Computation limits are most dramatic for traditional computers which are perhaps six orders of magnitude too slow for high performance vision. But as we consider VLSI implementations or biological structures, size, power, weight, and interconnection are important for mobile beings. As a consequence, there are whole worlds out there we cannot see. The human visual system makes use of regularity and smoothness, e.g. extended objects. The world seems well-described by our assumptions, but it must; we see only those things consistent with our assumptions. There could be things which change so fast we cannot follow, which have such fine detail, or are so disconnected that we cannot make sense of what we see. But we encounter some new things when we throw stones at them or throw planets near them or throw photons or fundamental particles, or use a microscope, or perform other experiments.

The design of perceptual image operators is a compromise among generality, limited computation, and limited information in an image. To an extent, perceptual parameters can be tuned to optimize performance. However, there is very little quantitative, analytic understanding of adaptation. Strong constraints on design of perceptual image operators follow from projection of scene structure and the process of observation. For example, the pixel sampling function, the intensity function (log, linear, or other), digitization function, and noise are properties of the sensor.

If we had special scenes or special knowledge about objects in the scene, we would sacrifice generality by choosing particular image operators specially tailored to those scenes. For example, if the scenes were planar sine wave patterns with a single known image frequency in a known direction, and if illumination and sensor sensitivity were constant, we might use sine and cosine functions of a single image frequency. Similarly, if scenes were black characters on white paper, or backlit objects, we might use thresholding.

The intent of image approximation is to approximate image intensities or to form a perceptually equivalent image. Much of MARR's work was based on the paradigm of image approximation, e.g. relating to LOGAN's theorem on reconstruction. The intent of feature detection is to estimate parameters of image elements which have been predicted, and to isolate image elements from background, under assumptions about background. An aspect of feature detection is to detect image features which correspond in especially simple ways to object features, e.g. image lines which are consistent with projections of vertical lines. We are concerned with stability in the image and stability on the object. In image approximation we aim to give a coherent, compact approximation of the image in terms of basis functions which may have no perceptual relevance. We believe that efficient basis functions really have perceptual relevance and that effective image coding or image approximation really closely resembles perception. The reason for this belief is that effective image coding must strongly discriminate sparse areas of rapid change from large areas of slow change; that process is similar to segmentation in perception. Image coding has been relatively ineffectual; a factor of 10 image compression is current. A sketch may represent a compression factor of 20,000. Typical approximation methods involve taking samples and interpolating, i.e. using splines. It is a useful rule of thumb in approximation that taking uniform samples is not efficient, that given a fixed number of samples, good approximations are obtained by sampling most densely where curvature is greatest. That sounds like segmentation and edge detection. Our prescription is to place knots at discontinuities of the intensity surface.

Feature detection aims for a fragmented, incomplete approximation of parts of an image in terms of selected functions which have perceptual relevance. An example of feature detection is determining a line drawing of image intensity discontinuities. A line drawing could also be used for image approximation. There is an implicit intention of relating image features to scene elements. For example, lines in a line drawing are interpreted as object edges, pigment boundaries, or illumination boundaries. Scene modelling and projection determine a one-one relation between scene elements and image elements, but a many-one inverse. Over a range of viewing conditions, some image elements may have few inverse images.

The set of operators on a sequence of image intensity surfaces in time should be chosen to satisfy the following constraints:

completeness;
limited complexity;
independence or uniqueness;
adjoined operators based on scene modelling;
information limits;
statistically efficient;
unbiased;
scale invariance; translational invariance; rotational invariance;
maximally stable under observation.

The arguments that follow can be made with varying precision, but they are not formalized, only phrased to encourage formalization. The chief problems lie in calculating complexity of sets of image operators, and in choosing optimality conditions.

We assume that perceptual operators form a complete subset of independent operators with limited complexity. Operators are added, based on scene modelling. Complete also means capable of modelling the image intensity surface arbitrarily well, i.e. to the limits of information. Independence is usually related to an orthonormal set, but statistical independence is also important. Those conditions do not specify a unique set of observables. An optimality condition does specify a unique set. We assume that observables satisfy optimality conditions (efficient estimation) which reflect limited information and which are based on scene models. As an example, where other conditions allow, edge operators are preferred which maximize signal to noise or which minimize transverse errors of edge locations. Invariance assumptions imply that operations must be evaluated at a dense set of scales, a dense set of positions, and a dense set of angles. In interpretation, any perceptual operator should depend only on observables, such as the gravity vector, direction and distance between a pair of points, time interval between events, etc.

The image stability condition reflects an approximate invariance condition. One form is that image observables should be as insensitive as possible to variation in illumination and other scene variations. We identify image observables with singularities in the intensity surface. We identify scene elements with constructional elements related to fabrication, and with geometric discontinuities. Estimation of image position of discontinuities of the image intensity surface like steps and peaks can be shown to have least variation in a neighbourhood under changes in illumination, sensor, and noise. Another form of invariance condition is that image observables should relate as closely as possible to scene observables. Estimation of object locations inferred from discontinuities of the image intensity surface also have the least variation in a neighbourhood under the same variations.

Information is limited. In detail, pixels are discrete, there are a finite number of them; an optical system has limited resolution because the point-spread function has non-zero width; the sensor receives a limited number of photons, thus it has intrinsic noise; the sensor may also have amplifier noise. Information limits are scene constraints ultimately. We believe that globally, computation limitations are more significant than information limits. We believe that locally, at the level of edge elements, information limits of the signal dominate. Information limits are related to the complexity of interpretations that we can make. For example, locally over a disk of a few pixels diameter, we can approximate the intensity surface with simple functions, e.g. a step along a straight line, or a corner. If we attempt to approximate the intensity surface with complex functions, there will be undetermined parameters. More globally, the more hypotheses we test, e.g. the more parameters, the greater the chance of accidental agreement with data.

Scene Modelling

Scene modelling is concerned with sets of observables which approximate objects and relations between objects. Observables should satisfy essentially the same constraints as image modelling. Observables are operators defined on representations of space/time or other domains such as function. We suspect that the only basic domain is space/time, i.e. representations of other domains ultimately reduce to space/time. To those conditions above, we add the following:

physical modelling;
physical partitioning;
maximal stability under variations within object class.

Modelling of depth data from projected light or direct ranging devices belongs more in this section than the last, since the essential determiner is the geometric class of the data.

We include space/time behaviour of physical systems, kinematic and dynamic behaviour, i.e. all physical phenomena. This does not necessarily mean that all of human knowledge, school physics, chemistry, biology, mathematics, psychology are necessary for every problem. But all problems require some knowledge, even if implicit. All of us have an informal model of the physical world and its function. In order to simplify problems, they are normally partitioned into systems. This partitioning has a semantic basis, i.e. a meaning which we give it, but which is not ordained by first principles. The partitioning does have a strong physical meaning. It reflects the boundaries we can draw around systems based on approximations of strengths of interactions. To put it another way, a system is defined depending on what observables we want to calculate, and relative contributions to those observables. For example, if we describe the motion of an object relative to the earth on a human scale, we need include only the object and the earth. If we describe the motion of an object in the solar system, we need to include the motion of the object relative to the earth and the motion of the earth with respect to the solar system, which brings in the masses of sun and planets. In choosing such partitionings or other simple descriptions, we rely heavily on body-centred coordinate systems based on physical vectors, e.g. the centre of mass system, or the gravity vector. As part of physical modelling, it is important to model the observation process, including illumination, sensors, and projective geometry. Modelling the observation process has received some attention, particularly from Horn. Modelling is especially important for exotic sensors such as SAR, acoustic imaging, and coherent infrared. It is the goal for ACRONYM to include mechanisms for describing the physical world mechanisms for choosing physical partitions of systems, mechanisms for choosing observables composed of physical vectors (as much as possible in basis-free form), and mechanisms for modelling observation.

The concept of maximal stability of object features deals with the usefulness of invariance assumptions, and the choice of features which are not sensitive to variations within an object class. For example, classical mathematical treatment of surfaces deals with surface discontinuities. In an estimation sense, discontinuities are the parts of the surface which can be localized most reliably.

Image modelling and object modelling may converge on identical perceptual elements, but they are not identical. For example, both approaches lead to considering 2-D step functions, either to describe image intensities or as images of edges of surfaces or surface markings. Object modelling gives special importance to the following elements: the shadow at the crack between objects which abut, and the dark line at the edge which is the image of the shadow; a vertical edge and its image; vegetation and green colour in an image (here we are making a naive assumption about colour constancy). They have little special meaning in image modelling.

Object modelling describes what is at once observable and simply interpreted. Observables are simply interpreted if they are invariant, or usefully quasi-invariant (i.e. if a sufficiently restrictive constraint can be imposed), and in addition, if image features have few possible interpretations. We can approximate images better using features obtained from object modelling. Object modelling contributes more to the interpretation of image features. Many assumptions in image modelling represent commitments to object, illumination, and spatial models. We tend to assume predominantly matte surfaces. With high resolution radar or acoustic imaging, the world is mostly very different, specularly reflecting, like a hall of mirrors.

System Model

We consider a hierarchy of operations based on a class of images which assume a scene model with piecewise smooth illumination, piecewise smooth surfaces, and piecewise constant pigmentation. As discussed previously, we identify image observables with discontinuities in the image intensity surface. We intend to relate image observables with discontinuities in illumination, with surface discontinuities, and with pigmentation discontinuities. Because of shading from illumination, surface curvature, sensor vignetting, we must consider discontinuities superimposed on a smooth intensity surface, not a flat surface. Thus, image operations compute the signal above the local average, which we call contrast functions. This is identical to lateral inhibition in biological systems.

Isotropic Operators

In this model all operators compute the contrast function, a difference of gaussian weighted averages over centre and surround. Operators differ only in size, ranging in powers of two from the smallest operators which have a centre which is a single pixel. In this model and in many devices, individual pixels have Gaussian sampling functions. In our implementation we used overlapping boxes instead of Gaussians. Scale invariance dictates a constant complexity condition, i.e. the sampling density is a constant fraction of operator size. The complexity of isotropic operators of sizes differing by a factor of two is obvious:

$$\frac{4}{3} D^2 = D^2(1 + 1/4 + 1/6 + 1/64 + \ldots) ,$$

where D is the diameter of the retina. This is an expansion in local functions in the spatial domain. We lack knowledge of their formal properties, independence and completeness. The contrast functions are not independent in absence of noise, but they may be statistically independent. Typical expansions in orthogonal functions have a global basis, e.g. the Fourier expansion. Gabor functions provide a local basis, similar to the quantum mechanical wave functions. We are currently also examining a local basis of step functions.

Features appear in several adjacent cells and at several scales. We encounter both step function and delta function signals. Estimating locations of delta function features involves estimating the location of the maximum of a peak of the contrast function, which is equivalent to estimating the location of a zero of the derivative of the contrast function [BINFORD 81]. This is slightly different from localizing an edge. Interpolation to zero gives estimates of location with variance that is easily estimated [BINFORD 81]. The linear approximation to the derivative has heuristic value, i.e. it is clear and efficient to implement. It is equivalent to interpolating the maximum of a quadratic approximation. Estimating location of step function features involves estimating the location of maximum directional derivative of the contrast function. This is equivalent to estimating the location of a zero of its second derivative.

Features appear at adjacent scales. The problem of combining results of multiple resolutions has been discussed by [MARR 79]. An ad hoc solution was presented. This model follows a fine-to-coarse regime, that is it segments out fine structure from the signal by a non-linear weighting, then processes the residual signal with coarse operators [BINFORD 81]. This is similar to cutting out small spots and estimating the properties of the background. We believe that this is much more effective than averaging the effects of small structure into large operators. The procedure is to segment at each scale what is detectable at that scale.

Only if we choose our input image carefully can we have large, localized patterns without having extended, directional features. We talk about isotropic operators and not isotropic features, i.e. non-directional. Large isotropic features would have to have low enough contrast that no part of the feature is separately detectable. For example, consider a circularly symmetric feature which is larger than pixel size. If contrast is moderate, edges of the feature can be detected at high resolution as directional, elongated features. In summary, there will be isotropic features, i.e. dots, which may be part of elongated features.

Consider designing an optimum detection operator. On the one hand, we have spatially-random noise. To minimize noise in its output, the optimal operator should average over large support. On the other hand, there is an unknown distribution of spatial patterns which are not random spatially. As we make operators larger, they have increased probability of including at least one pattern. Thus, there is a trade off between averaging out noise and introducing systematic errors by including very large 'noise' contributions from patterns. Depending on the distributions of random noise and patterns, there will be an optimum size of operator, which will typically be small. We want to find small patterns anyway. The random error is inversely proportional to the square root of the area, that is 1/radius, while the systematic error is linear with the area, or the square of the radius. The optimal radius will be roughly where the two expectation values are equal. There is a large penalty for systematic errors.

If we use that optimal operator, we will detect patterns which we can remove from the signal. If then we recurse to define optimal operators on the residual signal, a substantial part of the systematic 'noise' will have been removed, and the optimal operator will be roughly twice as large. Under reasonable assumptions about the distribution of sizes of operators, we argue that the recursive optimization of operators leads to results similar to those of the model.

This mechanism handles intensity data and feature data uniformly. Given a gray scale image which is really a pattern of dots, it first isolates the dots on a small scale, then groups them on a larger scale. Each operation produces an output and a residual signal. The residual signal goes to the next level operator. The output of the operator is continuous but quite nonlinear and approximates a thresholded quantity. The problem approaches one of clustering in dot patterns. The output goes into another level of operators which compute a local contrast function which compare weighted averages over a centre with weighted averages over a surround.

What quantities do they deal with? Computational complexity of these grouping operations is very sensitive to the choice of this function. The choice is a very simple non-combinatorial function which would exhibit severe limitations in discriminating structure within a region in order to discriminate its boundaries with another region. Designing test cases of texture discrimination is equivalent to an exercise in analysing natural strategies of camouflage. Uniformly, local operators determine a combined likelihood of systematic differences of local variables. One variable is size. Operators calculate the local average size over the centre and the local average size over the surround, then compare the difference with the local distributions of the variables (e.g. standard deviation) to determine a likelihood of a random assortment of size. Likelihood functions for individual variables like size are combined into a joint likelihood of a random assortment for that local area. We assume that local variables include size and contrast of individual isotropic features and the radial interval between neighbours. There will be an expanded list for directional features. Isotropic operators are simply a special case of directional operators with zero elongation. Here we use local to mean a few diameters of the centre of the size under consideration, including the size of the surround and the size of possible slightly larger regions over which distributions are computed.

Consider finding groups with sizes of elements differing by a factor of two, with spacings between elements differing by a factor of two, from fine to coarse, up to a fixed number of neighbours, a constant complexity condition. Consider size. We want to find peaks in size relative to background for all sizes. We assume that it is enough to find peaks with width a factor of two. The same is true of contrast. If there are n variables like size and contrast, then the complexity of grouping is approximately:

$$2^n \log(D)\,(4/3)\,D^2 \quad ,$$

where 2^n defines overlapping neighbourhoods, $\log(D)$ covers all spacings, and $(4/3)\,D^2$ is the complexity of isotropic features of intensity. For isotropic operators, we have $n = 3$. Within the isotropic restriction, pairs of dots should be indistinguishable from other features of similar size and contrast. This is not true for directional, extended features and not true in biological systems.

One process is estimating the properties of non-stationary distributions in order to answer the question, how likely is a fluctuation as great as the observed one or greater. A significant question is how we can make precise measurements of sparse statistical data in a way that encompasses both pattern texture and statistical texture. That is, we normally think of density as a fundamental observable. If we treat density as a Poisson distribution, if there are few dots, we have large errors in estimating dot density. That is, dot density is not a very good estimator unless there are lots of dots. We then would like to expand bin sizes for grouping to get large numbers of features, but we lose the ability to discriminate. We encountered this same problem in Fourier analysis of texture [BAJCSY 74]. Instead we estimate the local parameters of the distribution of intervals between elements. If features fall in a regular pattern, the distribution of intervals will be sharp. The interval distribution is a reasonable estimator also for features which are randomly placed. Pattern and statistical textures are equivalent in our scheme. They differ only in variance of certain variables. An equivalent estimation can be made with density, if the density estimate is made to fit the pattern, rather, than to fit the neighbourhood. An equivalent process is pitch synchronous estimation of frequency in speech signal analysis.

One of the main purposes of multiple resolution is adaptation to low light levels. The system uses an unspecified weighting function which effectively maximizes the spatial accuracy. This has the effect of using larger operators for low contrast and low light levels. The problem is the same as that of merging multiple resolutions, a problem still unsolved. The shift from fine to coarse resolution is automatic, driven by a shift in the level at which features are systematically detectable, a function of signal to noise.

Directional Features

The discussion of isotropic features was only a warm up for discussing directional features. We include extended neighbourhoods at a dense set of positions, elongations, and orientations. Isotropic operators are a subset with elongation zero. Along with extended features are their terminations which are point features, like intersections.

Extended features arise from intensity images and from linking features. We have a partial order on elongation from isotropic to maximum elongation, just as there is an order on diameter for isotropic operators. The partial order enforces a fine-to-coarse regime, at each stage removing fine extended features from consideration by more extended operators. The fine-coarse regime solves the problem of distinguishing a continuous linear feature from a broken structure. Finding terminations of features is an important operation in making that distinction.

Consider pairs of points. Colinearity is not relevant. The distance between points is relevant. The likelihood of two points being a random assortment based solely on their distance, depends on the distribution of distances. This is scale invariant; if two points are sufficiently close together that they lie far out on the tails of the distribution of interneighbour distances, it is unlikely that they are associated by random assortment. This generates what is sometimes called an 'affinity', but based on first principles. Grouping by distance is weak for pairs of points but not for more points. The comparison can be made using non-parametric statistics. Given a uniform spatial distribution with density d, what is the probability of random occurrence of a pair at a distance d or less? It is the integral from 0 to d of the product of the Poisson probability of no point within radius d and at least one point in the infinitesimal annulus with radii d and $d + dd$. The probability is roughly quadratic near zero.

At two points define a straight line. We must consider three or more points to find interesting alignments. Again consider randomly distributed points. Consider two points in the space and consider the distribution of a third point, in coordinates oriented along and normal to the line between the two points u, v respectively. The distribution is uniform in u and v. The likelihood of the nearest point falling at a normal distance $v1$ or closer is the integral from zero to $v1$ of the product of the Poisson probability for no points from zero to $v1$ and the Poisson probability for one or more points in the interval $v1, v1 + dv$. This analysis should be made symmetric and extended for more points.

We now calculate the complexity of computing directional operators for a dense sampling of scale, elongation, position, and orientation. First note that complexity is independent of elongation, because the number of angles increases proportional to l, but the number of samples along each direction decreases inversely proportional to l. Their product is constant. Also, there are few elongations for large scale, i.e. $\log(D) = k + l$; where D is the diameter of the retina, k is the scale and l is the elongation. Thus, the complexity for the first level, intensity, contains terms for scale = 1,2,... Complexity = $\log D \cdot D^2 + (\log D - 1) \cdot D^2/4 + \ldots$ which is less than $\log(D) \cdot (4/3) r^2 \pi$. The complexity of grouping directional features is given by

$$2^n (\log D)^2 \frac{4}{3} D^2 .$$

The ratio of complexity of directional grouping to isotropic grouping is 36. Here $n = 4$, the number of variables, which include contrast, length, width, and orientation. The list is not necessarily complete.

Figure 2 shows an example for which there are multiple peaks in the significance measure. They represent groupings at multiple levels.

Figure 3 demonstrates two groupings with the same standard deviation for a line. Calculation of isotropic significance is separated from calculation of significance of linearity.

Linear grouping has been applied to a dot pattern. Results are shown in figure 4 for clusters at one elongation at the 1% significance level.

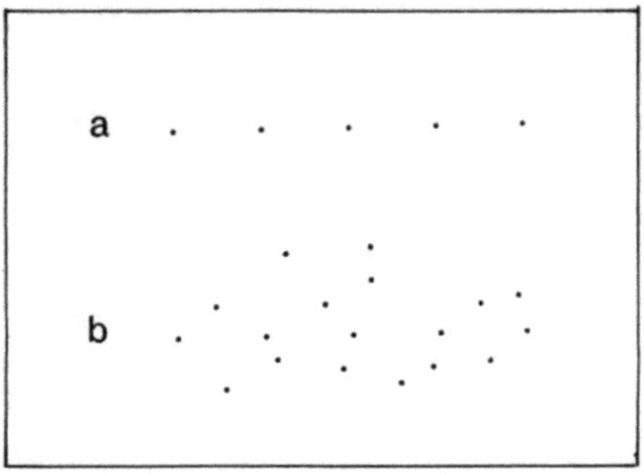

Fig. 1

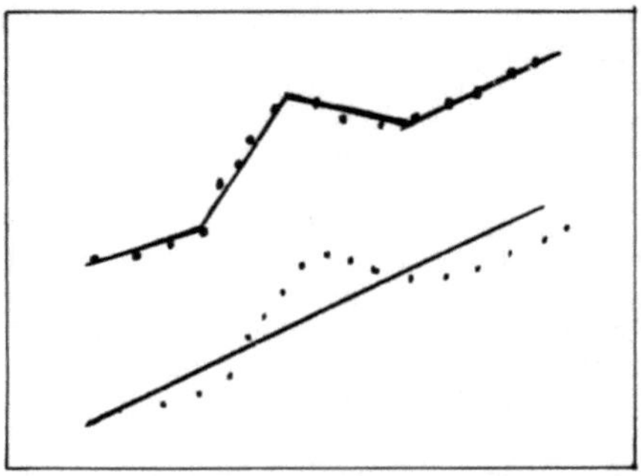

Fig. 2

Fig.1. The five dot pattern of 1a is not obvious in a background, even though it is highly non-random

Fig.2. Multiple peaks in likelihood of non-random assortment at different scales

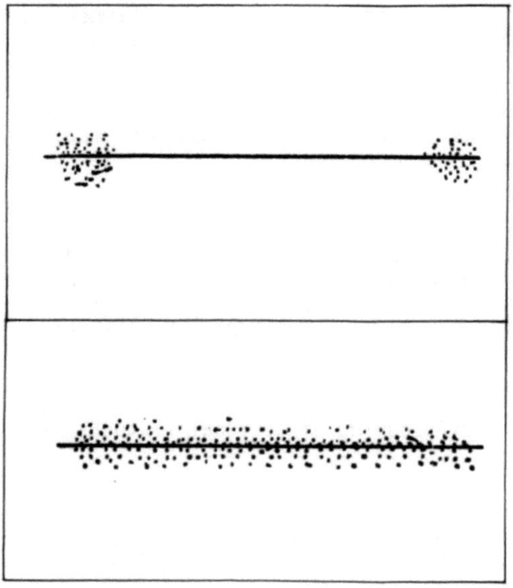

Fig. 3

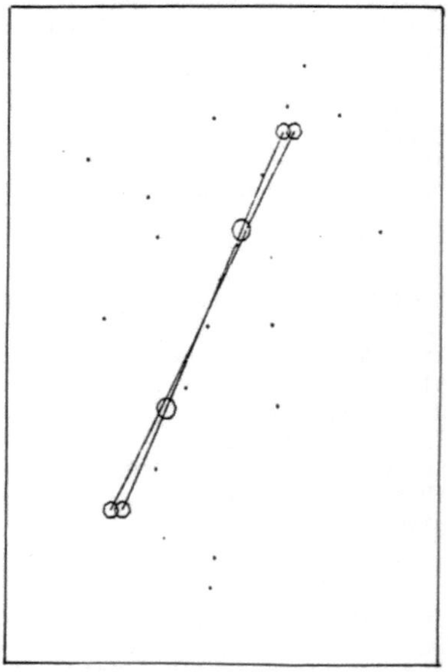

Fig. 4

Fig.3. Two sets of points which have the same standard deviations from least squares best fit lines but which have different significance as a linear feature

Fig.4. Dot clusters non-random at the 1% level

References

[Binford 81] T.O. Binford, *Inferring Surfaces from Images*, Artificial Intelligence Journal, July 1981

[Brooks 81] R. Brooks, *Symbolic Reasoning among 3-D Models and 2-d Images*, Artificial Intelligence Journal, August 1981

[Lowe and Binford 81] D. Lowe and T.O. Binford, *The Interpretation of Three-Dimensional Structure from Image Curves*, Proc.Int. Joint Conf. on AI, August 1981

[Lowe 82] David Lowe and T.O. Binford, *Segmentation and Aggregation: An Approach to Figure-Ground Phenomena*, Proc. IU Workshop, 1982

[Lowry 82a] Michael Lowry, *Reasoning between Structure and Function*, Proc. IU Workshop, 1982

[Lowry 82b] Michael Lowry and Allan Miller, *Analysis of Low-Level Computer Vision Algorithms for Implementation on a VLSI Processor Array*, Proc. IU Workshop, 1982

[Malik 82] J.M. Malik and T.O. Binford, *Representation of Time and Sequences of Events*, Proc. IU Workshop, 1982

The Role of Attention in Object Perception

A. Treisman

Department of Psychology, University of British Columbia,
Vancouver, B.C., Canada

Perception of complex everyday scenes appears introspectively to be immediate and effortless. Our first awareness is of recognizable objects organized coherently in a meaningful framework. This compelling impression is consistent with the Gestalt psychologists' view that perception directly encodes wholes, meanings, relations, which can, if necessary be analyzed into more elementary parts and properties. Other kinds of evidence should, however, give us pause. Why is it proving so difficult to programme scene analysis? What do we make of physiological and anatomical discoveries suggesting the existence of specialized feature detectors and even of spatially segregated areas of the brain responding selectively to different dimensions such as colour, orientation, stereoscopic depth, and movement [1]? The immediacy and directness of conscious experience are no guarantee that it reflects an early stage of information-processing in the nervous system.

Even psychological research, of a less phenomenological kind, suggests that some decomposition takes place into separable parts and properties. Selective adaptation can raise thresholds or produce after-effects which are restricted to particular aspects of a stimulus. GARNER [2] proposed several converging tests for inferring separable dimensions: dimensions are separable if they can be processed in parallel and without interference; if they can be selectively attended to; and if they contribute independently to judgments of similarity. Examples of dimensions which meet these criteria are the size of a circle and the orientation of a diameter in the circle. Even the wholistic quality of subjective impressions can be questioned: if we seriously test the nature of these impressions by flashing randomly jumbled objects or scenes, subjects perceive remarkably little—perhaps one object, a few colours or lines.

If we accept the possibility that early stages in vision decompose a scene along a number of separate dimensions, the question arises how these are recombined, how we avoid seeing a blue sun shining on yellow grass under a green sky. MINSKY [3] as early as 1961 posed a similar problem for artificial intelligence: how do we articulate a scene, grouping together the features that belong to the same object and separating those that belong to different objects? An obvious answer is through their common spatial origin; MARR's programme [4], for example, carries out local computations separately and in parallel for each small area. It then forms local groups and assigns them place-tokens in a "primal sketch". The results of the different local computations (registering orientation, intensity, etc.) are aggregated on a strictly local basis by reference to their coordinates in a single spatial map. BARROW and TENENBAUM [5] compute a number of separate "intrinsic images" for real world properties such as reflectance, orientation, distance and incident illumination. The images are "in registration" with the original input image and therefore with each other.

An alternative approach is to deal with each localized subset of features singly and serially. This is the possibility originally envisaged by Minsky, who suggested the "repeated application of a fixed set of pattern-recognition techniques." In human vision, the analogy might be to focus attention serially on different spatial points or areas. All the different features present in the current spotlight of attention could then be correctly integrated and intrusions of parts or properties from other objects could be prevented. This is our current hypothesis for which we have found considerable behavioural evidence [6,7,8,9].

The theory distinguishes two functional levels of perceptual processing. At the first, all the separate features are coded independently and in parallel to form different feature maps, using populations of feature detectors for such properties as colour, orientation, direction of movement and stereoscopic disparity. At this first preattentive stage, the different feature maps are organized into groups, textures, homogeneous areas — candidates for possible objects and events to be identified at the next stage using focused attention. This stage is similar to the preattentive level first proposed by NEISSER [10] and to MARR's primal sketch [4] except that it results in many separate feature maps rather than one global representation.

The second stage is a feature integration stage, where focused attention serially scans each location containing a feature group. Any features that are registered within the same central 'fixation' can be integrated into a single percept, their nature and relations determining the identity of the object. Thus the spatial spotlight of attention provides the 'glue' which puts objects together. Unlike the first preattentive stage which operates in parallel over the entire array, focused attention is a serial processor. Only one object can be produced at a time. The internal scanning process is at least partially independent of eye movements and visual fixations and it is very much faster. We may move our eyes every 250 msec, but the internal scanning rate that is estimated from the rate of visual search in complex displays is often 10 to 20 times faster.

What empirical predictions follow from this idea? (1) <u>Search</u>. If an object has a unique feature, we should be able to detect or identify it independently of the number of other objects which are present. So search for a feature target should be parallel across a display. If, on the other hand, an object has no unique feature and is defined only by a conjunction of features, identifying it should require focal attention to each object in turn, and should therefore force serial processing. Subjects showed the predicted pattern [7]. They searched for targets embedded in displays consisting of randomly mixed green X's and brown T's. When searching for targets defined by either of two disjunctive features (the colour blue or the letter S), they detected the target as quickly in a display of 30 distractors as in 1 or 5. When searching for a conjunction of two features (green and T) the pattern of results suggested a serial, self-terminating scan of each item in turn. If subjects check each item in turn until they find or fail to find the target, the search times should increase linearly with display size. Moreover, all the items would be checked on negative trials and on average only half the items would be checked before the target is found on positive trials. The slopes of search times against display size were in fact linear, and the time for a negative response increased twice as fast as for a positive with increasing display size (Fig. 1).

This serial scan for conjunction targets is quite a robust result. It is found both with easy and with difficult discriminations (see Fig. 2); it persists over a considerable number of practice sessions; it does not depend

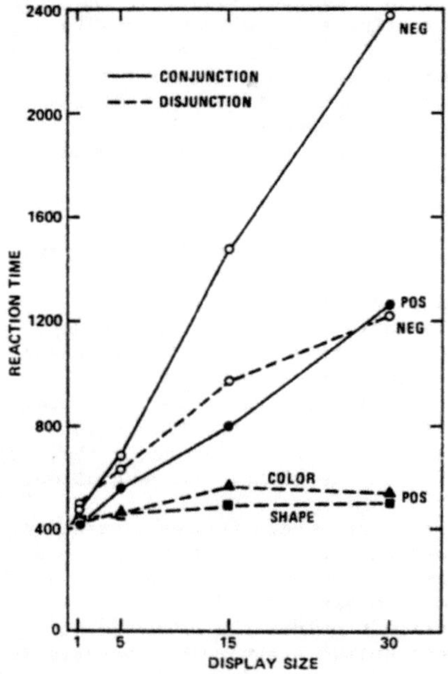

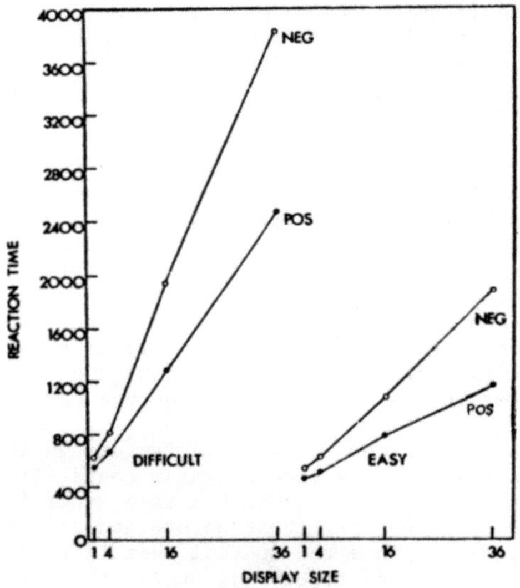

Fig. 1.
Search latencies for targets defined by disjunctive features or a conjunction of features

Fig. 2. Search latencies for conjunction targets defined by features that are difficult or easy to discriminate

critically on the visual angle of the display or on the haphazard or regular arrangement of the distractors, It is found not only with conjunctions of color and shape, but also with the local parts of more complex shapes. For example search for the letter 'R' is parallel in a background of P's and B's in which the diagonal line of the target R is a unique feature, but serial in a background of P's and Q's which between them contain all the components of the 'R' (Fig. 3).

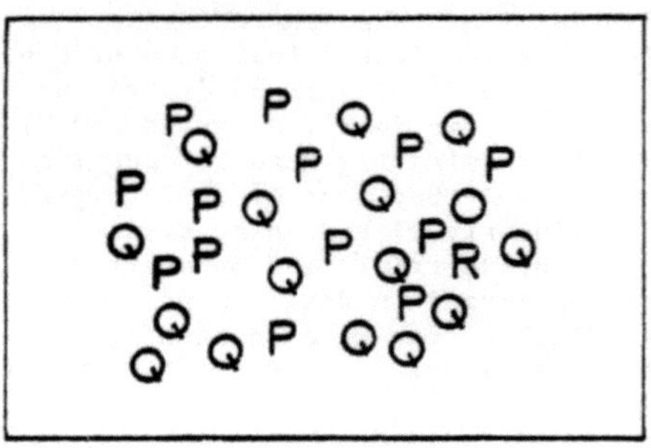

Fig. 3. Example of display containing target R in background of P's and Q's.

2) Texture Segregation. The next prediction concerns texture segregation and figure-ground grouping. It seems likely that these depend on pre-attentive, parallel processing. They should therefore be determined only

by separate features and not by conjunctions of features. So we should group two sets of items easily on the basis of colour (e.g. red curved or straight shapes vs. blue curved or straight shapes) or on the basis of a simple feature of shape (e.g. curved blue or red shapes vs. straight blue or red shapes) but not on the basis of conjunctions of these properties (red curved and blue straight shapes vs red straight and blue curved shapes). Experiments confirmed these predictions [9].

The theory makes quite a strong and surprising claim about early perceptual grouping. It implies that preattentive organization exists only within dimensions — within a colour map, within a shape map, within a map of movements or orientations — and that these maps are related to each other only where and when attention is focused. This suggests the possibility that we might effectively camouflage an object at the preattentive level by placing it at a boundary between two groups, each of which shares one of its features. We can choose an object which within either group alone would be quite salient, and see if adding the second group makes it harder to see (Fig. 4). In order to detect the presence of the red X, attention has to be narrowed down to exclude the adjacent red 'O's and blue X's and focused on the item itself. Yet in either group alone the red X would be easy to detect. If, on the other hand, the target has a unique feature, (for example the colour green or vertical lines) detection should be independent of grouping.

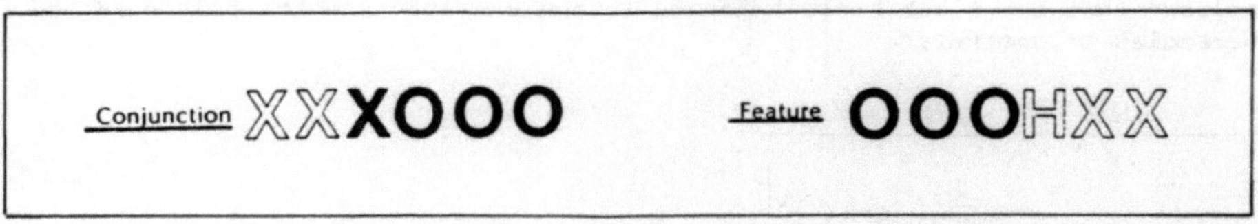

Fig. 4. Displays containing conjunction or feature target between two groups of distractors

An experiment confirmed the predicted camouflage of the conjunction target: subjects took 135 msec longer to find the conjunction target than the green or vertical feature targets. In fact they missed it altogether on 9% of trials, even though the display remained on until they responded. It seems that two competing ways of grouping this display exist — one within the preattentive colour map and one within the preattentive shape map. The conjunction target exists in neither of these maps, while the feature target is always unique in one of the two. Luckily, in normal life the preattentive boundaries of our multiple feature worlds are likely to agree. The features of real physical objects have highly correlated spatial boundaries. The edges of a dog co-exist and move together, whether we define them by their colour - brown, or by their texture - furry.

(3) Localization and Identification. Another function of focused attention in this view is to localize objects in perceived space. Different features that are separately perceived at the preattentive level may be free-floating relative to one another and to a common perceptual space. Objects, on the other hand, would be localized as a condition for being correctly perceived. It should be possible, then, to detect the presence of a simple feature (e.g. red or vertical) without locating it correctly, while for conjunctions, this should be impossible. We tested this idea, simply by

looking at the dependencies between the two types of judgement, identification and localization [7]. If we are unable to identify conjunctions without locating them, the dependency should be complete: we should never identify a conjunction without also knowing where it is. Subjects were asked to decide whether a target in a background of blue O's and red X's was a red O or a blue N. They were also to indicate its location in the matrix of 2 rows of 6 letters each. On trials on which the location was misperceived by at least two spaces, the identity judgment was at chance. However, when each target was defined by a single feature (orange or S), subjects identified about 70% correctly on trials on which they mislocated the target by two or more spaces. This is evidence that detection of a feature can occur not only without focused attention, but also without information about its spatial location. At some level we may have free-floating presence or absence information for features but not for conjunctions.

(4) Illusory Conjunctions. The next prediction is perhaps the most surprising. If features are free-floating with respect to one another when attention is prevented or overloaded, it follows that illusory conjunctions should be formed when more than one object is present, by wrongly recombining the features of different objects. For example a yellow chair and a green vase might be wrongly conjoined to give a yellow vase. Occasional anecdotes suggest that such hallucinatory couplings may actually occur: a friend walking in a busy street 'saw' a colleague and was about to address him when he realized that the black beard belonged to one passerby and the bald head and spectacles to another.

Fig. 5. Display used to investigate illusory conjunctions of colour and shape

To test the prediction in the laboratory, we used displays like those in Fig. 5 [8]. The digits were black and each letter was in a different colour. In order to ensure that attention was overloaded and would not be focused on any single item, the displays were flashed briefly and the primary task was to report the two digits correctly. Subjects were then asked to report all the coloured letters that they were reasonably confident they had seen. They made a large number of conjunction errors, in fact almost as many as correct reports. They averaged one illusory conjuntion in every two or three trials (0.39 per trial compared to 0.52 correct). Comparing these errors to those in which one feature was correct and the other an intrusion which had not appeared on the card, we found that conjunction errors were three times as likely as intrusions. It is worth mentioning an anecdotal observation. Many of the subjects stopped after a few trials and spontaneously made comments like 'Oh, you are tricking me. The numbers were colored that time.' No guessing or response bias or attempt to please the experimenter can explain that unrequested observation, since we had told subjects that all the digits were black and they were asked to report only their identity. At least some of these illusions appear to be genuine perceptual experiences rather than wrong guesses or memory failures.

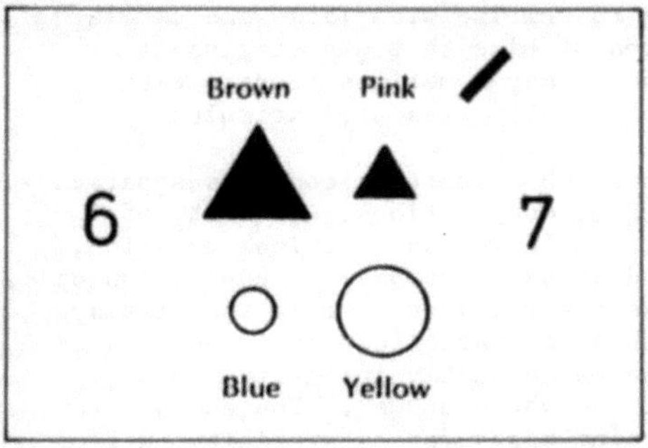

Fig. 6. Display containing heterogeneous items in four colours

The next question we asked was whether there are any constraints on the illusory recouplings. For example, can one take the red from a small, outline circle and use it to fill in the area of a larger triangle that was originally blue? Or are there limits to the sizes, shapes, areas and distances between which the exchanges can occur? We tried displays in which items varied in colour, size, shape and whether the colour was filled in or outlined (see Fig. 6). Some displays varied in only two features, some in three and some in all four. We also ran two different attention conditions. It is important to the theory that illusory conjunctions result from attention failures, and not from some other difficulty. We compared a divided attention with a focused attention condition. In the divided attention condition, subjects attended to and reported the digits as their primary task, and were cued only after the display was presented which coloured shape to report. In the focused attention condition, the digits could be ignored, and subjects were given a spatial cue 150 msec in advance of the display, telling them which coloured shape to attend to and report. We matched the overall accuracy in the two conditions by reducing the exposure duration in the focused attention condition. To summarize the results: (1) first in the divided attention condition, all four features were liable to switch. Subjects reported illusory exchanges of size and solidity as well as of shape and colour. All four dimensions seem to be separable by this criterion. (2) Illusory conjunctions were as frequent with heterogeneous displays as with more homogeneous ones. Colours were as likely to switch between items which differed on the other three features as between items which were otherwise identical. (3) Illusory conjunctions do seem to be linked to attention load rather than to task difficulty in general. When we changed from post-cued, divided attention to pre-cued, focused attention we found a big shift in the type of errors. With attention cued in advance and no primary load, subjects made very few conjunction errors; either they omitted the item altogether, or they were as likely to report a feature that was not presented as a feature from the wrong location in the display.

These last two experiments have confirmed the initial premise of the theory. They suggest a strong interpretation of feature "separability", and couple the physical stimulus with conscious experience much more loosely than we might intuitively suppose. The presence of each feature (e.g. 'red', 'outline', and 'curved') appears to be encoded quite independently, so that the colour red is registered without its spatial extent or distribution, and a circle is identified separately from the colour which fills it. These feature labels are then combined, either with or without focused attention, to generate the conscious impression which we experience. If attention is diverted, the synthesis of recombined features may generate a representation which differs significantly from the original display. For example, taking 'blue' from a small blue outline triangle to combine with large and circle must increase, alter and transpose the area of blue that was originally presented. Yet these illusory hybrid objects may sometimes be seen with as much confidence as those that match the physically presented stimulus.

So far, I have begged the question of which properties count as separable features and which have to be put together as conjunctions. In fact, of course, this is far from obvious, and is, I believe, an empirical question. My claim has been that attention is needed whenever we have no population of specialized detectors which directly sense the relevant item in the display. Does this create a vicious circle in which a separable feature is defined as one which requires attention before it can be correctly integrated with others? One answer is yes, but it is not the whole answer. The escape from the circle is through a variety of converging tests for separability, all of which, we hope, will pinpoint the same candidates for separable feature hood. The strategy is to choose two features which are most likely to be separable — for example we might use colour and line orientation — to establish across a variety of tasks two different behavioural syndromes typical of features on the one hand and conjunctions on the other. So, just as a physician uses spots, a fever and a sore throat to diagnose measles, we can use the feature syndrome— parallel search, location errors, illusory conjunctions, texture segregation — as new diagnostic criteria for featurehood. We predict that there should be correlations between these different tasks, both across stimuli and perhaps also across subjects. If this holds up, we may then be able to apply the same tests to help us decide, for example, whether more dubious features like closure or intersection or symmetry qualify as elementary features, whether faces or words are recognized as unitary wholes or built out of features. We can also test perceptual learning to see whether new detectors can be set up, registering conjunctions of previously separable features as unitary wholes.

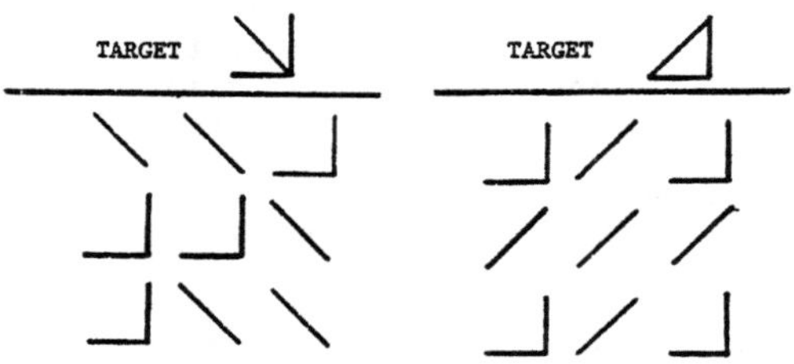

Fig. 7. Example of displays to test illusory arrows and triangles

With colour and shape, there may be built in constraints on unitization. We did not succeed in training parallel search for a conjunction over thirteen sessions of practice [7]. For component parts of shapes (e.g. curves, lines, angles, etc.) it would make biological sense to allow some flexibility. There would be advantages in being able to develop automatic detection of important faces or objects. Another difference between conjunctions of colour and shape and conjunctions of shape components is that conjoined shapes may produce emergent features [11]. For example, joining L with \ may produce a triangle ▷ with a new property, closure. Are these emergent properties also picked up by separate populations of feature detectors? If so, they could affect the probability of illusory conjunctions. People might be less willing to see a ▷, given an incorrect conjunction of L and \ , simply because the closure feature was missing in the physical stimulus. The more salient an emergent feature is, the more its absence should inhibit us from seeing the illusory conjunction. We tested this prediction by comparing two conjunctions of the same features, the diagonal line and the right angle — one which generates closure and one which does not (◺ versus ⦣ , see Fig. 7). We found significantly more illusory conjunctions with the arrow than with the triangle (an excess of 11.2% illusory arrows in a conjunction compared to a feature background, compared to only 3.6% illusory triangles). Closure may be more salient as an emergent feature than intersection.

But this generates another prediction: if there is an emergent feature of closure which is picked up by a separate population of detectors, it should also behave like a feature in other tests: it should allow parallel search and mediate texture segregation. So search should more often be serial with arrows than with triangles. This prediction was also confirmed. Search latencies increased more rapidly with increasing display size with arrows as targets than with triangles, and the mean ratio of positive to negative slopes was close to 0.5 for arrows (0.47), but not for triangles (0.23). Texture segregation was also slower with arrows than with triangles, averaging 926 msec versus 746 msec. So, across stimuli, the tasks seem to covary as predicted.

Another prediction can be made: there were quite marked individual differences in the number of conjunction errors, particularly with the arrows. This suggests the possibility that particular individuals may give different weight to different features in determining whether to 'see', for example, a triangle. For some people, three lines in the right arrangement might be enough, while for others closure may also be essential. This gives us another way of testing the theory; of seeing whether the different tasks I have discussed are related in the way I suggest. Just as different stimuli may generate performance patterns which covary across the tasks, so may individual subjects with any given stimulus. People who make illusory arrows out of their parts should also search serially for arrow targets in potential conjunction distractors, and should have trouble segregating textures which contain arrows and their parts. On the other hand, people who code arrows by some emergent wholistic property, for example the three-way intersection, should find an arrow target in parallel across search displays and should be able to use the presence or absence of arrow intersections to segregate one area from another.

We tested these correlations across tasks within each type of stimulus. The findings were that for the arrows, there were significant correlations across individuals between (a) the frequency of illusory conjunctions, (b) the difference in search slopes for arrow targets in a conjunction and in a feature background, and (c) the difficulty of texture segregation. For triangles, the correlations were not significant, perhaps because so few illusory conjunctions were seen. It looks, then, as if closure does function as a primitive feature in the same way as line orientation and colour. It also looks as if the three tasks I used are inter-related in the way the theory suggests.

We have recently tried to analyze further just what happens in the apparently wholistic perception of triangles. The suggestion was that triangles differ from arrows only because triangles have an additional emergent feature — closure. So at the feature level a triangle would activate detectors for lines, angles, and also a detector for closure, while an arrow would activate only detectors for lines and angles. An alternative account would be that triangles are perceived as integral units, Gestalts, and not decomposed at all. One way we might distinguish these two views is to see whether the lines of a triangle are less detachable than those of an arrow. Are they less available to make illusory interchanges with other shapes? If we ask subjects to look for a dollar sign in a display of S's and triangles (Fig. 8) are they less likely to conjoin a line from a triangle with an S to form an illusory dollar sign, than they are to take a line from an arrow? If triangles are registered only as unitary Gestalts, they should have no lines to give to an illusory target dollar sign. On the other hand, if triangles are seen when we conjoin their separately detected lines, angles and closure, there is no reason to expect triangle lines to be less available at the feature level than arrow lines. This seems to be the case. Illusory dollar signs were seen on 46% of arrow cards and 42% of triangle cards; the difference was not significant. A triangle seems to yield its separate lines with as little reluctance as an arrow. There is no evidence here for wholistic perception at the early preattentive stage.

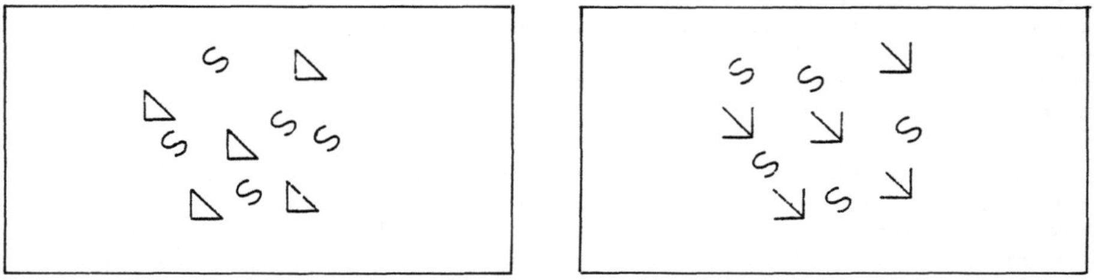

Fig. 8. Examples of displays generating illusory dollar signs

Would a single target line then be as hard to find in a display of triangles as in a display of arrows? If both consist of lines at the feature level, this might follow in theory. Unfortunately it did not in fact. Search for a target line was considerably slower when it was surrounded by arrows than when it was surrounded by triangles. This seemed like a set-back until it occurred to me that target lines and arrow lines both have a feature that triangle lines do not — they have free ends. If these are encoded by detectors other than those which register line orientations, subjects could solve the search problem in the triangle displays by looking for a free end. This may sound like an unbeatable theory — when in trouble invent a new feature — and that is certainly a cause for worry as well as for pride! It was good to hear that JULESZ' research [12] on texture segregation has led him to postulate two basic candidates for primitive feature detectors, one of which in fact encodes free ends (terminators).

In the laboratory, the theory has held up reasonably well so far. How would it fare in the context of everyday perception of objects and complex scenes? Can we reconcile what I have been claiming with the apparent speed and richness of information processing that we constantly experience? I suggest that this apparent wealth of detail at the level of objects and scenes is largely something we create, an informed hallucination. The preattentive registration of features along many dimensions occurs automatically and in parallel. But, when we test more carefully to see how accurate we are in putting features together to construct faces, places or words, the results tend to be less impressive. Scanning a school photograph for my daughter's face, among hundreds of other teenagers is a painstaking business. Proof-reading is also best done slowly and serially. In both these examples contextual redundancy is less useful than it normally is. Much of our peripheral or non-attentive 'seeing' may capitalize on our prior knowledge. In natural contexts there are many known constraints limiting which features we can sensibly combine: grass is normally green and the sky blue or grey.

Perception maps sensory data into expected 'frames', selecting the combinations of features that make sense. We may match expected features to actual features without checking how they are combined. If so, the 'wholes' or objects would still exist in our heads. They would be what we expect to see and normally end up seeing. Yet they may not be the initial code that registers the stimulus in its sensory form.

I suggest there may be three ways in which we can see whole objects; and we may not be aware which we have used in any given instance. 1) We may see them, as I have suggested, by integrating their features in the spotlight of attention. 2) We may see them by predicting their features in a familiar context and separately confirming that each feature is present. 3) Finally, in the absence of either prior information or focused attention, we may be reduced to random resyntheses which result in illusory conjunctions.

References

1. Zeki, S.M. The functional organization of projections from striate to prestriate visual cortex in the rhesus monkey. Cold Spring Harbor Symposia on Quantitative Biology, 1976, 15, 591-600.

2. Garner, W.R. The processing of information and structure. Potomac, MD.: Lawrence Erlbaum, 1974.

3. Minsky, M. Steps towards artificial intelligence. Proceedings of the Institute of Radio Engineers, 1961, 49, 8-30.

4. Marr, D. Vision: A computational investigation into the human representation of visual information. San Francisco: Freeman, 1982.

5. Barrow, H.G., & Tenenbaum, J.M. Recovering intrinsic scene characteristics from images. In A. Hanson & E. Riseman (Eds.), Computer vision systems. New York: Academic Press, 1978.

6. Treisman, A., Sykes, M., & Gelade, G. Selective attention and stimulus integration. In S. Dornic (Ed.), Attention and performance VI. Hillsdale, N.J.: Lawrence Erlbaum, 1977.

7. Treisman, A., & Gelade, G. A feature integration theory of attention. Cognitive Psychology, 1980, 12, 97-136.

8. Treisman, A., & Schmidt, H. Illusory conjunctions in the perception of objects. Cognitive Psychology, 1982, 14, 107-141.

9. Treisman, A. Perceptual grouping and attention in visual search for features and for objects. Journal of Experimental Psychology: Human Perception and Performance, 1982, 8, 194-214.

10. Neisser, U. Cognitive psychology. New York: Appleton-Century-Crofts, 1967.

11. Pomerantz, J.R. Perceptual organization in information processing. In M. Kubovy & J.R. Pomerantz (Eds.), Perceptual organization. Hillsdale, N.J.: Lawrence Erlbaum, 1981.

12. Julesz, B. Spatial nonlinearities in the instantaneous perception of textures with identical power spectra. In C. Longuet-Higgins, & N.S. Sutherland (Eds.), The psychology of vision. Philosophical Transactions of the Royal Society, London, 1980, 290, 83-94.

Cooperative Grouping and Early Orientation Selection[1]

S.W. Zucker

Computer Vision and Graphics Laboratory, Department of Electrical Engineering, McGill University, Montreal, Quebec, Canada

1. Introduction

The visual world is rich in fragments. Objects occlude one another, so their surrounding contours are only partially complete. Shadows obscure boundaries, and motion juxtaposes almost everything. The visual system must therefore have mechanisms available for completing these fragments, one prerequisite for which, in psychological terms, is grouping.

Grouping, as on organizational principle for visual data, was first studied by the Gestalt psychologists. Noting that the limiting case of a contour fragment is a collection of closely spaced dots, their studies of dot grouping become immediately relevant. In particular,when do collections of dots group into lines? into curves? How? Why? The Gestalt answer was formulated as a series of principles of perceptual organization, such as the tendency of dots to group with their nearest neighbors. But these principles were purely descriptive; from what are they a result? They are not explanatory, nor do they have direct predictive power. In this paper we shall re-examine these questions within the context of what appear to be certain very early--almost primitive--forms of dot pattern grouping. We shall be concerned with one kind of structure into which dots can group--oriented entities; as well as why they should, how they could, and when they ought to form these groupings.

The main results in this paper are psychological and computational. In psychological terms, two very different kinds of distinctions must be made. The first of these we refer to as a size/density constraint: the results of grouping processes differ drastically as a function of the spacing between dots. The second is that, even when the dots are closely spaced, there are two principally different types of grouping processes that are necessary for orientation selection. The first of these processes has very high spatial specificity while the second has relatively low specificity. The two psychological distinctions give rise to hard constraints on the processes that can accomplish dot grouping. An abstract mechanism is proposed for dot grouping that is based on an optimal interpretation of the responses of operators with asymmetric receptive fields by a relaxation network. The result shows that the difference between the two types of dot grouping processes is essentially one of interpolation. From a computer vision perspective, it also leads to a process for finding curves that performs robustly in the presence of noise. All of these results, and many related ones, are described more fully in ZUCKER [1982].

---

[1] This research was supported by NSERC Grant A4470 and by DSS Contract 01.SU.CD3217043 administered by DREA

2. On the Role of Early Grouping Processes

The need for two early types of grouping processes is suggested by the surface-inferencing task. Within this context, the most basic image feature is the 1-dimensional contour that arises from the intersection of different surfaces. This is almost always apparent, since it is likely that both reflectivity and orientation will be changing abruptly along it. For our purposes it can be viewed as a special case of any well—defined surface contour, because changes in the local orientation (or tangent) to the curve imply analagous changes for the surface on which it lies [BINFORD, 1981; PENTLAND, 1982]. Such contours will be most useful when they are very accurately positioned, because then they can serve as tight boundary conditions against which many different sources of surface discontinuity, depth (from stereo), and other information can be integrated. But it should be stressed that these contours are never perfectly continuous and non artifactual everywhere; see MARR and HILDRETH [1980; Fig. 8].

There are many circumstances, however, within which the above well-defined surface contours cannot exist, due to rapid variations of the material comprising the surface. Instead these rapid variations, or physical "microstructures", give rise to more complex image microstructures. Examples of such surfaces include those covered by grass or hair or fur--in short, surfaces rich in texture. In this case each entity would have a general orientation, but they would all overlap and sway to the extent that individual differences would be substantial. They will be roughly aligned, and of roughly the same length, but not precisely so. Images formed from this class of surface would reflect this random, discontinuous structure. Sections of the micro-entities that were properly aligned (with respect to the viewer and the light sources) would produce strong highlights, but generally these would be very short. No long contours would be seen; rather, the impression would be a more global one of a field that is "flowing" in a certain way. Information about the surfaces is available, however, although it is somewhat less precise than the first type above.

3. Type I and Type II Processes

The functional identification of Type I and Type II processes is clearly with the two sub-tasks discussed in the context of surface inferencing. Type I processes are responsible for the highly accurate boundary conditions, while Type II processes are responsible for the more coarse interior constraints. (Their functional roles may, however, become more intertwined than this; the essential point is the specificity difference.) The following demonstrations exhibit these two types of grouping with dot patterns.

3.1 Type I (Wertheimer) Grouping

Everyone is familiar with the manner in which closely-spaced dots can depict a smooth curve; see Fig. 1.

Figure 1. A smooth dotted curve

Additional demonstrations, such as those in WERTHEIMER's original paper [1923], together with considerations about three-dot acuity tasks [BECK and SCHWARTZ, 1978], lead us to the defining characteristics of Type I grouping: (i) high positional specificity; (ii) high orientational change specificity; (iii) high primitive specificity; (iv) the percept as a one-dimensional curve; and (v) the presence of endpoints and singular points.

The basic characteristic of Type I grouping is (i)--high spatial specificity. Properties (ii-v) follow from it.

3.2 Type II (Random Dot Moire Pattern) Grouping

Random dot Moire patterns (RDMP's), (GLASS [1969] and ANSTIS [1970]) provide examples of alternative grouping phenomena. A RDMP can be made by producing a random field of dots, making a copy of it, transforming the copy, and then displaying the transformed original and the copy superimposed. If the transformation is a rigid rotation, then the result is a pattern with a circular appearance. If the transformation is sinusoidal or triangular, then patterns such as the ones in Fig. 2 arise.

Fig. 2. A RDMP in which the top half is sinusoidal and the bottom triangular

Note the absence of any well-formed contours, and, furthermore, that there are no clear endpoints or discontinuities in curvature (within these limits).

The most striking difference between RDMP's and the Wertheimer groupings shown earlier is their random and incomplete character; they give an impression rather than the crisp, clear effect that would be present with individual contours. There is a much looser requirement on the underlying spatial structure, and the result is more of a field or a flow than a complete one-dimensional contour. Dots can be moved around, e.g., by making the pattern with a different underlying random distribution, and the overall effect does not change. Almost all of the above specificities, with the exception of orientation specificity, can be shown to be looser, and the requirements for discontinuities are much more severe.

The term specificity has been used above to stress the fact that the differences between Type I and Type II processes are distinct from those that would arise from differences in spatial resolution.

The difference between Type I and II patterns can be illustrated with a final demonstration, shown in Fig. 3. It consists of dots arranged in slope horizontally across the center. Fig. 3a, which permits Type I grouping, has the dots arranged in pairs randomly spaced along (imaginary) lines about 5 dot-diameters apart. The line of slope change is immediately apparent, as are the oriented fields on either side of it. The second

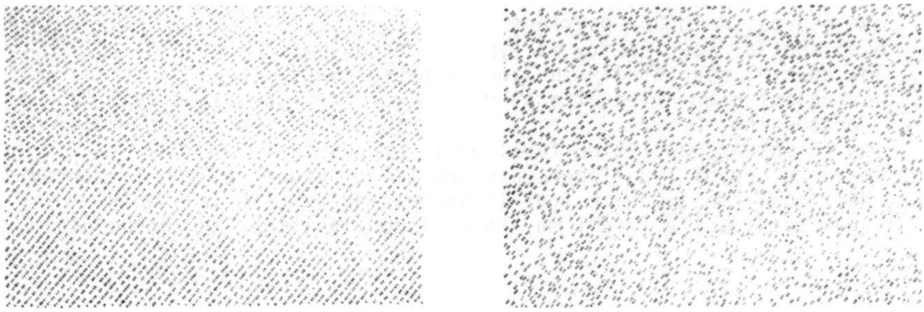

3a 3b

Fig. 3. Comparison of Type I(a) and Type II(b) patterns. All dot pairs are perfectly aligned in slope. The (a) pattern is spatially aligned along these curves, but the (b) pattern has a random positional variation within 3 dot diameters. The slope difference between the top and the bottom is 50%

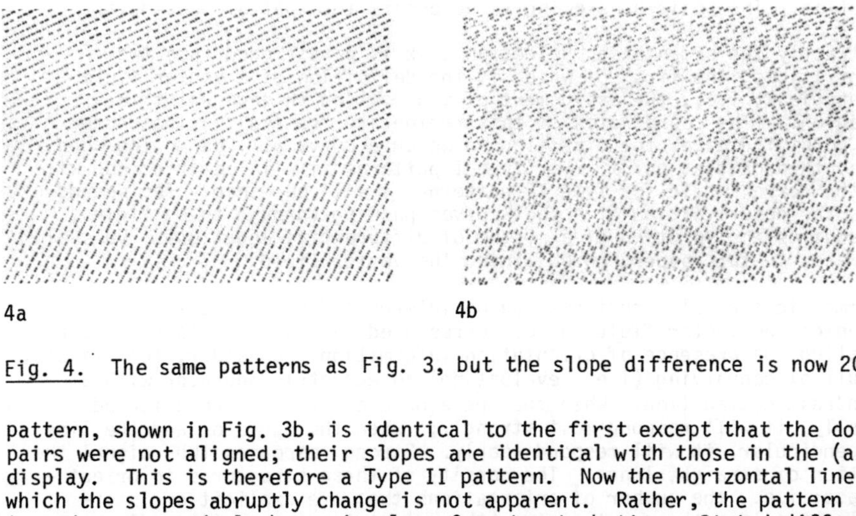

4a 4b

Fig. 4. The same patterns as Fig. 3, but the slope difference is now 200%

pattern, shown in Fig. 3b, is identical to the first except that the dot pairs were not aligned; their slopes are identical with those in the (a) display. This is therefore a Type II pattern. Now the horizontal line along which the slopes abruptly change is not apparent. Rather, the pattern seems to undergo a gradual change in slope from top to bottom. Stated differently, the discontinuity in the Type I pattern seems to have been smoothed away in the Type II pattern. A 200% change in slope is visible, however; see Fig.4.

The patterns in Fig. 3 and 4 illustrate another point. They were constructed so that all patterns have nearest neighbors that point in precisely the correct direction. If this information were usable, then the slope change would be visible in all patterns. The Type II process clearly ignores it even for 50% changes in slope, which suggests that nearest neighbors are not the basis for grouping (but compare STEVENS [1978]).

To close this section, we review the constraints obtained from the psychophysics. While not necessarily a complete list, any model for orientation selection must account for them.

C-1 non nearest neighbor pairings;
C-2 high spatial specificity in Type I processing;
C-3 low spatial specificity in Type II processing;
C-4 equivalent high orientational resolution for Type I and Type II processing;
C-5 low curvature resolution for Type II processing;
C-6 curve nature of Type I percepts and field nature of Type II percepts;
C-7 additional smoothing in Type II processing results in smooth percepts, while Type I may have discontinuities (and endpoints).

4. A Model For Orientation Selection

Differences between Type I and Type II grouping processes would seem to suggest that the mechanisms supporting them must differ drastically as well. This is not the case, however, as we attempt to show with a model for orientation selection. The model amounts to a (non linear) lateral inhibitory network running among differently oriented operators of different sizes, with the central locations of these operators skewed slightly in position. The orientation selection task can be formally posed as a variational problem, and the multiple-size operators provide curvature constraints necessary for its solution. The position skewing is necessary for spatial continuity.

Abstractly the orientation selection task is partitioned into two different stages, the first of which is the definition of a vector field of orientations; i.e., the assignment of a direction to each point in a fixed region. (The coordinates of this region are "retinal".) The vector at each point indicates the direction in which a curve will be moving when it passes through that point. For Type I patterns it will be thin, and will follow the curve, while for Type II patterns it will be dense. The second stage is the determination of actual curves passing through these fields, which requires the solution of a system of differential equations. We shall concentrate, in this paper, on obtaining the vector field.

Asymmetric spatial structures can be interpreted both as a framework within which the vector field can be represented, and as operators capable of signaling the presence of orientation information. Consider, for example, the result of convolving (i.e., evaluating) an eccentric operator with a high-contrast dotted line. When the operator and the line are oriented identically, it will respond most strongly; when the operator and the line are perpendicular, it will respond weakly. Strong responses thus signal the orientation of straight lines. The density of these structures in orientation space; i.e., the number of orientations that are explicit or interpolatable, provides the constraint on the orientation resolution. By constraint C-4, we know that this number should be about equal for Type I and Type II processes.

The choice of operators is constrained by (i) the asymmetry requirement; (ii) constraint C-1, which indicates more of a local arithmetic operation than a logical one (such as computing nearest neighbors); (iii) the surface inferencing framework, which connects second derivatives with surface curvature [Pentland, 1982]; and (iv) available biological machinery. Second directional derivative operators are the simplest operators that are consistent with all of these constraints, so we shall use them in this paper. They resemble what neurophysiologists have called "line detectors", although, as we shall see, significant further processing is necessary before any but the clearest lines can be detected.

Problems arise in the interpretation of the responses of second directional derivatives when the lines over which they are evaluated are not perfectly contrasted with the background, when noise is present, or when they curve--situations at the heart of many Type I, and virtually all Type II, patterns. The operators give intermediate responses, neither strong nor weak. How can these intermediate operator responses be interpreted? This question can be formally recast in terms of what we shall call the *response matching problem*. Since we know the structure of the operators, we can compute their expected response to known patterns. Then, given the response observed to arise from an unknown pattern, we assert that the unknown pattern is the one which, if present, would have given an expected response as similar as possible to the observed response.

The solution to this response-matching problem requires constraints as well as a mechanism for applying them. The constraints come from the operators, both by varying their orientation and their size. Small operators can only signal (locally) straight lines, as was implicitly shown above. But larger ones, when rotated, can actually signal curves. This is because the curve will systematically enter and leave the support, or "receptive field", of the operator as it is rotated; see Fig. 5. Intuitively, the small operators localize the curve, while the larger ones provide its context.

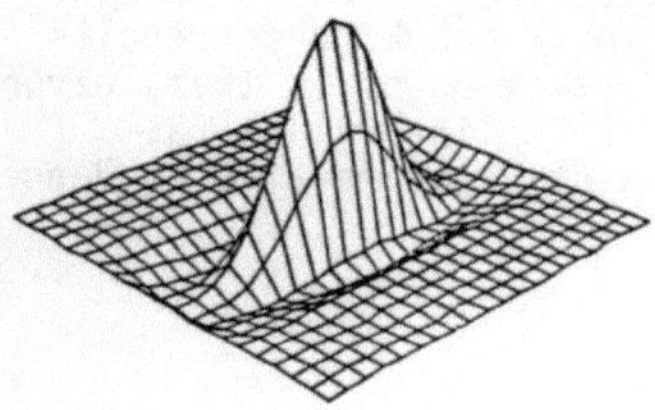

Fig. 5a. A second directional derivative operator presented in relief.

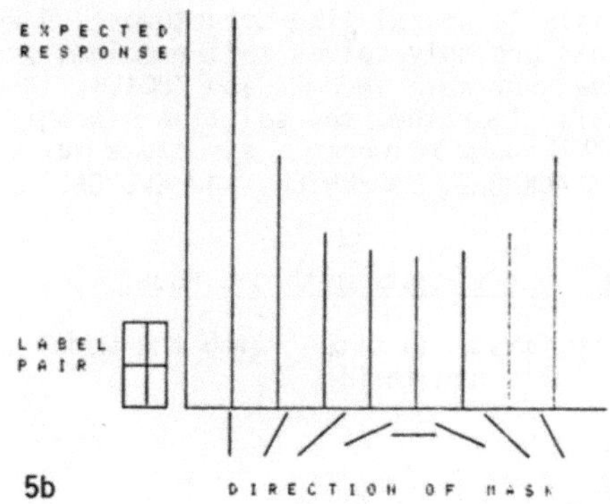

5b

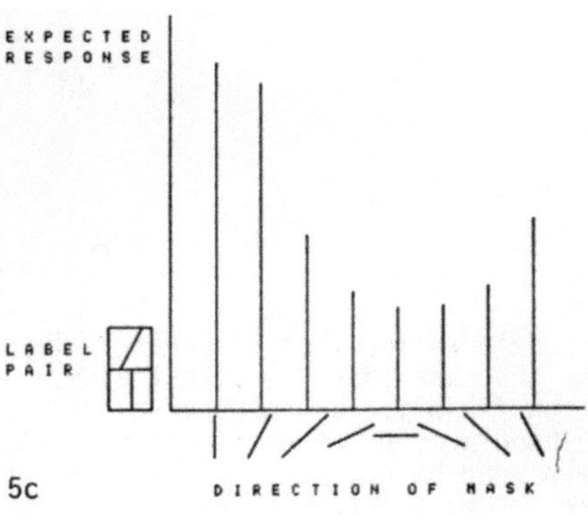

5c

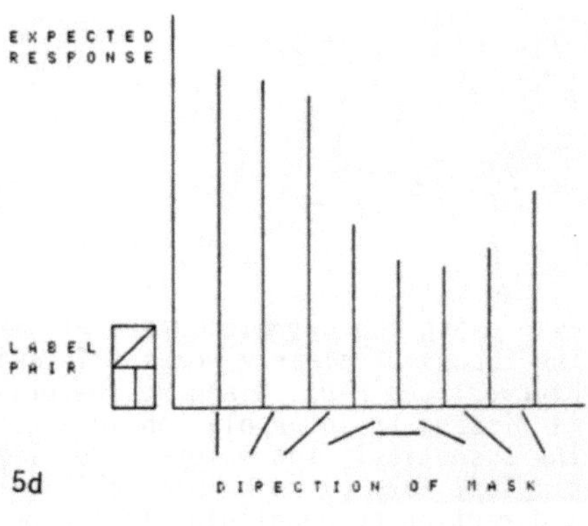

5d

Figs. 5b-5d. Expected responses for the operators shown in 5a over straight (5b) and curved (5c-d) line patterns

The search for the underlying pattern can be done cooperatively by allowing the responses to interact with one another. The solution to the response matching problem requires minimizing differences of the form:

$$\| R_{exp}(or;\ size) - R_{obs}(or;\ size) \|$$

under a suitable form. Intuitively this means that a short straight segment pointing in direction Θ should receive positive reinforcement from large operators at that same position that are also pointing in direction Θ as well as from similarly oriented small and large operators at neighboring positions (in the directions $+\ \Theta$). It should receive negative support from all of these operators oriented in directions $\Theta \pm \pi/2$, since these would indicate lines running in perpendicular directions. (The constraints are actually more complicated than this, however.)

This notion of support can be made precise, and it leads to a variational setting for the response interpretation problem. In fact, when the variational problem is completely developed, it can be shown that a functional exists with local maxima that coincide exactly with solutions. Thus there is a formal sense in which the solution to the response matching problem can be said to be optimal.

For the above solution to be feasible biologically, it must be implementable in neural-like structures. A relaxation algorithm can be formulated that provably solves this problem, and for which local convergence results are available [HUMMEL and ZUCKER, 1980]. The important point is that, given this algorithm, the solution strongly resembles a (non linear) lateral inhibitory network, a structure for which there exists some related evidence [BLACKMORE, CARPENTER, and GEORGESON, 1970].

5. Experiments with the Model

Our first experiment with the model is on the noisy curve in Fig. 6; the results are in Fig. 7.

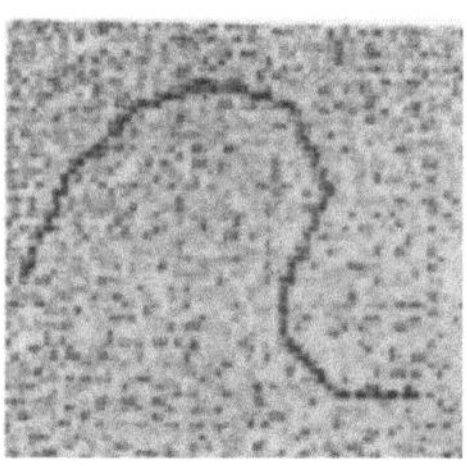

Fig. 6

Fig. 7

Even using the extreme quantizations of 8 orientations and 3 operator sizes, the technique clearly works. This is not the case for Type II patterns, however; see Figs. 8 and 9. The orientation field is far too sparse. What is missing is interpolation in a neighborhood around the operators so that the essentially incomplete Type II patterns can be completed. When even the simplest form of such interpolation is added, as a spreading of support in a direction perpendicular to the operators' orientation, the result is successful (Fig. 10). Thus we can now pose the essential difference between Type I and Type II patterns: the spatial specificity or, in computational

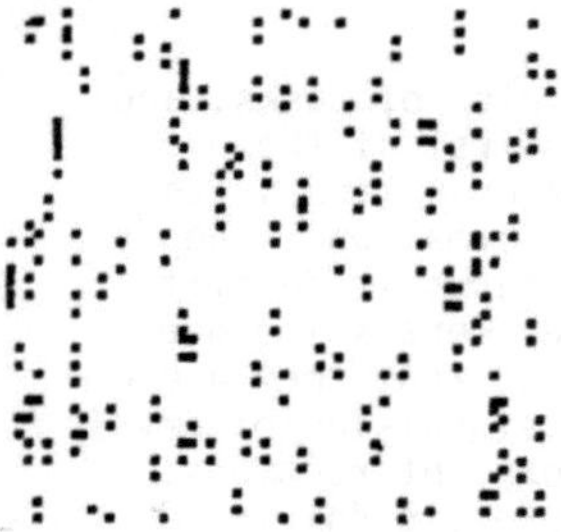

Fig. 8

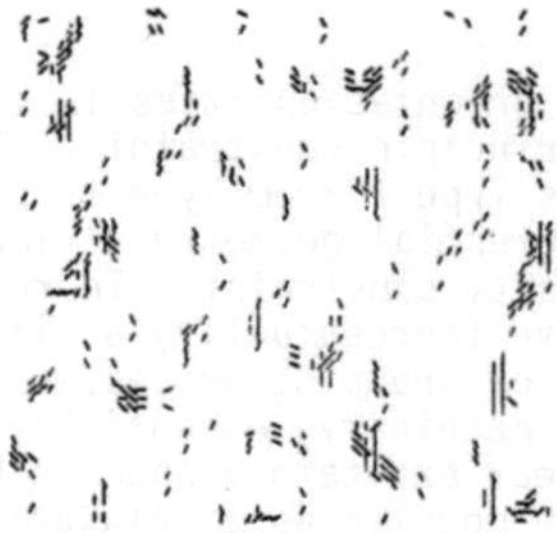

Fig. 9

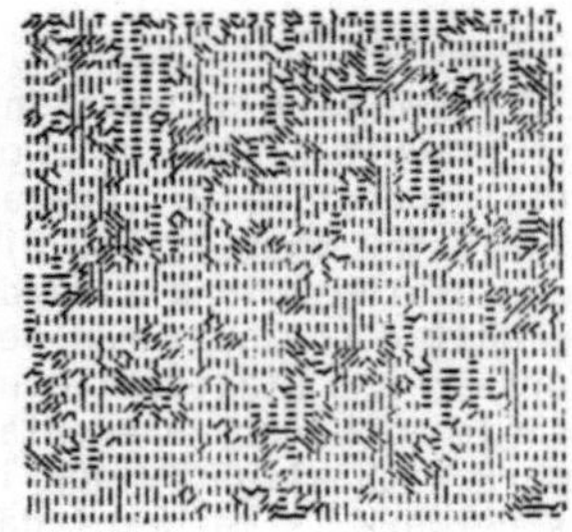

Fig. 10

Fig. 8. A section of a RDMP in which the displacement is vertical

Fig. 9. The results of Type I grouping on the pattern in Fig. 8

Fig. 10. The result when interpolation, or "thickening", is added to the Type I process

terms, the extra interpolation. It is this interpolation that "smoothed out" the abrupt changes in Fig. 3.

6. Size and Density Constraints for Very Early Grouping Processes

The model for grouping suggests very early, essentially syntactic processes that are data-directed and inflexible in their behavior. But such behaviour only works under well-defined, or delimited, conditions, two of which seem to be active for dot grouping. The first of these involves the size of the dots, or individual entities, and the second their density. Many demonstrations show that the organization of dot patterns changes as a function of their density, leaving other aspects of their geometry constant; see Fig. 11.

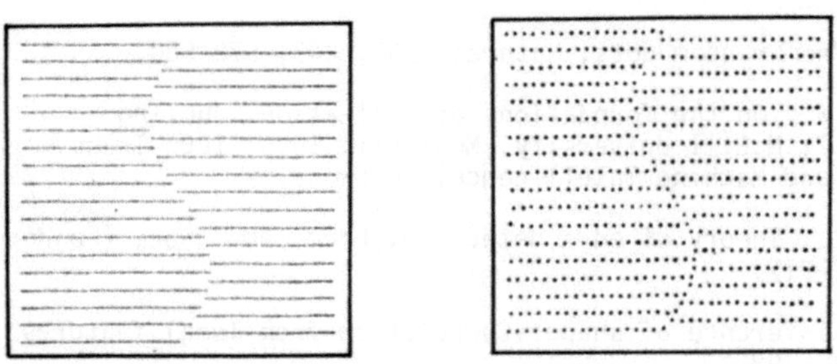

Fig. 11. Illustration of the size/density constraint for a subjective contour

The surprise is that there is a rather sharp value of a size/density ratio that delimits where this change occurs. Our experiments suggest that it lies between 3-8 dot diameters, for dots about a minute of visual angle in diameter. Such a constraint is consistent with the above model, because it constrains the size of the operators involved.

7. Conclusions

The grouping of dots into oriented entities is a complex of tasks, and their study has lead to two principle constraints. The first of these, spatial specificity, separates Type I from Type II patterns. It has its roots in the physics and differential geometry of image formation. The second of these is a size/density constraint. In rough terms it delimits when dotted contours can behave (perceptually) as if they were solid ones. A further computational study of grouping processes showed that they are feasible with very elementary machinery, essentially a lateral-inhibitory network with slightly non linear excitatory and inhibitory interactions. This network is mathematically modeled as a relaxation process solving a variational (in particular optimization) problem, and the dot grouping task is formulated with no constraints other than those available from second directional derivative operators. Thus while it is only a sufficient model, it is not biologically implausible.

Dot grouping was studied because dot patterns are a limiting case of many of the different kinds of fragments that necessarily arise in the processing of visual information. The results presented here have many implications for texture, symmetry, and other forms of early spatial information processing. They also have important analogues in time, or motion, that are currently under study. Together they seem to provide a foundation of "first guesses" upon which further processing can be profitably based.

References

Anstis, S., Phi movement as a subtraction Process, Vision Research, 1970, 10, 1411-1430.

Beck, J., and Schwartz, T., Vernier acuity with test dot objects, Vision Research, 1978, 19, 313-319.

Binford, T., Inferring surfaces form images, Artificial Intelligence, 1981.

Blakemore, C., Carpenter, R., and Georgeson, M., Lateral inhibition between orientation detectors in the human visual system, Nature (Lond.), 1970, 228, 37-39.

Glass, L., Moire effect from random dots, Nature, 1969, 243, 578-580.

Hummel, R., and Zucker S.W., On the foundations of relaxation labeling processes, TR-80-7, McGill University, Montreal, 1980; IEEE Trans. Pattern Analysis and Machine Intelligence, in press.

Marr, D., and Hildreth, E., Theory of edge detection, Proc. Roy. Soc. (London), B, 1980, 207, 187-217.

Pentland, A., The Visual inference of shape: Computation from local features, Ph.D. thesis, MIT, 1982.

Stevens, K., Computation of locally parallel structure, Biol. Cyber., 1978, 29, 19-26.

Wertheimer, M., Laws of organization in perceptual forms, Psych. Forsch., 1923, 4, 301-350; trans. in Ellis, W., A Source Book of Gestalt Psychology, Routledge and Kegan Paul, London, 1938, 71-88.

Zucker, S., Early orientation selection and grouping: Type I and Type II processes, TR-82-6, McGill University, Montreal, 1982.

Recognition of Shape in Visible Surfaces

H.K. Nishihara

M.I.T. Artificial Intelligence Laboratory, 545 Technology Square, Cambridge, MA 02139, USA

1. Introduction

Vision can be thought of as a process that produces from images of the external world a description that is useful to the viewer and not cluttered by irrelevant information. Answers to the question of what is useful and what is clutter depend on the task at hand. I will concentrate here on the problem of recognizing three-dimensional shapes.

Many kinds of visually derived information can play a role in recognition, like the color and texture of an orange or the rapid jittering motion that excites a cat. I will be concerned here only with shape which I define as the geometry of an object's physical surface. Thus two statues cast from the same mold will have the same shape even though different materials may have been used.

The study of even this more narrowly defined problem of recognizing shapes is easily confounded by the large body of non-visual information such as contextual constraints that people can and do use with facility to augment their performance. The challenge is to limit the problem treated without oversimplifying or denying the rich use of other knowledge. I will present an approach taken by D. MARR and myself [1] which focuses on the kinds of information relevant to recognition. This study led us to a way of conceptually subdividing processes in vision according to the type of information they use and produce.

2. Shape Representations

Early computational work on the problem of visual shape recognition dealt with polyhedra which have natural descriptions as two-dimensional line drawings (see fig.1). Research in understanding the problem of obtaining and interpreting such drawings was pioneered by ROBERTS [2], GUZMAN [3], WALTZ [4], HUFFMAN [5], CLOWES [6], and MACKWORTH [7]. The approach, however, is specialized to solids with planar faces and even for those shapes, line drawing descriptions are still not easily obtained from images.

At about the same time, another approach was developing in the work of BLUM [8], BINFORD [9], AGIN [10], NEVATIA [11], and HOLLERBACH [12] which focused on descriptions built about a central axis of the shape. These representations were tied more directly to the distribution of a shape's volume than to surface properties like corners and edges which were the basis of descriptions in the former approach.

A stick figure model like those shown in Fig.2 illustrates the emphasis of this alternate view. Just as line drawing descriptions were attractive

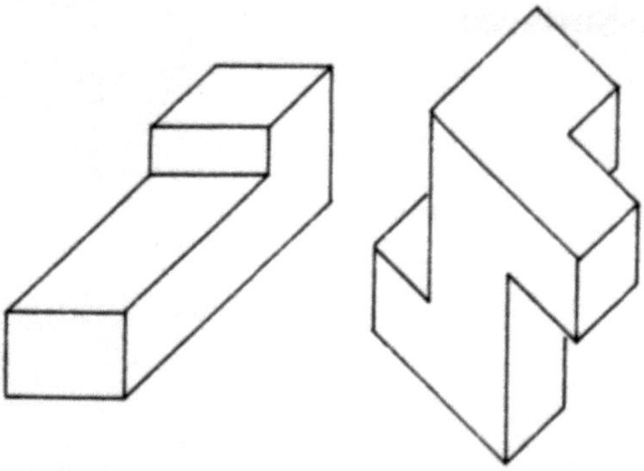

Fig.1. Line drawings of polyhedra

for their simplicity, the arrangement here of a few pieces of wire effectively portrays complex animal shapes. That they work so well is surprising considering the paucity of information in these models. What was especially revealing to us was that a shape representation does not have to reproduce a shape's surface in order to describe it adequately for recognition. The only information made explicit here is the arrangement and relative sizes of a small number of sticks standing symbolically for major components of the shape.

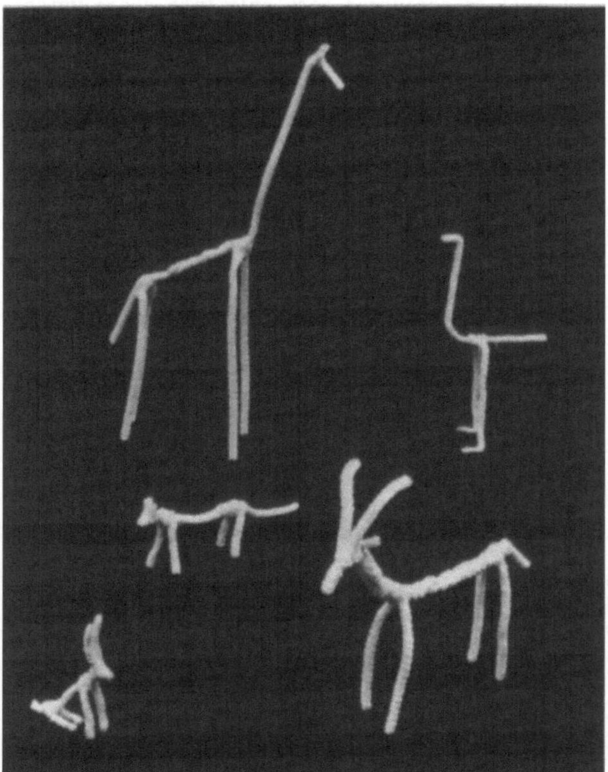

Fig.2. Pipe-cleaner models

A way to think about why these models might work, and more generally about what matters for shape recognition is to suppose we are in the business of designing a prosthesis for the blind, and ask what sort of limited device would be most helpful to a blind person trying to discriminate shapes. I will present a few possibilities as examples and then go into a more detailed examination of alternatives and criteria for comparing them.

The simplest device from a technological standpoint would present the user in some way with information about surface brightness. For example, a T.V. camera could be interfaced to a two-dimensional tactile array of mechanical transducers such as pins which either rise or vibrate according to the image intensity at the corresponding location in the T.V. image.

Figure 3 shows what a T.V. camera might obtain. The information is adequate for our visual cortex to interpret, but a blind user cannot look at the image as we do; he or she must acquire this information through another modality which will not have as great a data transfer rate as the early stages of our visual systems. Even a relatively coarse 100 by 100 pixel image has 10,000 distinct brightness values (each image in Fig.3 has 150,000 pixels). Even if this volume of information could be communicated in a timely manner, the data is about brightness, not shape, so it cannot be used directly. Extensive analysis is required to extract information about the physical surfaces imaged and only then can inferences about shape be made. This analysis might, for example, make use of texture and colour variations to identify regions likely to be part of the same physical surface or if---as in binocular vision---two images from slightly different vantage points are available, they can be compared and horizontal position

Fig.3. Stereo pair of intensity descriptions

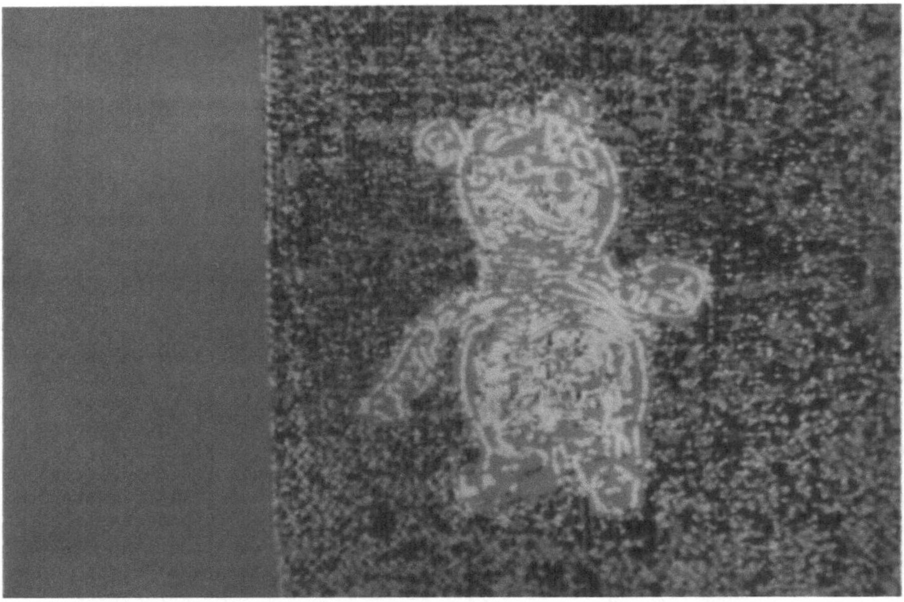

Fig.4. Near/far stereo disparity description

disparities translated into measurements of stereo depth at the corresponding point in the visual field.

Computer vision technology is advancing and it may be possible in a few years to accomplish some forms of this analysis automatically in a prosthesis so that data perhaps like that shown in Fig.4 could be presented instead the raw intensity data from a camera. What is shown in the figure is the output of a simple but fast stereo matching algorithm that I have been studying which follows the ideas of MARR and POGGIO [13]. The algorithm indicates only whether a region is likely to be nearer (white) or farther (black) than a given fixation plane in depth. Regions where no measurement was made by the algorithm are shaded grey. This information is very limited and it is somewhat noisy, but it is information about the geometry of the physical surfaces viewed and not about how they reflect light and this is a critical step. Many problems remain, however, particularly the large volume of data still present in the description and the dependence of the description on vantage point.

The progression we have been making through different categories of representation [14] is illustrated in Fig.5. All vision systems begin with intensity information which is the first category shown. From there information about physical surface properties that directly affect the image can be recovered and this makes the second category. Once obtained, this visible-surface information can be reformulated to best meet the requirements of a particular application---in our case for describing shape effectively for recognition.

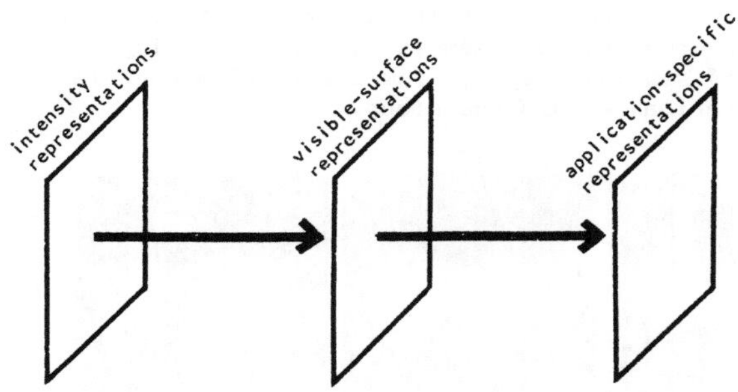

Fig.5. Processes in vision can be broadly classified according to the type of information they manipulate. Early visual processes are concerned primarily with computing whatever visible surface information they can from the intensity variations present in images at the retinas while later processes address the requirements of specific applications like recognition or navigation

3. Criteria for Judging the Effectiveness of a Shape Representation

There are many different aspects of an object's shape that could be taken as a basis for a representation, some more useful than others, and any one aspect can be described in a number of ways, for example using different coordinate systems. Although it is difficult to formulate a completely general classification of shape representations, we can set out criteria for judging the suitability of a given representation and basic design choices that have to be made when a representation is formulated.

One point we can raise as a criterion is the question of accessibility. Can the desired description be computed from an image, and can it be done inexpensively? There are fundamental limitations inherent in the information available in an image, for example there is no information about the back surfaces of a viewed shape so a representation requiring a specification of an object's complete surface would not satisfy this criterion.

Another question we can ask is about scope and uniqueness. What class of shapes is the representation designed for and do the shapes in that class have unique descriptions? For example, an edge based representation includes polyhedra within its scope, but shapes with smooth curved surfaces have no natural edges on which to base a description. Of course edges can be chosen arbitrarily to produce a polyhedral approximation to the shape, but the description will not be unique. Generally a shape will have many distinct descriptions in a given representation due to variations of non-shape parameters such as orientation relative to the viewer. To obtain consistent descriptions one needs to decouple the shape description from these non-shape details and this limits the representation's scope to a class of shapes for which the decoupling can be accomplished.

A third criterion is stability and sensitivity. Within the scope and uniqueness requirement, there lie questions about the continuity and resolution of a representation. To be useful for recognition, the degree of similarity between two shapes must be reflected in their descriptions, but at the same time even subtle differences should be expressible. These opposing conditions can be satisfied only if it is possible to decouple stable information that captures the more general and less varying properties within a group of similar shapes, from information that is sensitive to the finer variations within the group.

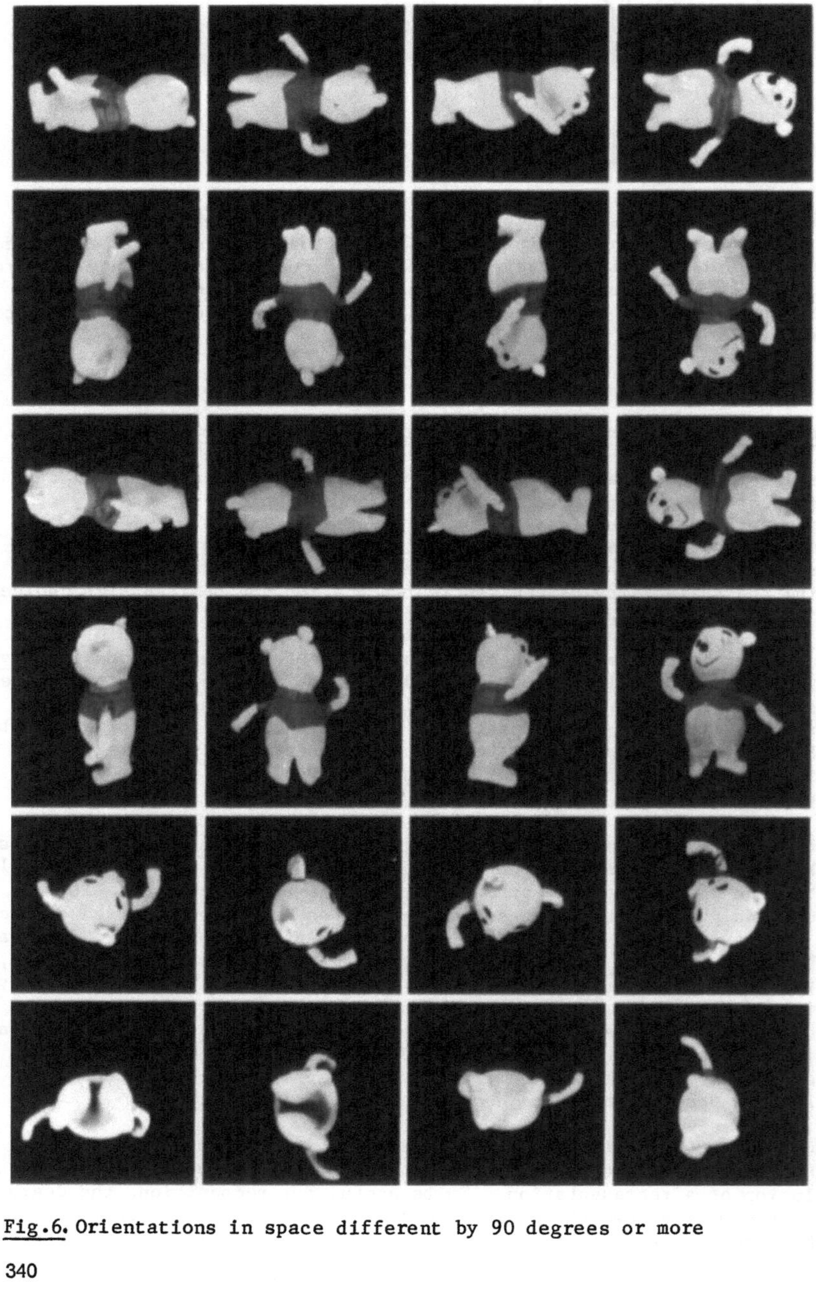

Fig.6. Orientations in space different by 90 degrees or more

4. Choices in the Design of a Shape Representation

We can now relate these three performance criteria to different choices in the design of a shape representation. Perhaps the most fundamental property of a representation is that it can make some types of information explicit, bringing the essential information to the foreground, and allowing the descriptions to be smaller and manipulated more quickly. Three aspects of a shape representation's design are particularly important: the coordinate system used to specify how a shape is arranged in space, its primitives which are the primary units of shape information used in a description, and the kind of organization imposed on the information in that description.

4.1 Coordinate Systems

The most important aspect of the coordinate system used by a representation is the physical reference that defines it. For vision the principle possibilities are viewer-centred frames in which spatial positions are specified relative to the viewer or some other base not tied to the shape itself, and object-centred frames in which locations are specified relative to the shape viewed.

The optical imaging process imposes viewer-centred coordinates on the representations in early visual processing and it would be preferable from the standpoint of the accessibility criterion to have subsequent representations use the same type of coordinate system. This, however, is not realistic for the case of shape recognition because of the large number of orientations a given shape can have in space.

Figure 6 illustrates the problem. If our prosthesis presents its user with surface information in camera-based coordinates, the user will have to be familiar with the range of orientation dependent "appearances" a given shape can have. Any pair of views in this set of 24 are at least 90 degrees apart in orientation. Even rotations of 30 degrees produce substantial changes in appearance for most shapes and there are some 500 distinct views differing by 30 degrees or more in orientation. Thus to make use of viewer-centred data for recognition, the user will have to be familiar with each of the possible appearances requiring a memory of between 24 and 500 views of each shape, or he will have to have some way of relating a new view of an object to a possibly different view he has stored in memory.

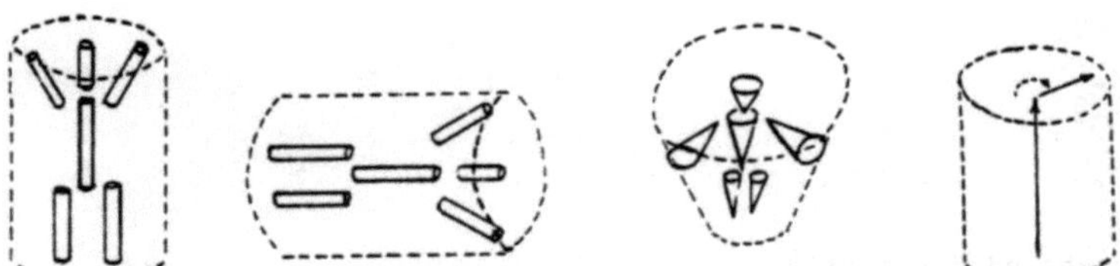

Fig.7. Object-centred coordinate frames

To compare two descriptions which may be of the same object in different orientations involves carrying out a coordinate transform to rotate one description so that it coincides with the other. If the appropriate rotation is not known, all possibilities would have to be exhaustively searched.

In many cases, however, the structure of the shape can be used to suggest one or a few standard orientations to check. For example, the longest axis or an axis of symmetry of the shape can be used to remove two degrees of freedom in determining a standard orientation. Fig.7 shows how a figure in different orientations can be associated with a cylindrical coordinate system aligned with the long axis of the shape. With that frame of reference, it is possible to describe the positions of the parts consistently in all three cases.

In this way, the effects of vantage can be decoupled from a description of an object's shape to a degree determined by the detectability of a standard object-centred coordinate frame for it.

4.2 Primitive Shapes

Another design issue is the type of shape primitive used by the representation. These are the indivisible shape properties or features out of which a shape description is built. For recognition, an important distinction is whether they are large or small relative to the whole shape. If small primitives are used, many elements will be required to give a complete description and the individual elements will specify a local property of the shape---usually its surface. Large primitives on the other hand will make explicit more global properties of the shape usually related to the distribution of its volume. We refer to the two classes of primitive types as surface-based and volumetric primitives.

Fig.8. A surface-based description

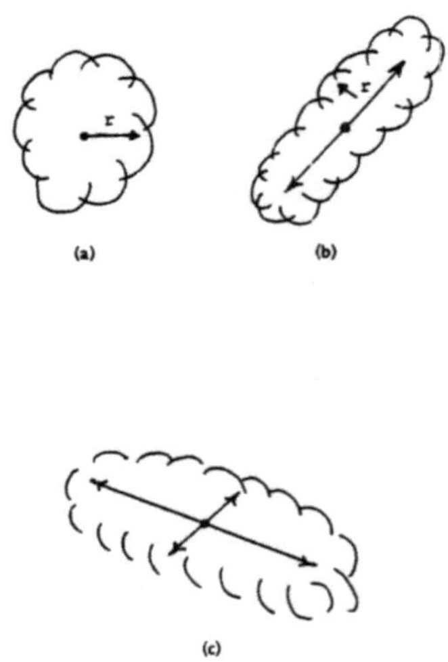

Fig.9. Volumetric primitives

Figure 8 is an example of a surface-based description. The small surface patches indicate local surface orientation. Fig.4 was another example where the local elements indicated the relative depth of the surface.

For shape recognition, small or surface-based primitives have two flaws. Descriptions using them are large and so difficult to communicate; and they perform poorly in terms of the stability and sensitivity criterion. Small primitive elements give prominence to fine structure at the expense of larger characteristics. For example, there are no explicit answers to basic questions about the size or elongation of the overall shape in Fig.8. What is explicit is information about the positions of small patches of surface. As a consequence, the relative accessibility of answers to queries about the shape described is inversely proportional to the size of the characteristic in question. This runs counter to the needs of recognition where global characteristics such as size or elongation are more significant than the orientation of a particular patch on the surface.

Volumetric primitives allow the opposite emphasis. Large shape properties such as size, position, and orientation or elongation can be made explicit in a description at the expense of fine detail as illustrated by Fig.9. It would be hard to tell much about a shape with this alone, but the same primitive information can be specified for concentrations of mass within the shape and this is essentially what the stick models in Fig.2 do.

Large volumetric primitives capture global information which tends to be more stable among similar shapes while progressively smaller primitives provide sensitivity to finer shape distinctions with the smallest primitives being surface based. A range of primitive sizes are thus necessary to maintain both stability and sensitivity.

4.3 Organization

The benefits of a range of primitive sizes is only achieved if the more stable elements of the description can be distinguished from the less stable but more sensitive ones. This brings us to the third design issue which concerns the organization in a description. In the simplest case no organization is imposed by the representation and all elements of the description have the same status. Fig.10 illustrates this for a stick figure description. All parts of the figure large and small are specified with primitive elements in the same way and no part is described at more than a single scale.

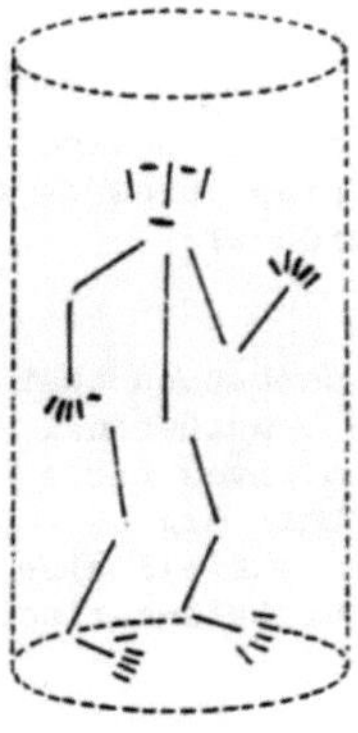

Fig.10. Uniform organization

It is also possible to describe a shape at several levels of detail simultaneously, maintaining information about the general orientation and structure of the shape at the coarsest level while also making the fine structure of smaller components explicit. A modular organization allows us to make explicit in a description the primitive relationships that are

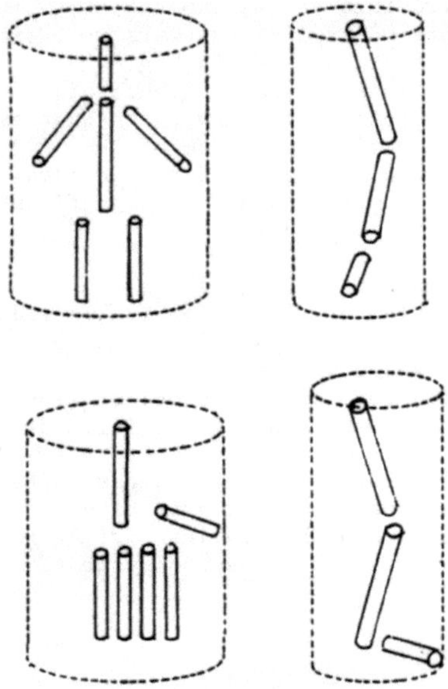

Fig.11. Modular organization

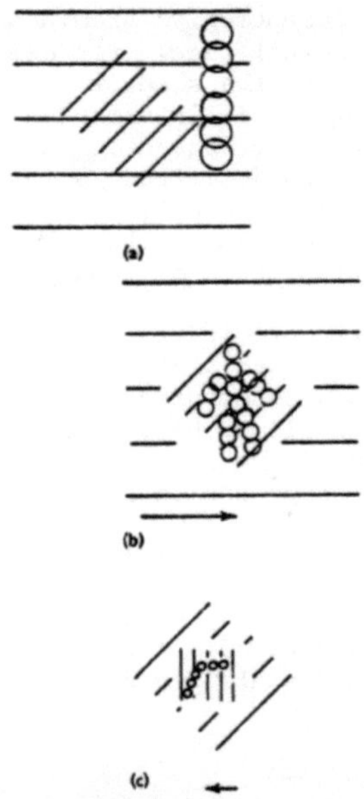

Fig.12. Prosthesis operation

meaningful. Fig.11 gives us an example. In a human shape the arrangement of major body parts with respect to each other is more important than the relationship between a foot and finger of the shape. Similarly the relative arrangement of the fingers of one hand has greater stability than the arrangement of fingers from both hands.

5. The 3-D model representation

Thus a modular organization which groups primitive elements close in size and proximity makes it possible to separate the more stable properties of a shape found at coarser resolutions from the finer details which are important for sensitivity.

In summary, the three design criteria favor a shape representation that is object-centred, volumetric, and modular. This suggests a hypothetical device which would help a blind user identify volumetric primitives for a single module at a time. Fig.12 illustrates a way in which this might operate. The user would manually point the device and choose a field size and it would indicate to him the presence of "bumps" in the visible-surface at a scale appropriate to the field size.

To keep the proposal practical for early studies, it would be left to the user to identify the groupings of elements that go together as a module in the sense of Fig.11 and it would be his responsibility to choose an appropriate coordinate system orientation. He is in a better position to do this than before, since the quantity of data requiring manipulation is now substantially reduced. Fig 12a illustrates what the output of this device would be like. The sensitive field is indicated by the horizontal line grid and the circles indicate loci of coarse surface bumps detected by the device. From this we know there is something to the left and it is vertically elongated, but nothing else. Pointing toward this object and scaling down the field size might yield the output shown over the diagonal line grid in Fig.12b. Note that the orientation discovered in Fig.12a is a sensible reference frame for relating the structure we find in Fig.12b. We can proceed to smaller field sizes inspecting the structure of the various elements found at the medium scale as shown in Fig.12c. The user would use the wide field information to locate interesting regions for more detailed examination and the orientations observed at that scale would suggest coordinate frames for relating part positions at the finer scales. After a few moments of examination, switching between coarse and fine fields the user might be able to build up an impression like that illustrated in Fig.13.

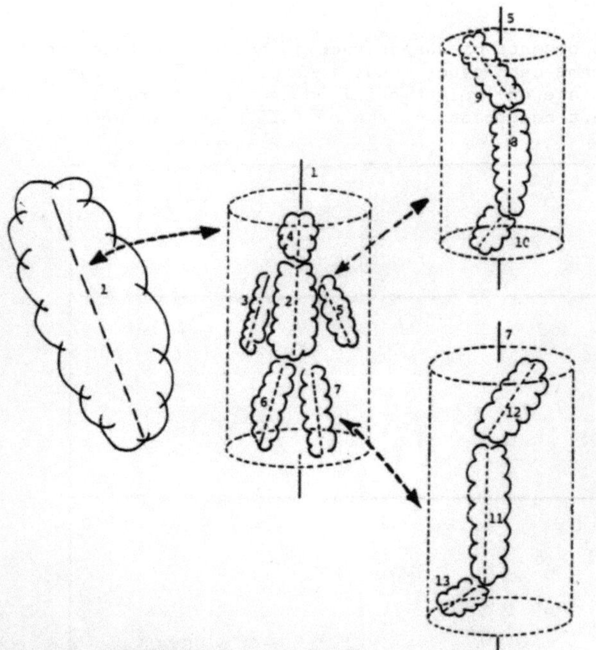

Fig.13. A 3-D model description

Each of the modules shown here is a complete unit of shape information and could stand on its own. D. MARR and I called these modules 3-D models. A description in the 3-D model representation would consist of a collection of related 3-D models like that illustrated in Figure 13.

6. Discussion

The 3-D model representation characterizes a class of shape representations responsive to the demands of recognition. I have attempted to structure this presentation around the problem of helping a blind person but much the same issues have to be addressed by any system attempting to recognize shape. In this section I will briefly survey several questions concerning the use and limitations of such representations.

If a consistent description can be produced for a shape each time it is seen, that description can be used as an index into a catalogue in memory containing information about shapes seen previously. Fig.14 illustrates a possible organization structured according to specificity. If the available shape information allows construction of a 3-D model indicating only position and orientation of the overall shape, that would index the catalogue entry at the top of the figure. If information is sufficient to count the number of major components and roughly estimate their placement within the shape, catalogue entries from the second level could be discriminated. Similarly if more precise relative size and orientation information were obtained, access could proceed down to the third and fourth rows of the illustration.

3-D models for the components of shapes such as hands and faces would also be indexed in the same catalogue. Thus a shape seen from a bad angle---such as a horse face on---might still be recognized from its head or legs even though important components of the overall shape are obscured.

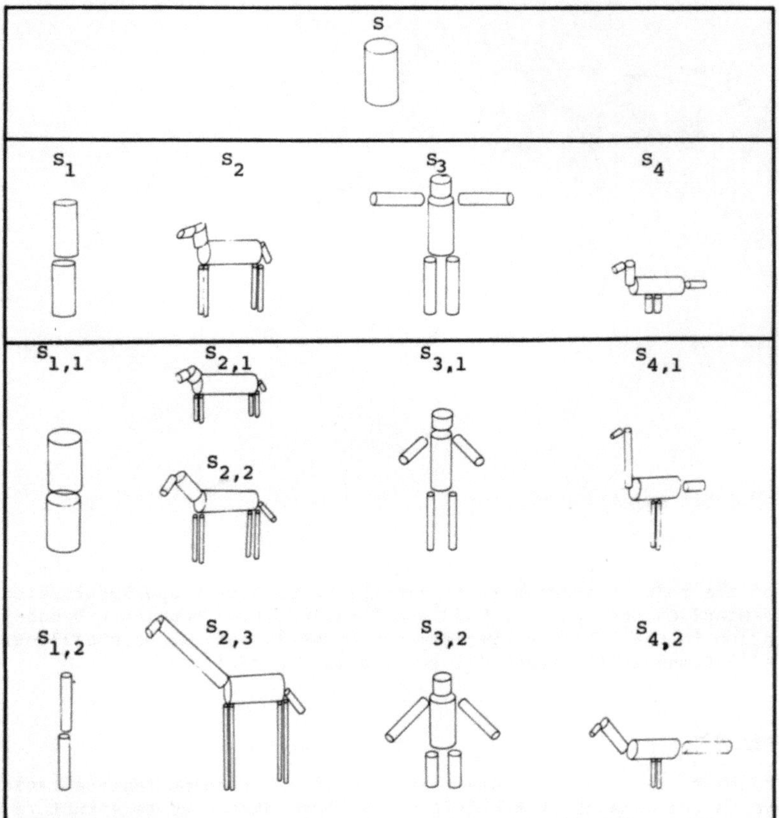

Fig.14. Organization of a shape catalogue

Implicit in this observation is a definition of object or shape in terms of describable stable structure rather than physical connectivity. This makes it possible to handle novel composite shapes such as a unicorn as a horse with a human bust even though it is neither. The simplicity of the individual 3-D models allows useful descriptions for indexing a catalogue to be constructed quickly. Further 3-D models can be assembled when there is time and this refinement can be guided---to significant details---by information made available from the initial catalogue access.

An object-centred, volumetric, and modularly organized representation will perform poorly, however, for shapes in which any of these three design parameters is weakly defined. For example, a piece of paper crumpled into a ball tends to have many small volumetric elements---the wrinkles---all at the same scale with little discernible organization into groupings of coarser scale structure and no clear orientation to the whole shape. The accessibility of this class of representations is also limited by occlusion problems, if a shape is seen from a poor vantage point where important components are hidden from view, or the major axis of the shape is foreshortened causing a minor axis to be chosen as the coordinate frame of the shape.

In many instances, there are ways to overcome these limitations, for example by noticing that a component or parent of the shape is recognizable or that as in the case of a sphere, the absence of a discernible coordinate frame and primitive elements is actually a useful characterization of the shape. Beyond this, these limitations are an important component of the theory we are developing [15]. They tell us what ought to be difficult shapes for recognition and if this turns out not to be the case, they probably will provide the strong clues to an improved theory.

The purpose of this study was to set out a framework for thinking about the problem of recognizing shape using visual information. We began this by considering what type of information we begin with and what type of information is appropriate to solving the problem. It appears that rather strong implications can be drawn from a consideration of a few basic criteria and design issues.

7. References

1. Marr, D. and Nishihara, H.K. Representation and Recognition of the Spatial Organization of Three-Dimensional Shapes, Proc. R. Soc. Lond. B., 200, 269-294 (1978).

2. Roberts, L.G. Machine Perception of Three-Dimensional Solids. In: Electro-optical information processing, (ed. J.T. Tippett et al.), 159-197, M.I.T Press. (1965).

3. Guzman, A. Computer Recognition of Three-Dimensional Objects in a Visual Scene. Artificial Intelligence Laboratory Memo AI-TR-228, Massachusetts Institute of Technology, (1968).

4. Waltz, D. Understanding Line Drawings of Scenes with Shadows. In The psychology of computer vision, (ed. P.H. Winston), 19-91, McGraw Hill (1975).

5. Huffman, D. Impossible Objects as Nonsense Sentences. In Machine Intelligence 6, (eds. B. Meltzer and D. Mitchie), 295-323, Edinburgh University Press (1971).

6. Clowes, M.B. On seeing things. Artificial Intelligence 2, 79-112 (1971).

7. Mackworth, A.K. On the Interpretation of Drawings as Three-Dimensional Scenes. Ph.D. thesis, University of Sussex (1974).

8. Blum, H. Biological Shape and Visual Science (Part I). J. Theor. Biol. 38, 205-287 (1973).

9. Binford, T.O. Visual Perception by Computer. Presented to IEEE Conference on Systems and Control, Miami (1971).

10. Agin, G.J. Representation and Description of Curved Objects. Stanford Artificial Intelligence Project, Memo AIM-173, Stanford University (1972).

11. Nevatia, R. Structured Descriptions of Complex Curved Objects for Recognition and Visual Memory. Stanford Artificial Intelligence Project, Memo AIM-250, Stanford University (1974).

12. Hollerbach, J.M. Hierarchical Shape Description of Objects by Selection and Modification of Prototypes. Artificial Intelligence Laboratory Memo AI-TR-346, Massachusetts Institute of Technology (1975).

13. Marr, D., Poggio, T. A Computational Theory of Human Stereo Vision. Proc. R. Soc. Lond. B., 204, 301-328 (1979).

14. Nishihara, H.K. Intensity, Visible-Surface, and Volumetric Representations. Artificial Intelligence, 17, 265-284 (1981).

15. Sutherland, N.S. The Representation of Three-Dimensional Objects. Nature, 278, 395-398 (1979).

Parallel Computers and Vision – Can Hardware Concepts Give Clues to "Brainware"?

A.H. Bond

Image Processing Institute, University of Southern California,
Los Angeles, USA

and

Artificial Intelligence Laboratory, Queen Mary College,
London University, London, England

1. The Two Fields

This conference brings together experts from two fields. On the one hand, we have researchers in the biology of vision in humans and animals. They study anatomy, physiology and psychology. And on the other, we have researchers in artificial intelligence and computer vision. They work in intelligent software for computer vision, in hardware approaches to computer vision and industrial applications of computer vision.

Although in roughly equal numbers at this meeting, in general it seems to me that there are probably about ten times as many workers in the biology of vision than there are in computer vision.

At the risk of insulting both fields let me characterise (or caricature) what they do and what they want from each other.

What They Do

The biologist is engrossed in experimental technique and experimental design. He is trying to discover the phenomena of vision in living systems. His main problem is to obtain good experimental data and to explain it. He is in a very exciting period of discovery at the moment.

The computer scientist is a compulsive model maker. In spite of an interest in living systems, he approaches vision as an engineering problem, de novo.

What They Want From Each Other

The computer scientist wants the biologist to describe models, to speculate but to produce a model which is well enough defined that it can be specified as an engineered mechanism. However the biologist does not want to speculate, his recent past is strewn with discarded concepts and assumptions. Any theory he might produce would be premature and naive, it would have to be thrown away three weeks later due to new experimental findings.

The biologist wants the computer scientist to produce concepts and mechanisms useful for the explanation of biological phenomena, so he can understand his last experiment and design his next experiment. However, the computer scientist does not know enough about biological mechanisms to produce models related to them. He needs exact specifications, but he is given vague specifications that he cannot construct or test like the ones he works with.

What We Are All Looking For

Is a set of concepts, a vocabulary and a language for describing, representing and discussing information processing in humans.

Attempts include: "boxes and arrows", Lisp and relational data structures.

2. Brain System Description

Computer System Design

Computers have seemed to offer many things to psychologists, physiologists and even anatomists, concepts for discussing information processing and a modelling tool allowing the testing of theories of information processing in humans.

However, the existing design of real computers has differed greatly from information processing systems considered as explanations of biological phenomena. Computers can hardly have differed more in their structure and performance from the kinds of information processing structure and performance needed to explain living phenomena. After all, computer design arose from two or three pressures: the technology of reliable switching mechanisms - hence binary coding, the commercial profit from business applications - hence fast arithmetic, and known programming techniques based upon von Neumann random access memory and expression as nests or sequences of functions.

We are however now in a new era in which computers are sufficiently inexpensive that experimentation in structure is feasible. LSI "chip" technology has reduced the design and construction problem by many orders or magnitude. Interfacing and communication between units has become understood and standardised. Hardware is now sufficiently reliable that it is not a major effort to keep it operational.

The computer architecture problem is to select from modules such as:

Memory Modules - random access addressable (RAM and ROM), associative, cache, shift registers (chip or bubble, etc.), etc.

Processing Modules - logical units, arithmetic units, central processing units.

Interfacing Modules - serial and parallel ports, direct memory access ports, multiplexers, encoders and decoders.

Communication Modules - serial and parallel lines, busses, broadcast carriers.

and to connect them together to produce a machine which can be programmed to produce a desired total computer system.

The space of computer architecture is very large and only about four types of idea have been so far explored:

Pipleline computers - with a small set of different special processors executing one program.

Array processors - with a large set of identical processors executing one program in a synchronised manner. This type includes the so-called associative processors.

Multiprocessors - with a small set of processors executing separate but strongly interacting programs asynchronously, this includes multimicro-processors.

Computer networks - with a loosely coupled set of autonomous processors executing separate programs which interact occasionally, mainly by sharing data.

Computer System Description And Brain System Description

The total computer system design problem is to specify an information processing system with required functionality, input and output channels, and performance. For example, a single user computer for word processing, having to handle certain types and quantities of information at certain reliability and cost, or a controller for a robot with certain mechanics and sensors to enable it to navigate through certain types of environment and carry out certain time-dependent tasks.

A solution of the design problem is a system structure with a specified physical implementation, defining the hardware and its associated software.

Computer systems are extremely complex, and their design and understanding would not be possible without a partitioning of their description into a heirachy of layers. Fig. 1 shows a typical layered description scheme. The layers are defined by specifying the interfaces between them. A program or process at level n calls functions at the n-1 interface and provides functions at the n interface.

An interface usually constitutes a language providing functions, methods of combining functions and control structures, and often having a syntax for an external linguistic description.

Program in High Level Language

——— e.g. Fortran

Assembly Code Program

——— Assembly language

——— Supervisor calls (Utilities, I/O)

——— Executive (Scheduling of processes, Resource sharing, error and interrupt handling)

——— Hardware (Interprets machine code directly)

Computer Architecture

——— Architecture modules

Module Design as Circuits

——— Gates

——— Physical implementation (As TTL or VLSI, Electrical loads, Spatial layout)

Fig. 1. Computer system description

The reason we have worked through all this is to earn the ability to propose, by analogy, a layered approach to brain system description, with interfaces which provide a hierarchy of languages for describing the structure and performance of the system at each level. In Figure 2, we give such a simple layered scheme.

Description of process in terms of rules and processes running on a machine	Psychology
Description of machine in terms of modules	Brain Architecture
Description of modules as hardware units, e.g., design of a cortex in terms of abstract neurones	Module Design
Description of small neuronal structures	Circuit Design
Implementation Spatial Layout Redundancy Fault Tolerance Energy Considerations	Anatomy

Fig. 2. Brain system description

Questions About The Brain

When I started thinking about the set of modules which could form the architecture of the brain, and comparing them with computer concepts, I realised that there were many basic distinctions and concepts in computing that might have to be modified in dealing with living systems.

Here are some questions I should very much like biologists to answer for me:

1. Is the division of an information processing system into program and machine a useful concept for living systems?

A major conceptual breakthrough both in practice as well as later in theory was the description of information processing systems as consisting of two elements - the program and the machine. The operation of the system was described by the program being executed by the machine. This concept allowed much simpler description of systems than had been previously possible.

The program as controller seems appropriate for brainware. However, other properties of the computer concept of program may not be. We are used to being able to process programs as data, to move them around, make copies, modify them from another program, etc. These properties flow from the von Neumann machine concept. We can still have a program and machine concept without these von Neumann properties. It amounts to a two-layer hierachical control description with the top layer being the program controlling the bottom layer, the machine. We tend to think in serial terms of a single point of control active at any one time in the program, this may not be the case.

The program - machine distinction is separate from the software - hardware distinction, which is an implementation issue. A program could be implemented in hardware, running on a hardware machine. It would be a controller selecting which elements of the hardware machine are active, which data paths are to be active and which data is to flow.

2. Is the memory versus logic distinction useful?

In computer system design, one traditionally distinguishes between memory ("sequential") circuits which can change their action and state permanently or at least stably, as a result of inputs received, and logic ("combinatory") circuits which do not change permanently and always perform the same function. This distinction was reinforced by two other concomitant properties. Memory was slow and logic fast in action, and memory was expensive and logic cheap. These days these distinctions are becoming blurred. One has mixed units, memory with additional logic, modifiable logic circuits, etc. Is such a distinction present in biological systems?

3. Associative memory presumably is to be used, but what kind?

What is it like to design a system with only associative memory and no addressable memory or sequential memory? If there is no addressing at all, then presumably there is no storage management problem, no need to have processes for storage allocation and no need to compactify, etc. What is the width of data items normally stored in living associative memories, how wide are typical retrieval keys and how general is the process of association of the key into the memory? The stimuli and other symbolic items are presumably coded in a highly parallel form, and probably highly redundant. What kind of parallel encoding of stimuli into signals is used?

4. Distributed memory and copying of information.

First, does the brain copy information? As a stimulus moves through the system, does it get transferred by a copy being made, or is there destructive readout destroying the previous record and maintaining the uniqueness of the instance of the event record? A unique copy might be economical in energy, since energy could be transferred to the new copy upon destruction of the old.

Second, does the brain make multiple copies of a record? By this I really mean similar copies of a record, where each copy might record some aspect of the total event. If as seems likely there are multiple copies then one has to be able to retrieve them together. The danger of multiple copies comes if and when you modify them, in which case all other copies, similar copies and partial copies have to be modified as well, to maintain consistency.

Along with copying of information goes the idea of modifying an existing memory trace. Does this happen or is a new complete modified version installed?

5. Slow memory and parallel processing.

Since the long-term memory is rather slow, about 300ms per chunk, this must mean that speed is attained by a small sequence of highly parallel actions. Is this a reasonable conception and if so, how can this be described?

6. Are memory protection mechanisms used?

That is, as part of the processing, can whole areas of memory be inhibited from access by keys? Could data typing be used, where data fall into classes (or "types") which are processed in ways and by processes specific to their particular class?

7. Are their special memory structures?

There are two special memory structures, which if they existed could make a big difference to the action of the system. First, are their special structures for spatial maps? Second, is there a "tape recorder" type episodic memory always recording what is happening and allowing free associative access?

Send your answers please to Bond, California.

If we had answers to these questions, we could go ahead and produce practical architectural designs for the brain and also go on to a higher level psychological description in terms of programs running on, in or through this architecture.

3. Our Work With A Parallel Computer

The Distributed Array Processor

We want now to describe our work on computer vision using a particular parallel machine, the ICL Distributed Array Processor (the DAP). We will first outline the architecture of the DAP and its operation. It is perhaps best thought of as an intelligent parallel memory machine with a single serial controller.

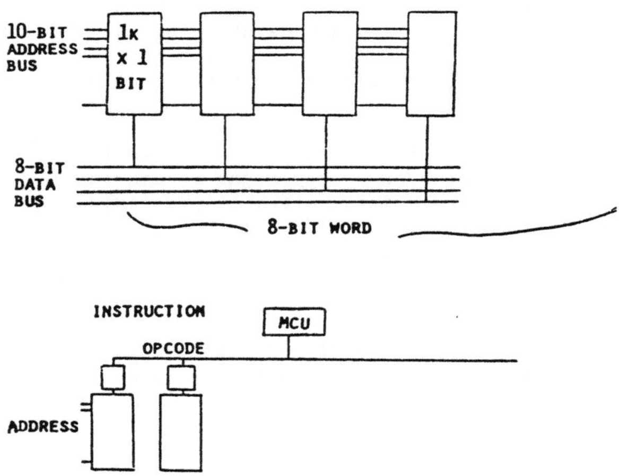

Fig. 3. Intelligent memory

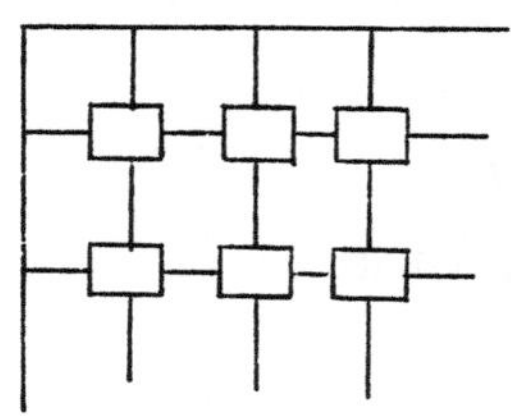

Fig. 4. Square array and NSEW neighbours

In Fig. 3, we diagram first normal addressable memory as used say in an 8 bit microprocessor, in which a set of 8 memory chips receives a 12 bit address (giving 4096 different addresses) and each chip outputs a 1 bit value, 0 or 1. The set of 8 bits produced is the computer word that has been retrieved. The DAP uses intelligent memory obtained by adding some logic to each memory chip. The logic is very simple and provides only boolean operations on bits. In addition, 3 one bit registers, call Q, C and A, are added to each chip. The intelligent memory chip will be called a PE (processing element).

The DAP control unit reads through a serial program. It decodes each instruction and broadcasts the operation code and operand address to all the PEs, so they all perform exactly the same instruction at the same time. These identical operations are of course performed upon different data stored in each PE. In addition, a PE has the choice of ignoring the currently broadcast instruction and doing nothing. This is controlled by A register (the activity bit).

This type of parallelism with a set of identical processes is often called "uniform parallelism".

There are a large number of PEs, namely 4096. They are arranged in a square 64 x 64 array, see Fig. 4, and there are connections from each PE to its N, S, E and W neighbours. The meaning of these connections is that data from the neighbours' Q, C and A registers or store can be used in operations. Each DAP operation uses one, two or three one bit operands and produces a one bit result. The operand can be either the contents of a PE register, i.e., Q, C or A, or the contents of the selected address in the PE store, i.e., one of the 4K bits, S, see Fig. 5.

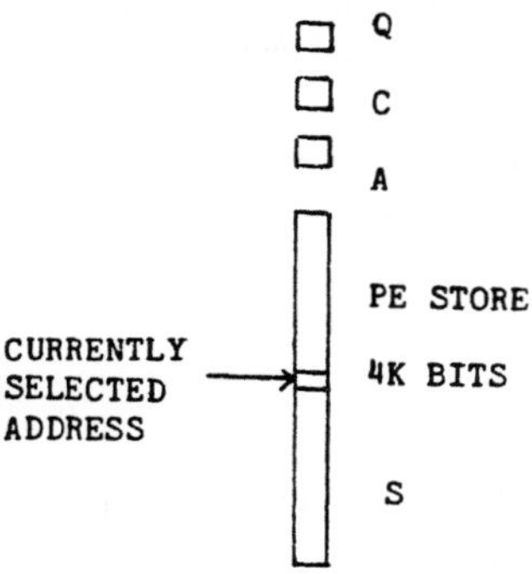

Fig. 5. DAP operands

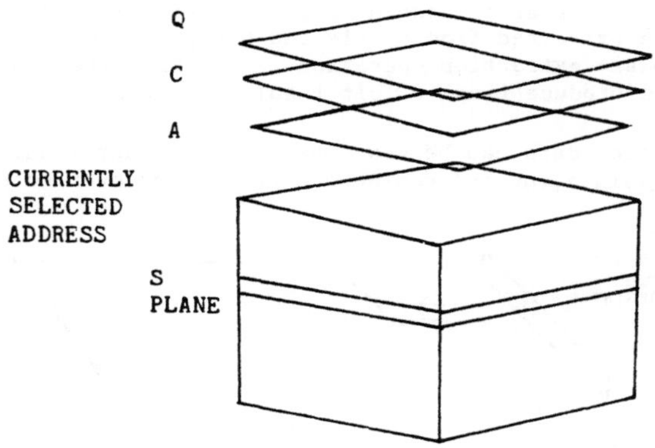

Fig. 6. Bit planes

The result is placed in a register or the selected PE store address. A basic DAP PE machine operation takes the one, two or three selected bits, which can be Q, C, A or S, does a boolean operation and puts the (one bit) result into Q, C, A or S.

The set of selected store bits over the whole array forms a "bit plane", i.e., a 64 x 64 array of bits. Likewise, the Q, C and A registers form the Q-, C- and A- bit planes. Since any operation is replicated over the whole array in parallel, the whole array of PEs is operating on one, two or three planes of bits to produce a resultant bit plane, see Fig. 6.

Any instruction can specify that the operands are to be taken from the N, S, E or W neighbour. Since all PEs do the same thing, this is usually thought of as a shift operation combined with the basic operation, see Fig. 7.

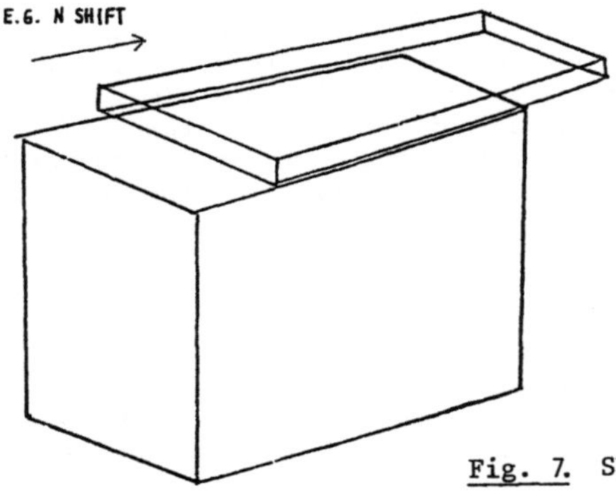

Fig. 7. Shifts

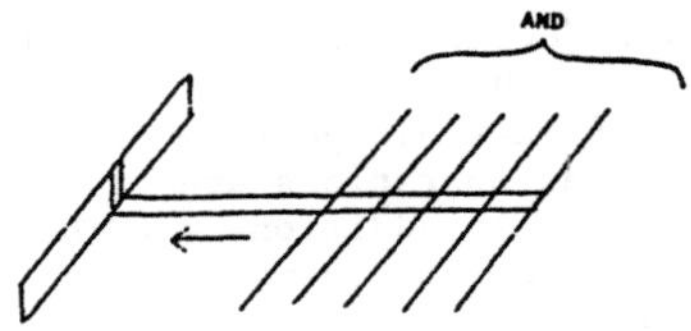

Fig. 8. Data extraction

The control unit for the DAP is like an ordinary serial computer. It has eight 64 bit registers, can do logic and arithmetic, etc. in these registers and it can branch conditionally upon their contents.

In order to branch upon the data computed in the array of PEs, data must be extracted from a selected bit plane into a control unit register. The basic extraction operation is to AND corresponding bits along rows or columns to produce final 64 bit result, see Fig. 8.

Also, data can be broadcast from a control unit register to the array by replacating its contents over rows or over columns, see Fig. 9.

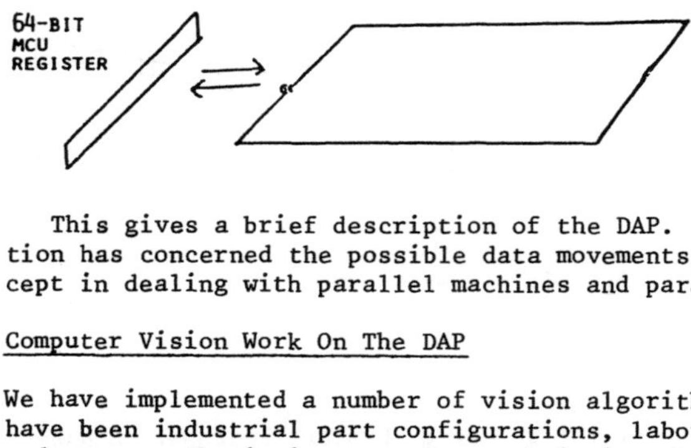

Fig. 9. Data broadcasting

This gives a brief description of the DAP. Note that most of the description has concerned the possible data movements, and this is an important concept in dealing with parallel machines and parallel algorithms.

Computer Vision Work On The DAP

We have implemented a number of vision algorithms on the DAP. The images used have been industrial part configurations, laboratory scenes, blocks worlds and some synthetic images. Our work is just beginning to investigate the software design issues in this area [1].

Local Operators

Local operators of the type discussed by M.J. Duff in this symposium are easy to program on the DAP and compute quickly. We have implemented gradient operators and Walsh function edge feature detectors [2]. The Marr-Hildreth operator has been run for diameters up to 128 pixels on 256 x 256 8 bit images giving timings of a few seconds.

My colleague, HILARY BUXTON has implemented a space-time generalisation of the Marr-Hildreth operator, which finds moving edges [3]. This has been run for 40 x 40 x 40 windows in 52 image temporal sequences of 64 x 64 8 bit images timings of a few seconds.

Other interesting local operators that we intend to investigate are texture measures and local shape from shading methods.

Histograms And Region Extraction

My colleague, PHILIP MARKS, has reported work on the parallel extraction of intensity histograms [2], using the basic extraction operation of the DAP. From this, intensity buckets can be extracted and used to extract regions by their differing intensities. The bucket interval is broadcast to the image and regions projected out in parallel.

Curve Fitting

In a similar way, a straight line or known curve can be fitted to a set of points in parallel by broadcasting the curve's description and having all the PEs compute their distance in parallel, see Fig. 10.

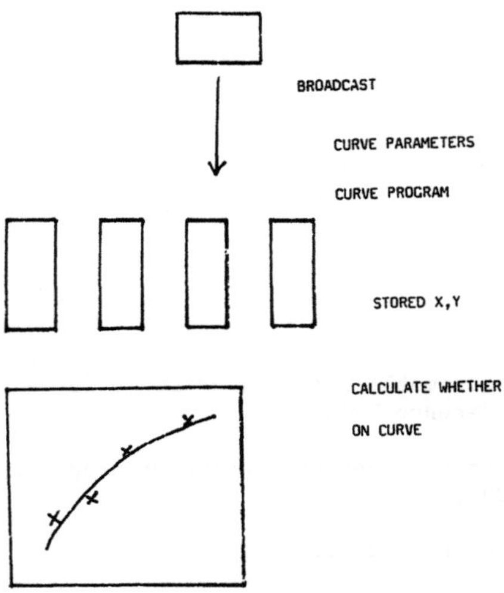

Fig. 10. Global curve fitting

Many To Many Matching

The problem of matching two sets of items with scalar properties to each other can be performed in parallel, see Fig. 11. Each set can be represented as a vector, then one can be replicated over rows, the other over columns and the two matching in parallel, giving the fit of each item in one set to each item in the other set.

Relaxation

We do not yet know whether two graphs can be matched easily in a fully parallel way. MARKS [4] implemented a Waltz relaxation for the special case of line drawings, however he was only able to constrain in parallel all the arcs from one vertex at a time.

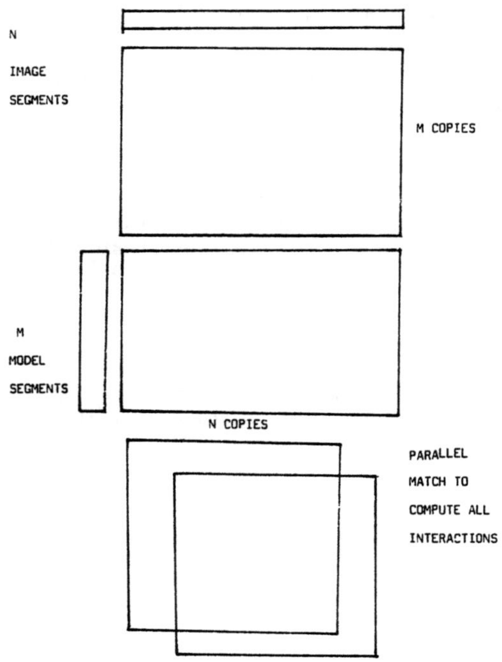

Fig. 11. Many - Many matching

References

1. Bond A.H. and Buxton H. "The use of programmable parallel hardware for industrial vision" Proceedings SPIE Conference 1982.

2. Marks P.C. "Low level vision using an array processor" Computer Graphics and Image Processing, Vol. 14, pp. 281-292, 1980.

3. Buxton B. and Buxton H. submitted to Proc. Roy. Soc., 1982.

4. Bond A.H. and Marks P.C. "Scene analysis using the ICL Distributed Array Processor" Technical Report, Artificial Intelligence Laboratory, Queen Mary College, London, 1981.

Image Processing in Perception and Cognition

G. Sperling, M. Pavel, Y. Cohen, M.S. Landy, and B.J. Schwartz

New York University, New York, N.Y. 10003, USA

1. Overview

In this paper we briefly survey theories and ideas about image processing, with some illustrative examples, taken mostly from the Human Information Processing Laboratory at N.Y.U. First we develop the concepts of *multiple stable states* and *path dependence* in a basic visual-motor task (vergence of the eyes) and show how these can be encompassed in *potential theory*. We then consider two examples of human information extraction from complex, dynamic, visual displays: (1) the extraction of the shape and the motion of 3D wire objects from 2D images, and (2) the extraction of meaning from displays of deaf signers communicating in American Sign Language (ASL).

In extracting 3D shape, the human perceptual system uses *fallible heuristics* which can lead to obvious perceptual errors (i.e., powerful perceptual illusions). Further, in ambiguous stimuli, the induced perceptual state depends strongly on immediately preceding stimuli *(path dependence)*. The theory of the "logic of perception" proposed to account for these observations is that information about a stimulus--derived concurrently by many different processes--is combined in a non-linear competition/cooperation network to achieve a perceptual decision.

In cognitive information processing tasks, such as extracting meaning from continuous ASL communication, there are two levels of processing: first, parsing the continuous sequence of images into a sequence of signs (analogous to parsing spoken language into words or cursive script into letters) and second, extracting meaning from the sign sequence. The parsing problem is characterized by the extremely restrictive structure of the information received. For example, the number of different signs in ASL is only on the order of thousands. Therefore, experienced signers can extract the reference form of a sign from very reduced images and from very different contexts (even though performance of the same sign may vary considerably with context). A practical consequence is that we have been able to produce legible ASL images of very low bandwidth (on the order of 10 kHz) by various image-coding schemes: spatial low-pass filtering, binary intensity encoding, and cartooning by edge detection or zero-crossing algorithms. *Feature templates* and *intentions* are the operative concepts for parsing; *schemas* are the most hopeful approach to the problem of extracting meaning.

2. Image Processing in Perception

2.1. Multiple stable states and path dependence

2.1.1. Multistability of vergence. We consider first a classical demonstration of HELMHOLTZ [17] that may seem specialized but is prototypical of human perception. A subject views a binocular stereogram, that is, a stimulus that produces images that are identical--or nearly identical--at corresponding points in each eye (Fig. 1). To normal subjects, such a stimulus appears as a single object, that is, the images in the two eyes are said to be perceptually fused. Suppose, at this point, the lines of sight of the two eyes are parallel. Now, let the left and right half-images of the stereogram be slowly moved apart in physical space so that, in order to maintain fusion, the lines of sight of the two eyes must diverge. Helmholtz found that he was able to cause his eyes to diverge by eight degrees before fusion was lost and his eyes returned to their resting position. To restore fusion, Helmholtz had to bring the stereo half-images much closer together than the point at which fusion was lost.

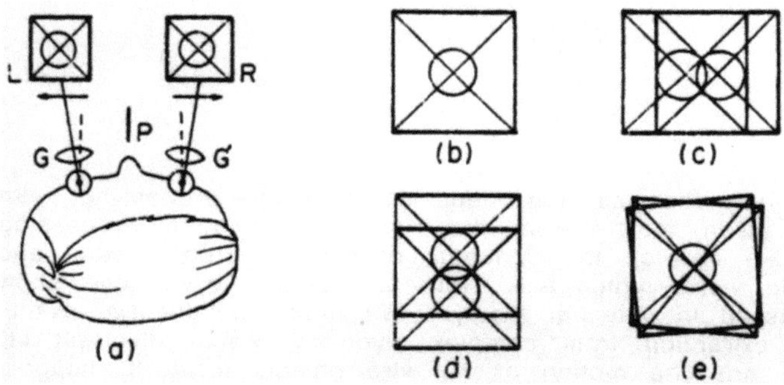

Fig. 1. (a) A stereogram showing the left *L* and right *R* half-images viewed by the left and right eyes, respectively. The lenses *G*, *G'* enable the observer to focus at the short viewing distance. The partition *P* isolates the views of the two eyes. The arrows indicate the direction in which the stimuli are moved in Helmholtz's horizontal vergence demonstration. The broken lines indicate parallel lines of sight; the solid lines indicate the actual lines of sight when the observer is able to verge accurately at the angle of divergence shown. (b) Representation of the fused perception when the eyes are correctly verged. The perception after vergence fails for (c) horizontal vergence, (d) vertical vergence, and (e) torsional vergence. (After Sperling [42])

Some people can control horizontal vergence voluntarily, and this will complicate any demonstration. However, analogous results to those described above are obtained with vertical and with torsional displacements. That is, when the stereo half-fields are displaced vertically the eyes can be forced to diverge vertically by up to about eight degrees, and with torsional rotation of the fields, the eyes can be forced to rotate torsionally around their axes (Figs. 1d,e). To the best of our knowledge, no one has yet succeeded in *voluntarily* making vertical or torsional eye movements.

2.1.2. Multiple stable states, path dependence, hysteresis. These demonstrations from Helmholtz illustrate path dependence for responses to a certain range of stimuli. When a stereogram requires an ocular divergence of, say, one degree, then divergence always occurs and fusion is always achieved, independent of the recent history of stimuli. With a stereogram requiring a divergence of twelve degrees, fusion is never achieved. With a stereogram requiring a vergence of eight degrees, fusion will occur if the immediately preceding stimulus requires a divergence seven degrees and the eyes were fused on it but not if the preceding stimulus divergence was one or twelve degrees. There are two stable states, fused or unfused, in response to precisely the same stereogram. We say there are *multiple stable states* and there is *path dependence.* Insofar as we restrict ourselves to deterministic theories, multiple stable states and path-dependence are two sides of the same coin: different states can be reached only by different paths (sequences of previous stimuli) and, path-dependence implies that the same stimulus induces different states depending on the path taken. *Hysteresis* is the name given to path dependence in which a state tends to perseverate even after the inducing conditions have changed. This is the kind of path dependence demonstrated above.

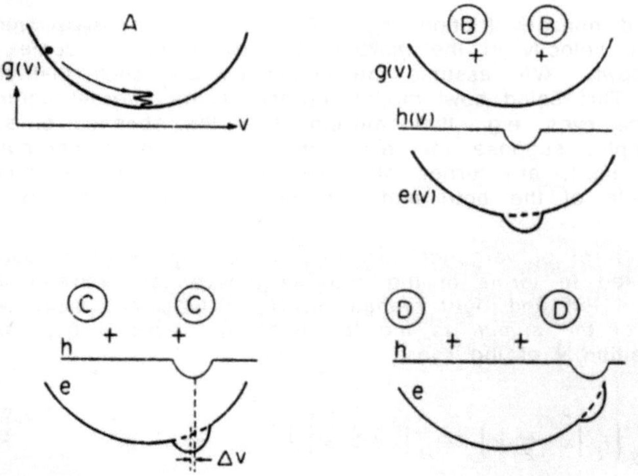

Fig. 2. Potential theory model. (A) Vergence displacement energy *g(v)* as a function of vergence angle *v*. The eye's vergence position is represented by the projection onto the abscissa of a marble rolling on the surface. This surface governs the eyes' return to their neutral vergence position when they are somehow displaced from it; a typical path is indicated. (B) Two halves of a stereogram *(B,B)*, fixation points are indicated by +. Vergence displacement energy *g(v)* adds to image-disparity energy *h(v)* to produce (net) vergence energy *e(v)*, which has a single minimum slightly displaced in the direction of divergence. (C) Two halves of a stereogram requiring a greater divergence than (B) to achieve fusion; *g(v)* is same as in (B) and therefore omitted; the corresponding *h(v)* is shifted to the right. The e(v) surface has two minima. The arrows indicate "vergence disparity" Δv: the displacement of the minimum of *e(v)* [which determines the actual vergence position] away from the minimum of *h(v)* [which minimizes *L/R* image disparity]. (D) A stereogram which is at the limit of vergence fusion. The *e(v)* surface has only one minimum, which corresponds to the neutral vergence position. (After Sperling [38])

2.2. Potential theory interpretation of path dependence.[2]

Let the vergence position, v, of the eyes as a function of time, t, be given by v(t). Let the composite of all forces tending to alter vergence position be dg/dv, and let

$$dg/dv = -k_1 v'' - k_2 v' \tag{1}$$

where ′ denotes differentiation with respect to t. Equation 1 is an elementary, linear differential equation in which the coefficient k_1 can be thought of as describing the mass of a moving body and k_2, the friction of the medium in which it moves. A simple, concrete realization of (1) would be a salad bowl filled with oil in which a marble rolls under the influence of gravity. The shape of the bowl is described by g(v) (Fig. 2a). The vergence position of the eyes is represented by the horizontal coordinate of the marble. In electrical potential theory, g represents field potential, dg/dv represents force; here, g is called *vergence displacement energy.*

For bowl-shaped g(v) and nonzero friction ($k_2 > 0$), it is evident that, whatever the starting position and velocity of the marble, it must eventually come to rest at the bottom of the bowl. (We assume the conditions are such that the marble remains in the bowl.) This salad bowl model represents the internal control of vergence movements of the eyes, e.g., their movement in the absence of any external stimulus. For example, suppose the eyes are verged on a particular stereogram and suddenly the lights are turned off. This is represented by positioning the marble at the side of the bowl and allowing it to roll freely to its resting position (Fig. 2a).

External factors in the control of vergence are represented by image disparity energy, h(v). h(v) is expressed in terms of the squared differences between the illuminance distributions on the left and right *retinas*, which in turn, are expressed in terms of the luminances of the *stimuli* to the left and right eyes, $I_L(x,y)$ and $I_R(x,y)$, and the vergence position v of the eyes:

$$h(\nu) = -\iint_{x,y} \left[I_L\left[x-\frac{\nu}{2},y\right] - I_R\left[x+\frac{\nu}{2},y\right] \right]^2 dx\ dy. \tag{2}$$

One should think of *h* as the error in image registration--the square of the difference between the two eyes' images. The critical point is the assumption that h(v), or some other error function like it, can be computed not only for the eyes' present vergence position but for other values of vergence. It is easy to imagine several ways in which neurons might compute h(v) in the neighborhood of the current vergence, v [38]. Insofar as h(v) can be computed in a neighborhood of v by the visual system, it would know in which direction to move the eyes to reduce the registration error h(v), i.e., to increase the correspondence of the two left and right retinal images. Figure 2b illustrates a typical h(v).

In the potential-theory model, vergence is controlled by e(v) = g(v) + h(v), that is, by the sum of internal and external factors, i.e., the marble rolls on the e(v) surface. How this works is illustrated for three cases. When a stereogram requires only a small amount of divergence for fusion, there is a single minimum in e(v) and consequently only one stable state: fusion of the object. When the stereogram requires a greater amount of divergence, there are two stable states: resting position and fused on the stereogram. From e(v) in Fig. 2c it is clear how, by moving the minimum of h gradually from the position represented in 2b to that in 2c, the marble will stay in the minimum corresponding to the fused state. Figure 2d illustrates the case of too great a disparity for fusion, the perturbation in e(v) caused by h(v) is too shallow to hold the marble.

Figure 2 illustrates the potential theory representation of the multiple stable states of vergence. There is also a catastrophe theory representation (SPERLING [42]) which is more succinct but omits any suggestion of the dynamics of the system. The vergence system is an especially attractive one in which to investigate dynamics because all the intermediate states of the system as well as the stable states are observable.

2.3. Neural model for multiple stable states

Like vergence, higher level perceptual processes in humans (and presumably, in all biological systems) also are characterized by path dependence and multiple stable states. The functional description of these processes is quite similar to the case of vergence, but the actual mechanism involves, we believe, a very specialized neural network for making decisions. In horizontal vergence, the eyes can be verged on only one vertical plane at any one time, and are verged on exactly one plane because of physical constraints.

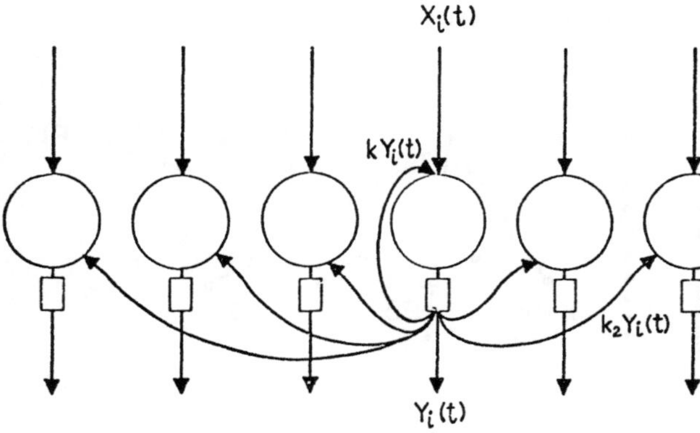

Fig. 3. A monoactive column of model neurons. Connections are shown for one neuron, i. It receives an external input $x_i(t)$ that sums with an output-produced feedback excitatory input $k_1 y_i(t)$. It sends shunting inhibitory signals $k_2 y_i(t)$ to all other cells. The small rectangular box in the output path represents a threshold: when its input is y, $y > 0$, its output is $\max(y - e, 0)$. (After Sperling [42])

An analogous neural network is the "monoactive" network of n neurons, only one neuron can be "active" at one time and exactly one neuron is active at every time. A proposed wiring diagram for a monoactive net is shown in Figure 3. Neurons are labeled i, inputs are $x_i(t)$, outputs are $y_i(t)$, each neuron feeds back its output onto itself as an excitatory input and every neuron sends its output to all the other neurons in the net as a strong inhibitory input. Simple inhibitory networks are not monoactive. Three properties make this inhibitory network monoactive: (1) the output range of neurons is bounded; (2) neurons have a threshold, ϵ, which their net input must exceed in order for there to be an output; (3) each neuron has positive self-feedback. The first property holds for all neurons, the second for nearly all, and the third is unusual. A network with these properties was proposed by SPERLING [38], and independently by GROSSBERG [15, 16] who developed powerful mathematical analyses of such systems. A monoactive network is necessary where decisions need to be made between alternatives that cannot or should not be combined. For example, when there are several widely-spaced objects competing for attention in the visual field, the eyes point at each of the objects in turn, not at the mean position, which may be blank. When an animal is hungry, thirsty, and sleepy, and must satisfy these drives at different locations, it goes to each location in turn, not to an in between location.

2.4. Cooperation/Competition

We consider here three processes that occur in the human cyclopean image, that is, the perceived image of the world produced by the two eyes acting together. The thrust of these examples is that various decisions are made about each point in the cyclopean field. Each decision mechanism can be represented as a monoactive column of cells centered over x, y. Although the interaction within a column is entirely competitive, neighboring columns are assumed to interact cooperatively. That is, a decision reached in one column favorably influences the analogous outcome in its neighbors (i.e., the active neuron provides an excitatory input to neurons on the same level in neighboring columns). Between-column interactions are local, but effects can propagate widely.

A

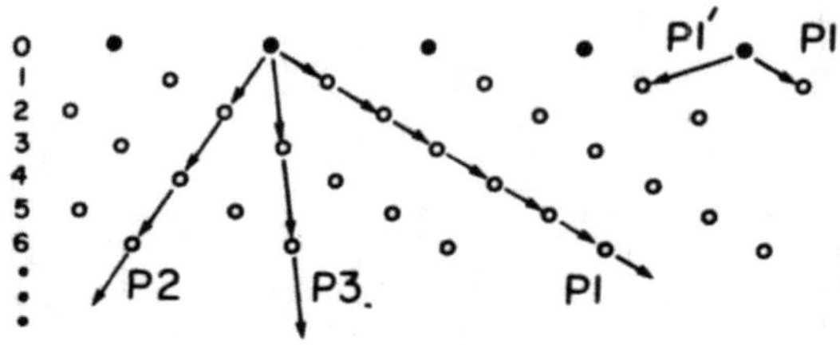

B

Fig. 4. Two examples of ambiguous displays. (A) Static figure/ground ambiguity, Rubin's faces/vase. (B) Ambiguous motion display. Dot row zero is flashed first, followed by row 1, 2, and so on. Motion can be perceived along any of the paths P1, P2, P3; which of these paths dominates in this configuration depends on the time t and the distance d between successive points along the path. Although P1′ is a potential pathway, it is not perceived because it has the same t as P1 but a larger d, and is therefore (theoretically and empirically) always weaker. (Based on Burt & Sperling [4].)

2.4.1. Stereoscopic Depth. At any point x,y in our cyclopean image we perceive one and only one value of depth. That is, we cannot see two fine details at different depths at precisely the same point in the cyclopean field. Even in looking at a landscape through a lace curtain, each point will be seen to belong either to the landscape or the curtain. In stereograms containing large regions of ambiguous depth, the perceptual resolution of these regions can be determined by a few points of unambiguous depth.

2.4.2. Binocular rivalry. When conflicting features (e.g., lines at different orientations) are presented to corresponding points of the two eyes, binocular rivalry is usually observed--that is, a location x, y of the cyclopean field will contain the feature of one eye or the other, but not both. Large ambiguous (rivalrous) regions are easily influenced by small unambiguous regions, and tend to resolve as wholes.

2.4.3. Figure-ground. In the classical ambiguous figures (Fig. 4a), every point in the cyclopean image is labelled as "figure" or "ground". Any region that is unambiguously figure tends to expand to its limits, as does any region labelled ground. In the classical ambiguous figures, there are two equally valid, symmetrically opposite labelling schemes. The figure-ground labelling process probably is a manifestation of a more general object labelling process that assigns each point of the visual field to an object or to the background.

Since SPERLING [38] (proposed cooperation/competition networks to account for phenomena of stereoscopic depth perception, they have been widely adopted by theorists in similar contexts (e.g., JULESZ [22], DEV [11], NELSON [30], MARR & POGGIO [28]).

2.5. Perceptual Selection: The Dominance Problem

The examples up to this point have dealt largely with the mechanism by which perceptual decisions are made and have not considered the nature or situation-specific content that enters into these decisions. To open this discussion, let us consider a three-dimensional wire cube rotating about a vertical axis. When viewed monocularly, such a cube is ambiguous; it can be perceived either *veridically* (correctly) or *reversed*, (i.e., the front face appears in the rear, the rear face in front, and the apparent direction of rotation is reversed). To avoid the influence of incidental cues (shadows, texture and imperfections in the wires, the slight increase in angular subtense of the wire edges as they approach the observer, etc.) it is best to represent the cube dynamically on a CRT screen. Ideally, this CRT display would produce the same monocular retinal image as a real wire cube except that the CRT's visual properties are more flexibly controlled than those of real wire cubes. Thus, it is not surprising that a well-constructed 2D CRT display can look virtually as solid, as 3D, and as convincing as a real wire cube. When our subjects' visual perception of a CRT display spontaneously "flips" from the veridical to reversed state (i.e., the direction of motion reverses), it is so convincing that they mistake the perceptual state change for an objective change in direction.

There are two critical questions relating to the perceptual reconstruction of 3D wire figures from their 2D projections. (1) How are the various alternative perceptual states computed? (2) What determines the relative dominance of the perceptual alternatives? We consider the question of dominance first.

2.5.1. Control of dominance by proximity luminance covariance (PLC). In the case of parallel projection of a rotating wire object (Fig. 5a), the veridical and the reversed figures are mirror images of each other and appear to rotate in opposite directions. There is no physical basis for discriminating between the two alternatives. In experiments, subjects report the two alternative states with equal frequency. Suppose we introduce a positive PLC into the display, that is, we make edges that represent portions of the 3D cube close to the observer brighter than those edges that represent distant portions (Figs. 5c, g). Now we find (SCHWARTZ & SPERLING [37]) that 96% of the time the observers report the perceptual alternative with brighter edges forward; only 4 percent of the time are bright edges seen in the rear. In the case of parallel projection, making *distant* parts of a cube brighter (a negative PLC) does not produce a different result because, by symmetry, the stimulus is objectively identical to that obtained by making near parts brighter. (Only the direction of rotation is reversed.)

To test the strength of PLC as a cue to shape, we can pit it against other factors, such as rigidity. Consider a polar projection of a cube, such as that obtained by viewing a cube from close up, say, from a vantage point whose distance to the center of the cube is only 2.25 times the width of a side. This 2D perspective view is illustrated in Fig. 5e. When such a wire cube is rotated, the shape ambiguity is removed. There is one, and only one, interpretation as a rigid object. Nevertheless, as with parallel projection, the cube can easily be perceived in two states: veridical and reversed. With polar projection, only the veridical state is perceived as rigid; the reversed state is perceived as a nonrigid, rubbery object whose sides distort enormously, with nonuniform expansions and contractions throughout the rotation. One might suppose that subjects have a strong tendency to perceive the object in the rigid mode, but empirically, Schwartz and Sperling found this not to be the case.

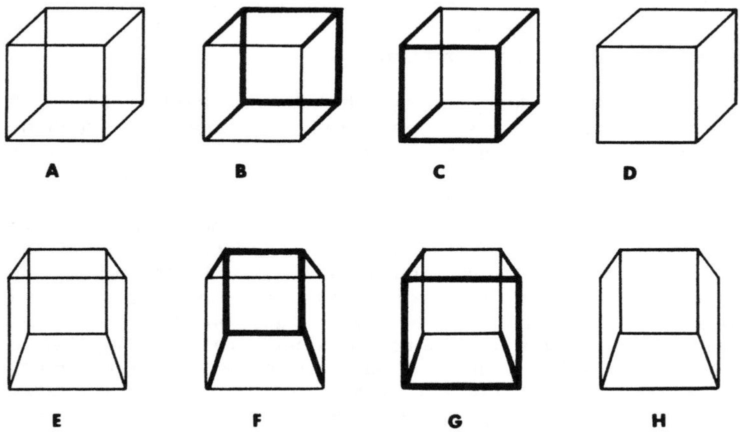

Fig. 5. Frames from rotating Necker cubes. (A) Necker cube. (B) Positive *proximity luminance correlation* (PLC). (C) Negative PLC. (D) Hidden lines (occlusion). (E) Perspective cube. (F) Positive PLC. (G) Negative PLC. (H) Hidden lines (occlusion). (Based on Schwartz & Sperling [37])

There is a modest tendency, for the rigid mode dominates slightly over the nonrigid mode, but this seldom exceeds about 70:30 and depends irregularly on such factors as polar projection distance and subject, with some subjects perceiving the nonrigid mode more often. It is not surprising then, that PLC overwhelms rigid interpretation (in linear perspective) as a determiner of dominance. With a positive PLC that favors the veridical mode (Fig. 5g), the rigid percept occurs on 94% of the presentations. With a negative PLC that favors the reversed mode (Fig. 5f), the *nonrigid* percept occurs on 92% of the presentations. The significance of these demonstrations of the power of PLC is that PLC is a fallible cue to depth that nevertheless consistently overrides unequivocal geometrical cues. It might be well to design machines to use more reliable cues to depth than humans do.

2.5.2. Other selection cues. Proximity luminance covariation is the most powerful cue we have discovered in wire cubes. By reducing the PLC (so that the brightness ratio of near points to far approaches 1.0), the strength of the PLC cue can be made arbitrarily small, and equated to any other cue, such as linear perspective that may be pitted against it. This balance method, and the corresponding analytic techniques of conjoint measurement [24] [12] can be used to accurately scale the strength of cues that determine the choice of perceptual states, and ultimately to construct an additive scale for the strength of any combination of factors. For example, *occlusion* (the removal of lines that would be hidden if the sides of the cube were opaque) is a weak cue to the perception of reversed states because, in polar projections, very little is actually occluded (Fig. 5h), but a strong cue to rigid perceptions because much is removed (not shown, cf. Fig. 5d). On the other hand, by painting opaque blotches on the side of a figure, occlusion can be made overwhelming (BRAUNSTEIN [3]). Linear perspective, as we noted above, is a relatively weak cue. The effect of a preview of a static field is quite strong. Context (adjacent or surrounding cubes) is not particularly effective.[4]

2.5.3. Weighing the evidence. The theory that we derive from experiments of this sort, is that there are a small number of discrete perceptual states, and that the evidence in favor of the states is simply weighed linearly, with the most favored state dominating in proportion to the weight advantage of its evidence over that of the strongest contender(s). Two comprehensive recent studies take a similar approach.

(1) The late Frank RESTLE [34] introduced a mathematical strength theory to the study of motion in his seminal analysis of JOHANSSON'S [19] fascinating motion stimuli. In a typical stimulus configuration, the movement of a small number of dots along interrelated linear and elliptical paths yields a rich variety of possible percepts (2D and 3D interpretations). Restle's strength function counts the number of parameters that are necessary, in his system, to describe the trajectories of dots in each of the candidate perceptual modes. The smaller the number of parameters--the simpler the description of the perceptual state--the greater its strength.

(2) BURT and SPERLING [4] studied ambiguous motion in stimuli composed of rows of dots (Fig. 5b). They found that the time t and distance d between dots determined the path, among several, along which motion was perceived. They derived a concise additive representation for the strength S of any candidate path of the form

$$S = te^{-t/\lambda}/d\,, \qquad (3)$$

where λ was about 20 msec. (This representation becomes additive by taking logarithms of both sides of (3)). By functional analysis, Burt and Sperling were able to prove that (3) was a unique description of their data.

In the past, investigators have not looked for additive representations for the weight of sources of evidence in favor of perceptual hypotheses. On the other hand, linear estimates are the simplest, and in many instances the most powerful statistical estimates of the significance of evidence, and almost certainly among the easiest to implement in machines. The monoactive neural network suggests that biological implementation also may be simple. Indeed, we may be entering an epoch of convergent evolution of psychological theories, biological theories, and machine implementations.

2.6. Perception Selection: The Selection Problem

How does the perceptual system arrive at the candidate perceptual states for which it seeks evidence? In the case of rotating wire figures, there are three proposals to consider.

(1) ULLMAN [46] demonstrated that by identifying (the same) five points of a 3D object in three different 2D projections, the 3D configuration of the points could be computed algebraically. This is a global computation of depth in the sense that no bit of information in any single frame conveys any 3D information; the 3D emerges only when the whole computation, involving all the five points under consideration, is complete. So far, this algorithm has not been elaborated to deal with errors, even small errors, but it is constructive and demonstrably works well with error-free data.

(2) TODD [45] proposed a theory quite analogous to RESTLE'S [34] in which motion is analyzed in terms of projections of ellipses in 2D representing conic sections in 3D. This approach is promising but lacks generality.

(3) We propose that local cues provide inputs to an interactive competition/cooperation network whose state tends toward the most self-consistent percept. Examples of local cues are shrinking or expanding lines, changes in direction of movement, etc. in dynamic views, and vertices, converging lines, etc., in both static and dynamic views.[4]

Although in formal models, the selection and dominance problems can be treated separately, in biological systems, selection and dominance undoubtedly are two manifestations of the same computation.

3. Image Processing in Cognition

There are no hard and fast differences between perception and cognition: we use the term *cognition* to refer to complex situations in which learning, language, and meaningfulness play a greater role.

3.1. Bandwidth of American Sign Language (ASL)

Although we think of language communication as one of the more complex of cognitive tasks, there is an old English saying that belittles language relative to images; it translates into modern English roughly as

$$\textit{An image has the utility of } 10^3 \text{ speech tokens.} \qquad (4)$$

In the next sections, we investigate the extent to which (4) is indeed true.

In the U.S.A., the bandwidth of AM radio transmission is about 4.5 kHz. That is, radio transmits acoustic signals from about 0.3 to about 4.8 kHz. American television was designed for a bandwidth of over 4 MHz [31] so that we find Eq. (4) verified. Similarly, the bandwidth of a telephone line in the U.S.A. is approximately 3 kHz [47]. American Picturephone and British Viewphone, systems for face-to-face video telecommunication, were designed with bandwidths of over 1 MHz [9], [18]. Again, Eq. (4) is near the truth. Unfortunately, today the cost of a communication line is closely related to its bandwidth, so that video communication costs orders of magnitude more than speech communication.

The high cost of video communication is of little concern to hearing persons who normally use telephones for communication. It is a tragically insurmountable obstacle to congenitally deaf persons who habitually communicate in sign language; they can not use telephones and could not afford video telephones. In this regard, studies of proficient users of American Sign Language (ASL) show that they communicate information at about the same rate as in spoken language [1], so that the signing population could use a video telephone as effectively as the hearing population uses telephones. No one had ever measured the bandwidth actually required to communicate in ASL by means of raster encoded images, and when no one responded to a suggestion to do so [39], the first author set about to do it himself.

The ordinary television screen encompasses a nominal bandwidth of 4 MHz. By dividing the screen into a large number of rectangles, like a sheet of postage stamps, a different ASL conversation could be carried in each postage-stamp rectangle. By varying the size of the stamps, and calculating the fraction of total bandwidth allocated to each, SPERLING [40][41] was able to show that with a bandwidth of 21 kHz, ASL sentences and ASL word lists were communicated at 90% of the intelligibility of the control condition and finger spelling was at 70% of control. The most experienced deaf subjects achieved intelligibility scores of 40% to 50% at a bandwidth of 4.4 kHz. This is about what would be obtained with hearing subjects listening to voice communication over a voice communication channel with a bandwidth of 1.5 kHz [13]. Thus, while a television picture may use 1000 times the bandwidth of a telephone, ASL requires only a few times more bandwidth than voice communication. There are some obvious next questions. (1) Is it possible to use image processing methods to reduce the picture bandwidth so that ASL could be communicated over ordinary telephone lines, or, if not, what is the requirement for an ASL communication line? (2) How is it possible for experienced signers (or speakers) to use degraded signals so effectively?

3.2. Image processing of ASL

The intelligibility of degraded images indicates that a sequence of good quality ASL images contains a large number of redundancies. That is, some of the information present at a given point of the image is also present elsewhere. A human observer is capable of taking advantage of the redundancies and extracting the important information from a quite impaired image. Solution of the transmission problem requires a machine to take advantage of these redundancies to transmit an efficient code. There are three potential sources of redundancies:

(1) within-frame statistical and deterministic dependencies

(2) between-frame statistical dependencies

(3) source constraints.

These three cases overlap. The first two cases involve assumptions concerning only the images or train of images without any additional higher level knowledge. The last case involves assumptions regarding the physical, semantic and pragmatic constraints. That is, the signer is limited by her anatomical structure combined with the laws of physics, by the rules of the language, and by the message to be conveyed. For example, Fig. 6 shows a selection of frames from the sign "tomato". In "tomato" the index finger first touches the lips (indicating taste), then rotates out and imitates a slicing movement. There are large relatively homogeneous areas within all the frames of Fig. 6; some areas hardly change from frame to frame; when the head or body moves, it moves very slowly; details in the right hand are important for conveying meaning primarily when it is still or moving very slowly; details in the left hand are unimportant, and so on. These facts could and should be embodied in an efficient coding scheme.

3.2.1. Nondestructive Coding. The studies of the parameters of raster coding of ASL [32][40][41] did not enable us to answer the most interesting questions, and it was necessary to establish a digital image processing facility--a substantial technical undertaking (see Acknowledgements and LANDY, [25]; COHEN, [6][7]). Our basic digitized images of ASL were 96 pixels (picture elements) high by 64 pixels wide and quantized to seven significant bits of intensity (128 discriminable grey levels). There were a maximum of 30 full frames per second, with interlace. In this paper we concentrate on the preliminary results rather than the technical details. Our first step in the search for a more efficient image representation is

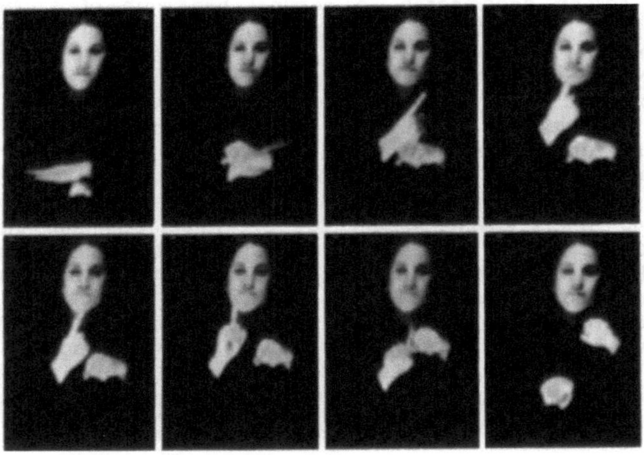

Fig. 6. Frames from a 1.7 sec (50 frame) production of the ASL sign "tomato". Beginning at upper left, the frame number in sequence is 3, 13, 17, 21, 28, 36, 45, 49. The frames are 64x96 pixels, with 7 bits of intensity resolution in the original display. The signer is wearing a dark sweater and viewed against a black background in order to facilitate discrimination of the hands. The white dots in the corners are fiducial marks placed just outside the corners of the picture area

nondestructive coding. Here, we represent the images by a code which is shorter for frequently occurring symbols or which combines a number of symbols into one.

3.2.1.1. *Huffman codes.* One such code, based on Huffman sequences, treats each pixel individually. Each value of a grey level is converted into a binary word whose length is inversely proportional to the relative frequency of occurrence of that level. Thus, the frequently occurring values of luminance are represented by only a few bits and the rare values by many bits. Significant reduction of information is achieved in those cases where the distribution (histogram) of the gray levels differs considerably from a uniform distribution. When a coding of this type is applied to the ASL images, the amount of data compression is relatively small. The theoretical limit of such coding schemes may be estimated by computing the entropy of individual pixels for the ASL images. For our sample of the ASL images, the entropy indicated maximum possible compression to be 38% of the original value.

3.2.1.2. *Run length codes.* Another nondestructive code which appeared promising is the run-length code. Instead of representing each pixel, this code represents any sequence of pixels of the same value of gray level by the level and the length of the sequence. Consequently, the run-length code is particularly powerful when the images contain considerable patches of the same luminance level. In the case of the ASL images with full grey scale, the savings were nominal; i.e., 70% of bits were required.

3.2.1.3. *Hierarchical codes.* There are many other, more sophisticated, methods for information compression. An attempt was made to apply some of them, such as hierarchical coding [23], but the amount of compression achieved was small in comparison to the target reduction. Therefore, we proceeded to investigate methods which are effective at the expense of losing some information.

3.2.2. Nonspecific Within-Frame Coding. The codes considered in this section take advantage of the fact that not all the information in the original images has to be transmitted. There are two separable--but not independent--coding problems: the *image* code and the *transmission* code. We are concerned primarily with the image code, but we need to also specify a transmission code in order to evaluate the transmission requirement of the image code.

3.2.2.1. *Low-pass filtering (postage stamp experiment).* Among the reasons that allow us to transmit less information are the limitations of the human visual system and the fact that not all details are required for an observer to be able to recognize the message in the image. For example, in the "postage stamp" experiment described above, information is lost because of subsampling and low-pass spatial frequency filtering (imperfect resolution by the television display). A digital approximation to the analog "postage stamp" stimuli (with all stimuli scaled to the same size) is shown in Figs. 7a through 7d.

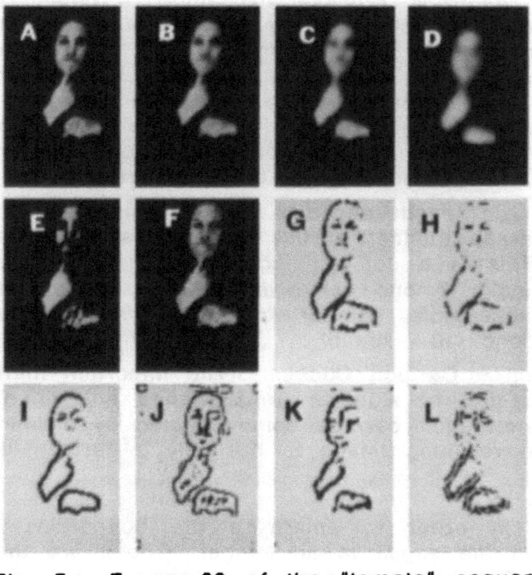

Fig. 7. Frame 20 of the "tomato" sequence, 64x96 pixels, subjected to 11 image transformations. The image coding scheme, and the average number of bits (1000 bits = kb) per frame in "tomato" sequence are indicated for each panel. The transmission code used to compute the bit rate is DPCM for frames (A-D) and hierarchical coding (Cohen [8]) for (E-L). (A) Original (18.4 kb). (B, C, D) Low-pass spatial filtering (8.7, 3.5, 1.3 kb). (E) Binary intensity, 6% black (0.9 kb). (F) Block truncation (3.7 kb). (G) Positive Laplacian, 6% black (1.3 kb). (H) Positive Laplacian, 3% black (1.0 kb). (I) Edge detection mask (1.3 kb). (J) Zero crossings 6% black (2.0 kb). (K) Negative Laplacian, 6% black (1.3 kb). (L) Union of positive and negative 3% black Laplacians (1.5 kb)

3.2.2.2. *Block truncation codes.* There are many other quite powerful methods for data compression at the cost of eliminating some (hopefully unimportant) information. Block truncation coding [29] is among the most efficient image codes that utilizes the full range of grey scale values (see Fig. 7f). In this coding scheme, each frame is divided into blocks that are transmitted separately. The code for a block depends upon the variance of the pixels inside the block. When applied to the ASL images, the code required approximately 0.6 bits per pixel for intelligibility, representing a net reduction of the amount of information by a factor of 20 relative to the original image. We plan to study an elaboration of the code in which blocks that change between frames are allotted more bits of information than unchanging blocks.

Even though the block truncation method is a quite elaborate code for full grey scale, the effective transmission rate for ASL is still 55,000 bits per second (96x64 pixels x 0.6 bits/pixel x 15 fps). This rate is many times higher, we suspect, than that essential for sign language communication. More complex methods are required, methods that are adapted, at least to some degree, to the idiosyncratic features of ASL communication.

3.2.2.3. *Binary intensity code.* The conversion of analog images into digital form without compression usually results in at least a five-fold increase in the required transmission bandwidth. However, the quantizing process also offers a means of compression by a reduction of the number of quantizing levels. The ultimate minimum number of intensity values in an informative image is two: black and white. Figure 7e shows a binary image. ASL sequences of these images are surprisingly comprehensible. In part, this is because the signer is wearing a black sweater and is viewed against a black background, so that the hands and face are the only nonblack image components. These conditions, obviously, are ideal for enhancing intelligibility of ASL, and particularly, of binary images of ASL.

Binary images lend themselves to run length encoding (for transmission) and to various hierarchical codes. The average number of bits per picture needed to encode these images is about 900. This is a coding efficiency of 0.15 bits per pixel, a 50 fold saving over the original's 7 bits per pixel. At 15 frames per second (approximately the minimum acceptable number for ASL), the required channel capacity is 13,500 bits per second (13.5 kbaud).

3.2.3. ASL-Specific Within-Frame Coding – Cartoon Codes. The important information in ASL is conveyed by the position of signer's arms, hands and fingers relative to the face and body. This information can be communicated by outlining the boundaries of the extremities. The resulting image is basically a cartoon-like representation of the signer.

3.2.3.1. *Edge detection by masks.* In order to determine the boundaries of objects, it is convenient to find most of the *edges* in an image. These are then connected to form closed curves. Fortunately, there has been considerable theoretical and empirical work done on the problem of edge extraction. By and large, the techniques applied involved various modification of gradient operators and masks. One method [35] selects the maximum of four different 3 by 3 masks applied at each point and retains the points with the largest resulting values. In the example of Fig. 7i only the largest 6% of the values were retained.

3.2.3.2. *Edge detection by Laplacian operators.* MARR and HILDRETH [27] proposed this method, but we are using several simpler variations on their algorithms that we have developed. Our preliminary observations indicated that a 5x5 linear operator applied to 64x96 pixel frames, followed by a thresholding to preserve only the strongest edges, was quite effective. Setting the threshold such that only 6% of the most prominent edge pixels were retained (Fig. 7g) results in a net reduction of the amount of information to 0.16 bits per pixel using a hierarchical coding scheme [8]. The case when only 3% of the edges were kept is illustrated in Fig. 7h. Detection of the positive side of edges is illustrated in Figs. 7g, h, the negative side of edges in Fig 7k, and the union of negative and positive edges in Fig. 7l. Positive and negative Laplacians reveal somewhat different aspects of images; both are about equally effective for this subject matter. The union is included to illustrate this.

3.2.3.3. *Zero crossings.* In addition to detecting the inside and outside of edges by the extreme values of the Laplacian operators, Marr and Hildreth proposed finding the midpoint of edges by locating the Laplacian zero crossings. We use a different algorithm to locate zero crossings; Figure 7j illustrates zero crossings based on the same Laplacian operator described above.

Most of the segmentation schemes based on edge detection require a method for combining the detected edges into a continuous boundaries. However, in the case of the sign language images, we found an interesting effect: Since the ultimate receiver is a human observer and the individual frames are presented in a rapid succession, the combination process occurs in the visual system of the observers. Even when each frame contains only disconnected[5] edges, a sequence of frames appears to represent the boundaries quite effectively.

The application of cartoon schemes followed by hierarchical coding enabled an information reduction from the original image by a factor of about 30. However, the resulting image still requires about 1300 bits/frame [8]. Intelligibility is maintained at frame rates as low as 15 frames per second, resulting in a transmission rate of about 20 kilobaud for 64x96 cartoon-coded ASL. The 20 kbaud rate for cartoons and the 13.5 kbaud rate for 1-bit intensity quantized images could be reduced somewhat by using a coarser pixel grid (fewer pixels) and by refinements in coding procedures to bring it below 10 kbaud, which is available on ordinary switched telephone networks. Substantial further reduction in transmission rate of binary images--or the transmission of grey scale pictures at this rate--may well require incorporation into the code of the physical constraints imposed by the construction[6] of a human body, and the knowledge of ASL (signs, syntax and semantics).

3.3. Information processing strategies

How is it that signers can do so well with such reduced visual information? In fact, their performance is not so different from other skilled human performance: listeners can understand speech in incredible amounts of noise; experienced military pilots have landed airplanes in virtually zero visibility; and chess masters can accurately reconstruct complicated chess positions from memory when their viewing time of the chess position is restricted to a few seconds.

The memory performance of chess masters provides perhaps the purest prototype for all these skilled performances. There are two studies of the memory of of chess masters for chess positions. DE GROOT [10] found that after a five second look at a chess board, his master could reconstruct it, correctly placing 24 or 25 of 25 pieces on an empty chessboard. CHASE and SIMON [5] found that their master could correctly place about 20 of 25 pieces from comparable positions. Their class A chess players could correctly place only half as many pieces as the master, and a novice could place only half as many again. The extraordinary performance of chess masters obtained only when the pieces were in the kinds of positions reached by skilled players in actual games. When chess positions were constructed by placing chess pieces randomly on the board, both the novice and class A player actually recalled the positions better than the master!

3.3.1. Templates, intentions. The extraordinary memory of chess masters for chess positions (together with other aspects of their performance) has been interpreted in terms of what we here call memory *templates*. These presumed templates are memories for configurations of chess pieces that have been learned--stored in long term memory--during many thousands of hours of viewing chess positions. Obviously, templates are not literal representations of the chess pieces themselves, they are higher-order representations of chess configurations. In other contexts, it might be appropriate to call them *propositions*. It is estimated--but not known for certain--that templates typically deal with half a dozen pieces and that a chess master has acquired on the order of tens of thousands of templates. A master analyzes a novel position into its component templates, the deviations of particular pieces from these, and the connections between templates. It is reasonable to suppose, but not known, that templates themselves are organized into hierarchies of templates of templates.

We suppose that skilled ASL communicators have built up vast reservoirs of templates for hand and body configurations, and that speakers of a language have acquired sound templates of words and of typical sound sequences. Again, we suppose that such templates are represented in a more abstract feature space, where variations in the angle of view or individual differences between signers do not affect the representation.

3.3.1.1. *A gedanken experiment.* There is an additional difficulty in the study of language, relative to chess. Actual ASL signing and actual speech proceed so rapidly that the target positions of articulators are never quite reached. As a gedanken experiment, imagine a chess game being played so fast that the pieces never quite settle down. Before a piece being played by one player has actually touched the board, the other player has, perhaps, already begun to remove it and simultaneously has initiated the movement of another piece. We could simulate such a chess game by representing the chess board on a computer display. Players enter moves by keying a code at a terminal. This causes the displayed representation of the piece to move as though it were a massive body moving through an extremely viscous medium, taking many seconds to reach its destination. After a period of familiarization with the dynamics of the system, players could initiate new moves long before old ones had settled. In a rapid game between experienced players, photographs of the display board would never show all the pieces in resting positions and would be quite difficult to interpret. The gedanken chess board, we claim, represents the first-order situation in language communication, in which the target positions of the articulators--vocal tract or hands--are virtually never reached in actual discourse. Speakers and signers, like experienced chess players in the simulator, recognize the *intended* positions and reply in kind. In fact, in rapid discourse, speakers do not simply speed up--making the same articulatory gestures as in slow discourse at a faster rate; there are complex changes, including modifying their articulatory gestures so as to make their vocal intentions more discernible to the listener. The analogous modification of rapid movements probably occurs in ASL but it has not been reported.

3.3.1.2. *Intentions.* By *intention*, we mean an inference of the intended target position of a chess piece in the simulated game or of an articulator in spoken or signed language communication. To parse spoken or signed discourse we have to discover these intentions. According to this view, continuous discourse can be parsed into a sequence of target positions. Sequences of these target positions are represented as templates in the listener's memory, representing words or syllables. For ASL, the number of templates is presumed to be on the order of thousands, or at most, a few ten thousands. The reason a skilled signer can communicate with reduced visual input is that he does not attempt to parse the input into an arbitrary sequence of movements, but uses it only as evidence for choosing among templates. The number of eligible templates is further constrained by context and pragmatics as has been pointed out above.

3.3.2. Analysis by synthesis, schemas. It is not necessary to be able to make signs in order to be able to understand ASL, nor is it necessary to be able to speak in order to understand speech. What is necessary is an understanding of the dynamics--the physical and biological constraints--of the articulators in order to derive the intention of a movement trajectory. A dog knows where to look for a stick when his master throws it or pretends to throw it. The dog can derive the intended trajectory from his master's arm and body movement even though a dog can not throw sticks. On the other hand, to utilize the constraints in a signed or spoken message requires a very sophisticated understanding of language, and of meaning. One proposed way of utilizing these constraints is to construct the alternative messages and to evaluate the incoming signals in terms of its evidential weight for these alternatives. The difficulties inherent in such procedures will be discussed in other papers here.

To represent meaning in a message SCHANK and ABELSON [36] and many others have proposed *schemas* as an appropriate means of representing meaning. A schema for representing a kitchen scene, for example, would include a floor, walls, cabinets, stove, refrigerator, utensils, a person, etc. A picture of a particular kitchen scene would be represented by filling details for the variables listed above, with default values representing the most typical scene. Evidence that people remember meaningful scenes in this way comes from their good recall of unusual details, and the nature of the errors they make, the errors being presumed to reflect default values. A schema can be viewed as a higher-order template whose components are variables, with values to be derived from the input.

To return to ASL, it is an article of faith that the representation of meaning of ASL sequences will not differ from the representation of meaning of spoken sentences. On the other hand, the surface form of ASL is quite different from spoken language; the grammar and semantics of ASL are quite different from spoken languages. Certainly, despite its name, American Sign Language bears little relation to English. The major advantages of studying a visible language like ASL is that the articulations--the hands and body--are continuously visible, and their physical properties can be readily measured. The inaccessibility of the articulators for speech, and hence the extraordinary difficulty of obtaining a dynamic description of their movements in speech [14] has been one of the stumbling blocks to an understanding of the speech process. ASL does not have this problem and seems ideally suited to the study of language processing at all levels by monkeys [44], or humans, or machines.

4. Acknowledgements

The preparation of this article and the work on motion perception was supported by U. S. Air Force, Life Sciences Directorate, Grant No. AFOSR-80-0279; the work on image processing of American Sign Language was supported by National Science Foundation, Science and Technology to Aid the Handicapped, Grant No. PFR-80171189.

The authors wish to acknowledge the technical, engineering assistance of Thomas Riedl and Robert Picardi, the advice of Dr. Nancy Frishberg, who along with Ellen Roth, also served as signer, and O. R. Mitchell, who made available his computer programs for block truncation codes.

5. Notes

1. When the two eyes are correctly pointed, we say they have achieved motor fusion. In this article, it will not be necessary to distinguish perceptual from motor fusion.

2. SPERLING [38] contains a fully detailed elaboration of the potential-theory model; see SPERLING [42] for a summary.

3. As defined, h represents a comparison in image space $I(x,y)$. As we shall observe later, comparisons (or template matches) are best made in feature space, after the most relevant, invariant stimulus properties have been abstracted. Thus h is better defined in terms of $T{:}I(x,y)$, where T represents the transformation to feature space ([38], p. 470; [26]), but T is an unnecessary complication at this juncture.

4. BRAUNSTEIN [2] proposes a similar theory (heuristics are used to disambiguate motion), and an informative review of motion cues.

5. The perceptual connection of dynamic segments into boundaries is analogous to JOHANSSON'S [20][21] classical demonstrations of the perceptual inference of moving objects (such as persons and bicycles) from several luminous points painted on their surfaces. POIZNER, BELLUGI and LUTES-DRISCOLL [33] tested the intelligibility of ASL communication when images were composed of lights placed on fingers, arms, and shoulders. Of the various arrangements tested, lights on the finger tips were most useful. TARTTER and KNOWLTON [43] demonstrated that ASL could be communicated by viewing 13 luminous points on each hand and one on the nose. In dynamic point-light displays, the reconstruction of hands occurs in the visual system of the viewer.

6. Since this article was prepared, the authors have become aware of two similar attempts to reduce the required transmission bandwidth of ASL by means of image encoding: (1) Pearson, D. E. & Six, H. Low data-rate moving-image transmission for deaf communication. *International Conference on Electronic Image Processing, 1982*, 204–208. (2) Abramatic, J. F., Letellier, P., & Nadler, M. A narrow-band video communication system for the transmission of sign language over ordinary telephone lines. *Conference Report*, August, 1982.

6. References

1. Bellugi, U., & Fischer, S. A comparison of sign language and spoken language. *Cognition*, 1972, **1**, 173–200.
2. Braunstein, M. L. *Depth Perception Through Motion.* New York: Academic Press, 1976.
3. Braunstein, M. L., Andersen, G. J., & Riefer, D. M. Distance perception by monocular observers: Conflicting dynamic cues. *Investigative Ophthalmology and Visual Science*, ARVO Supplement, 1982, **21**, 273.
4. Burt, P., & Sperling, G. Time, distance, and feature trade-offs in visual apparent motion. *Psychological Review*, 1981, **88**, 171–195.
5. Chase, W. G., & Simon, H. A. The mind's eye in chess. In W. G. Chase (Ed.), *Visual Information Processing*. New York: Academic Press, 1973, pp. 215–281.
6. Cohen, Y. Measurement of Video Noise. *Technical Report, Human Information Processing Laboratory, N.Y.U.* February, 1982a.
7. Cohen, Y. The HIPL Picture Processing Software. *Technical Report, Human Information Processing Laboratory, N.Y.U.* July, 1982b.
8. Cohen, Y. Hierarchical coding of binary images. *Technical Report, Human Information Processing Laboratory, N.Y.U.* August, 1982c.

9. Crater, T. V. The Picturephone system: Service standards. *Bell System Technical Journal*, 1971, **50**, 235-269.

10. de Groot, A. *Thought and Choice in Chess*. The Hague: Mouton, 1965.

11. Dev, P. Perception of depth surfaces in random-dot stereograms: a neural model. *International Journal of Man-Machine Studies*. 1975, **7**, 511-528.

12. Falmagne, J-C. Random Conjoint Measurement and Loudness Summation. *Psychological Review*, 1976, **83**, 65-79.

13. French, N. R., & Steinberg, J. C. Factors governing the intelligibility of speech sounds. *Journal of the Acoustical Society of America*, 1947, **19**, 90-119.

14. Fujimura, O. Modern methods of investigation in speech production. *Phonetica*, 1980, **37**, 38-54.

15. Grossberg, S. Contour enhancement, short term memory, and constancies in reverberating neural networks. *Studies in Applied Mathematics*, 1973, 52, 217-257.

16. Grossberg, S. Competition, decision, and consensus. *Journal of Mathematical Analysis and Applications*, 1978, **66**, 470-493.

17. Helmholtz, H. L. F., von. *Treatise on Physiological Optics*, (3rd ed.), J. P. C. Southall (Trans.). Rochester, N. Y.: Optical Society of America, 1924. (Reprinted, New York, N. Y.: Dover Publications, Inc., 1962.)

18. Hillen, C. F. J. The face to face telephone. *Post Office Telecommun. Journal*, 1972, **24**, 4-7.

19. Johansson, G. *Configurations in Event Perception*. Stockholm, Sweden: Almqvist & Wiksell, 1950.

20. Johansson, G. Visual perception of biological motion and a model for its analysis. *Perception and Psychophysics*, 1973, **14**, 201-211.

21. Johansson, G. Visual motion perception. *Scientific American*, 1975, **232**, 76-88.

22. Julesz, B. *Foundations of Cyclopean Perception*. Chicago: University of Chicago Press, 1971.

23. Klinger A., & Dyer C. R. Experiments on picture representation using regular decomposition. *Computer graphics and image processing* 1976, **5**, 68-105.

24. Krantz, D. H., Luce, R. D., Suppes, P., and Tversky, A. *Foundations of Measurement*. New York: Academic Press, 1971.

25. Landy, M. The HIPL Picture/Header Format Standard. *Technical Report, Human Information Processing Laboratory, N.Y.U.* March, 1982.

26. Marr, D. *Vision: A Computational Investigation into the Human Representation and Processing of Visual Information*. San Francisco: W. H. Freeman and Company, 1982.

27. Marr, D., & Hildreth, E. Theory of edge detection. *Proceedings of the Royal Society of London*, **207**, 187-217.

28. Marr, D., & Poggio, T. Cooperative computation of stereo disparity. *Science*, 1976, **194**, 283-287.

29. Mitchell, O. R., & Delp, E. J. Multilevel graphics representation using block truncation coding. *Proceedings of the IEEE*, 1980, **68**, 868-873.

30. Nelson, J. I. Globality and stereoscopic fusion in binocular vision. *Journal of Theoretical Biology*, **49**, 1-88.

31. Pearson, D. E. *Transmission and Display of Pictorial Information*. New York: Wiley, 1975.

32. Pearson, D. E. Visual communication systems for the deaf. *IEEE Trans. on Communications*, 1981, **29**, 1986-1992.

33. Poizner, H., Bellugi, U., and Lutes-Driscoll, V. Perception of American Sign Language in Dynamic Point-Light Displays. *Journal of Experimental Psychology: Human Perception and Performance*, 1981, **7**, 430–440.

34. Restle, F. Coding theory of the perception of motion configurations. *Psychological Review*, 1979, **86**, 1–24.

35. Robinson, G. S. Edge detection by compass gradient masks. *Computer Graphics and Image Processing*, 1977, **6**, 492–501.

36. Schank, R. C., & Abelson, R. P. *Scripts, Plans, Goals and Understanding*. Hillsdale, N. J.: Lawrence Erlbaum Assoc., 1977.

37. Schwartz, B. J., & Sperling, G. Nonrigid perceptions consistently elicited by rigid object stimuli. Manuscript submitted for publication, 1982.

38. Sperling, G. Binocular vision: a physical and a neural theory. *American Journal of Psychology*, 1970, **83**, 461–534.

39. Sperling, G. Future prospects in language and communication for the congenitally deaf. In L. Liben (Ed.), *Deaf children: Developmental perspectives*. New York, N. Y.: Academic Press, 1978. Pp. 103–114.

40. Sperling, G. Bandwidth requirements for video transmission of American Sign Language and finger spelling. *Science*, 1980, **210**, 797–799.

41. Sperling, G. Video transmission of American Sign Language and finger spelling: present and projected bandwidth requirements. In A. Habibi and A. N. Netravali (Eds.), *IEEE Transactions on Communication* [Special Issue on Picture Communication Systems]. New York: IEEE Communications Society, 1981a, 1993–2002.

42. Sperling, G. Mathematical models of binocular vision. In S. Grossberg (Ed.), *Mathematical Psychology and Psychophysiology*. Providence, Rhode Island: Society of Industrial and Applied Mathematics–American Mathematical Association (SIAM–AMS) Proceedings, 1981b, **13**, 281–300.

43. Tartter, V. C., & Knowlton, K. C. Perception of sign language from an array of 27 moving spots. *Nature*, 1981, **289**, 676–678.

44. Terrace, H. S. *Nim*. New York, N. Y.: A. Knopf, 1979.

45. Todd, R. Visual information about rigid and nonrigid motion: A geometric analysis. *Journal of Experimental Psychology, Human Perception and Performance*, 1982, **8**, 238–252.

46. Ullman, S. *The Interpretation of Visual Motion*. Cambridge, Massachusetts: The MIT Press, 1979.

47. Technical Staff, Bell Labs. *Transmission Systems for Communications* (Rev. ed. 4), Western Electric Co., Winston-Salem, N. C. Tech. Pub., 1971.

Part VII

Postscript

Image Interpretation: The Way Ahead?

A. Sloman

Cognitive Studies Programme, University of Sussex,
Brighton, United Kindgom

Abstract

Some unsolved problems about vision are discussed in relation to the goal of understanding the space of possible mechanisms with the power of human vision. The following issues are addressed: What are the functions of vision? What needs to be represented? How should it be represented? What is a good global architecture for a human-like visual system? How should the visual sub-system relate to the rest of an intelligent system? It is argued that there is much we do not understand about the representation of visible structures, the functions of a visual system and its relation to the rest of the human mind. Some tentative positive suggestions are made, but more questions are posed than answers.

A.1. Introduction

The human visual system is the most powerful information-processing system known. It very rapidly processes a continuous stream of information, from millions of receptors. It copes with an enormous variety of formats, and many kinds and degrees of image degradation. It improves itself over time, and it can be used for purposes as varied as sight-reading music, diagnosing diseases, peeling a banana and enjoying a ballet. Explaining how it all works is a mammoth task, and no sane person should claim to be able to see the way ahead. But having been asked, I shall try to peer into the mists.

Assuming that a criterion for understanding is the ability to design a working model, there are many things we still don't understand, despite the progress of the last few years, mostly concerned with 'low level' processing. (See HANSEN and RISEMAN [12], BRADY [6], MARR [25], recent Artificial Intelligence conference proceedings, e.g. IJCAI 1981, and this volume. A lot of this research was originally inspired by the work of HORN, e.g. [22], who showed that more information about scenes could be extracted from images in data-driven fashion than was previously thought possible. Some of the work is very technical, and I cannot claim to have understood it all.)

I take it that our aim is to understand not just human vision but general principles relevant to the design of visual systems, especially those which approximate the power of the human visual system. Science is not just the study of what is, but also of what is and is not possible and why. (See my [30] chapters 1 and 2.) This aim may not be important for special-purpose applications of image processing, but in the long term, the design of flexible and robust robots will require a deep understanding of general principles, including the geometry and physics of scenes and image formation, and also computational principles concerned with how best to represent information, how to cope with ambiguity and degraded information, how to combine multiple sources of information for a single task, how to maximise speed or trade space and time, how to improve the system's abilities over time, etc. In the short run, for commercial reasons, most research and development funding is likely to be directed to special-purpose systems which are heavily model-driven, and cannot cope with arbitrary scenes, unlike human beings and many other animals, which are not restricted to seeing known sorts of objects.

There is a great discrepancy between the kinds of tasks that can be performed by existing computer models and the experienced richness and multiple uses of human vision. This is not merely a quantitative difference which might easily be overcome by the use of better hardware. There are too many limitations in our theoretical understanding for technological advances to make much immediate difference. Given computers many times faster and bigger than now, and much better TV cameras, we still would not know how to design the visual system for a robot which could bath the baby or clear away the dinner things, let alone enjoy a ballet.

A.2. The main problems

The literature cited includes descriptions of both achievements and detailed unsolved problems, especially problems of interpreting local image features in terms of scene features or fragments. This is an area in which much more progress has been made, and will be made, than used to be thought possible. However, there are some equally important problems not receiving so much attention:

(1) What are the functions of a visual system?

(2) What needs to be represented, and how?

(3) What sort of global architecture can enable a system to perform those functions?

In an attempt to elucidate the nature of these problems I shall show that common assumptions about the functions of vision are too restrictive, and that representations used in current models are not adequate, even for such apparently simple things as straight lines. I shall offer some speculations about the architecture required by a human-like visual system and the way in which it should relate to the rest of the information processing mechanism. In particular, it should not be restricted to producing descriptions of the geometry and motion of the environment and it should have several subsystems which are themselves linked not only to each other but also to other mental mechanisms.

A.3. Methodological note

Being a philosopher and a programmer, my approach to an attempt to understand the space of possible computational systems is primarily top down: attempting to derive structure from functions and constraints. Analysis of function involves studying the purposes for which vision can be used and the circumstances in which it can achieve those purposes. Such things as required reaction times, error rates, type of degradation in various circumstances (e.g. occluded objects, poor lighting, mist, blizzards, loss of one eye, rapid motion, etc.) can be thought of as more detailed specifications of function. (Strangely, few text books on vision seem to discuss the functions of vision -- e.g. the relation between vision and action. HOCHBERG [20] mentions action on the last page!)

Although it is very important to understand human vision we can fruitfully aim for a higher level of generality, based only on assumptions common to natural and artificial systems: assumptions concerning the nature of the environment, the nature of the tasks of a visual sub-system of an intelligent system, and the need for rapid decisions relative to the available processing speed. We need to understand the space of possible visual systems and how they relate to the functions.

But for the crucial speed constraint, a visual system might systematically generate all possible 3-D scenes from all possible viewpoints, project them onto a representation of a retina, and compare with actual retinal stimulation. (Moving images would require representation of changing scenes.) Subject to a suitable finite quantisation of the search space and 'tolerant' matching, the selection of the best match could be done after an exhaustive search. The fact that problems and theories are formulated in such a way as (rightly) to rule out consideration of such absurd strategies shows that separating the theory of the domain from the design of algorithms and mechanisms may not always be as easy, or as useful, as MARR suggests (e.g. [25]). See also section C.8 below.

B.1. The functions of vision

A full survey of the functions of vision is not possible here, but it obviously provides information about the environment, which can be used for searching for things, controlling and monitoring actions, finding one's way, forming plans, reacting to opportunities and dangers, making predictions, understanding mechanisms, testing theories, interpreting communications, making pictures, building databases for future use, and even improving visual abilities. Vision can also trigger desires (e.g. sexual), emotions or reflex actions.

How can all this be done? It is often assumed that animals, and robots, need a visual subsystem whose function is to take in retinal stimulation and produce descriptions of (possibly changing) three-dimensional objects, their properties and relationships, thus:

```
     r    ___________    _________    __________
-> e ->| Visual     |-->|3 or 4-D|-->|Non-visual|
-> t ->|mechanisms|-->|descrip |-->| sub-     |
-> i ->|           |-->| -tions |-->|systems   |
-> n ->|___________|-->|________|-->|__________|
     a
```

I shall challenge the assumption that such a limited interface is compatible with the common uses of vision. I shall try to indicate the need for visual mechanisms to communicate more than just the geometry and motion of 3-D scenes.

B.2. Is there only one kind of output?

Information from lower levels is often useful, for instance when 2-D information, not 3-D information, is needed. Putting a finger on the edge of a table (usually) requires information about the three-dimensional location of the edge. However, being able to run one's eye along the edge (e.g. looking for defects in the workmanship) requires only a two-dimensional representation of the location of the edge within the visual field. You can run your eye along a horizontal wire looking for kinks even if you cannot judge the distance of the wire. Similarly, moving a paint brush smoothly along an edge may require only the 2-D representation for monitoring the motion (once it has started) and providing adjustments if the hand is going too far up or down, in situations where the depth information comes from a non-visual source, such as touch, or is not needed because the geometry of the arm constrains the motion. Using disparity in a 2-D image to control movement may be far easier than using the 3-D disparity, for instance where there is insufficient time, or depth information, to compute the 3-D disparity. 2-D disparity will be specially useful when the only corrections to position of the arm are essentially 2-dimensional, e.g. merely raising or lowering.

2-D image structure is useful for deciding which way to move to see more of a partially hidden object. In some cases local information near an occluding edge (especially when there's motion) suffices to determine the best direction, whereas in others more global computation is required. Compare the problems of how to move to see more of A or more of B in the figures. Absolute depth information is not needed.

```
                           -----                   -----
 ----------               |     \                 /     |
|         |----           |      \               /      |
|         |    |          |       \-------------/       |
|         | A  |          |        \     B     /        |
|         |    |          |         \         /         |
|         |----           |          \-------/          |
 ----------                ------------\     /----------
```

2-D image structure could also be used by higher levels in: planning a route across a room, finding space for a new package on a cluttered floor, anticipating a trajectory, working out what someone is looking at, keeping to the centre of a path, sighting with a gun, moving out of sight of an enemy. A 2-D image map may help to constrain searching, both by providing rapid access to information about nearness, collinearity, inclusion etc, and by constraining proposals for solutions. How exactly such maps can be used remains a research problem. FUNT [11] deals with some simple cases. (See ch 7 of [30] for a discussion of pros and cons of 'analogical' and 'applicative' representations.)

Some of these tasks merely extract information from the image representation, whereas others require an ability to manipulate the representation, as in FUNT's program which computes collision points of arbitrarily shaped objects by sliding and rotating images. This could use the mechanism MARR ([24],[25]) postulates in claiming that 'place tokens' representing structures inferred by higher level processes can be added to the primal sketch). An empirical question is what sorts of manipulation people can do. In the following figure it is much easier to see where A will first touch B on approaching it than to see where A' will first touch B':

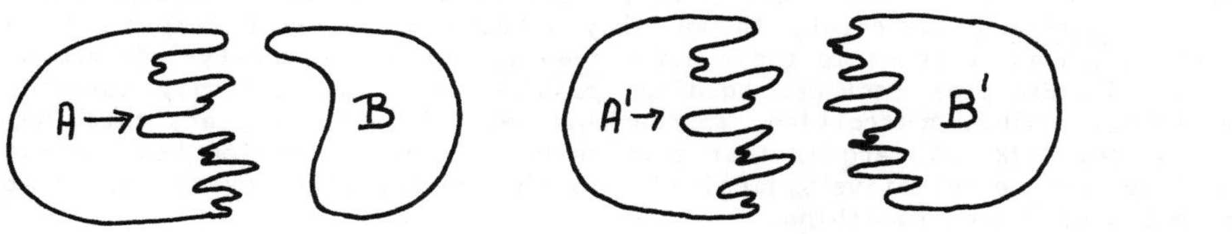

In a computer it is simple to define array manipulations which will find both tasks equally easy. Is the human brain a poor array processor? Or is there some important generally useful design feature of the human visual system which is responsible for its limitations in this case? For many tasks array manipulations are not general enough, e.g. detecting the kind of approximate symmetry exhibited in many living organisms.

Different tasks require access to different layers of interpretation, i.e. the results of different sub-modules of the visual system. For some of the tasks 3-D structure could be used. For others, e.g. drawing a realistic picture, it is essential to use image relationships. Much more detailed information about the original image is needed for a realistic oil-painting than for a sketchy line-drawing. The latter and the guidance examples could use relatively high level intermediate results of visual processing: two-dimensional yet quite abstract.

B.3. Not the retinal image.

The 2-D structure accessed from outside the visual system should not necessarily be a representation of the retinal image. That will constantly change as different parts of a scene are fixated while a problem is being solved. Something less transient is needed, such as a representation of the available 'optic array' at a viewpoint (DRAPER [9]), possibly built up over a period of time from different samples. Or it might be a map of what BARROW and TENNENBAUM [3] call 'intrinsic scene features' - closely related to what MARR called the two and a half D sketch. Several such representations are needed, dealing with different levels of abstraction, all with array-like qualities and some sort of registration to enable information at different levels to be combined easily when appropriate (compare BALLARD [1]).

The fact that (especially with the aid of relative motion), we can easily perceive two scenes superimposed in a television screen, or one seen through a window and another reflected in it, suggests that different retinal flow elements may be separated into different arrays, and then two (or more?) sets of scene interpretations built on them in parallel. This may be related to the ability to see through mist or muddy water. Stereoscopic vision reverses the process, merging two arrays. (Recent experiments suggests that locations of different sorts of features even of the same object might be stored in different maps in human vision, requiring some time for inter-map relations to be computed TREISMAN [34]).

B.4. Is only geometry represented?

Vision does not produce only geometric information. Consider a view of a busy workshop. At any moment there will be complex changing retinal images, representing 3-D structures composed of walls, floor, furniture, machines, tools, materials with changing shapes, and human bodies in a variety of changing postures. Among the relationships we can see are many which are not purely geometrical. For instance, seeing a trestle as supporting a table involves seeing it as applying an upward force and as preventing downward motion. This interpretation can play an important role in assessing the consequences of moving the trestle, or predicting the effects of placing a very large object on the table. We can see many other relationships which are not purely geometrical, though they involve geometrical components (a fact which is relevant to their being seen as opposed to merely inferred or believed). Examples include: holding, pushing, pulling, cutting, turning, moulding, using, controlling, approaching, avoiding, looking at, catching, and so on. MARR [25] argues that this is not visual perception because the ability can be selectively impaired by brain damage. I'll try to show that he may have missed something.

This is not merely a semantic quibble. The claim is that the kinds of representations which are used for geometric structures can also be used for non-geometric information which maps usefully onto geometric structure, and that the same processes can operate on them. In diagrams, we often find it useful, or even essential, to combine representations of non-geometrical properties or relationships with the representation of geometrical structure. Fields of force, lines of possible movement, causal relationships, can all be usefully represented this way, for problems in which spatial relationships are important. For similar reasons an intelligent system can benefit from integrating geometrical and non-geometrical information in combined representations.

There are many different detailed ways this can be done. For instance, in a computer it could be done by storing a variety of different sorts of information in 2-D arrays in registration with an array storing information about the structure of the current field of view. Alternatively a single array could have lists of pointers to different types of descriptions. Switching attention between classes of tasks or subtasks would be facilitated by having different 'maps' (array-like structures) storing different sorts of information, but with a mechanism for rapid access of corresponding locations in different maps, so that relations like contiguity and collinearity may be detected readily even among entries in different stores.

So non-geometrical concepts may generate geometrical problems, and it would be useful to design a visual system to include interpretations in terms of such non-geometrical concepts in order to facilitate practical problem solving.

B.5. Seeing mind in matter

Even mental states of other agents may usefully be represented in registration with spatial information. An animal may find it as important to see what another animal is looking at or aiming to pick up, as to see whether it is big, or coming closer. The computation of relations between another's field of view and objects he might react to can benefit from an integrated 'analogical' representation. Problems about the intersection of objects, or trajectories, with another's view field may make use of similar representations to those used for detecting overlap of two physical objects, or for controlling the motion of the hand, mentioned above.

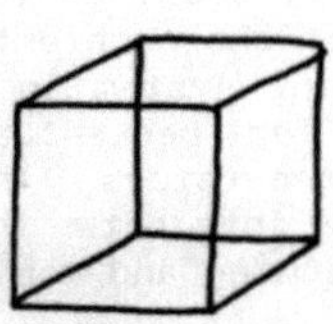

When a Necker cube drawing 'flips', you see different sets of geometrical relations. But many visual ambiguities go beyond geometry. The vase/face figure includes both geometrical ambiguities concerning relative depth and also abstract ambiguities of grouping: what 'goes with' what. But when you look at the duck-rabbit picture, the flips feel just as visual, though the change is not geometrical. Instead you see different *functional* components (eyes, ears, etc.) and even more importantly the *direction the animal is facing* changes. Abstract functional descriptions like "front" and "back" are involved. The duck or rabbit is seen as *looking* in one direction or another. Here "front", "back", "looking" are not just arbitrary labels, but imply links to elaborate systems of concepts relating to the representation of something as an *agent*, capable of moving, of taking in information, of making decisions, etc. For instance "front" indicates potential forms of motion and also the direction from which information is obtained about the environment. The attribution of 3-D slope and relative depth to parts of the Necker cube involves going beyond what is given. So does interpretation of natural images in terms of surface orientation or curvature, convexity of edges, occlusion, etc. Given the need to be able to interpret 2-D structures in terms of these quite different structures, why stop at mapping onto geometrical structures? We shall see that it can be useful for a visual system to map arbitrary image features onto arbitrary structures and processes, including direct triggering of actions.

B.6. Not only descriptions

Vision need not be restricted to the production of descriptions, whether of geometrical or non-geometrical, structures and processes. It may, for example, involve the direct invocation of action. In situations of potential danger or opportunities it would be useful for a disturbance in a peripheral part of the visual field to trigger an eye-movement to re-direct attention. A tendency to react quickly to a global pattern of optical flow produced by a large object moving rapidly towards the viewer would be very useful. A fencer or boxer who does not have time for processes of planning and deciding can also benefit from direct coupling between scene fragment detectors and the invocation of actions. Where speed of response is crucial, visual feature detectors should be able directly to trigger stored action routines without the mediation of a decision maker examining a description of what has been detected. (This seems to be the only way some animal visual systems work). The triggering could make use of general-purpose associative mechanisms which seem to be needed for other purposes. Visual learning wotuld then include the creation of new detectors (TREISMAN and GELADE [34] refer to a process of 'unitisation') and the creation of new links between such detectors and other parts of the system. Some of the links might provide new feed-back loops for fine control of actions. Some of the reflex responses may turn out to be misguided because the central decision making system is not given time to take context into account.

This use of special-purpose processors to provide 'direct coupling' could also provide the basis for many more abstract visual skills, such as fluent reading (including sight-reading music). It may also be very relevant to the fact that visual experiences can be rich in aesthetic or emotional content. It is argued in SLOMAN and CROUCHER [33] that in an intelligent system in a complex and partly unpredictable world, it is necessary for perceptual processes to be able to activate dormant motives, motive generators, and thereby generate processes which may be emotional. How to integrate the interrupt function of vision with its use in more relaxed planning and monitoring of actions is an important research issue.

B.7. What makes it VISION?

To sum up so far, instead of an output interface for only one type of result of visual processing, there is a need for different sorts of communication between visual sub-processes and other sub-systems. Sometimes descriptions of three or four-dimensional structures may be useful, sometimes only two-dimensional. And sometimes the descriptions needed are not purely geometrical, but include extra layers of interpretation, involving notions such as force, causation, prevention, function, or even the mental states of agents.

The suggestion that vision involves much more than the production of descriptions of three-dimensional structures, at least in the higher animals, conforms with the common-sense view that we can see a person looking happy, sad, puzzled, etc. Seeing into the mind of another, seeing an object as supporting another, seeing a box as a three-dimensional structure with invisible far sides, may all therefore use the same powerful representations and inference mechanism as seeing a line as straight or curved, seeing one shape as containing another, etc.

Is there then no difference between vision and other forms of cognition, for instance reasoning about what is seen? A tentative answer is that the difference has to do with whether the representations constructed are closely related to 'analogical' representations of a field of view. This is a different boundary from the one postulated by MARR, between processes which are essentially data-driven and use only very general information about the physics and geometry of the image-forming processes, and processes which may be guided by prior knowledge, and which make inferences about non-geometric properties and relations.

Research issues arising out of this discussion include the following. What range of tasks can vision be used for? Is there a useful taxonomy of such functions? What range of pre-3-D structures is useful and how are they useful? What sorts of non-geometrical information can usefully be extracted from images and embedded in visual representations? What sorts of operations on these representations can play a useful role in reasoning, planning, monitoring actions, etc.?

B.8. Problems for a neural net implementation

How should the addressing be done in a neural net? It is relatively easy to use arrays on a conventional computer, with locations represented numerically and neighbourhood relations represented by simply incrementing or decrementing co-ordinates. This enables any subroutine to 'crawl' along the array examining its components, using the number series as an analogical representation of a line, following Descartes. On a multi-processor (e.g. neural net) representation, where location is represented by location in the network, problems of access may be totally different. Will each module accessing the array have to have physical links to all elements of the array? If so how will it represent those links and their properties and relations? Moreover, there is a risk of a 'combinatorial explosion' of connections. Would it be useful to have a single 'manager' process through which all others communicate with the array, using symbolic addresses? I suspect that even these questions are probably based on too limited a view of possible forms of computation. There is a need for further investigation of possible models [15].

Assuming that a temporary memory is needed in which information about the optic array can be accumulated, how should a neural net cope with changing scenes? E.g. if you swivel slowly round to the right does the stored structure really 'scroll' gradually to the left as new information is added at the right edge? This would happen automatically if the whole thing were constantly being re-computed from retinal stimulation -- but we need something more stable, built up from many fixations. Scrolling an array-like representation in a computer can be done easily without massive copying, merely by altering a table of pointers to the columns, provided that all access goes via symbolic addresses. When the representation is embodied in a network of active processors, simultaneous transmission across network links could achieve a similar effect, but the problems of how other sub-systems continue to be able to access information are considerable. Do neurones use symbolic addresses: using these rather than physical links might solve the problem. Some recent work explores the possibility that alternative mappings from one structure to another might be represented explicitly by different active 'units' which would mutually inhibit one another. (See HINTON [17,18,19], BALLARD [1]). This seems to require an enormous explosion of physical connections, to accommodate all possible translations, rotations, etc.

C.1 Problems of representation: what and how?

Even if there were some perfect mechanism which analysed retinal stimulation and constructed a detailed representation of all visible surface fragments, their orientation, curvature, texture, colour, depth, etc., this would not solve the problems of vision. This unarticulated database would itself have to be analysed and interpreted before it could be used for the main purposes of vision. And right now we don't know very much about what such an interpretation process would be like. In particular, what structures should be represented for different purposes, and how should they be represented? HINTON [16] demonstrates that different representations of so simple a structure as a cube can profoundly influence the tasks that can be performed. We know little about the ways of representing (for instance) a face to facilitate recognition from different angles or with different facial expressions, or to facilitate interpretation of subtle mood changes, or lip-reading.

I am not talking about the problems of detecting structures, (in the world or in images) but about how to represent them in a useful way.

Even if we consider only two-dimensional geometric structures, such as the visible outlines of objects and the visible textures and markings on surfaces, we find a richness and variety that defeats existing representational schemes. Representations must not merely be mathematically adequate: they must also be epistemologically and heuristically adequate - i.e. including all the information required for a variety of tasks and facilitating computation in a reasonable time [23].

A representational scheme has two main components, the representation of primitives and the representation of composition. For instance, in many systems for representing images, primitives are local quantitative measures (e.g. of intensity gradients, or optical flow), and composition is merely the embedding of such measures in a two-dimensional array. Often a hierarchical mode of composition is more useful, e.g. using a relation "part of" to define a tree structure as used by linguists and extended by MINSKY [27a] to vision (elaborated, for example, by MARR and NISHIHARA [26]). However, a hierarchichal tree structure is often not general enough to capture perceivable relationships in a useful way. For instance, the choice of a top-level node may be arbitrary, and computing relations between the 'tips' of the tree (e.g. computing relations between a person's finger and the tip of his nose from a hierarchical body representation) may be difficult and time consuming. Yet we often see such relationships apparently effortlessly (hence letters to the press complaining about drivers who pick their noses whilst waiting at traffic lights). Moreover, many objects do not have a tree-like topology, for instance a wire-frame cube. So, instead, a network is often used, with links representing a variety of relationships, e.g. "above", "touches", "supports", "same-size", "three feet away from", etc. (See WINSTON [39], MINSKY [27]). This allows both global relations between major components, and useful information about arbitrary sub-components to be represented in a quickly accessible form. The network may still be hierarchical in the sense that its nodes may themselves be networks, possibly with cross-links to other sub-nets. One problem with such networks is their sensitivity to change. Relatively simple movements may drastically change the structure of the net. Sometimes parts of an object are not represented in terms of their mutual relationships, but in terms of their relationship to a frame of reference selected for the whole object. This can facilitate recognition of rigid objects using a generalised 'Hough transform' and cooperative networks of processors (BALLARD [1], HINTON [19]). It reduces the problem of representing changed states, since each part merely changes its relation to the frame of reference. However, relational networks seem to be more suited to non-rigid objects which preserve their topology rather than their metrical properties, like a snake or a sweater. (The Hough transform uses what BARLOW [2] calls 'non-topographical' representations, i.e. mapping objects into abstract spaces other than physical space.)

C.2. On perceiving a line

Despite the richness and complexity of such representational schemes, many percepts do not seem to be captured adequately by them. For instance, a circle might be represented approximately as made of a number of straight or curved line segments, or by parameters for an equation. But neither representation does justice to the richness of the structure we perceive, which has the visible potential for decomposition in indefinitely many ways, into semi-circles or smaller arcs, or myriad points, etc. The decomposition perceived may change as the surrounding figure changes or as the task changes. Yet, there is also a unchanging percept: we see a persistent continuous structure. (Phenomenal continuity is an interesting topic requiring further analysis. We certainly do not see what quantum physicists tell us is really there.)

This perception of continuity through change can also occur when an object changes its shape. If a circle is gradually deformed, by adding dents and bumps, the mathematical representation in terms of its equation suddenly becomes grossly inadequate, but we can see a continuous change. We can see the identity of a continuously changing object. A 'chain' encoding in terms of length and orientation of many small segments may be less sensitive to change, but will not capture much of the global structure that is perceived. It also fails to capture the perception of the space surrounding the line which is also seen as continuous. For instance it makes it hard to discover the closeness of two diametrically opposite points after the circle has been squashed to a dumbbell shape.

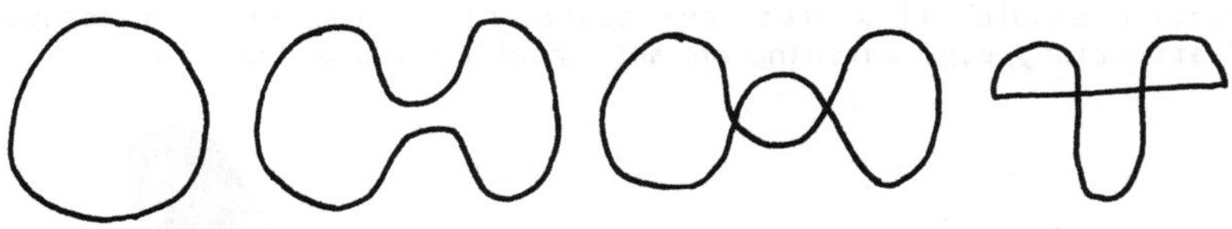

The line may gradually be deformed into a very irregular shape with many changes of curvature, sharp corners, lines crossing, etc. The algebraically representable mathematical properties will change drastically and discontinuously, and so will any network representing the more obvious decomposition based on discontinuities and inflection points. Yet we easily see the continuing identity. We don't have to switch into totally different modes of perception as the figure changes (though there may be sudden recognition of a new shape or relationship, emerging from the basic perception of the layout of the line in space). We need some account both of the relatively unchanging perception of the line in a surface as well as the awareness of higher level patterns and relationships which come and go.

C.3. The special case trap

Many computer programs make use of representations of straight lines, circles, and other mathematically tractable shapes. One obvious moral is that even when we can represent certain special cases in a computer model, we may have totally failed to explain how they are actually represented in human or animal vision. Living systems may deal with the simple cases using resources which are sufficiently powerful to cope with far more complex cases. Seeing a straight line is probably just a special case of seeing an arbitrary curve. How to represent arbitrary curves in a general purpose visual system remains an unsolved problem.

Similarly, theorems about perception of smooth surfaces may tell us little about a system which can see a porcupine and treats smooth surfaces as a special case. Many existing programs represent static configurations of blocks, but cannot cope with moving scenes. Perhaps a human-like system needs to treat such static scenes as special cases of scenes with arbitrary patterns of motion? This would imply that much work so far is of limited value, insofar as it employs representations which could not cope with motion. Of course, this does not prove that the simple models are totally irrelevant: continuous non-rigid motion may perhaps be seen in terms of locally rigid motion, which in turn may be represented in terms of a succession of static structures, just as arbitrary curves may be approximated by local straight lines. However, it needs to be shown that such a representation is useful for tasks like unfolding a garment in order to put it on, or trying to catch a rabbit which has escaped from its hutch. It may be possible for a visual system to use many special modules which deal only with special cases, when they are applicable. If so, our account of the global organisation of a visual system needs to allow for this.

C.4. Criteria for a satisfactory representation

How can we tell when we have found a satisfactory 'general purpose' representation for lines in 2-D or 3-D space? Any answer will ultimately have to be justified in relation to the tasks for which it is used. But our own experience provides an initial guide. The representation should not change in a totally discontinuous fashion as the shape of the line changes, for we sometimes need to see the continuity through change. We also need to be able to examine the neighbourhood of the line, for instance looking for blemishes near the edge of a table, so we need to be able to represent the locations in the space surrounding the line as well as the locations on the line: the 'emptiness' of other locations may be significant for many tasks. The representation should allow arbitrary locations on the line to become the focus of attention, e.g. watching an ant crawling along the line.

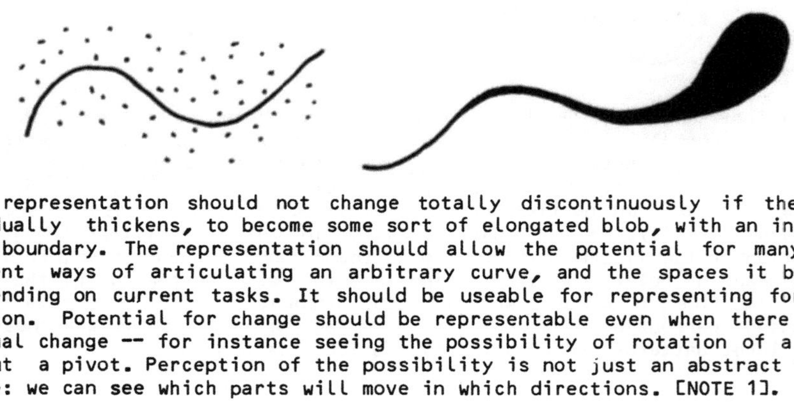

The representation should not change totally discontinuously if the line gradually thickens, to become some sort of elongated blob, with an interior and boundary. The representation should allow the potential for many different ways of articulating an arbitrary curve, and the spaces it bounds, depending on current tasks. It should be useable for representing forms of motion. Potential for change should be representable even when there is no actual change -- for instance seeing the possibility of rotation of a lever about a pivot. Perception of the possibility is not just an abstract inference: we can see which parts will move in which directions. [NOTE 1].

C.5. An 'active' multi-processor representation?

The discussion so far suggests that it would be useful to represent arbitrary structures both by projecting them onto 'analogical' representations in registration with a representation of the optic array, and also by more abstract symbolic networks of relationships, some of them object-centred, some scene-centred. There is also a need for a large number of independent processors all simultaneously attempting to analyse these structures in a variety of different ways and offering their results to other sub-systems. As a perceived structure changes, so will the pattern of activity of all the processors accessing the arrays. At lower levels the changes will be approximately as continuous as the geometrical changes. At higher levels, new processes may suddenly become active, or die away, as shapes and relationships are recognised or disappear. Hence the impression of both continuity and discontinuity through change.

Here we have a representation not in terms of some static database or network of descriptions, but in terms of a pattern of processing in a large number of different sorts of processors, including for instance some reporting on the "emptiness" around a perceived line. It is not going to be easy to build working models.

C.6. The horrors of the real world

For limited purposes, such as recognition or guiding a simple automatic assembly system, some simple cases, e.g. rigid, plane-sided objects, with simple forms of motion (e.g. rectilinear or circular) can be represented using conventional mathematical techniques.

But when it comes to the peeling of a banana, the movements of a dancer, the ever changing visible patterns on the surface of a swiftly flowing river, we have to look for new ideas. Even for static scenes, conventional AI representations tend to make explicit only singularities in space -- edges, vertices, surfaces of objects -- and not the visible continuum (or apparent continuum) of locations in which they are embedded, a problem already mentioned in connection with the simpler 2-D case.

The use of integrative array-like analogical representations described above may not be feasible for three and four-dimensional structures, owing to prohibitive storage and connectivity requirements. (This is not obvious: it depends both on brain capacity and on what needs to be represented.) Perhaps the desired integration, for some purposes, can be achieved by projecting three and four-dimensional representations into 'place tokens' [24] in one or more changing two-dimensional analogical representations in registration with the optic array. Some arrays might represent surface structure more closely than retinal structure, for instance, if a square table top seen from the side is represented by a square array, not a trapezium. Array representations would help to solve some problems about detecting relationships between arbitrary components of a scene but would still leave the necessity for articulation, and a description in terms of recognised objects their properties and relationships, in a manner which is independent of the current viewpoint.

C.7. What sort of "construction-kit" would help?

We have suggested a two-tier representation of spatial structures, using both projection into a family of array-like structures and networks representing structural and functional relationships between parts. There seems to be a very large "vocabulary" of recognisable visual forms which can be combined in many ways. The attempt to reduce them all to a very small set of 'primitives' such as generalised cylinders MARR [26] does not do justice to the variety of structures we can see. We seem to need a vocabulary of many sorts of scene-fragments including: surface patches -- concave and convex, corners of various sorts, surface edges, lamina edges, tangent edges (relative to a viewpoint), furrows, dents, bumps, ridges, rods, cones, spheroids, laminas, strings, holes, tubes, rims, gaps between objects, etc. Besides such shape fragments, there may be a large vocabulary of process fragments - folding, twisting, moving together, coming apart, entering, flowing, splashing, etc. [NOTE 1]. Compare the "Naive Physics Project" of HAYES [13]. Larger structures might then be represented in terms of a network of relationships between these "primitives". (Christopher Longuet-Higgins, in a discussion, suggested the analogy of a construction kit.) Do we also need primitives for entities with fuzzy boundaries and indefinite shapes like wisps of smoke or a bushy head of hair?

Primitives are not enough: we also need to represent their composition into larger wholes, and once again there are gaps in existing techniques. How is the relation between a winding furrow in a field and the field itself represented? Conventional network representations allow a relatively small number of 'attachment' points to be represented. But we see the furrow embedded _along its whole length_. Is this adequately catered for by combining a conventional network-like description with a projection back into a shared array-like representation? Even for a block structure like an arch, the conventional network representation (e.g. [39]) of one block as "above" or "supported by" another does not adequately represent the perceived line of contact, which can be seen as a continuum of possibilities for inserting a wedge to separate the blocks. How is a happy expression represented as embedded in a face?

At a lower level the primitives themselves would need a representation which permitted them to be recognised and their relationships characterised in detail. If this were based in part on analogical representations both of the image forms and of the 3-D structures, then this might provide a means for linking percepts together by embedding them in a larger map-like representation, e.g. linking 3-D edge descriptions to 2-D image features detected by edge-image recognisers. Could some generalisation of this help to explain the perception of a composite object as a continuous whole, unlike the relational network representation? (I am not saying that the representation is continuous, like a drawn map, only that it may be sufficiently dense to represent continuity, especially if the resolution can be changed as required.) At a very general level both forms of representation are equivalent to collections of propositions: the network is equivalent to propositions about object parts and their relationships, the map is equivalent to propositions about locations and their occupants. But they have different heuristic power.

C.8. Why the system will be messy

The use of a large number of different visual primitives for describing scenes is from a mathematical point of view redundant: for instance the geometrical structure of any scene made of objects with non-fuzzy boundaries can be represented with as much precision as required in terms of suitably small plane surface fragments. But that representation will not be useful for many tasks, such as recognition of a non-rigid object. The use of a large number of mathematically inessential primitives is analogous to the use of a BECKER's 'phrasal lexicon' [4], instead of a non-redundant grammar in a language understanding system. It is also analogous to a mathematician's learning many lemmas and rules which are redundant in the sense that they can all be derived from a more basic set of axioms. The use of the redundant system can constrain searching in such a way as to save considerable amounts of time, since searching for the right derivation from a non-redundant set of general axioms can founder on the combinatorial explosion of possible inference steps. Imagine trying to find a derivation of the theorem that there is no largest prime number from Peano's axioms.

The process of compiling the strictly redundant rules would to a considerable extent be influenced by experience of dealing with successively more complex cases, and storing the results for future use. In that case the total system at any one time, instead of having a neat mathematically analysable mode of operation, might be a large and messy collection of rules. The rules could even be partly inconsistent, if mistakes have been made and not all the rules have been thoroughly tested. The theory of how such a system works could not satisfy a craving for mathematical elegance and clarity. In MARR's terms this is a 'Type 1' theory explaining why no 'Type 1' theory can account for fluent visual processing. (I believe this is a very general point, which applies to many forms of intelligence, including language understanding and problem solving.)

C.9. Further problems.

There are many open research issues associated with the discussion so far. Which sorts of representation primitives and modes of composition are useful for seeing different sorts of environments and for performing different tasks? I've made some tentative proposals above, but they need further study.

Ethologists may be able to provide some clues as to which animals make use of which primitives. This may help in the design of less ambitious computer models and help us understand our evolutionary history.

A visual system which is not allowed to take millions of years of trial and error before it becomes useful as a result of 'self-organising processes' needs some primitives to be built in from the start, as the philosopher Immanuel Kant argued a long time ago. What needs to be built in depends in part on how much time is available for the system to develop before it has to be relied on in matters of life and death. We still don't know what would need to be built in to a general purpose robot, nor how it could synthesise new primitives and new modes of composition.

Investigation of these issues needs to be guided by a number of constraints. The representations must be useable for the purposes of human, animal, or robot vision, such as controlling actions, making plans, making predictions, forming generalisations, etc. They need not be effective for all possible goals, such as recognising the number of stars visible on a clear night, or matching two arbitrarily complex network structures. Is there some general way of characterising the boundary between feasible and over-ambitious goals?

The representations must be useful for coping with known sorts of environment, but they need not be applicable to all physically possible environments. In particular, it may be the case that the design of a visual system as powerful as ours must use assumptions about what I've called the 'cognitive friendliness' of the environment [32]. (More on this below.)

The representations must be processable in a reasonable time. This may prove to be a very powerful constraint on theories of animal vision, given the slow speeds of neuronal processing.

The representations must be capable at least in part, of being developed through some sort of learning. This allows adaptation to new environments with different properties.

They must be capable of representing partial information, for instance when objects are partially obscured, the light is bad, vision is blurred, etc., or when lower-level processing is incomplete but decisions have to be taken. (Partial information is not to be confused with uncertainty.)

More specific constraints need to be considered in relation to specific tasks. When the task is recognition, the representation should be insensitive to non-rigid deformations of the object, changes of view-point, etc. When the task is manipulation of a fragile object, the reverse, great sensitivity, is required. (Compare [26]).

D.1. The architecture of a visual system

The space of possible computational architectures is enormous, and only a tiny corner has been explored so far. It is hard to make sensible choices from a dimly visible range of options. Far more work is needed, on the properties of different forms of computation, especially multi-processor computations and their relations to the functions of vision. I shall try to show how design issues can usefully be related to different forms of 'cognitive friendliness' which may be present or absent in the environment.

Architectures, like representational schemes, may be discussed in terms of primitives and modes of composition, at different levels. Given basic limitations in processor speed, low-level visual processing needs a highly parallel organisation, in order to deal with massive amounts of information fast enough for decision-making in a rapidly changing environment. This is the biological solution, and, for the lower levels, may be the only physically possible solution given constraints on size and portability of the system. Of course, mere parallelism achieves nothing. Quite specific algorithms related to the physics and geometry of scenes and image production are required. There is plenty of work being done on this, and I'll say no more about it.

One of the main differences between different computational architectures concerns the way in which local ambiguity is handled. Some early programs used simple searching for consistent combinations of interpretations, but were defeated by the combinatorial explosion, e.g. [8], as well as a rather limited grasp of scene or image structures. Speed is crucial to an animal or robot in an environment like ours and rules out systems based on combinatorial searching. Co-operative models, such as 'Waltz filtering' [36] and relaxation use a parallel organisation to speed up the search enormously, when local features, together with their remaining interpretations, can reduce the ambiguity of their neighbours, in a feedback process. However, Waltz filtering is subject to what HINTON [14] called the 'gangrene' problem: eliminating local hypotheses which violate constraints can ultimately cause everything to be eliminated, in cases where the only good interpretation includes some local violations. Relaxation methods get round this, but it is not easy to apply them to networks of modules which need to be able to create new hypotheses on the basis of partial results of other modules. (MARR [25] argues that co-operative methods are too slow, if implemented in human neurones - I don't know whether this argument is sound. It depends on how neurones encode information.)

New forms of co-operative computation based on massive parallelism (BALLARD [1], HINTON [18,19]) seem to be potentially very important for visual processing. BALLARD calls them 'unit/value' computers, HINTON 'mosaic' computers. They replace combinatorial search in time with massive spatial connectivity, and an open question is whether the combinatorial explosion can be controlled for realistic problems by a suitable choice of representations. The combinatorial possibilities are not so great at the lower levels, so perhaps they are restricted to the detection of relatively simple, relatively local, image features.

It is not so clear whether higher levels need a parallel organisation, and to what extent the processes are essentially serial and unable to benefit from parallelism. Our discussion of functions and representations suggests that it would be useful to have a number of sub-processes dealing with different aspects of analysis and interpretation. I shall show that allowing different processes to be relatively independent, so that they can operate in parallel, makes it possible to take advantage of certain forms of cognitive friendliness of the environment, in order to compensate for unfriendliness in other dimensions.

My discussion bears a superficial resemblance to claims which used to be made about the need for 'heterarchic' control. The heterarchy/hierarchy distinction played an important role in the early 1970's in relation to the problems of designing systems to run on a single essentially sequential processor. (See WINSTON [37], SHIRAI [29], BRADY [5].) In that context, 'heterarchy' was hard to distinguish from 'anarchy'. The restriction to a serial processor generated problems of control which no longer seem relevant. If many modules can operate in parallel we need not argue about how they transfer control to one another, and there is no worry that one of the modules may assume control and never relinquish it. Instead the important questions are: what information flows where, and when, and how it is represented and processed. We attmpted to address such questions in the POPEYE project [31,32].

D.2. <u>The relevance of the environment to architecture</u>

Cited work by Horn, MARR, Barrow and Tennebaum, and others, has shown how prior assumptions about the general nature of the environment can reduce search spaces by providing local disambiguation: for instance assuming that surfaces are rigid, or continuous, and have clear boundaries, or that illumination is diffuse. These are examples of 'cognitive friendliness' of the environment. Another example is the assumption that there is adequate short wavelength illumination and a clear atmosphere. The availability of space to move, so that parallax and optical flow can be used for disambiguation is another. The relative infrequency of confusing coincidences, such as edges appearing collinear or parallel in an image when they are not in the scene is another (often described as the 'general viewpoint' assumption). The reliability of certain cues for directly triggering predatory or evasive action (section B.6.) is another form of cognitive friendliness, in an environment which may be unfriendly in other respects. An oft-noted form of friendliness is limited independent variation of object features, implying that the space of possible scenes is only sparsely instantiated in the actual world, so that scenes and therefore images have redundant structures, which can be useful for disambiguation. (E.g. BARLOW [2]). (The assumptions of rigidity and continuity of objects are special cases of this.) The 'phrasal lexicon' strategy sketched above in C.8. presupposes this sort of friendliness - limited variation implies re-usability of the results of computations.

Some of the general assumptions can be 'hard wired' into some of the processing modules, such as edge detectors, detectors of shape from shading or shape from optical flow. More specific assumptions, e.g. concerning <u>which</u> object features tend to co-occur in a particular geographical region, would be learnt and represented symbolically, like knowledge of common plant forms.

But the degree of friendliness can vary. Implicit or explicit assumptions about constraints can prove wrong, and if HINTON's 'gangrene' is to be avoided the organisation used needs to allow for local violations, if that provides a good global interpretation of an image, e.g. in perceiving camouflaged objects, or coming across a new sort of plant.

A common form of temporary unfriendliness involves poor viewing conditions (bad light, mist, snow storms, intervening shrubbery, damaged lenses, etc) which can undermine the performance of modules which work well in good conditions. A traditional way of dealing with this is to allow incomplete or unreliable data to be combined with previously stored information to generate interpretations. This implicitly assumes that not all forms of cognitive friendliness deteriorate at the same time: creatures with novel shapes don't suddenly come into view when the light fades. (Feeding children false information about this can influence what they see in dim light.)

The idea that intelligent systems need to degrade gracefully as conditions deteriorate is old. However, it is often implicitly assumed that the main or only form of cognitive unfriendliness is noise or poor resolution in images. There are several dimensions of cognitive friendliness which need to be studied, and we need to understand how visual systems can exploit the friendliness and combat the unfriendliness. Human vision seems to achieve this by great modularity: many independent modules co-operate when they can, yet manage on their own when they have to. Binocular stereo vision is certainly useful, but normally there is no drastic change if one eye is covered -- driving a car, or even playing table tennis, remain possible, though with some reduction in skill. Similarly loss of colour information makes little difference to the perception of most scene structure, though various specialised skills may be degraded. Motion parallax and optical flow patterns are powerful disambiguators, yet a static scene can be perceived through a peep hole. We can see quite unfamiliar structures very well when the light is good, but in dim light or mist when much disambiguating information is lost, we can still often cope with relatively familiar objects.

D.3. Speed and graceful degradation.

Previously, it was argued that non-visual subsystems need to obtain information from different visual subsystems. It can also be useful to have information flowing into visual data-bases not only from other parts of the visual system, but also from other submechanisms, including long-term memory stores. For instance, if some data-bases have to deal with incomplete information or possibly even incorrect information, because of some form of cognitive unfriendliness in the environment, then it will be useful to allow prior knowledge to be invoked to suggest the correct information. A more complex use of prior knowledge is to interact with partial results to generate useful constraints on subsequent processing. PAUL [28] showed how the layout of dimly perceived limb-like structures could interact with knowledge of the form of a puppet to specify which are arms and which legs, indictating roughly where the head is, and even suggesting approximate 3-D orientation. This sort of process seems to be inconsistent with the first of BARLOW's two 'quantitative laws of perception', which states that information is only lost, never gained, on passing from physical stimuli to perceptual representations [2].

In good viewing conditions this sort of mechanism is not necessary, and a modular design can allow what is found in the data to dominate the interpretation (though it doesn't always in humans, for instance in proof-reading). When very rapid decisions are needed, higher levels may start processing more quickly, and if lower levels have not completed their analysis, decisions may have to be based on data which are as bad as when viewing conditions are bad. The experience of the 'double take', thinking you've seen a friend then realising that it was someone else, could be explained in this way.

So both speedy decision making, and graceful degradation, can be facilitated in related ways. If modules have to be able to cope with incomplete information by using prior knowledge of the environment, then sometimes a high-level decision can be taken before all lower level analysis has been completed, either because part of the field of view has been processed fully, revealing an unambiguous detail, or because coarse-grained global analysis has quickly provided information about a large scale structure, e.g. the outlines of a person seen before fine details have been analysed. (The presence of visual modules which process sketchy, incomplete information, indexed by location relative to the optic array, may account for the ease with which children learn to interpret very sketchy drawings.)

Decisions based on partial information are, of course, liable to error. In an environment where different forms of cognitive friendliness do not all degrade simultaneously, errors will be comparatively rare. This liability to error coupled with tremendous power and speed, is indeed one of the facts about human vision which requires explanation. It points to the importance of designs which may not be totally general and may not always find optimal solutions, but which achieve speed and robustness in most circumstances.

An animal visual system need not be guaranteed to be error-free, or even to find the best interpretation, so long as it works well most of the time. The 'good is best' principle states that in an environment with limited independent variation of features any good interpretation is usually the only good interpretation, and therefore the best one. So designs guaranteeing optimal interpretations (e.g. [40]) may not be relevant to explaining human perception. An open question is whether task constraints will often require a guarantee of optimality to be sacrificed for speed, even for robots. This could have implications for how robots are to be used and controlled. A lot depends on how friendly the non-cognitive aspects of the environment are, i.e. what the consequences of errors are.

Besides factual information, it may sometimes be useful for information about goals to flow into visual sub-modules. What sorts of interactions are desirable? An omnivore's goal of finding food could not interact directly with edge-detectors -- but what about the goal of looking for cracks in a vase? Higher level goals could not normally be fed directly into visual modules. Nor can they normally be translated into specific collections of lower level goals, except in special cases (e.g. finding vertical cracks requires finding vertical edges). However, goals may be able to constrain processing by directing fixations, and possibly by influencing which stores of prior information are used by certain modules. An example might be the use of different discrimination nets linked into some object-recognising module depending on the type of object searched for. If you are searching for a particular numeral on a page, it may be useful to use a different discrimination net from one relevant to searching for a particular letter, even though, at a lower level, the same set of feature detectors is used. In that case, at higher levels, the process of searching for the digit '0' will be different from the process of searching for the letter 'O' despite the physical identity of the two objects.

D.4. Recapitulation

I have tried to illustrate the way in which task analysis can precede global design or theorising about mechanisms. I've suggested that a visual system should not fit into the mental economy like a black box computing some function from 2-D image structures to 3-D scene structures. Instead, we have a sketch of the visual system as a network of processes feeding many sub-databases which may be linked to different non-visual subsystems. (The general form of this sketch is not new, and not restricted to vision: e.g. [10].) Among the visual databases will be a subset which have an array like structure (with location representing location, and relationships implicit). For indexing, and problem-solving purposes, these should be mapped onto each other and a representation of the field of view, possibly built up over several fixations. The contents of higher level data bases, with more explicit represention of relationships, can be projected back into these array-like structures. Some of the databases should be able to trigger actions which bypass the central decision making process. Some may include amodal abstract, representations shared with other sensory subsystems. (This might allow some essentially visual modules to be used for spatial reasoning by the blind, even if they get no information via the eyes.) The central decision-making process needs to have access to a large number of the visual databases, though it will not be able simultaneously to process everything. (If the information available is not used it may not be stored in long term memory. So inability to recall does not prove that something has not been processed visually.)

The enormous redundancy in such a system makes empirical investigation a difficult and chancy process. For without being able to switch modules on and off independently it will be very hard to observe their individual capacities and limitations. Perhaps the effects of brain damage, combined with performance in very (cognitively) unfriendly situations, will provide important clues. Perhaps it won't.

It is likely that the design goals can be achieved in more than one way. However, there may be an interesting class of constraints, including the nature of the environment, the tasks of the system, and the maximum speeds of individual processors, which determine unique solutions (apart from the sorts of individual variations we already find between humans).

D.5. Further research

We need to explore in more detail the different dimensions of cognitive friendliness / unfriendliness of the environment, and how exactly they affect design requirements. Which sorts of friendliness can only be exploited by hard-wired design features and which can be adapted to through learning processes?

Given the nature of the environment, and the needs of an animal or the purposes of a robot, what kinds of data-bases are likely to be useful in a visual system, and what should the topology of their interconnections be? Can we get some clues from comparative studies of animals? I've made tentative suggestions about some of the sorts of data-bases which could play a role in human vision, and how they are interconnected. Could experimental investigations shed more light on this?

A problem not faced by most computer models is that in real life there is not a single image to be processed, nor even a succession of images, but a continual stream of information [7]. The problem of representing motion was mentioned in C.7. How constantly changing information is to be processed raises other problems. Once again we don't have a good grasp of the possible alternatives. As remarked in section C.3, it may be that only a system which is good at coping with changes will be really good at interpreting the special case of static images. The lowest levels of the system will probably be physical transducers which react asynchronously to the stream of incoming information. Is there a subset of data-bases which makes use of the "succession of snapshots" strategy? What are the trade-offs? Should higher level modules be synchronised in some way, or are they best left to work at their own speeds? If the environment is cognitively friendly in that most changes are continuous, and most objects endure with minimal change of structure, this provides enormous redundancy in the stream of information. The architecture of the system, and the representations used, could exploit this, avoiding much recomputation.

Much current research is aimed at finding out how much can be achieved by totally data-driven processing. We have seen that integration of prior knowledge with incoming data can provide speed and graceful degradation. We need to find out exactly which kinds of long-term memory need to be able to interact with which temporary visual databases.

My discussion has stressed the modularity and redundancy of a visual system. We need to explore in more detail the ways in which different sorts of failure of individual modules or connections between modules would affect total performance. There may be failures due to lack of relevant information or internal failures due to physical malfunction, or programming errors.

Our discussion has implications concerning the relationship between vision and consciousness. As usual, many questions remain unanswered. In particular, what exactly determines which databases should be accessible to consciousness?

D.6. Conclusion

I said at the beginning that I would present more questions than answers. I have outlined an approach to studying the space of possible visual mechanisms, by relating them to functions and properties of the environment. The study of the functions of possible mechanisms can have many levels. I have mostly stuck to a level at which it is indifferent whether the modules are embedded in brains or computers. As many AI researchers have pointed out, it's the logical not the physical nature of the representations and manipulations thereon that we need to understand initially. However, we cannot try to build realistic models of the type sketched here until we know a lot more about what should go into the various data-bases. This requires finding out more about what needs to be represented and how it can be represented usefully.

This top-down research strategy is only one among several: we can learn from many disciplines and approaches. But analysis of function can provide a useful framework for assessing relevance. However, we must always bear in mind that our attempts to derive structure from function are inherently limited by our current knowledge of possible forms of representation and computation. The way ahead includes increasing this knowledge.

Acknowledgements

I have learnt from talking to Frank Birch, Margaret Boden, Steve Draper, John Firsby, Bob Hawley, Gerald Gazdar, Geoffrey Hinton, David Hogg, Christopher Longuet-Higgins, John Mayhew, Julie Rutkowska, Mike Scaife, Frank O'Gorman, David Owen, John Rickwood, and the late Max Clowes who set me thinking about these problems. Prof. Barlow made useful comments on an early draft of this paper. Between drafts I read most of and learnt from Marr's book on vision. Alison Sloman drew the pictures. Alison Mudd and Judith Dennison helped with production of the paper. The work has been supported in part by SERC grant GR-A-80679.

[NOTE 1]
The motion primitives referred to in C.7 may be used to link static scene descriptions, E.g. the description of shut scissors may be linked via a

description of relative rotation to a description of open scissors. A description of a ball may be linked via descriptions of a squashing process to descriptions of disks and cylinders. Such linking of static and non-static concepts may both facilitate prediction and account in part for the experienced continuity as scenes change, referred to in C.4. MINSKY makes similar suggestions in [27]. If such links are accessible while static scenes are perceived, this could account for the perception of 'potential for change' referred to in C.4, which seems to play an important role in planning, understanding perceived mechanisms, and solving problems.

BIBLIOGRAPHY

1 Ballard, D.H. 'Parameter networks: towards a theory of low-level vision' in Proceedings 7th IJCAI, VOL II, Vancouver, 1981.

2 Barlow, H.B. 'Perception: what quantitative laws govern the acquisition of knowledge from the senses?' to appear in C. Coen (ed) Functions of the Brain, Oxford University Press, 1982.

3 Barrow, H.G. and Tenenbaum J.M. 'Recovering intrinsic scene characteristics from images', in [12] 1978

4 Becker, J.D. 'The Phrasal Lexicon' Theoretical Issues in National Language Processing. Eds. R.C. Schank and B.L. Nash-Webber. Proc. Workshop of A.C.L., held at M.I.T. June 1975. Arlington, Va.: Association for Computational Linguistics.

5 Brady, J.M. 'Reading the writing on the wall', in [12] 1978

6 Brady, J.M. (ed) Special Volume on Computer Vision Artifical Intelligence, 17,1, 1981. North Holland.

7 Clocksin, W.F. 'A.I. theories of vision: a personal view', AISB Quarterly 31 1978

8 Clowes, M.B. 'On seeing things', in Journal of Artificial Intelligence vol 2, no. 1 1971.

9 Draper S.W. 'Optical flow, the constructivist approach to visual perception, and picture perception: a reply to Clocksin', A.I.S.B. Quarterly, 33, 1979.

10 Erman L.D. and V.R. Lesser 'A multi-level organization for problem solving using many diverse cooperating sources of knowledge' in IJCAI-4, M.I.T 1975.

11 Funt, Brian V. WHISPER: A Computer Implementation using Analogues in Reasoning. Technical Report 76-09, Dept. of Computer Science University of British Columbia, Vancouver, 1976 (Summarised in IJCAI-77.)

12 Hanson, A. and Riseman E. (Eds) Computer Vision Systems Academic Press, New York, 1978.

13 Hayes, P.J., 'The naive physics manifesto' in D. Michie (ed) Expert Systems in the Microelectronic Age, Edinburgh University Press, 1979.

14 Hinton G.E. 'Using relaxation to find a puppet', in Proceedings A.I.S.B. Summer Conference, Edinburgh 1976.

15 Hinton G.E. and Anderson J.A. (eds) Parallel models of associative memory Hillsdale, NJ, Erlbaum, 1981

16 Hinton, G.E. 'Some demonstrations of the effects of structural descriptions in mental imagery', Cognitive Science, 3, 231-250, 1979.

17 Hinton, G.E. 'The role of spatial working memory in shape perception', in Third Annual Conference of the Cognitive Science Society, Berkeley 1981.

18 Hinton G.E. 'A parallel computation that assigns canonical object-based frames of reference', Proceedings 7th IJCAI, VOL II, Vancouver, 1981.

19 Hinton G.E. 'Shape representation in parallel systems' Proceedings 7th IJCAI, VOL II, Vancouver, 1981.

20 Hochberg, J.E., Perception, (Second edition) Prentice Hall, 1978.

21 Hogg D, 'Model based vision: a program to see a walking person', in preparation, University of Sussex, 1982.

22 Horn B.K.P, 'Obtaining shape from shading information', in [38] 1975.

23 McCarthy, J and Hayes P, 'Some philosophical problems from the standpoint of Artificial Intelligence' in Machine Intelligence 4, eds B. Meltzer and D. Michie, Edinburgh University Press, 1969.

24 Marr, D. 'Early processing of visual information', in Philosophical transactions of the Royal Society of London, pp. 483-519 1976.

25 Marr, D. Vision, Freeman, 1982

26 Marr, D. and Nishihara, H.K., 'Representation and recognition of the spatial organisation of three-dimensional shapes.' Proc. Royal Society of London, B. 200, 1978

27a Minsky, M.L. 'Steps towards artificial intelligence' reprinted in Feigenbaum and Feldman (eds) Computers and Thought 1961

27 Minsky, M.L. 'A framework for representing knowledge', in [38] 1975.

28 Paul, J.L, 'Seeing puppets quickly' in Proceedings A.I.S.B. Summer Conference, Edinburgh 1976.

29 Shirai, Y. Analysing intensity arrays using knowledge about Scenes, in [38] 1975.

30 Sloman A. The Computer Revolution in Philosophy: Philosophy Science and Models of Mind, Harvester Press and Humanities Press, 1978.

31 Sloman A, and D. Owen, G. Hinton, F. Birch, 'Representation and control in vision' in Proceedings ASIB/GI Conference, Hamburg, 1978.

32 Sloman Aaron, and David Owen, 'Why visual systems process sketches', in Proc. AISB Conference, Amsterdam, 1980.

33 Sloman A. and Croucher M, 'Why robots will have emotions', in Procedings 7th IJCAI, Vancouver 1981.

34 Treisman A.M. and Gelade G., 'A feature-integration theory of attention' Cognitive Psychology, 12, 97-136, 1980.

35 Treisman A.M. and Schmidt H., 'Illusory Conjunctions in the Perception of Objects' Cognitive Psychology, 107-140, 1982.

36 Waltz, D. 'Understanding line drawings of scenes with shadows', in [38] 1975

37 Winston, P.H. 'The M.I.T. Robot' in Machine Intelligence Vol 7 ed D. Michie and B. Meltzer, Edinburgh University Press, 1972.

38 Winston, P.H. (ed) The Psychology of Computer Vision, McGraw-Hill 1975.

39 Winston, P.H. 'Learning structural descriptions from examples', in [38] 1975.

40 Woods, W.H. 'Theory formation and control in a speech understanding system with extrapolations towards vision', in [12] 1978.

Index of Contributors

Springer Series in

Synergetics

Series Editor: **H. Haken**

Volume 1: **H. Haken**
Synergetics
An Introduction. Nonequilibrium Phase Transitions and Self-Organization in Physics, Chemistry and Biology. 2nd enlarged edition. ISBN 3-540-08866-0

Volume 3:
Synergetics
Far from Equilibrium. Proceedings of the Conference Far From Equilibrium: Instabilities and Structures, Bordeaux, France, September 27–29, 1978
Editors: **A. Pacault, C. Vidal**
ISBN 3-540-09304-4

Volume 4:
Structural Stability in Physics
Proceedings of Two International Symposia on Applications of Catastrophe Theory and Topological Concepts in Physics, Tübingen, Federal Republic of Germany, May 2–6 and December 11–14, 1978
Editors: **W. Güttinger, H. Eikemeier**
ISBN 3-540-09463-6

Volume 5:
Pattern Formation by Dynamic Systems and Pattern Recognition
Proceedings of the International Symposium on Synergetics at Schloß Elmau, Bavaria, April 30–May 5, 1979
Editor: **H. Haken**
ISBN 3-540-09770-8

Volume 6:
Dynamics of Synergetic Systems
Proceedings of the International Symposium on Synergetics, Bielefeld, Federal Republic of Germany, September 24–29, 1979
Editor: **H. Haken**
ISBN 3-540-09918-2

Volume 7: **L. A. Blumenfeld**
Problems of Biological Physics
ISBN 3-540-10401-1

Volume 8:
Stochastic Nonlinear Systems
in Physics, Chemistry and Biology. Proceedings of the Workshop, Bielefeld, Federal Republic of Germany, October 5–11, 1980
Editors: **L. Arnold, R. Lefever**
ISBN 3-540-10713-4

Volume 9:
Numerical Methods in the Study of Critical Phenomena
Proceedings of a Colloquium, Carry-le-Rout, France, June 2–4, 1980
Editors: **J. Della Dora, J. Demongeot, B. Lacolle**
ISBN 3-540-11009-7

Volume 10: **Y. L. Klimontovich**
The Kinetic Theory of Electromagnetic Processes
ISBN 3-540-11458-0

Volume 11:
Chaos and Order in Nature
Proceedings of the International Symposium on Synergetics at Schloß Elmau, Bavaria, April 27–May 2, 1981
Editor: **H. Haken**
ISBN 3-540-11101-8

Volume 12:
Nonlinear Phenomena in Chemical Dynamics
Proceedings of an International Conference, Bordeaux, France, September 7–11, 1981
Editors: **C. Vidal, A. Pacault**
ISBN 3-540-11294-4

Volume 13: **C. W. Gardiner**
Handbook of Stochastic Methods
for Physics, Chemistry and the Natural Sciences
ISBN 3-540-11357-6

Volume 14: **W. Weidlich, G. Haag**
Concepts and Models of a Quantitative Sociology
The Dynamics of Interacting Populations
ISBN 3-540-11358-4

Volume 15: **W. Horsthemke, R. Lefever**
Nonequilibrium Transitions Induced by External Noise
ISBN 3-540-11359-2

Volume 16: **L. A. Blumenfeld**
Physics of Bioenergetic Processes
ISBN 3-540-11417-3

Volume 17:
Evolution of Order and Chaos
in Physics, Chemistry, and Biology
Proceedings of the International Symposium on Synergetics at Schloß Elmau, Bavaria, April 26–May 1, 1982
Editor: **H. Haken**
ISBN 3-540-11904-3

Springer-Verlag
Berlin
Heidelberg
New York

MIX
Papier aus verantwortungsvollen Quellen
Paper from responsible sources
FSC® C105338

If you have any concerns about our products, you can contact us on
ProductSafety@springernature.com

In case Publisher is established outside the EU, the EU authorized representative is:
Springer Nature Customer Service Center GmbH
Europaplatz 3, 69115 Heidelberg, Germany

Printed by Libri Plureos GmbH
in Hamburg, Germany